Switching Arc Phenomena in Transmission Voltage Level Vacuum Circuit Breakers

Zhiyuan Liu · Jianhua Wang · Yingsan Geng · Zhenxing Wang

Switching Arc Phenomena in Transmission Voltage Level Vacuum Circuit Breakers

西安交通大学出版社
XI'AN JIAOTONG UNIVERSITY PRESS

Springer

Zhiyuan Liu
Department of Electrical Engineering
Xi'an Jiaotong University
Xi'an, Shaanxi, China

Jianhua Wang
Department of Electrical Engineering
Xi'an Jiaotong University
Xi'an, Shaanxi, China

Yingsan Geng
Department of Electrical Engineering
Xi'an Jiaotong University
Xi'an, Shaanxi, China

Zhenxing Wang
Department of Electrical Engineering
Xi'an Jiaotong University
Xi'an, Shaanxi, China

ISBN 978-981-16-1397-5 ISBN 978-981-16-1398-2 (eBook)
https://doi.org/10.1007/978-981-16-1398-2

Jointly published with Xi'an Jiaotong University Press
The print edition is not for sale in China Mainland. Customers from China Mainland please order the print book from: Xi'an Jiaotong University Press.

This Springer imprint is published by the registered company Springer Nature Singapore Pte Ltd.
The registered company address is: 152 Beach Road, #21-01/04 Gateway East, Singapore 189721, Singapore

To Professor Jimei Wang (1922–2014)

Preface

In China, the research on vacuum arc started in 1958. At that time, Prof. Wang Jimei of Xi'an Jiaotong University, Xi'an, China, set up a research group studying on vacuum arcs and applications. He built up a research team between Xi'an Jiaotong University and Xi'an High Voltage Switchgear and Power Rectifier Factory to develop the first vacuum interrupter in China. Fortunately, the first vacuum interrupter was born after 6 months' efforts. The first vacuum interrupter interrupted a current of 5 kA at a voltage of 4kV in a synthetic circuit in Xi'an Jiaotong University. In 1960, Xi'an High Voltage Apparatus Research Institute, Xi'an, China, developed a single phase vacuum switch with 6.7kV and 500A. Thereafter Xi'an Jiaotong University developed a three-phase vacuum switch with 10 kV and 1500A in 1961. This was the first three-phase vacuum switch in China so that it attracted great attention from many Chinese switchgear factories. Five years later, Xi'an High Voltage Apparatus Research Institute developed a single phase vacuum switch with 10 kV and 2000A in 1969. In 10 years from 1966 to 1976, it was a turmoil period in China. There was almost no scientific research on vacuum arc theory and its applications. Since 1976, China restarted the research of vacuum arc and its applications. At that time, three major contributors in this field in China were Xi'an Jiaotong University, Xi'an High Voltage Apparatus Research Institute and Shanghai Electrical Apparatus Science Research Institute. After 1978, there were more than ten Chinese factories producing various vacuum switchgear products. Two major manufactures in vacuum interrupters in China were Baoguang Electronic Tube Factory, Baoji, China and Huaguang Electronic Tube Factory, Jinzhou, China. Baoguang Electronic Tube Factory licensed technology and production facilities from Siemens Co., Germany in 1984. Hauguang Electronic Tube Factory licensed technology and production facilities from Westinghouse Co., USA, in 1985. Since then, vacuum switchgear including vacuum interrupters, vacuum switches, vacuum circuit breakers, vacuum reclosers, vacuum disconnectors, ring main units, etc., entered a mass production in China. There was a fast expansion of vacuum switchgear from a few percent to more than 90 percent in the Chinese medium voltage market in a short period from 1988 to 1999. Since then, vacuum switchgear proved popular in the medium voltage market in China. For example, vacuum circuit breaker occupied over 98% of 12-kV circuit breaker market and over 80% of 40.5-kV circuit breaker market in China. In the year 2020, China is producing 60% of vacuum interrupters and vacuum switchgears of the world.

Vacuum circuit breakers are stepping into higher voltage field in China. In 1989, a research team was set up including Beijing Switchgear Factory, Beijing Dongfang Vacuum Tube Factory and Xi'an Jiaotong University. The team developed an outdoor type double-break 126-kV/1250-A/31.5-kA vacuum circuit breaker prototype in 1990. Since 2003, the research focused on developing 126-kV single-break vacuum circuit breaker. There was a close cooperation between Xi'an Jiaotong University, Shaanxi Baoguang Vacuum Device Company Ltd., Zhejiang Ziguang Electrical Apparatus Company, and Shaanxi Sirui Advanced Materials CO., LTD Electric Company Ltd. et al. Under the cooperation, a 126-kV/2500-A/40-kA single-break vacuum circuit breaker was developed and it passed type test in Xi'an High Voltage Apparatus Research Institute, Xi'an, China in 2013. Higher voltage vacuum circuit breakers were still under development in China. This book was based on the above-mentioned research work. At present, besides Xi'an Jiaotong University, Xi'an High Voltage Apparatus Research Institute, Shaanxi Baoguang Vacuum Device Company Ltd., Xi'an SHIKY High Voltage Electric Company Ltd., Pinggao Group, Dalian Technology University, Huahzhong University of Science and Technology, etc., are significant contributors in the development of high voltage level vacuum circuit breakers and vacuum arc theory In China.

Obviously, this book is based on teamwork. Most of this material is gratefully chosen from the published papers of the Ph.D. and Master's students in a research group in Xi'an Jiaotong University. The authors appreciate the contributions of the following people: Dr. Guowei Kong and Dr. Hui Ma et al. (Sect. 1.2); Dr. Hui Ma and Dr. Zaiqin et al. (Sect. 1.3); (Dr. Hui Ma and Dr. Yunbo Tian et al. (Sect. 1.4.1); Dr. Yunbo Tian, Miss Yanjun Jiang and Dr. Hui Ma et al. (Sect. 1.4.2); Dr. Hui Ma, Dr. Guowei Kong and Dr. Yunbo Tian et al. (Sects. 1.4.3 and 1.4.4); Dr. Xiaoshe Zhai et al. (Sect. 2.1); Dr. Haoran Wang and Mr. Jiankun Liu et al. (Sect. 2.2); Dr. Haoran Wang and Dr. Zhipeng Zhou et al. (Sect. 2.3); Dr. Haoran Wang, Dr. Yunbo Tian and Dr. Zhipeng Zhou et al. (Sect. 2.4.1); Dr. Wenlong Yan, Miss Yanjun Jiang et al. (Sect. 2.4.2); Dr. Jiangang Ding et al. (Sect. 2.4.3); Dr. Yingyao Zhang et al. (Sects. 3.1.1 and 3.1.2); Dr. Shimin Li et al. (Sect. 3.1.3); Dr. Yingyao Zhang and Dr. Xiaofei Yao et al. (Sect. 3.2.1); Dr. Haomin Li and Zihan Wang et al. (Sect. 3.2.2); Dr. Hui Ma et al. (Sect. 3.2.3); Dr. Jiang Yan and Mr. Sheng Zhang et al. (Sect. 3.3); Dr. Li Yu and Dr. Liqiong Sun et al. (Sect. 4.1.1); Dr. Xiaofei Yao et al. (Sect. 4.1.2); Dr. Haomin Li and Zihan Wang et al. (Sect. 4.1.3); Dr. Li Yu et al. (Sects. 4.1.4, 4.2.1, 4.2.2, and 4.2.4); Dr. Xiaofei Yao et al. (Sect. 4.2.3); Dr. Liqiong Sun et al. (Sect. 4.2.5); Dr. Xiaoling Yu, Mr. Yijiang Wei, Mr. Yafei Chen and Miss Qian Lv et al. (Sect. 4.4); Dr. He Yang et al. (Sect. 5.1.1); Dr. Yongxiang Yu and Dr. Yun Geng et al. (Sects. 5.1.2, 5.2.1, and 5.3.1); Dr. Haoqing Wang et al. (Sects. 5.1.3 and 5.1.5); Dr. Feng Zhao, Dr. Yingyao Zhang and Dr. He Yang et al. (Sect. 5.1.4); Dr. Feng Zhao et al. (Sect. 5.2.2). Dr. He Yang and Mr. Yonghui Li et al. (Sect. 5.3.2); Dr. He Yang, Dr. Xiaoshe Zhai, and Dr. Yingyao Zhang et al. (Sect. 5.4); Dr. Feng Zhao, Mr. Biao Hu, Mrs. Ling Li and Dr. He Yang et al. (Sect. 5.5).

The authors also appreciate great help and contributions from numerous friends. They include senior engineer Chongming Hu, Mr. Yugui Qian, Dr. Jianjun Yao, Mr. Chen Yu, Dr. Jing Yan, Profs. Xiaoling Yu, Zhongyi Wang, Shixin Xiu, Wanliang Fang, Dr. Xiaoshe Zhai, Dr. Li Yu, Dr. Yingyao Zhang, Dr. Shaoyong Cheng, Dr. Yuesheng Zheng, Mr. Yijiang Wei, Mr. Yafei Chen, Mr. Heming Zhou, Mr. Fuxing Yuan, Mr. Ping Liu, Mrs. Liping Yan, Mr. Chunhong Shen, Mr. Chuanchuan Wang, Mr. Zhenle Nan, Mrs. Xiaofeng Zhuang, Mr. Yanchao Zhang, Mrs. Zhe Yang, Mr. Shaocheng Tong, Mr. Peiji Gu, Mr. Feng Zhou, Mr. Yuanwang Lei, Mr. Quan Wang, Mrs. Guangli He, Mrs. Dongli Bi, Mr. Jianfei Lin, Mr. Yali Zhang, Mr. Peng Li, Mr. Yong Chen, Mr. Wenbin Wang, Mr. Gang Li, Mr. Xiaojun Wang, Mr. Shiping Wei, Mr. Zhaogui You, Mr. Shumo Han, Mr. Junhui Wu, Mr. Guohui Han, Mrs. Shuping Sun, Mrs. Xiaoqin Wang, Dr. Guiquan Han, Mr. Xiaomin Zhao, Mr. Yinghua Bi, Mr. Yonglin Li, Mr. Yong Chen, Mr. Shinian Wang, Mr. Jian Chen, and numerous people involved in the related work. The authors appreciate the great support from Mrs. Ying Li of Xi'an Jiaotong University Publishing House. The authors thank the support from the National Natural Science Foundation of China under grants of 50777050, 51177122, and 51937009.

We dedicated this book to Prof. Jimei Wang (1922–2014), who was the founder in the field of vacuum arc theory and vacuum switchgear in China. He marked the beginning of the development of vacuum switchgears in China. He promoted the progress of vacuum arc theory in China. He developed the first three-phase vacuum switch and the first 126 kV double-break vacuum circuit breaker in China. He also promoted the first 126 kV single-break vacuum circuit breaker in China. He trained two generations of engineers in the field vacuum switchgears in China. Before he passed away in 2014, Prof. Jimei Wang witnessed the 126-kV/2500-A/40-kA single break vacuum circuit breaker passing the type test in Xi'an High Voltage Apparatus Research Institute, Xi'an, China. He also promoted and witnessed a new 126 kV/40 kA synthetic circuit was put into use in Xi'an Jiaotong University in 2013. This book is dedicated to Prof. Jimei Wang, the founding father of vacuum switchgears in China.

Xi'an, China

Zhiyuan Liu
Jianhua Wang
Yingsan Geng
Zhenxing Wang

Introduction

In 2009, a CIGRE working group WG A3.27 was approved and built up. The title of the working group was *The impact of the application of vacuum switchgear at transmission voltages.*[1] Efforts have been made to study the impact of the application of vacuum switching technology at voltages above 52 kV. The formal technical brochure of the workgroup was published in July 2014, as No. 589. The technical brochure is considered as a milestone in the field of transmission voltage level vacuum circuit breakers.

The transmission voltage level vacuum circuit breakers are developing very fast recently worldwide, remarkably in Japan, Germany and China. According to technical brochure No. 589, there were approximately 10,000 units of 52 kV and above vacuum circuit breakers were delivered to utilities and industries in Japan. In 2015, Meidensha company developed the world-first 145 kV gas-insulated switchgear incorporated a single-break vacuum circuit breaker (V-GIS), as shown in Fig. I.1.[2] The Meidensha V-GIS has a nominal rated current of 2000 A and a nominal rated

Fig. I.1 Meidensha company developed the world-first 145 kV GIS incorporating single-break VCB in 2015

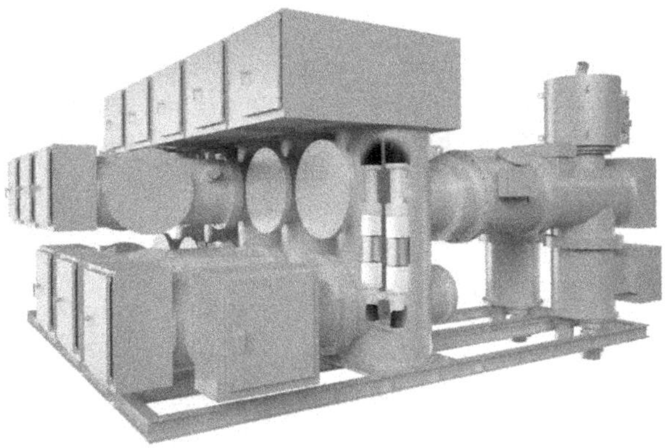

[1]CIGRE technical brochure No. 589, working group A3.27, The impact of the application of vacuum switchgear at transmission voltages, July 2014.

[2]https://www.meidensha.com/products/energy/prod_01/prod_01_03/prod_01_03_05/index.html.

Fig. I.2 Siemens company launched a 145 kV blue GIS using high voltage vacuum interrupters in 2016

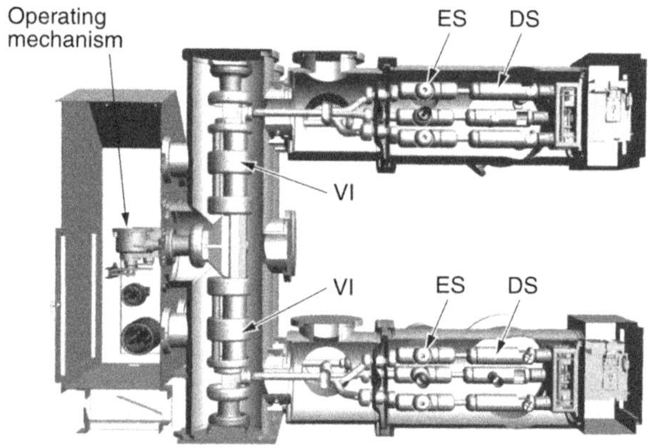

Fig. I.3 Meidensha company developed a 204 kV double-break V-GIS

short circuit breaking current of 31.5 kA. In 2016, Siemens company launched a 145 kV gas-insulated switchgear using high voltage vacuum circuit breaker (blue GIS) in Cigré session in Paris, as shown in Fig. I.2.[3] It adopted vacuum interrupter technology and so-called clean-air technology. The Siemens blue GIS has a nominal rated current of up to 3150 A and a rated short-circuit breaking current up to 40 kA. In 2018, Cigré session exhibition in Paris, Siemens company exhibited a 170 kV single-break vacuum interrupter and a 245 kV single-break vacuum interrupter prototype. In 2018, Meidensha company developed a 204kV dead-tank double-break vacuum circuit breaker and they successfully developed a 204 kV V-GIS, as shown in Fig. I.3.[4] In 2020, Pinggao group developed a 126 kV GIS with single break vacuum interrupter and using CO_2 as insulation gas, as shown in Fig. I.4. The mentioned efforts evidenced recent intense research and development work in the field.

In China, a research team was built up by Prof. Wang Jimei to develop transmission voltage level vacuum circuit breakers in 1989. This team included Xi'an Jiaotong University, Beijing Switchgear Factory, and Beijing Dongfang Vacuum Tube

[3]https://new.siemens.com/global/en/products/energy/high-voltage/transmission-products/blue-high-voltage-products.html.

[4]Harada H, 204kV Gas-Insulated Switchgear (GIS), Meiden Review, No.174, No.3, 2018, pp. 2–6.

Fig. I.4 Pinggao group developed a 126 kV single-break V-GIS in 2020

Factory. A double-break 126-kV/1250-A/31.5-kA vacuum circuit breaker prototype was developed by this team. In 2003, a new research team was built up to develop a 126-kV single-break vacuum circuit breaker, with the cooperation of Xi'an Jiaotong University, Shaanxi Baoguang Vacuum Device Company Ltd., Zhejiang Ziguang Electrical Apparatus Company, and Shaanxi Sirui Advanced Materials CO., LTD Electric Company Ltd. The team developed a 126-kV/2500-A/40-kA single-break vacuum circuit breaker and it passed type test in XIHARI, Xi'an, China, in 2013. The efforts inspired more 126 kV vacuum circuit breaker products flourishing in China, which were significantly commercialized by China XD group, Pinggao group, Baosheng electric company, etc.

This book is a summary of the research work in the development of the 126kV single-break vacuum interrupter and vacuum circuit breaker in China. The book includes five chapters. Chapter 1 deals with the high-current vacuum arc phenomena, which is critical for successful short circuit current interruptions, especially at transmission voltage level. Chapter 2 discusses the dielectric recovery properties after current interruption in vacuum. An insight into this phenomena helps to prevent restrikes in high-current interruptions. Chapters 3 and 4 deal with the technical challenges in developing vacuum interrupters and vacuum circuit breakers at transmission voltage level. An important finding is that the opening and closing displacement curve of transmission voltage level vacuum circuit breakers should match the vacuum arc characteristics to achieve high short circuit current breaking capacity and good anti-welding performance. Chapter 5 discusses the capacitive current switching in vacuum. Experiments evidenced that the capacitive current switching duty was a specific technical challenge for vacuum circuit breakers, especially at transmission voltage levels.

Contents

Chapter 1
High-Current Vacuum Arcs Phenomena at Transmission Voltage Level

1.1 Introduction

Vacuum circuit breakers (VCBs) are widely used to protect power distribution systems because of high interrupting capacity, low maintenance, long operating life, and eco-friendly [1]. In the field of transmission voltage levels high voltage circuit breakers, SF_6 circuit breakers dominate. Since SF_6 gas is regarded as a strong greenhouse effect gas with an extremely high global warming potential (GWP), which for a 100 year time horizon is estimated to be ~22,800 times larger than the GWP of CO_2 [2], it seems to be a good candidate to solve the problem of the strong greenhouse effect of SF_6 gas by developing vacuum circuit breakers from distribution voltage levels to transmission voltage levels. That is why transmission voltage levels vacuum circuit breakers are drawing an intense research and development efforts in the world.

Vacuum interrupters are the quenching chambers of the vacuum circuit breakers. Vacuum arcs are initiated and extinguished in vacuum interrupters in current interruptions. During the arcing period, a vacuum arc, which is a metallic vapor arc in vacuum, forms in the vacuum interrupter. In a low current mode, the vacuum arc operates in the diffuse mode, as shown in Fig. 1.1 [3]. In the diffuse arc mode a few moving spots appear on the cathode, while a diffuse glow originates at the cathode and reaches across the gap to the anode. Cathode spots that sustain the vacuum arc represent the main source of metallic vapor [4, 5]. The anode is often regarded as a negative collector of the cathodic plasma [6]. After the current reaches zero and the arc extinguishes, metallic atoms dissipates from the gap, and the dielectric strength is recovered, which typically achieves a successful current interruption [7].

However, when the arc current exceeds a threshold level, an anode spot comes to form suddenly, which is a luminous or bright region on the anode, as shown in Fig. 1.2 [8]. High-current anode spot can severely heat and grossly melt the anode in the arcing period, which is schematically shown in Fig. 1.3a. In addition, the anode in arcing period becomes cathode in post arcing period. The post arc cathode with the local molten region starts emitting metallic atoms through evaporation, as shown in Fig. 1.3b. If a transient recovery voltage is applied, it deforms the local molten

Z. Liu et al., *Switching Arc Phenomena in Transmission Voltage Level Vacuum Circuit Breakers*, https://doi.org/10.1007/978-981-16-1398-2_1

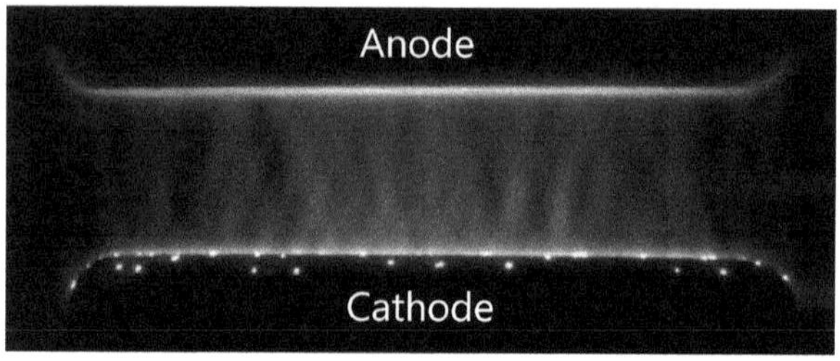

Fig. 1.1 Diffused vacuum arc [3]

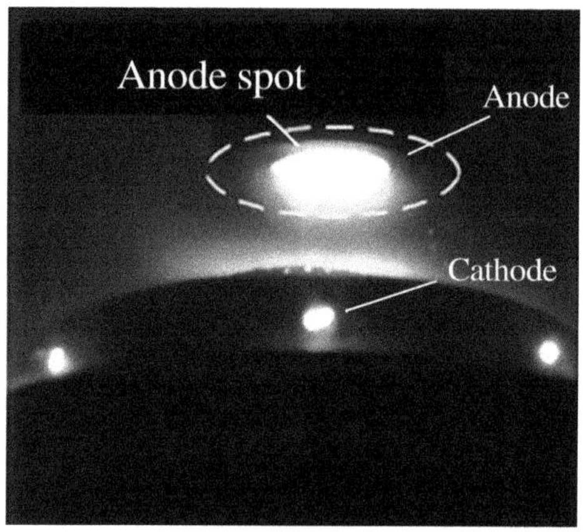

Fig. 1.2 Anode spot appearance in vacuum arc [8]

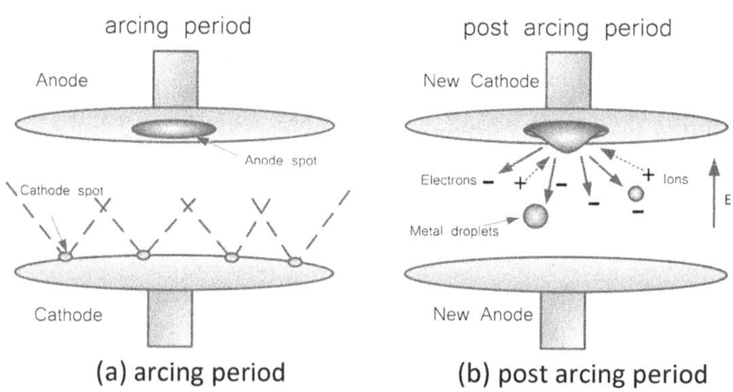

Fig. 1.3 A schematic diagram to show high-current anode spots severely heat and grossly melt the anode in the arcing period and the melted anode region starts emitting metallic atoms through evaporation in post arcing period

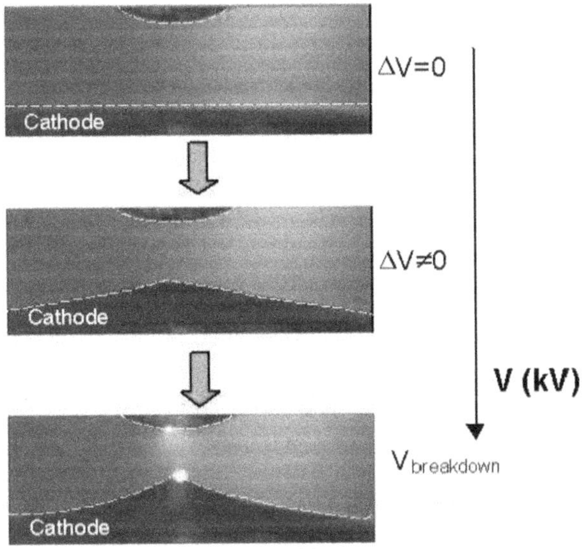

Fig. 1.4 Deformation of molten metal under an application of transient recovery voltage [9]

region on the post arc cathode and a Taylor cone would form, as shown in Fig. 1.4 [9]. In such case, a breakdown may occur under the transient recovery voltage and the current interruption becomes unsuccessful [10]. So that a formation of anode spots in the vacuum arcs significantly reduces the high-current interruption capabilities of vacuum interrupters [11], which suggests a successful high-current interruption in vacuum is severely influenced by anode phenomena of the vacuum arc. Miller gave excellent reviews about anode modes in vacuum arcs [11, 12–14].

Miller suggested to classify vacuum arc anode modes into four modes, which are diffuse arc, footpoint, anode spot and intense arc, as shown in the anode diagram in Fig. 1.5a [13]. In a Miller's new work on the classification of vacuum arc anode modes, a change appeared to the classical anode modes due to the work of Batrakov and colleagues, which identified new modes of high-current vacuum arcs, including anodic plume and anode spot type 2 [15]. The plume and type 2 anode spot modes can be seen in Fig. 1.5b, which may occur just for certain materials, since they have only been seen with CuCr contacts. In addition to the anode modes investigations of vacuum arc, Heberlein and Gorman observed five modes of vacuum arc columns: diffuse arc, diffuse column, constricted column, jet column, and anode jet [16]. Based on the classification work of anode modes [13] and column modes [16], and anode discharge modes observed in the authors' lab, three vacuum arc modes are determined according to their influence on successful current interruption, which are diffuse arc mode, high-current anode mode, and intense arc mode. Only diffuse arcs mode is grateful in current interruptions by vacuum. High-current anode mode and intense arc mode should be avoided because of the negative influences on successful current interruptions. Table 1.1 shows a correlation between the three vacuum arc modes (diffuse arc mode, high-current anode mode and intense arc mode) and the

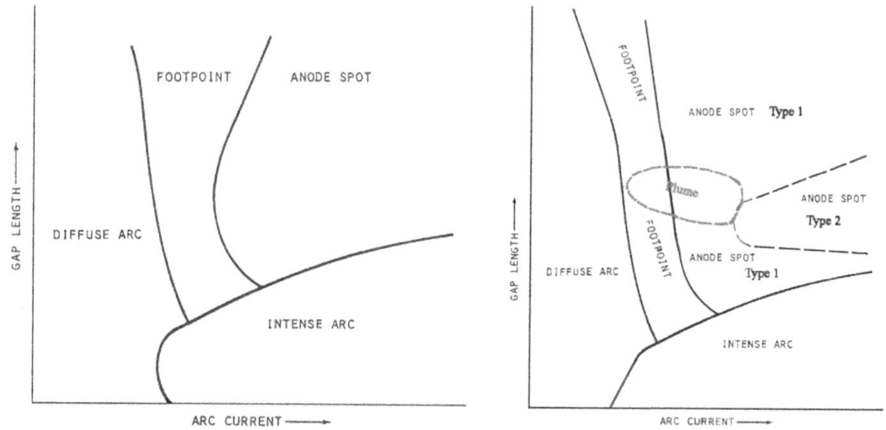

(a) four classical anode modes (b) new classification of anode modes

Fig. 1.5 Anode modes as a function of current and gap length [13]

Table 1.1 Vacuum arc discharge modes

Anode modes	Anode modes by Miller [13]	Column modes by Heberlein and Gorman [16]
Diffuse arc	Diffuse arc	Diffuse arc
High-current anode	Footpoint	Diffuse column
	Anode spot	Anode jet
Intense arc	Intense arc	Constricted column
		Jet column

anode modes proposed by Miller and arc column modes proposed by Heberlein and Gorman. The characteristics of these modes are described as follows.

(1) ***Diffuse Arc mode***: The behavior of the vacuum arc is controlled by cathode phenomena. The anode is basically passive or has only slight effect on the overall behavior of the arc [12]. At higher current, some constriction of the interelectrode plasma may occur [12].

(2) ***High-current anode mode***: There is a luminous or bright region (footpoint or anode spot) on the anode often with high arc voltage noise component [12]. The arc column appears as follows.

   (a) *Diffuse column*: A transition vacuum arc mode, characterized by the appearance of one or more luminous spots on the anode [12]. The arc column has diffuse boundaries.

   (b) *Anode jet*: There is a large anode bright region. This mode occurs at larger contact gaps. The anode jet separated from the cathode jet and with the appearance of strong anode domination [16].

(3) ***Intense Arc***: Also, a high current arc mode, where the anode is very active. The intense arc mode occurs at shorter gap lengths than does the high-current anode mode. The arc column appears as follows.

   (a) *Constricted column*: The arc boundaries and arc roots at the anode and cathode are well defined. The column is cylindrical with a slight constriction in front of the anode [16].
   (b) *Jet column*: It occurs at relatively small gap with a more intense column compared with constricted column. But the column is wider at the contact ends and the column has an appearance of two cones meeting in their apexes [16].

Therefore, from the prospective of successful current interruption, we define the arc current at which a high-current anode mode first appeared as a threshold current I_{th}. The intense arc is not specificly discussed in this book because it is typically related to a short arcing time in current interruptions. So far there is no enough knowledge on how to achieve successful current interrupitons with a short arcing time by controling or avoiding an intense arc.

There are many influence factors for high-current anode mode in vacuum arcs, such as contact materials, contact diameter, contact separation gap, magnetic fields [11–14], and contact opening velocity [4], etc. It is widely accepted that high-current anode mode formation threshold current is closely related to contact geometry parameters, including contact diameter and contact separation gap [17–21]. The contact geometry factor probably operates through its effects on both the voltage drop near the anode and the average current density. The threshold current I_{th} increases with anode diameter and decreases with contact separation. In transmission voltage level vacuum circuit breakers, the contact separation gap is typically higher than its counterpart in distribution voltages, because increased insulation levels should be met. For example, the contact gaps for 12 and 40.5 kV vacuum circuit breakers are typical ranging between 8 and 10 mm and between 18 and 20 mm, respectively. However, the 72 kV and above vacuum circuit breakers have contact gaps typically over 30 mm. For instance, a 126 kV vacuum circuit breaker reported a contact gap of 60 mm [22]. The higher contact separation gap in transmission voltage level vacuum circuit breakers means the anode spot formation occurs more easily than in distribution voltage level vacuum circuit breakers. In addition, the transient recovery voltages (TRVs) at the transmission voltage level circuit breakers are much higher than that in the distribution voltage level ones. This makes it more difficult for successful current interruption in transmission voltage level vacuum circuit breakers. Therefore, it is meaningful to quantitatively determine the threshold current I_{th}, which would be useful for designing vacuum interrupters of the transmission voltage level vacuum circuit breakers.

The principle of high current interruption in vacuum is using arc control technology. Arc control technology in vacuum is accomplished by the application of a transverse magnetic field (TMF) or an axial magnetic field (AMF) on a vacuum arc. The transverse magnetic field is applied perpendicular to the arc column in a direction that forces it to run rapidly around the periphery of the contacts. The axial magnetic field is applied parallel to the arc column to keep the arc in diffuse mode up to high current levels. Both transverse magnetic field and axial magnetic field arc control technologies aim to distribute arc heating homogeneously over the surfaces of the contacts, to avoid overheating at single locations. In transmission voltage level vacuum circuit breakers, axial magnetic field is preferred because the contact gap is high. With such high contact gap, transverse magnetic field would be too weak to control the vacuum arcs. Therefore, this book focuses on the axial magnetic field arc control technology.

This chapter deals with high-current vacuum arcs phenomena at transmission voltage level from the following four aspects. First, it discusses how contact geometry parameters and axial magnetic field influence the high-current anode mode formation threshold current. Second, the anode spot current density is determined. As a benchmark, the diffuse arc current density distribution on anode is given. Third, it shows how the high-current anode mode vacuum arcs changes the anode surface conditions, including anode erosion patterns, anode crater formation, and the anode surface temperature after current zero. Fourth, the anode spot formation mechanisms are discussed.

1.2 Threshold Current of High-Current Anode Mode Formation

This part discusses how contact geometry parameters and axial magnetic field influence the high-current anode mode formation threshold current I_{th} for transmission voltage level vacuum circuit breakers [23, 24]. The experimental results showed a linear relationship between I_{th} and a dynamic solid angle of the contacts subjected to an axial magnetic field. It also revealed a power function between the critical axial magnetic field strength per unit arc current needed to avoid anode spot formation and the contact diameter.

Compared with anode diameter or contact separation gap alone, solid angle is often a more important geometrical parameter to investigate the high-current anode mode formation threshold current I_{th} [11]. Solid angle is subtended by the anode from the cathode center or by a ratio of the anode diameter to a fixed contact gap [11]. The interactions between I_{th} and solid angle were evidenced by the results in the past papers, which revealed that I_{th} would increase with solid angle [25–28].

In addition, it is well known that I_{th} is significantly influenced by an axial magnetic field in the vacuum arcs [29–33]. An axial magnetic field contributed to improve threshold current level of anode spot formation significantly [30, 31]. With an axial magnetic field applied, positive ion in vacuum arc plasma is prevented from escaping from the contact gap. This effect caused an increase of plasma toward anode surface. The region of plasma plumes overlap in the interelectrode was reduced, which lead to an arc voltage decrease. Then such effects resulted in an enhancement of anode spot threshold current [29–31].

Although a relationship between I_{th} and the solid angle was evidenced in the past work, the results scattered because of various experimental conditions. There seems a necessary to investigate the I_{th} influenced by solid angle and axial magnetic field in combination under a set of unified experimental conditions. We investigated the I_{th} under the unified experimental conditions as following.

Experiments were performed in a $L-C$ discharging circuit, where the capacitors were pre-charged. The current frequency was 50 Hz. The applied arc current was up to 30 kA rms. The charging arc current was adjusted according to experimental conditions for high-current anode phenomena formation. The experimental setup is shown in Fig. 1.6. The experiments were operated in a cylindrical demountable vacuum chamber which was evacuated and sustained to 10^{-3}–10^{-4} Pa, as shown in Fig. 1.6a. Through an observing window located on the chamber, a high-speed charge-coupled device video camera recorded the vacuum arc evolution. The camera recording speed was set as 10,000 frames/s with the aperture fixed at four and the exposure time was set as 2 μs. To generate a uniform axial magnetic field, a Helmholtz coil passing through a direct current (dc) from a dc source was installed coaxially with a pair of butt type contacts in the chamber. As shown in Fig. 1.6b, the Helmholtz coils were 80 turns with an average radius of 80 mm and with a separation of 80 mm from a coil's center to another coil's center. The dc source was turned on ~100 ms before arcing and it kept turned on through the whole arcing process. The applied axial magnetic flux density B_{AMF} can be adjusted in a range of 0–110 mT. The axial magnetic field direction was from the upper contact to the lower contact in the chamber. Butt type contacts with a thickness of 5 mm were used in the experiment, and the contact material was CuCr25 (25% Cr) with contact diameters of 12, 25, 40, 60, and 80 mm, respectively. The contact separation length l as arc extinguishes was adjusted by an opening velocity using a permanent magnet operating mechanism.

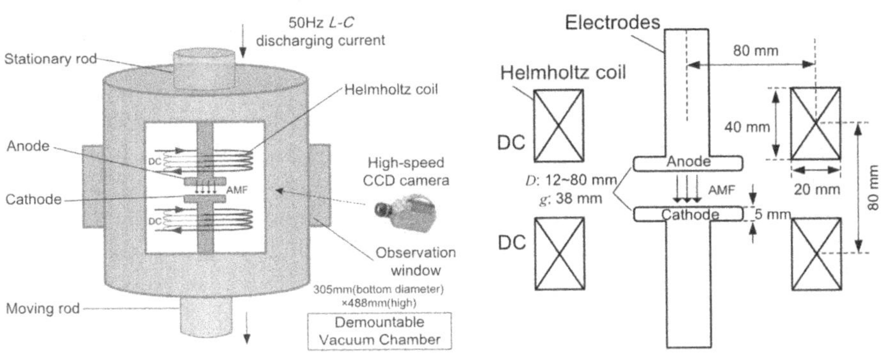

(a) Experimental setup

(b) Contacts and external Helmholtz coil

Fig. 1.6 Diagram of the experimental setup for the high-current anode phenomena research

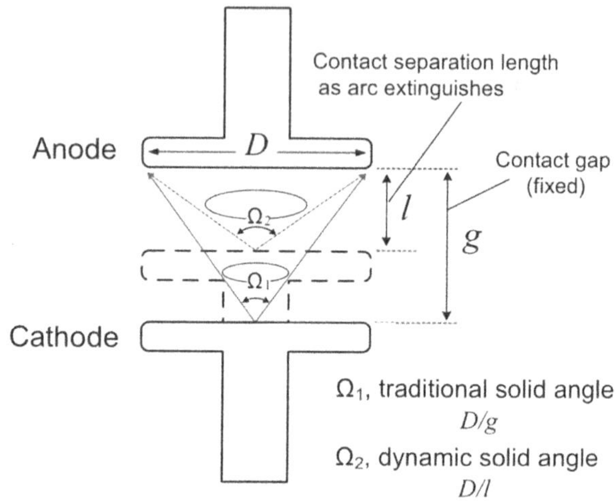

Fig. 1.7 Contact dynamic solid angle and traditional solid angle in a drawn vacuum arc

The arcing time was controlled at ~10 ms. An average opening velocity was defined as an average velocity within the contact separation during the arcing time, which was set as 1.2, 1.8, and 2.4 m/s, respectively. Thus, the contact separation length l as arc extinguished was 12, 18, and 24 mm, respectively. The full contact gap was set as 38 mm.

As mentioned above, solid angle (unit sr) is subtended by the anode from the cathode center or by a ratio of the anode diameter to the contact gap. However, once drawn vacuum arcs are initiated, one can observe two types of solid angles. As shown in Fig. 1.7, one solid angle Ω_1 is subtended by a ratio of the anode diameter to a fixed contact gap (D/g), which was as the same definition as that in the past work. While another solid angle Ω_2 was subtended by a ratio of the anode diameter D to the contact separation length l as arc extinguishes in a drawn vacuum arc (D/l). Ω_2 was defined as a dynamic solid angle, which was dynamically related to arcing time and contact opening velocity. In a vacuum arc evolution, it is a dynamic process during arcing. The experimental results proved dynamic solid angle Ω_2 to be more appropriate to reveal the arcing process in a drawn vacuum arc than the fixed solid angle Ω_1. Therefore Ω_2 was used in a drawn vacuum arc, i.e. the ratio of the anode diameter D to the contact separation length l at arc extinguishes, to understand high-current anode phenomena in vacuum.

Figure 1.8 shows an example that how high-current anode mode formation threshold current I_{th} was determined. Figure 1.8a compares the anode discharge mode for the arc current I_{arc} (rms) of 9.7, 11.0, and 12.8 kA, respectively, which was subjected to an axial magnetic flux density of $B_{AMF} = 37$ mT. The contact diameter D was 60 mm. The contact separation length at arc extinguishes l was 24 mm. The opening velocity was 2.4 m/s. As shown in Fig. 1.8a, when the arc current I_{arc} increased from 9.7 to 12.8 kA, the first high-current anode phenomenon appeared at the arc current 11.0 kA with which an anode jet (AJ) appeared during the arcing period of 5–6 ms. There was a bright region on the anode and the anode jet was

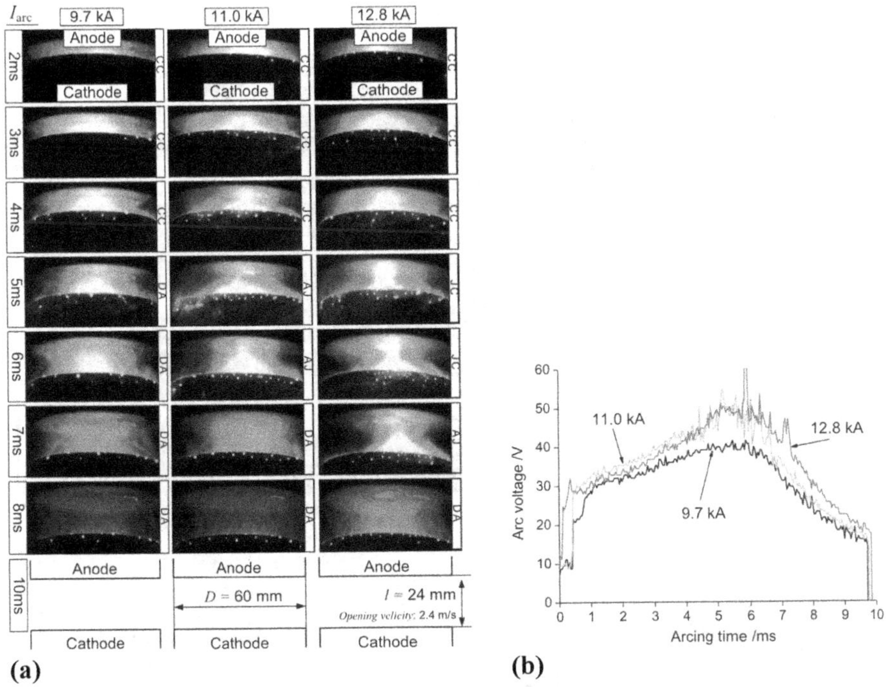

Fig. 1.8 Determination of high-current anode mode formation threshold current I_{th}. Contact material CuCr25 (25% Cr), arcing time = 10 ms, contact diameter $D = 60$ mm, contact separation length $l = 24$ mm, axial magnetic flux density $B_{AMF} = 37$ mT. **a** Anode discharge photos at arc current I_{arc} of 9.7, 11.0, and 12.8 kA, respectively. Anode spot first formed at the arc current $I_{arc} = 11.0$ kA. Arc modes are labeled beside each photo, where DA: diffuse arc, DC: diffuse column, AJ: anode jet, CC: constricted column, and JC: jet column. **b** Arc voltage characteristics at the arc current shown in **a**

significant. At a lower arc current of 9.7 kA, no anode spot was found yet. At the higher arc current of 12.8 kA, it showed a stronger anode spot mode. Therefore, the anode spot modes formation threshold current I_{th} was determined as 11.0 kA under the experimental conditions. Figure 1.8b shows the corresponding arc voltage waveforms corresponding to the arc current in Fig. 1.8a. As shown in Fig. 1.8b, there was an obvious arc voltage noise component starting at the arc current of 11.0 kA.

Figure 1.9 shows the anode discharge modes at three I_{th} at 24.3, 22.1, and 8.2 kA, respectively, with different contact diameter D and contact separation length l. The applied axial magnetic flux density B_{AMF} was 110 mT. Based on the experimentally determined I_{th}, one can find how I_{th} is influenced by dynamic solid angle and axial magnetic field.

I_{th}
24.3 kA
22.1 kA
8.2 kA
Anode
Cathode
2ms
3ms
4ms
5ms
6ms
7ms
8ms
10ms
CC
JC
DC
AJ
DA
l = 18 mm
Opening velocity: 1.8 m/s
l = 24 mm
Opening velocity: 2.4 m/s
D = 60 mm
D = 25 mm

◀**Fig. 1.9** Anode discharge modes at three high-current anode mode formation threshold current I_{th} with different contact diameter D and contact separation length l. Contact material CuCr25 (25% Cr), arcing time = 10 ms, and axial magnetic flux density B_{AMF} = 110 mT. Arc modes are labeled beside each photo, where DA: diffuse arc, DC: diffuse column, AJ: anode jet, CC: constricted column, and JC: jet column

1.2.1 *Influence of Dynamic Solid Angle on the Threshold Current*

Figure 1.10 shows a dependence of high-current anode mode formation threshold current I_{th} on dynamic contact solid angle D/l. In Fig. 1.10a D/l is adjusted by contact diameter D and in Fig. 1.10b D/l is adjusted by contact separation length l as arc extinguishes. Figure 1.10 shows that I_{th} is directly proportional to D/l. The linear relationship is significantly influenced by the applied axial magnetic flux density B_{AMF}. Stronger axial magnetic field corresponded to a higher I_{th}. However, the influence of D and l on I_{th} is different. The slope of I_{th} to D/l is significantly higher when D/l is adjusted by D than that of the case when D/l is adjusted by l. The results imply that D played a more significant role on I_{th} than l.

The experimental results that D played a more significant role on I_{th} than l agreed with previous findings, such as Mitchell [17] and Londer and Ulyanov [34]. Mitchell pointed out that a starvation current for an anode spot formation is inversely proportional to a contact separation but is directly proportional to the anode diameter by a power of 1.2. A calculation by Londer and Ulyanov confirmed that a critical current

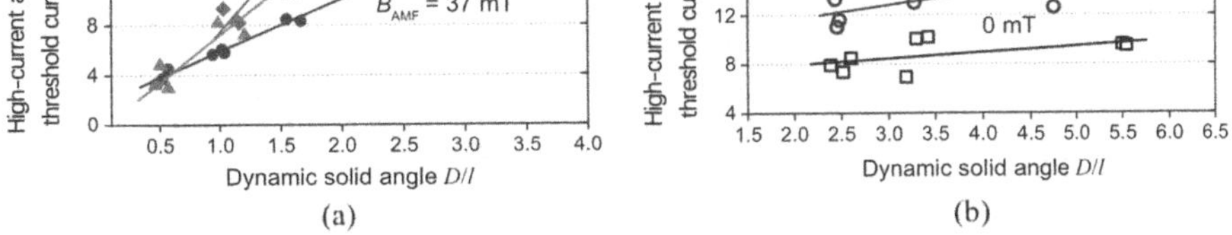

Fig. 1.10 Dependence of high-current anode mode formation threshold current I_{th} on dynamic contact solid angle D/l. Axial magnetic flux density is 0, 37, 74, and 110 mT, respectively. Contact material is CuCr25 (25% Cr). Arcing time is 10 ms. **a** Dynamic solid angle D/l is adjusted by contact diameter D of 12–80 mm. Contact separation length l as arc extinguishes is 24 mm. **b** Dynamic solid angle D/l is adjusted by contact separation length l of 12, 18, and 24 mm, respectively. Contact diameter D = 60 mm

for plasma flow crisis, correlated with an anode spot formation, is dependent more significantly on contact diameter than that on contact gap.

The phenomenon may be explained by the work of Dyuzhev and Shkol'nik et al. [33, 35]. They suggest that an anode spot comes to form at the case when a current density on the anode J_{anode} needed to maintain a stable arc exceeds the random current density on the anode J_{random} furnished by the adjoining plasma. A higher J_{random} is corresponding to a higher I_{th}. It is assumed that both a larger D and a smaller l may lead to a higher J_{random}. However, D could play a more significant role to enhance J_{random}, compared with l. Simultaneously, a larger D may easily lead to a smaller J_{anode} because of an increase of contact area. Therefore, D is more effective to enhance I_{th} than l.

While previous work found I_{th} was related to the fixed solid angle following an expression of D/g [11], $D^{1.2}/g$ [17] or $(D/g)^{1.1}$ [36], where g was the full contact gap, Fig. 1.10 shows a linear relationship between I_{th} and the dynamic solid angle D/l. An equivalent dynamic solid angle $(D/l)_{eq}$ can unify the two cases shown Fig. 1.10, where D/l is adjusted by D or by l. The equivalent dynamic solid angle $(D/l)_{eq}$ is proposed as follows:

$$(D/l)_{eq} = D/l + k \times D \tag{1.1}$$

where k is a constant. Figure 1.11 shows a dependence of I_{th} on $(D/l)_{eq}$, in which the experimental data comes from Fig. 1.10. As shown in Fig. 1.11, there is a linear relationship between I_{th} and $(D/l)_{eq}$ subjected to an axial magnetic field, in which k is 0.2. Therefore the equivalent dynamic solid angle unified the two cases in the experiments.

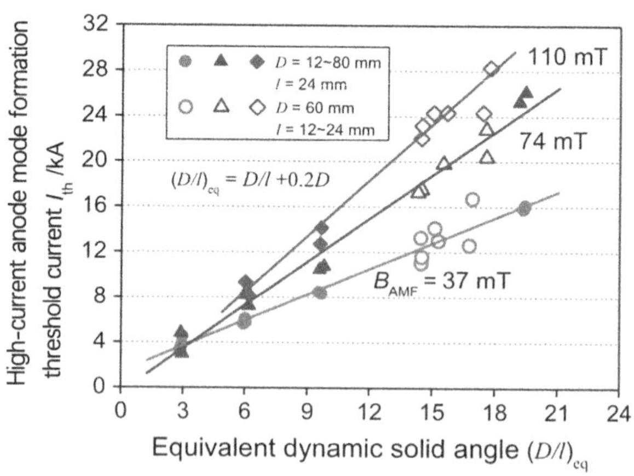

Fig. 1.11 Dependence of high-current anode mode formation threshold current I_{th} on equivalent dynamic solid angle $(D/l)_{eq}$. The experimental data comes from Fig. 1.10

1.2.2 Influence of Axial Magnetic Field on the Threshold Current

Axial magnetic field has a significant effect on anode spot formation in high-current vacuum arcs [30, 31, 33, 37]. A strong axial magnetic field keeps the vacuum arcs in diffuse mode up to high currents. An axial magnetic field also significantly increases the threshold current for anode spot formation [32, 38]. An axial magnetic field prevents positive ions in the vacuum arc plasma from escaping from the contact gap and thus increases the plasma transport to the anode surface. In addition, the plasma plumes overlap less in the interelectrode region, which reduces the arc voltage.

Extensive studies have considered the role of axial magnetic field contact design in vacuum interrupter to prevent anode spot formation in high-current vacuum arcs. Heberlein et al. [39] proposed that a linear function of critical axial magnetic field $B_{\rm crit}$ versus peak arc current $I_{\rm p}$ existed that would prevent anode spot formation. This linear function could be expressed as $B_{\rm AMF}^{\rm crit} \geq 3.1(I_{\rm p} - 10)$, where $B_{\rm AMF}^{\rm crit}$ is in mT and $I_{\rm p}$ is in kA. The constant of 3.1 has a unit of mT/kA. Their results were based on a pair of 100 mm-diameter Bruce profile contacts with an external axial magnetic field coil. Similar results were measured in Schulman and Bindas' work [32], in which they used the same contact structure and diameter. They found that $B_{\rm AMF}^{\rm crit}$ for prevention of anode spots in a high-current vacuum arc followed a function of $B_{\rm AMF}^{\rm crit} \geq 3.2(I_{\rm p}-9)$, where the units were same as in the previous equation.

To assess the effectiveness of an axial magnetic field to control vacuum arcs, an effective area criterion was proposed [40, 41]. Fenski et al. [40] proposed that the axial magnetic flux density covering an effective area that arcs tend to occupy should exceed 4 mT/kA to prevent arc constriction. Stoving and Bestel [41] proposed that $B_{\rm AMF}^{\rm crit}$ must fall in a 15–125 mT target range for a 12 kA (rms) current level, which was defined as an effective area reference.

However, the axial magnetic field criteria described above seems not satisfactory for vacuum interrupter designers, because a wide variety of contact geometries, contact gaps, and contact materials are in use. The criteria should correspond to the axial magnetic field contact design parameters, including the contact diameter, the arc extinguishing contact gap, and the contact materials et al. Previous work mentioned above indicated that $B_{\rm AMF}^{\rm crit}$ was related to the axial magnetic field contact design parameters. However, the data seems insufficient to quantify a relationship between $B_{\rm AMF}^{\rm crit}$ and the axial magnetic field contact design parameters. The following part would determine a criterion for $B_{\rm AMF}^{\rm crit}$ as a function of the contact design parameters in axial magnetic field vacuum interrupters, including the contact diameter, the contact separation gap as arc extinguishing, and the contact materials.

Figure 1.12 shows a dependence of $I_{\rm th}$ on $B_{\rm AMF}^{\rm crit}$ with a variation of D and of l. The contact material was CuCr25 (25% Cr). Figure 1.12a shows the case that dynamic solid angle is adjusted by D, in which D is from 12 to 100 mm while l is fixed at 24 mm. The applied axial magnetic flux density $B_{\rm AMF}$ was 0, 20, 37, 56, 74, and 110 mT, respectively. Figure 1.12b shows the case that dynamic solid angle is adjusted by l, in which l is 12, 18, and 24 mm, respectively while D is fixed at 60 mm. The

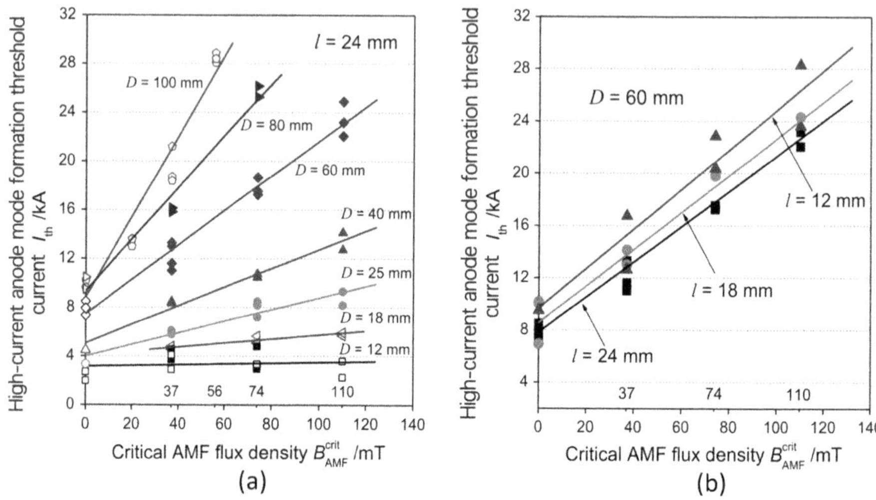

Fig. 1.12 Dependence of high-current anode mode formation threshold current I_{th} on $B_{\mathrm{AMF}}^{\mathrm{crit}}$. Contact material is CuCr25 (25% Cr). Arcing time is 10 ms. **a** Dynamic solid angle *D*/*l* is adjusted by contact diameter *D*. *D* is 12, 18, 25, 40, 60, 80, and 100 mm, respectively. Contact separation length *l* as arc extinguishes is 24 mm. Applied axial magnetic flux density B_{AMF} is 0, 20, 37, 56, 74, and 110 mT, respectively. **b** Dynamic solid angle *D*/*l* is adjusted by contact separation length *l*. *l* is 12, 18, and 24 mm, respectively. *D* is 60 mm. Applied axial magnetic flux density B_{AMF} is 0, 37, 74, and 110 mT, respectively

applied axial magnetic flux density B_{AMF} is 0, 37, 74, and 110 mT, respectively. When the external applied axial magnetic field flux intensity B_{AMF} exceeded $B_{\mathrm{AMF}}^{\mathrm{crit}}$, no high-current anode mode was observed in the arcing period at the given arc current. Otherwise, formation of high-current anode modes did occur. Figure 1.12a indicates that I_{th} is directly proportional to $B_{\mathrm{AMF}}^{\mathrm{crit}}$ for a variety of *D*. It shows axial magnetic field performed a much greater influence on I_{th} for a larger diameter contact than a smaller diameter contact. For a small *D*, the linear relationship between I_{th} and $B_{\mathrm{AMF}}^{\mathrm{crit}}$ tends to be parallel to the horizontal axis. At a *D* of 12 mm, even a very strong axial magnetic field cannot increase I_{th}. In contrast, for a large *D*, the slope of I_{th} increased with $B_{\mathrm{AMF}}^{\mathrm{crit}}$ rapidly. The higher *D* is, the more steepness of the slope is. Figure 1.12b indicates that I_{th} is directly proportional to $B_{\mathrm{AMF}}^{\mathrm{crit}}$ with various *l* of 12, 18, and 24 mm. It is found that the slopes of the three linear relationships were close to each other. This weak dependence of the slopes on *l* may be related to the constant and uniform applied external axial magnetic field. Otherwise, the gap distance would have a major effect on the axial magnetic field and thus on I_{th}.

Figure 1.13 shows a dependence of I_{th} on $B_{\mathrm{AMF}}^{\mathrm{crit}}$ for three contact materials: oxygen free high conductivity (OFHC) Cu, CuCr25 (25% Cr) and CuCr50 (50% Cr). *D* was 60 mm and *l* was 24 mm. The applied B_{AMF} values were 0, 37, 74, and 110 mT, respectively. Figure 1.13 indicates that there is a direct linear relationship between I_{th} and $B_{\mathrm{AMF}}^{\mathrm{crit}}$ for the three contact materials. It also shows that I_{th} was highest for the OFHC Cu contact material when it compared with the two Cu–Cr contact materials.

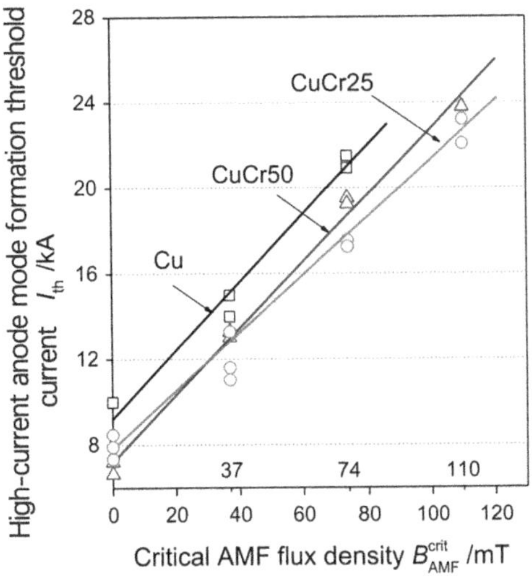

Fig. 1.13 Dependence of high-current anode mode formation threshold current I_{th} on B_{AMF}^{crit} for three Cu-Cr contact materials with different Cr contents by weight. Contact diameter D is 60 mm and contact separation gap l as arc extinguishing is 24 mm

The slopes of the linear relationships between I_{th} and B_{AMF}^{crit} for the three contact materials were close each other.

One can find a linear function that I_{th} is dependent on B_{AMF}^{crit}, as shown in Figs. 1.12 and 1.13. The linear function can be expressed as (1.2).

$$I_{th} = s \times B_{AMF}^{crit} + I_0 \tag{1.2}$$

where I_{th} is the threshold current for high-current anode mode formation (kA); I_0 is the I_{th} without applied axial magnetic field (kA); B_{AMF}^{crit} is the critical axial magnetic field to prevent high-current anode mode formation (mT); s is the slope of the linear relationship.

From (1.2), one can derive the following relations:

$$B_{AMF}^{crit} = k \times (I_{th} - I_0) \tag{1.3}$$

$$k = 1/s \tag{1.4}$$

where k is the critical axial magnetic field strength per unit arc current needed to avoid anode spot formation in a vacuum arcing process with a unit of mT/kA. In the case of $D = 100$ mm in Fig. 1.12, k is 3.0 mT/kA. In addition, I_0 is 9 kA. Therefore, (1.3) becomes (1.5).

$$B_{AMF}^{crit} = 3.0 \times (I_{th} - I_0) \tag{1.5}$$

Table 1.2 summarizes the k values for each contact diameter shown in Fig. 1.12a.

Table 1.2 Dependence of k on contact design parameters. k is the critical axial magnetic field strength per unit arc current needed to avoid anode spot formation in a vacuum arcing process with a unit of mT/kA

D (mm)	l (mm)	k dependent on D (mT/kA)						
		12 mm	18 mm	25 mm	40 mm	60 mm	80 mm	100 mm
12–100	24	308.6[a]	60.5	20.9	13.1	7.1	4.7	3.0

[a]denotes an infinite value shown in Fig. 1.14a, and it is excluded from the fitting in Eq. (1.7)

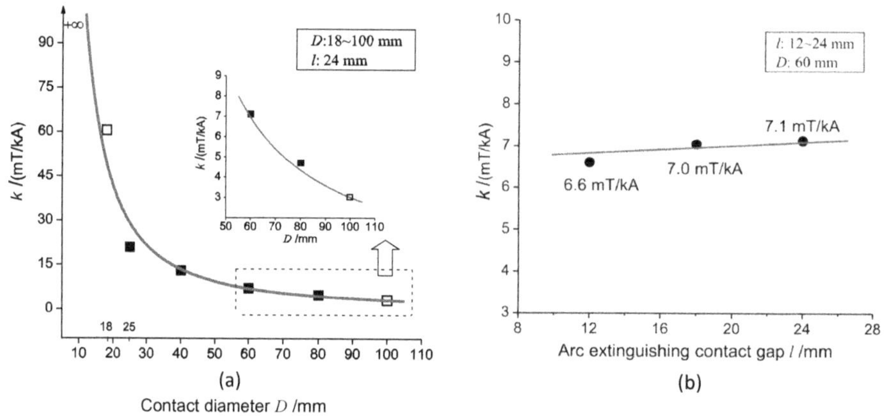

Fig. 1.14 Dependence of k on contact diameter D and arc extinguishing contact gap l. k is the critical axial magnetic field strength per unit arc current needed to avoid anode spot formation in a vacuum arcing process. Its unit is mT/kA. **a** A power dependence of k on contact diameter D. Contact material is CuCr25 (25% Cr). contact diameter $D = 18$–100 mm, and arc extinguishing contact gap $l = 24$ mm. **b** A linear dependence of k on the arc extinguishing contact gap l. Contact material is CuCr25 (25% Cr). contact diameter $D = 60$ mm, and arc extinguishing contact gaps $l =$ 12, 18, and 24 mm

Based on Table 1.2, Fig. 1.14a shows a dependence of k on D. It shows that when D increases from 18 to 100 mm at $l = 24$ mm, k decreases from 60.5 to 3.0 mT/kA, following a power function in the form of:

$$k = qD^{-\alpha} \tag{1.6}$$

where D is the contact diameter (mm) and q and α are the constants. Using the data given in Table 1.2, the power function of shown in (1.6) is fitted as

$$k = 4956.2D^{-1.6} \tag{1.7}$$

The experimentally measured value of k was 308.6 mT/kA at $D = 12$ mm. The data are so high that it is denoted as infinite in Fig. 1.14a and it is excluded from the fitting curve in (1.7). Therefore (1.7) is applicable for $D \geq 18$ mm, which is the case

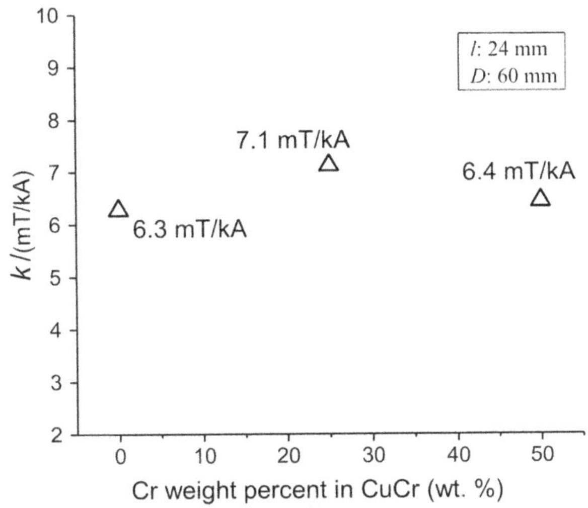

Fig. 1.15 Dependence of k on three contact materials of Cu, CuCr25 (25%), and CuCr50 (50%). Contact diameter $D = 60$ mm and contact separation gap l as arc extinguishing = 24 mm

in most commercial vacuum interrupters. The power function shown in Fig. 1.14a suggests the axial magnetic field has a more significant effect on I_{th} for a larger contact diameter than for a smaller contact diameter in axial magnetic field vacuum interrupters.

Figure 1.14b shows a dependence of k on l based on the data given in Fig. 1.12b. Figure 1.14b shows that k increased linearly but slowly with l, from 6.6 to 7.1 mT/kA when l increased from 12 to 24 mm with $D = 60$ mm. One can find that comparing with the strong influence of D, k is changing insignificantly with l in the range of 12–24 mm.

Based on the data given in Fig. 1.13, Fig. 1.15 shows a dependence of k on the three contact materials. It shows that k are quite close to each other for the three contact materials of Cu, CuCr25 (25% Cr), and CuCr50 (50% Cr). The maximum k of 7.1 mT/kA occurs with the CuCr25 contact material. However, this value is close to those for the OFHC Cu and CuCr50 contact materials, which are 6.3 and 6.4 mT/kA, respectively. Comparing with the influence of D shown in Fig. 1.14a, the influence of different Cr contents on k seems insignificant.

The results shown in Figs. 1.14 and 1.15 suggest that k is mainly determined by D. The l and the Cr content in the Cu–Cr contact materials do not play significant roles. Note, however, that this is valid only for the experimental conditions described above, where the external uniform axial magnetic field is applied, because the axial magnetic field has a significant effect on k.

Figure 1.14a shows that the dependence of k on D follows a power function. As contact diameters decrease, k tended to increase rapidly. There seemed that at a small enough contact diameter k approached infinity. This contact diameter can be defined as the minimum contact diameter. The minimum contact diameter was found to be 12 mm under the given experimental conditions, at which k was as high as 308.6 mT/kA. Figure 1.16 is an enlarged figure that focuses on the data for $D = 12$ mm in

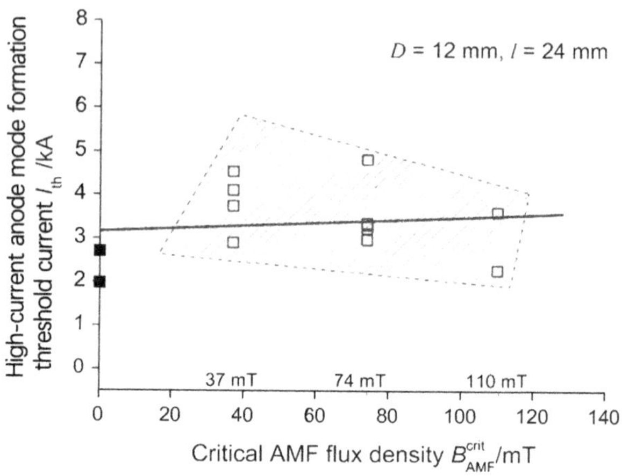

Fig. 1.16 Dependence of I_{th} on $B_{\text{AMF}}^{\text{crit}}$ at the minimum contact diameter at which k approaches infinity. The minimum contact diameter D is 12 mm in this case

Fig. 1.14a. Among the fluctuating I_{th} data, the fitting curve almost paralleled to the horizontal axis, while $B_{\text{AMF}}^{\text{crit}}$ was in a range of 0 to 110 mT. The physical meaning of the minimum contact diameter is that even an infinite high axial magnetic field cannot prevent a formation of high-current anode mode with so small contact diameter. This is probably because, at so small contact diameters, the overall anode power loading is so high that any redistribution by the external axial magnetic field does not help. In contrast, at larger contact diameters, the overall anode power loading can be easily redistributed by the external axial magnetic field.

The power function shown in Fig. 1.14a may help to design compact vacuum interrupters. Modern vacuum interrupters must be of compact size and must have a specific short circuit current interruption capacity. A design criterion would be related to high-current anode mode formation. When the axial magnetic field generated by the designed vacuum interrupter contacts exceeds $B_{\text{AMF}}^{\text{crit}}$, no high-current anode mode will appear, which is helpful for successful high- current interruptions.

A compact vacuum interrupter design may be achieved as following. From Figs. 1.14 and 1.15, one can find that k is not sensitive to l or to the three kind of contact materials. Therefore, k is mainly determined by D, which follows a power law that is shown as (1.6). The contact diameter determines the size of the vacuum interrupter. If a vacuum interrupter designer wants to reduce the contact diameter from a large contact diameter D_{A} to a small contact diameter D_{B}. From (1.6), he can get the required k_{B} at the small contact diameter D_{B} as the follow equation.

$$\frac{k_A}{k_B} = \left(\frac{D_A}{D_B}\right)^{\alpha} \tag{1.8}$$

where k_{A} and k_{B} are the k value at D_{A} and D_{B}, respectively. The α is a constant. It means that a compact vacuum interrupter can be achieved using the smaller contact diameter D_{B} with increasing k_{A} to k_{B}, at which there is no high-current anode mode formation occurring in the vacuum arc with the small contact diameter D_{B}.

Equations (1.6)–(1.8) can provide a semi-empirical information for compact axial magnetic field vacuum interrupter design. For example, if one needs to reduce the contact diameter from 60 to 40 mm, he should increase k from 7.1 to 13.5 mT/kA to ensure the absence of high-current anode mode. It is interesting that one can find the effective area criterion of 4 mT/kA proposed by Fenski et al. [40] is corresponding to a contact diameter of 80 mm shown in the enlarged figure of Fig. 1.14a. That means the effective area criterion of 4 mT/kA is not a general rule for axial magnetic field contacts design because it fits only to a specific contact diameter. If the vacuum interrupter designer wants to consider a more general rule, the rule described by Eqs. (1.6)–(1.8) seems to be a good candidate.

We should also note that these equations cannot be used directly as a basis for vacuum interrupter design, because the experiments used the Helmholtz coil to generate a uniform axial magnetic field, which is not the case in commercial vacuum interrupters. The contact gap length also plays a major role in commercial vacuum interrupters, because it affects the axial magnetic field amplitude and distribution between the contacts greatly. Also, in commercial vacuum interrupters, the effective axial magnetic field area is normally limited by the inner contact diameter and not by the external diameter. In addition, the axial magnetic field in a commercial vacuum interrupter is generated from the applied current itself through the contacts. Finally, a phase shift is always observed between the current wave and the generated axial magnetic field in commercial vacuum interrupters. More work is needed to translate the rule described by Eqs. (1.6)–(1.8) to be suitable for the design of compact commercial axial magnetic field vacuum interrupters.

1.3 Current Density of Anode Spots Subjected to Axial Magnetic Field

The anode current density distribution is a significant influencing factor for the formation of anode spot [42, 43]. It is well known that the anode current density distribution well reflects that of the anode energy flux. So that the anode current density distribution is a significant influencing factor for the formation of the anode spot. Moreover, the anode current density distribution is significantly influenced by an axial magnetic field in vacuum arcs [26, 29–32, 37, 44–46]. The axial magnetic field has an influence on both the positive ions in the vacuum arc plasma and the plasma-plume overlap region between contacts. As a result, the arc voltage first decreases as axial magnetic field increases until reaching a minimum. Subsequently, the arc voltage increases as axial magnetic field further increases.

The anode current density distribution is an important physical quantity in vacuum arcs, which is particularly associated with the anode energy flux density. Therefore, considerable attention has been focused on measuring the anode current density distribution in vacuum arcs. Mitchell used a 7-element anode to measure the anode current distribution without the application of a magnetic field [17]. The seven anode elements were spaced apart, and the current of each element was recorded from the potential of an element tube, which was used as a shunt resistor. Sherman et al. measured and calculated the arc current distribution using a magnetic probe at arc currents without an application of external magnetic field [47]. Schellekens measured the arc current distribution based on a ring-shaped contact configuration with an application of axial magnetic field [48]. The results revealed the influence of the axial magnetic field on the arc current distribution at an arc current of 5 kA. Drouet et al. measured the distribution using a multi-ring anode in a diffuse vacuum arc for several amplitudes of the arc current with different contact gaps [49]. Moreover, many studies on simulating vacuum arcs with an external axial magnetic field have been conducted [34, 42, 50, 51, 52, 53, 54, 55]. The works performed by Londer and Ulyanov [34], Izraeli et al. [53] and Wang et al. [54] focused on the effect of the axial magnetic field on high-current vacuum arcs. Numerical simulations conducted by Schade and Shmelev [50] and Wang et al. [55] revealed the physical behaviors and energy flux into the anode in high-current vacuum arcs.

In the following, we will experimentally determine the current density of anode spots subjected to axial magnetic field. However, the anode current density distribution in diffuse vacuum arcs subjected to axial magnetic field will be discussed first as a benchmark.

1.3.1 Anode Current Density Distribution in Diffuse Vacuum Arcs

Drawn arc experiments were conducted in a demountable vacuum chamber under an external and adjustable axial magnetic field. The experimental results provide evidence on the effect of the axial magnetic field and arc current on anode current density distribution in diffuse vacuum arcs. The experimental results are quantitative, so it can verify vacuum arc models.

The experiment was conducted using a *LC* discharging circuit with an external axial magnetic field generation circuit, as shown in Fig. 1.17. The arc current reached 20 kA (rms) with a current frequency of 50 Hz. The arcing time was controlled as approximately 10 ms. The average opening velocity was set as 1.8 and 2.4 m/s, which led a contact separation length as arc extinguishing of 18 and 24 mm, respectively. The full contact gap was 35 mm. The contacts were installed inside the demountable vacuum chamber, which was evacuated and maintained at 10^{-3}–10^{-4} Pa. A high-speed charge coupled device (CCD) video camera was used to record the evolution of the vacuum arc through an observation window of the demountable vacuum chamber. The camera recorded at 10,000 frames/s. Its aperture was fixed at 4 and the exposure time was set at 2 μs. The adjustable axial magnetic field was generated by an external, DC-excited Helmholtz coil, mounted coaxially with the contacts in the chamber. The Helmholtz coil had 80 turns and its average radius was 80 mm. A DC current source for the Helmholtz coil was turned on approximately 150 ms before contact separation.

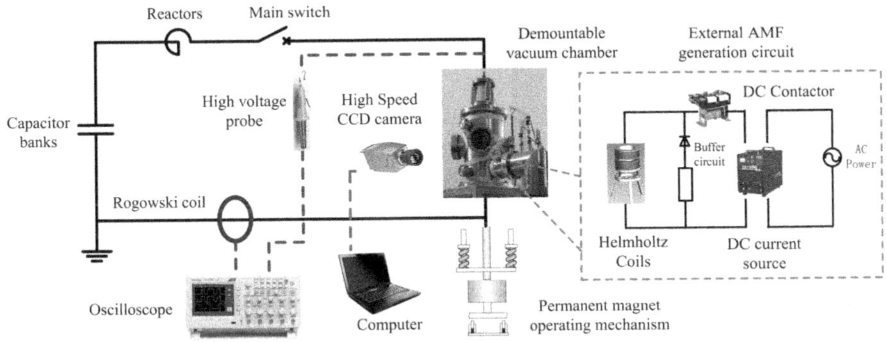

Fig. 1.17 An experimental system of *LC* discharging circuit with an external axial magnetic field generation circuit

B_{AMF} between the contacts was uniform and it can be adjusted from 0 to 110 mT. The generated axial magnetic field was in a direction from the upper contact to the lower contact.

Figure 1.18 shows an experimental setup for determination of anode current density distribution in diffuse vacuum arcs. An axial symmetrical four-area split contact and a butt-type contact were selected as the anode and the cathode, respectively. Under an application of the external axial magnetic field, it was experimentally found that the arc columns tended to concentrate in the central area of the anode. That is why an axisymmetrical, four-area split contact was selected as the anode, aiding in the measurement of the anode current density, by which one can easily focus on the current density of the anode central area. The contact material for both anode

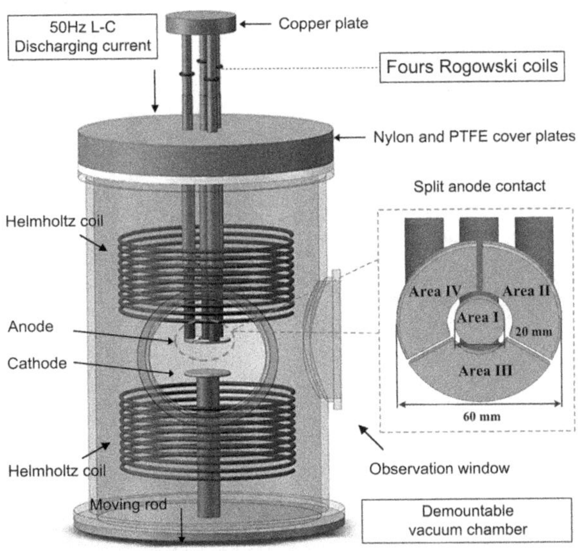

Fig. 1.18 An experimental setup for determination of anode current density distribution in diffuse vacuum arcs. It included vacuum chamber, Rogowski coils, Helmholtz coils, and the split four-area anode

and cathode was CuCr25 (25% Cr). The contact diameter of anode and cathode was same, which was 60 mm. The central area of the four-area split anode was labelled as Area I, and it had a diameter of 20 mm; the annular area was divided evenly into three annular arcs which were labelled Area II, Area III, and Area IV, respectively. There was a separation gap between the central area and the three peripheral areas and the gap was 2 mm. The central area was 314 mm^2, and the three annular areas were 778 mm^2 each.

The current in the four anode areas were measured by four Rogowski coils, which located outside of the demountable vacuum chamber. Four copper rods were connected to the four anode areas for current flowing. There are two cover plates enclosing the top and bottom side of the demountable vacuum chamber's cylinder body. The two cover plates were composed of insulating materials, with a combination of a nylon layer and a polytetrafluoroethylene (PTFE) layer. The two cover plates separated the current paths of the four split anode areas. The magnetic field induced by the current in nearby cables has an influence of the current density distribution to be measured. Figure 1.19 shows a schematic of the cable arrangement in the experiments. The distance from contacts to the current cable was 1.5 m. Area II was located at a further position to the cable. Furthermore, Area II was set at a nearest position to the camera.

Figure 1.20 shows typical experimental results for the anode current density distribution measurement with the split-anode, in which the arc column remained in the diffuse arc mode. The arc current was 10 kA (rms). The applied external B_{AMF} was 74 mT. Figure 1.20a shows the arc appearance at the current peak. In the figure,

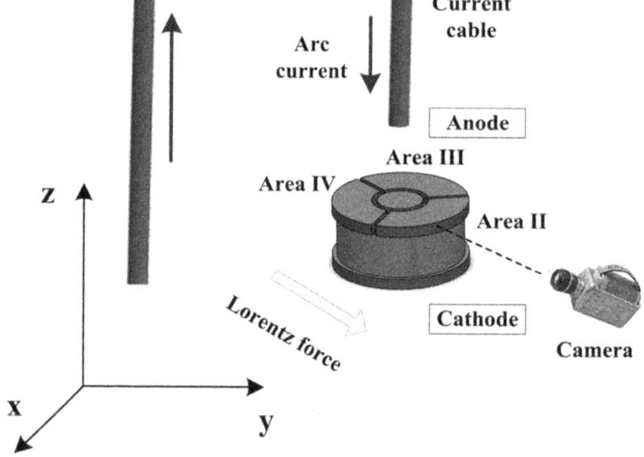

Fig. 1.19 Schematic of the cable arrangement in the current feeding circuit and experimental contacts

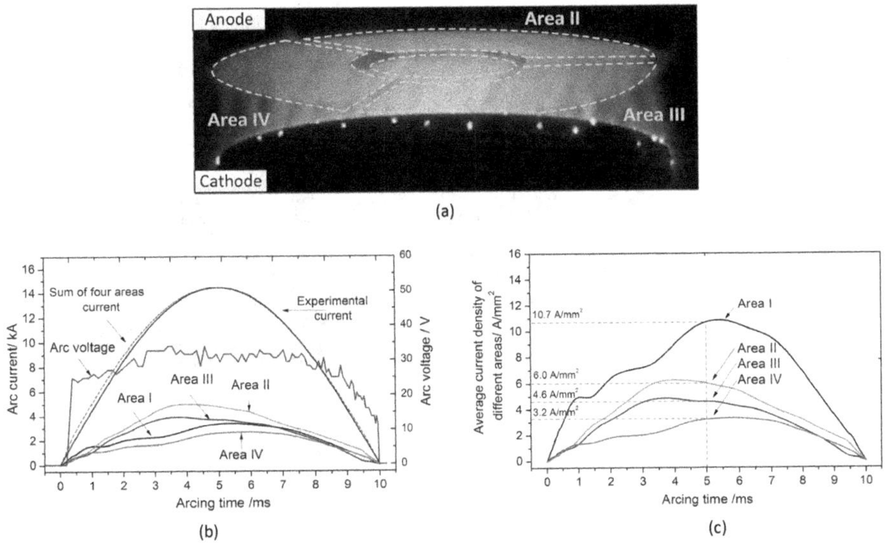

Fig. 1.20 Typical experimental results with a split-anode and a butt-type cathode in a diffuse vacuum arc. Arc current is 10 kA (rms); The arcing time is ~10 ms. B_{AMF} was 74 mT. **a** Diffuse vacuum arc appearance at the current peak; **b** Arc voltage and arc current in each of the four anode areas. The sum of the currents for the four areas (dashed line) approximated the experimental arc current (solid line); **c** Average anode current densities in areas I–IV

the split anode areas are outlined with dashed lines, and the three annular areas are labeled. It indicated a diffuse vacuum arc. No erosion activity was observed on the anode surface. The brightness of the inter-electrode plasma did not significantly vary in the axial direction. However, in the radial direction, the arc intensity in the central area was brighter than in the peripheral areas. Figure 1.20b shows the measured current waveforms of the four areas and the arc voltage. It shows a significant difference between the current waveform of each anode area. The sum of the four currents, shown as a dashed curve, well approximated the total arc current, which was shown as a solid line. The arc voltage reflected a typical diffuse vacuum arc. It ranged between 14 and 30 V and it was smooth during the arcing period. Figure 1.20c shows the average anode current densities in the four anode areas I–IV. The average anode current density in each area was determined as a ratio of the arc current in the area I to its area S: $J = I/S$. It shows the average current densities in the four anode areas at the arc current peak were 10.7, 6.0, 4.6, and 3.2 A/mm^2, respectively. It revealed that the current density distribution on the anode surface was not uniform in the diffuse arc mode subject to axial magnetic field, but rather the current concentrated in the central area. The current density in the central area was significantly higher than that in the three peripheral annular areas. Furthermore, the average current densities in the three peripheral annular areas differed significantly from each other after 1 ms.

What is the influence of axial magnetic flux density on the anode current density distribution? Figure 1.21 shows an effect of axial magnetic field on the vacuum arc

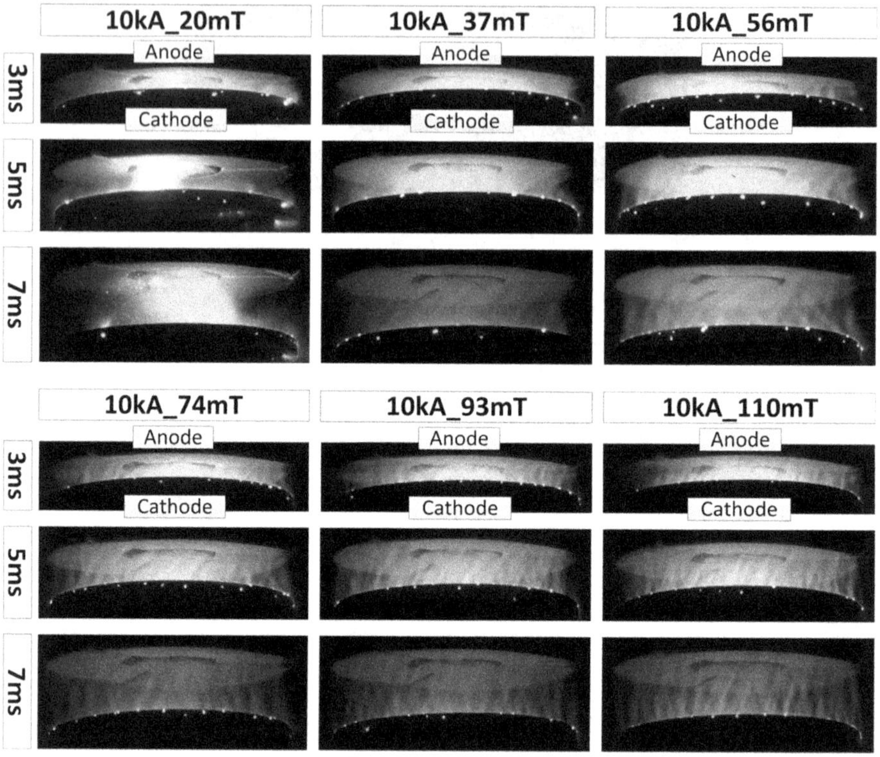

Fig. 1.21 Effect of axial magnetic field on the vacuum arc discharging process. Photographs of arc discharging at arcing times of 3, 5, and 7 ms with B_{AMF} of 20, 37, 56, 74, 93, and 110 mT Arcing current was 10 kA (rms)

discharging process. It shows a series of photographs of the vacuum arcs subjected to axial magnetic fields of 20, 37, 56, 74, 93, and 110 mT, respectively. Typical arc photographs are presented at arcing time of 3, 5, and 7 ms. Arc current was 10 kA (rms). It showed that, with a B_{AMF} of 20 mT, a bright column appeared in the central arcing region at 3 ms. At 5 ms (arc current peak), the arc column constricted in the central region, which emitted an intense light. At 7 ms, the arc still constricted with an intense light emission. Erosion can be observed in the central anode area after arcing. When B_{AMF} increased to 37 mT, at 5 ms, the arc column became more uniform than that of 20 mT. The anode acted as a collector of the arc current without any significant activity. When B_{AMF} increased to 56 mT, the arc appearance at current peak was like that of 37 mT. When B_{AMF} further increased to 74 mT and higher, the arc appearances at current peak seemed unchanged to that of 74 mT.

Figure 1.22 shows a dependence of the average current density in the central anode area on the applied axial magnetic field. The arc current was 10 kA (rms). It shows that the average current density in the central anode area at the current peak $J^{\mathrm{Peak}}_{\mathrm{Area\,I}}$ decreased as B_{AMF} increased. As B_{AMF} increased from 20 to 56 mT,

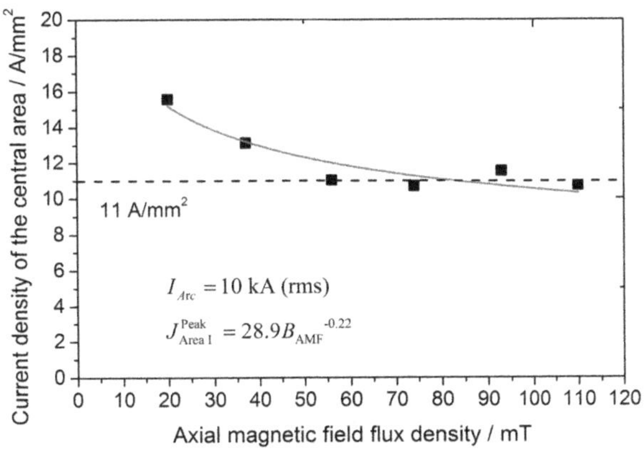

Fig. 1.22 Dependence of the current density in the central area on axial magnetic field. Arc current is 10 kA (rms). Solid line-the fitted curve of the power dependence of $J^{Peak}_{Area\,I}$ on B_{AMF} in Eq. (1.10)

$J^{Peak}_{Area\,I}$ rapidly decreased from 15.6–11.0 A/mm^2. As B_{AMF} further increased to 110 mT, $J^{Peak}_{Area\,I}$ remained almost a constant at approximately 11.0 A/mm^2. Figure 1.22 suggests that the relationship between $J^{Peak}_{Area\,I}$ and B_{AMF} follows a power function as shown Eq. (1.9), where *k'* and α are constant. Based on the data shown in Fig. 1.22, the power function of the $J^{Peak}_{Area\,I}$ dependence on B_{AMF} is fitted as Eq. (1.10).

$$J^{Peak}_{Area\,I} = k' B^{\alpha}_{AMF} \tag{1.9}$$

$$J^{Peak}_{Area\,I} = 28.9 B^{-0.22}_{AMF} \tag{1.10}$$

where B_{AMF} was the axial magnetic flux density in mT ranging from 20 to 110 mT, and $J^{Peak}_{Area\,I}$ was in A/mm^2. The fitted equation is shown as the solid curve in Fig. 1.22.

Figure 1.23 shows an effect of the axial magnetic flux density on the average current densities of the three peripheral annular areas at the arc current peak. The three peripheral annular areas were labeled Area II, Area III, and Area IV in Fig. 1.20a.

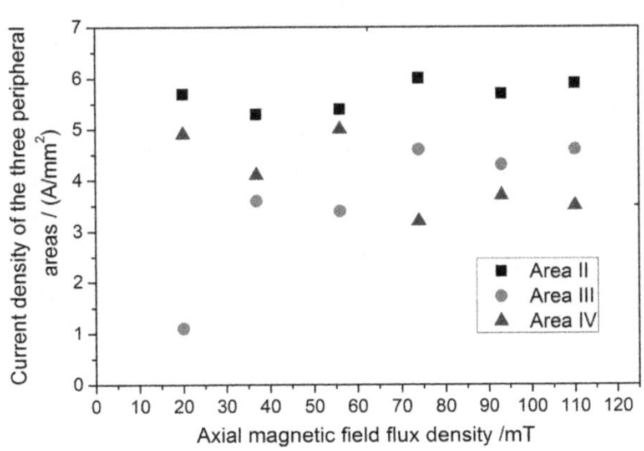

Fig. 1.23 Dependence of the average current densities in the three peripheral areas. B_{AMF} was 20, 37, 56, 74, 93, and 110 mT, respectively; Arc current was 10 kA (rms)

B_{AMF} was 20, 37, 56, 74, 93, and 110 mT, respectively. Arc current was 10 kA (rms). It shows the distributions of the average current densities in the three peripheral areas were quite uneven. The average current density in Area II (~5.8 A/mm^2) was higher than that in the other two areas. When the axial magnetic field was between 20 and 56 mT, the average current density in Area IV was higher than that in Area III, even though there was a significant fluctuation. When the axial magnetic field was between 74 and 110 mT, the average current density in Area III was higher than that in Area IV. As the axial magnetic flux density increased, the average current densities in the three peripheral areas became stable. The anode current density distribution stabilized with an increasing of axial magnetic flux density.

Figure 1.24 shows an effect of the arc current amplitude on the vacuum arc appearance. Three typical arcing instants are shown, which were 3, 5, and 7 ms, respectively. The arc current was 4, 6, 8, 10, 12, and 14 kA (rms), respectively. B_{AMF} was 74 mT. At 5 ms (arc current peak), the arc remained diffuse as arc current increased from 4 to 10 kA. As the arc current further increased to 12 and 14 kA, the arc column became brighter and it tended to constrict, as indicated by the intense luminance. At

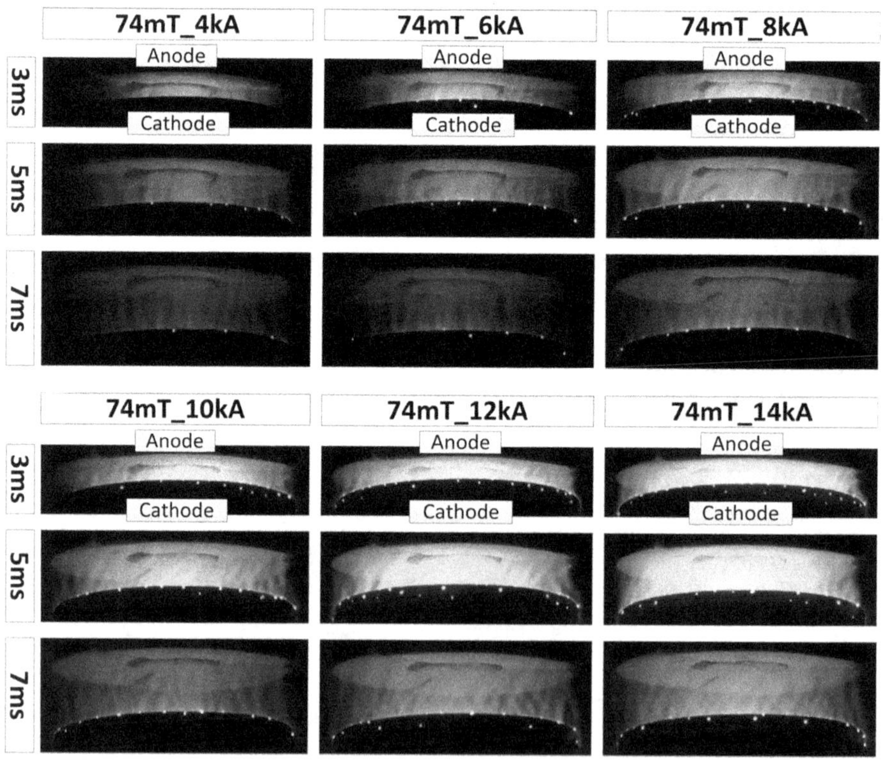

Fig. 1.24 Effect of the arc current amplitude on the vacuum arc appearance. The arcing time are 3, 5, and 7 ms, respectively. Arc current is 4, 6, 8, 10, 12, and 14 kA (rms), respectively. B_{AMF} is 74 mT

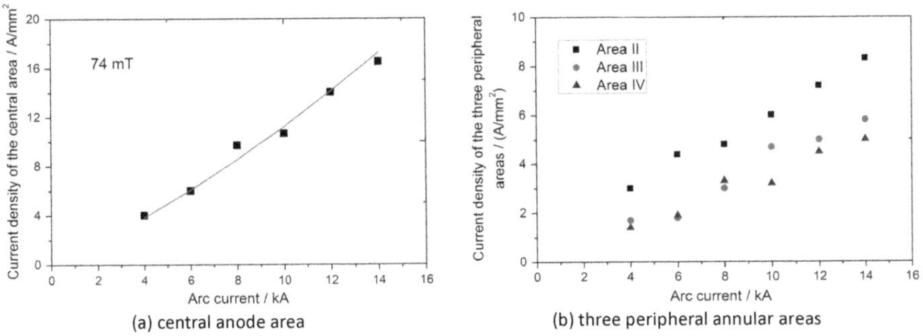

Fig. 1.25 Dependence of the average current density of each split-anode area on the arc current. B_{AMF} is 74 mT. The solid line shows the power law dependence of $J_{Area\ I}$ at the current peak on I_{ARC} in Eq. (1.12)

7 ms, at all the arc current the columns were diffuse. The anode acted as a passive collector of the arc current and there was no erosion observed on the anode surface.

Figure 1.25 shows a dependence of the average current density of each split-anode area at current peak $J_{Area\ I,II,III,IV}$ on the arc current. B_{AMF} was 74 mT. Figure 1.25a shows $J_{Area\ I}$ of the central area and Fig. 1.25b shows $J_{Area\ II,III,IV}$ of the three peripheral annular areas. Figure 1.25a indicates that $J_{Area\ I}$ increased as arc current increased from 4 to 14 kA. At arc current of 4 kA, $J_{Area\ I}$ was 4.0 A/mm^2. At arc current of 14 kA, $J_{Area\ I}$ increased to 16.5 A/mm^2. Figure 1.25b shows $J_{Area\ II,III,IV}$ increased as the arc current increased. $J_{Area\ II}$ was higher than that in the other two peripheral areas.

Figure 1.26 shows the average current density in the central anode area as a function of the axial magnetic field and arc current. The arc current was 4, 6, 8, 10, 12 and 14 kA (rms), respectively. For arc current of 4–10 kA, B_{AMF} of 0–110mT was applied. For arc current of 12–14 kA, B_{AMF} of 56–110 mT was applied. For each arc current, $J_{Area\ I}^{Peak}$ decreased with an increasing of B_{AMF}, and this decrease followed a power function. Table 1.3 shows a dependence of the coefficient *k'* in Eq. (1.9) on arc current. The dependence of *k'* on I_{ARC} was fitted to a non-linear Eq. (1.11).

$$k' = mI_{Arc} + nI_{Arc}^2 \tag{1.11}$$

where m and n are constants. Based on Table 1.3 and Fig. 1.26, it was fitted that $m = 2.2$ and $n = 0.069$. According to Eqs. (1.9) and (1.11), the dependence of $J_{Area\ I}^{Peak}$ on B_{AMF} (in the range of 20–110 mT) and on I_{ARC} (in the range of 4–14 kA (rms)) can be fitted to Eq. (1.12).

$$J_{AreaI}^{Peak} = \left(2.2I_{Arc} + 0.069I_{Arc}^2\right)B_{AMF}^{-0.22} \tag{1.12}$$

where B_{AMF} was in mT, $J_{Area\ I}^{Peak}$ was in A/mm^2, and I_{ARC} was in kA. The solid lines in Fig. 1.26 are the fitted curves of the power function of $J_{Area\ I}^{Peak}$ on B_{AMF} shown in Eq. (1.12) at various I_{ARC}. Figure 1.27 shows a dependence of $J_{Area\ I}^{Peak}$ on the arc

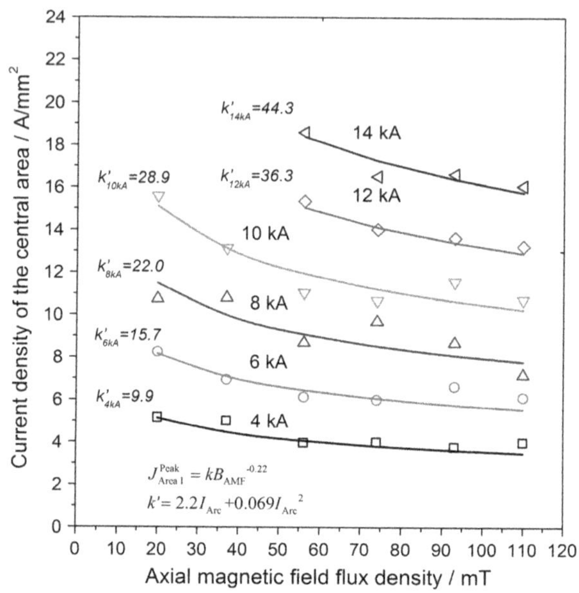

Fig. 1.26 Average current density of the central anode area as a function of the axial magnetic field and arc current. Arc current is 4, 6, 8, 10, 12, and 14 kA (rms), respectively. The fitted curves are the power dependence of $J_{\text{Area I}}^{\text{Peak}}$ on B_{AMF} in Eq. (1.12) (solid line)

Table 1.3 A dependence of k' in Eq. (1.9) on I_{ARC}

I_{ARC}/kA	k' dependence on I_{ARC} (rms)					
	4 kA	6 kA	8 kA	10 kA	12 kA	14 kA
4–14 kA	9.9	15.7	22.0	28.9	36.3	44.3
B_{AMF}/mT	20–110 mT				56–110 mT	

currents under B_{AMF} within a range of 20–110 mT, which is based on the same data shown in Fig. 1.26. It shows $J_{\text{Area I}}^{\text{Peak}}$ increased with the arc current. With a higher B_{AMF}, $J_{\text{Area I}}^{\text{Peak}}$ tended to be lower.

The experimental result was similar to that of Schellekens [48]. He found that the current density in the central area on an anode was approximately 3 A/mm^2 with an arc current of 5 kA subjected an axial magnetic flux density of 200 mT. The diameter of the anode in Schellekens' experiments was 60 mm, which was as same as the split anode. By using Eq. (1.12), one can get the average current density in the central anode area at current peak $J_{\text{Area I}}^{\text{Peak}}$ was approximately 4 A/mm^2 with an arc current of 5 kA subjected to an axial magnetic field of 200mT.

The experimental results also revealed an uneven current distribution characteristic on the three peripheral annular areas in diffuse arcs subjected to an external axial magnetic field. Figures 1.23 and 1.25b show that the average current density in Area II was higher than that in the other two peripheral areas. This phenomenon might be caused by an influence of the magnetic field generated by the current flowing in cables connecting the contacts. Figure 1.19 shows the schematic of the cable arrangement in the experiments. The vacuum arc column was driven by the Lorentz force generated

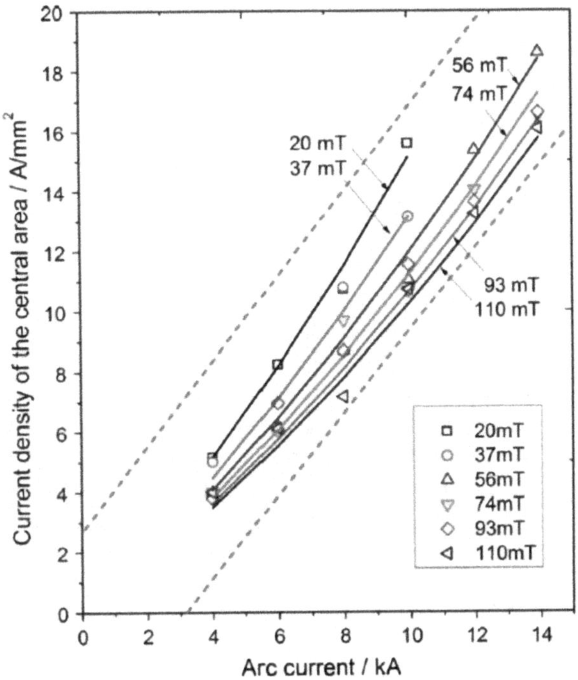

Fig. 1.27 Dependence of the average current density of central anode area on the arc current under various axial magnetic field. Hollow symbols—$J_{\text{Area I}}$ at the current peak at various B_{AMF} (20, 37,56, 74, 93, and 110 mT). Solid line—the power dependence of $J_{\text{Area I}}$ at the current peak on B_{AMF} given by Eq. (1.12)

by the current flowing through the cable. As Area II was located at a further position to the cable, more current was pushed to Area II by the magnetic field. So that the arc current in Area II was higher than that in Area III and Area IV. Thus, one can find a uneven distribution of the anode current density in the three peripheral annular areas.

The quantitative experimental results of the anode current density subjected to various axial magnetic fields may be used to validate the simulation results of vacuum arc models. One most related work was done by Schade and Shmelev [50]. They summarized the maximal heat flux densities at the centre of the anode predicted by numerical simulations, in which the heat flux densities were determined at constant ratios of B_{AMF} to arc current. The experimental results are also helpful for establishing appropriate boundary conditions for the vacuum arc models, which is quite meaningful for studies of the vacuum arc and the anode spot phenomenon [56].

1.3.2 Anode Spot Current Density

Now it is a right point to discuss the current density of anode spots subjected to axial magnetic field. This part will experimentally determine anode spot current density distribution by the following way. Once an anode spot forms, it occupies a local anode area, on which an eroded region can be observed after arc extinguishes. The locally anode spot occupied area, which corresponds the eroded anode region, increases with an increase of arc current. If one increases the arc current from a low current to a high current step by step, a diffuse arc will first appear. Then a small anode spot appears with further increasing of arc current. Finally, the central anode region (Area I shown in Fig. 1.18) will be entirely covered by a large anode spot and its erosion. At this point, the anode spot current density can be determined as the average current density of Area I. If one varies the size of Area I, more data can be obtained by the same way. Therefore, the anode spot current density was measured ranging from 31.9 to 37.7 A/mm^2. The measured results should contribute a basic experimental parameter to analyze the thermal processes of the anode in the vacuum arcs, during which the energy flux density input to the anode would be regulated by the anode current density and anode sheath voltage.

The experimental conditions were kept almost as same as in Sect. 1.3.1, except the axial magnetic field here was supplied by a cup-type axial magnetic field contact. So that the measured anode spot current density should reflect more closely to that of commercial vacuum interrupters than an external applied uniform axial magnetic field. The experiment was conducted with the same *LC* discharging circuit as shown in Fig. 1.17. The experimental setup for anode current density measurement included a pair of contacts installed inside a demountable vacuum chamber. The demountable vacuum chamber and the measuring method was as same as Fig. 1.18. Figure 1.28 shows the designs of contacts for anode spot current density measurement. It included a four-area split anode and a cup-type axial magnetic field cathode. The materials for both contacts was CuCr25 (25% Cr). The diameter of both anode and cathode is 60 mm. Three different diameters of Area I was designed as 12, 18, and 20 mm, respectively. The width of the gap between Area I and Area II, III, IV was 2 mm.

Figure 1.29 shows a typical axial magnetic field distribution on the split-anode surface. The current was 1 kA (rms). The contact separation length of 9 mm was set by taking account at current peak with an opening velocity of 1.8 m/s and an arcing

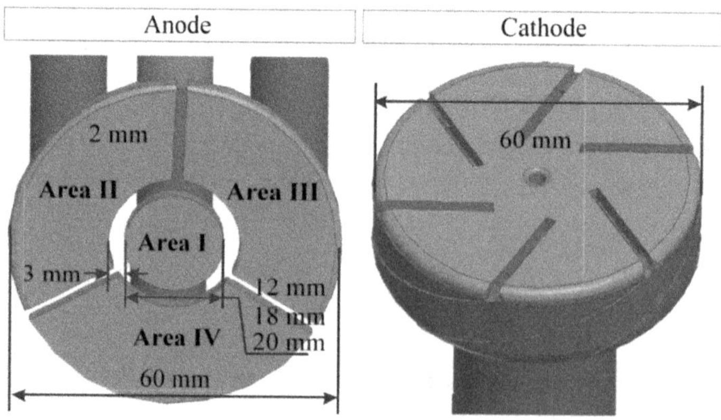

Fig. 1.28 Design of a split four-area anode and the cup-type axial magnetic field cathode

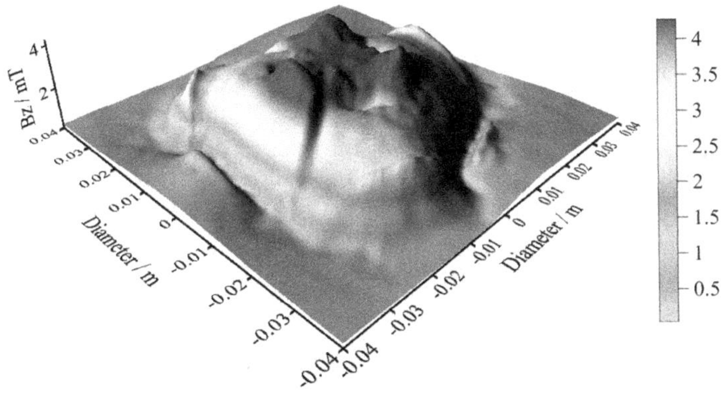

Fig. 1.29 Axial magnetic field distribution on the split-anode surface at current peak. The arc current is 1 kA (rms). The contact separation length is 9 mm. The contact configuration is shown in Fig. 1.28

time of 10 ms. The axial magnetic field at current peak (arcing time of 5 ms) was shown, which typically corresponded to an occurrence of the anode spots. The arc plasma was substituted by a conductor with the same diameter as the contacts; its conductivity was 1579 S/m. The arc conductivity was determined by the arc voltage and the arc current in the experiment. Figure 1.29 shows that the axial magnetic field distribution with a configuration of contacts shown in Fig. 1.28 is slightly different with that of the traditional cup-type axial magnetic field contact.

Figure 1.30 shows the axial magnetic field distributions on the cup-type cathode surface, the intermediate plane, and the split-anode surface at current peak. The

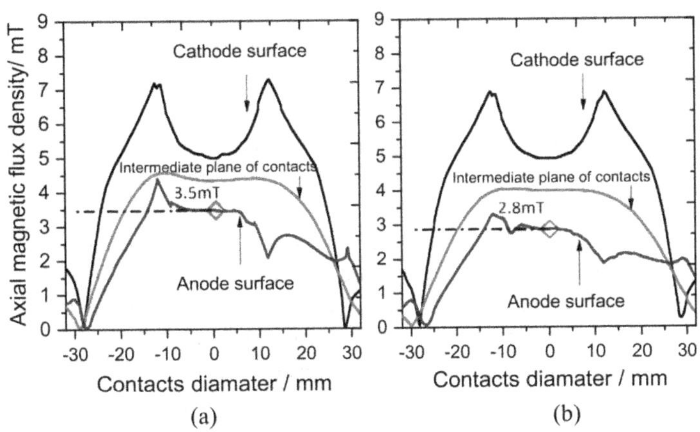

Fig. 1.30 Axial magnetic field distributions on the cathode surface, intermediate plane, and anode surface at current peak. The arc current is 1 kA (rms). **a** Contact separation length was 9 mm with an opening velocity of 1.8 m/s. **b** Contact separation length was 12 mm with an opening velocity of 2.4 m/s

current is 1 kA (rms). The arcing time was 10 ms. The opening velocity was 1.8 m/s. The contact separation length at current peak (5 ms) was 9 mm. In Fig. 1.30a, the axial magnetic field at the anode center point was 3.5 mT. In Fig. 1.30b, the opening velocity was 2.4 m/s. The contact separation length at arc current peak (5 ms) was 12 mm. The axial magnetic field at the anode center point was 2.8 mT. Moreover, despite different opening velocities, the axial magnetic field on the cathode surface were nearly equal.

Figure 1.31 shows the anode discharge measurement results at an arc current of 8 kA (rms). Figure 1.31a shows the time-sequence photographs of anode discharge evolution. It shows that the arc remained in the diffuse arc mode. The arc emitted a dim light. The anode collected and absorbed the particles produced by the cathode. No significant activity was observed on the anode surface. Furthermore, within the time interval of 4–7 ms, the arc intensity in the central region was brighter than that in the peripheral region, indicating a difference in current density distribution on different anode area. Between the time interval of 8 and 9 ms, arc brightness weakened and disappeared. During the arcing period, there is no visible erosion of the anode surface. Figure 1.31b shows the arc voltage and arc current waveforms of the four split anode areas. It indicates that in first few milliseconds after contact separation, the arc current was quite unevenly distributed over the four split anode areas. Initially, it concentrated in Area II. This was because an initial bridge column arc formed in Area II, which was reflected by an erosion region in Area II that was observed in a post-test examination. As the arc further expanded by an aid of axial magnetic field, Areas III and then Area I subsequently shared more current. The sum of the current of the four split anode areas (dashed line) approximately matched the total current (solid line). The arc voltage was low, which ranged between 14 and 30 V and it appeared smooth during the arcing period. Figure 1.31c shows the average anode current densities for each of the four areas. It shows that the maximum value of the average current density in peripheral areas reached 7.8 A/mm^2, occurs at 1.5 ms. The maximum value of the average current density of the central area reached 14.4 A/mm^2 at 4 ms. It indicated that, the current density in each area exhibited large discrepancies although the arc remained in the diffuse arc mode.

Then the arc current increased to 10 kA (rms). Figure 1.32 shows the anode discharge measurement results. The time-sequence photographs of anode discharge evolution are shown in Fig. 1.32a. It shows that between 4 and 6 ms, a bright region appeared on the central region of the anode surface. It indicated that an anode spot formed. In this period, the arc column tended to constrict. Between 7 and 8 ms, the brightness of the arc column weakened, but the central region of the anode remained bright. Moreover, an erosion on the anode was clearly observed during this period. At 9 ms, the arc column returns diffused; an erosion area was seen on the central region of anode. Figure 1.32b shows the arc voltage and arc current waveforms of the four split anode areas. It shows that in the first few milliseconds after contact separation, the arc current distributed very unevenly over the four anode areas. Initially, it concentrated in Area III. With an increasing of the current, the arc expanded. After 4 ms, the arc current mainly concentrated in Area I. The sum of the current of the four split anode areas (dashed line) approximately matched the total current measured in the

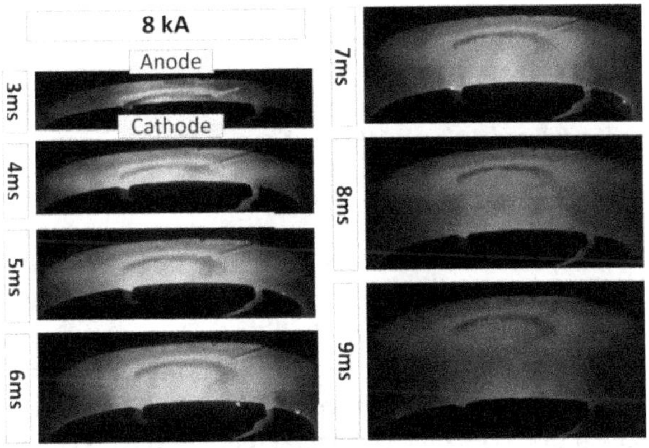

(a) Time-sequence photographs of anode discharge evolution

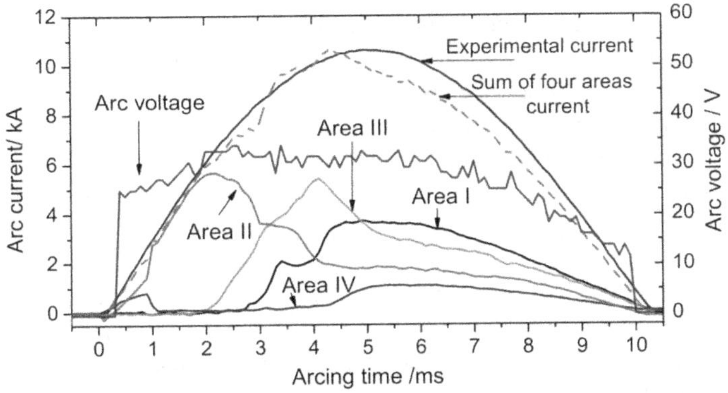

(b) Arc voltage and arc current of the four split anode areas

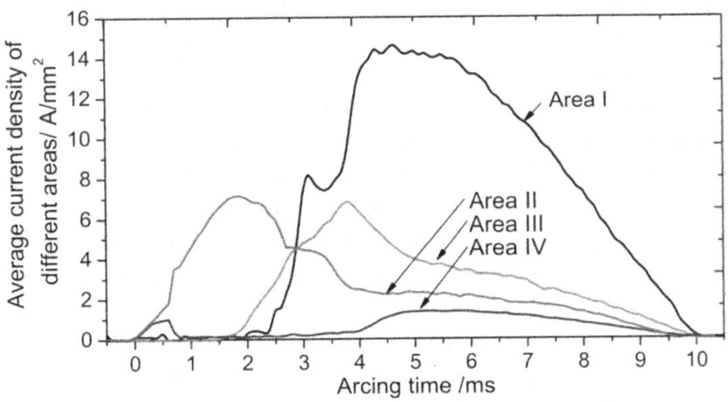

(c) Average anode current densities of the four split anode areas

◂**Fig. 1.31** Anode discharge measurement results at an arc current of 8 kA (rms). Contact configurations see Fig. 1.28. Both contact diameters are 60 mm. The diameter of Area I is 18 mm. Arcing time is 10 ms. The average opening velocity is 2.4 m/s. The contact separation length as arc extinguishing is 24 mm. The full contact gap is 35 mm. The sum of the current of the four split anode areas (dashed line) approximates the experimental current (solid line)

experiment (solid line). The arc voltage remained in a range of 10–50 V with some noise component. Figure 1.32c shows the average anode current densities for each of the four split anode areas. It shows that in the central area the current density was much higher than that of the peripheral regions. During arcing period of 3–5 ms, the current density in the central area increased rapidly, reaching a maximum value of 23.3 A/mm^2 at 4.5 ms. While the peripheral regions, the maximum value reached 7.7 A/mm^2 at 3 ms. Combined with the photographs shown in Fig. 1.32a, the results indicate that the anode current constricted more with an anode spot formation, and the distribution of anode current density became more uneven. As the erosion region caused by the anode spot did not cover the entire Area I, the measured average current density of 23.3 A/mm^2 should be less than the current density of an anode spot.

With a further increasing of arc current to 14 kA (rms), it was finally found that the erosion region caused by an anode spot did cover the entire Area I. Figure 1.33 shows the anode discharge measurement results at an arc current of 14 kA. The time-sequence photographs of anode discharge evolution are shown in Fig. 1.33a. It shows that the erosion region caused by an anode spot did cover the entire Area I, which was different with the discharging at a lower current of 10 kA (see Fig. 1.32a). When the arc current increased to 14 kA, the arc column emits luminous light and it began to constrict. The anode became active. The anode erosion area was expanding by a formation of a bright anode spot observed with a significant anode jet in the arcing period of 5–6 ms. Between the arcing period of 8 and 9 ms, the brightness of the arc weakened, and an eroded region was clearly seen in the Area I. Figure 1.33b shows the arc voltage and arc current waveforms of the four split anode areas. The arc current waveforms of the four split anode areas showed that the current concentrated in Area II initially. With an increasing of current, the arc expanded. After 4 ms, the arc current mainly concentrated in Area I. The sum of the currents of the four split anode areas (dashed line) approximately matched the total current measured in experiment (solid line). The arc voltage remained in a range of 35–70 V with a significant noise component. Figure 1.33c shows the average anode current densities for each of the four split anode areas. It indicated that the average current density of the central area is much higher than that of the peripheral areas. The average current density of Area I reached the maximum value of 37.7 A/mm^2 at 6 ms. The maximum value of the peripheral regions reached 7.2 A/mm^2 at 1.7 ms. Referencing to the corresponding arc photographs shown in Fig. 1.33a, when the anode became active with an appearance of anode spot and the arc became more constricted, the difference of average current density between Area I and Area II, III, IV was much larger, which indicated that the anode spot formation distributed the anode current more unevenly. As the erosion region caused by anode spot covered the entire Area

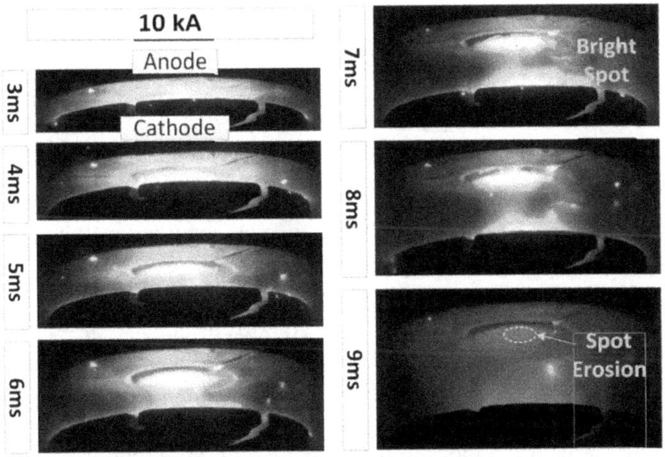

(a) Time-sequence photographs of anode discharge evolution

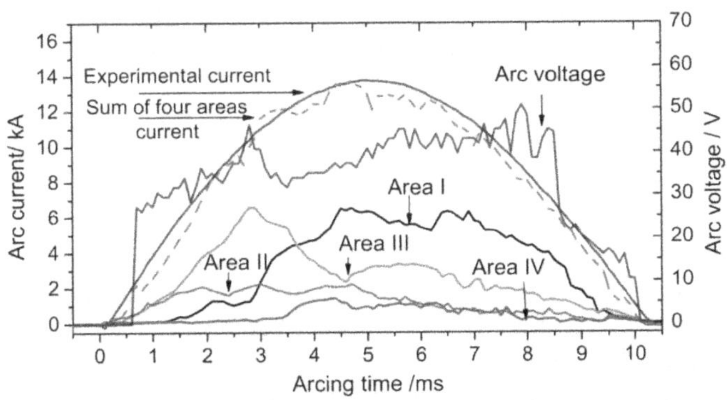

(b) Arc voltage and arc current of the four split anode areas

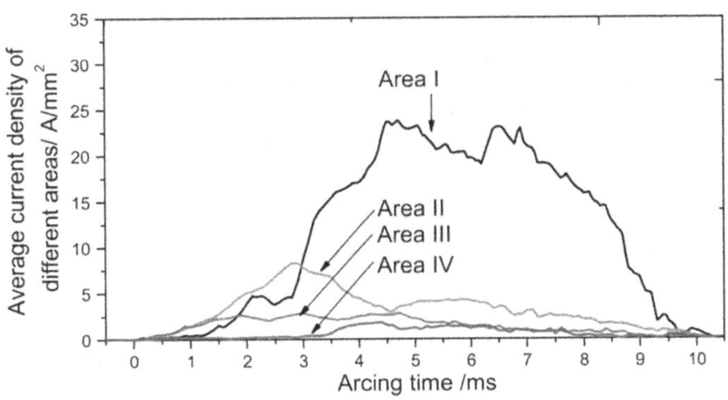

(c) Average anode current densities of the four split anode areas

◂**Fig. 1.32** Anode discharge measurement results at an arc current of 10 kA (rms). Contact configurations see Fig. 1.28. Both contact diameters are 60 mm. The diameter of Area I is 18 mm. Arcing time is 10 ms. The average opening velocity is 2.4 m/s. The contact separation length as arc extinguishing is 24 mm. The full contact gap is 35 mm. The sum of the current of the four split anode areas (dashed line) approximates the experimental current (solid line)

I, the maximum current density of 37.7 A/mm^2 was regarded as the current density of the anode spot.

In the determination of anode spot current density, the criteria were to estimate the anode spot current density as the average current density of Area I when the anode spot erosion area entirely covered the whole area of Area I. Therefore, the diameter of Area I was important to determine the anode spot current density. In the above measurement of anode spot current density, the diameter of Area I was 18 mm. Now, we repeat the measurement by changing the diameter of Area I. Two other diameters of Area I were chosen, which was 12 and 20 mm, respectively. The anode spot formation and whether the anode spot covered the whole area of Area I were distinguished by a combination of the optical observation, the arc voltage, and the erosion region checked by post-test examinations.

Figure 1.34 shows the anode mode photos at an arcing instant of 6 ms under various arc current at the three diameters of Area I, which were 12, 18, and 20 mm, respectively. The arc current ranged between 6 and 14 kA. Figure 1.34a shows the anode mode with the 12 mm diameter of Area I. It shows that at a small arc current of 6 kA or 8 kA, no anode spot appeared evidently. The anode spot first appeared at the arc current of 10 kA. A bright region appeared on a local area of the anode. Its arc voltage remained in a range of 20–50 V with some noise component, like the arc voltage waveform shown in Fig. 1.32b. After arc extinguishing, an erosion spot was observed on Area I. However, at 10 kA the anode erosion area occupied only part of Area I. At an arc current of 12 kA and above, anode spot appearing as a bright anode region started to cover the entire Area I. In Fig. 1.34b, with an 18-mm-diameter of Area I, an anode spot appeared at the arc current of 10 kA. The eroded anode region expanded with an increasing of arc current. At the arc current of 14 kA, the anode spot erosion area expanded to occupy the entire Area I. In Fig. 1.34c, the diameter of Area I was 20 mm. The anode spot appeared at the arc current of 10 kA. With the arc current increased to 14 kA, a constricted arc column was observed between the contacts. An accompanied eroded region was also found in Area I. However, it was found that the anode spot eroded region did not occupy Area I entirely until the arc current of 14 kA, as shown in Fig. 1.34c.

Figure 1.35 shows a dependence of average current density of Area I on arc current, which indicates an influence of the three diameters of Area I. The average current densities corresponded to the instant of arc current peak. It shows that the average current density of Area I increased with the increasing of arc current. At arc current of 6 kA, the average current density of Area I was around 8.5 A/mm^2 with the three kinds of contact diameters of Area I. At arc current of 8 kA, the average current density of Area I increased to above 10 A/mm^2, and the difference enlarged

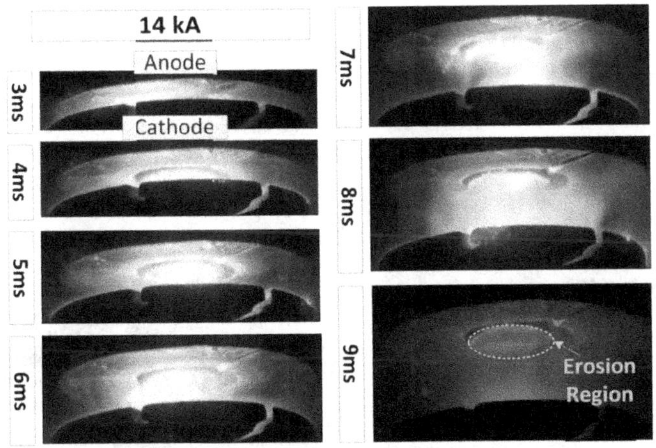

(a) Time-sequence photographs of anode discharge evolution

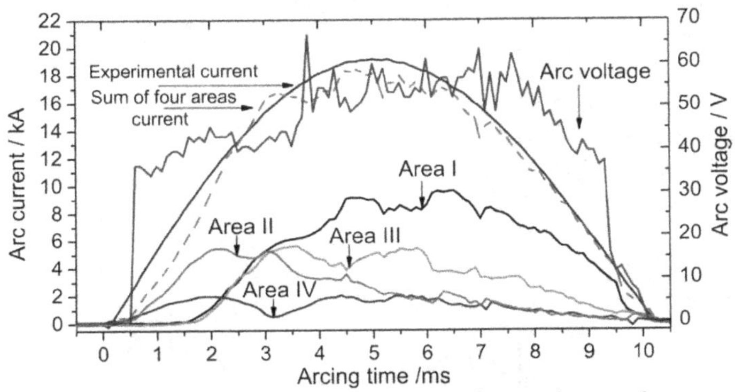

(b) Arc voltage and arc current of the four split anode areas

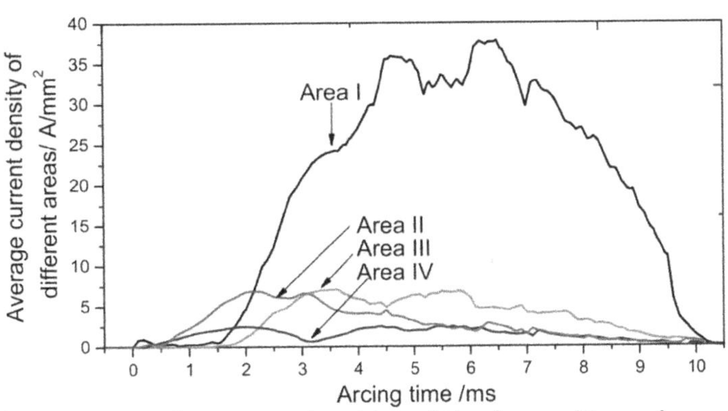

(c) Average anode current densities of the four split anode areas

◀**Fig. 1.33** Anode discharge measurement results at an arc current of 14 kA (rms). Contact configurations see Fig. 1.28. Both contact diameters are 60 mm. The diameter of Area I is 18 mm. Arcing time is 10 ms. The average opening velocity is 2.4 m/s. The contact separation length as arc extinguishing is 24 mm. The full contact gap is 35 mm. The sum of the current of the four split anode areas (dashed line) approximates the experimental current (solid line)

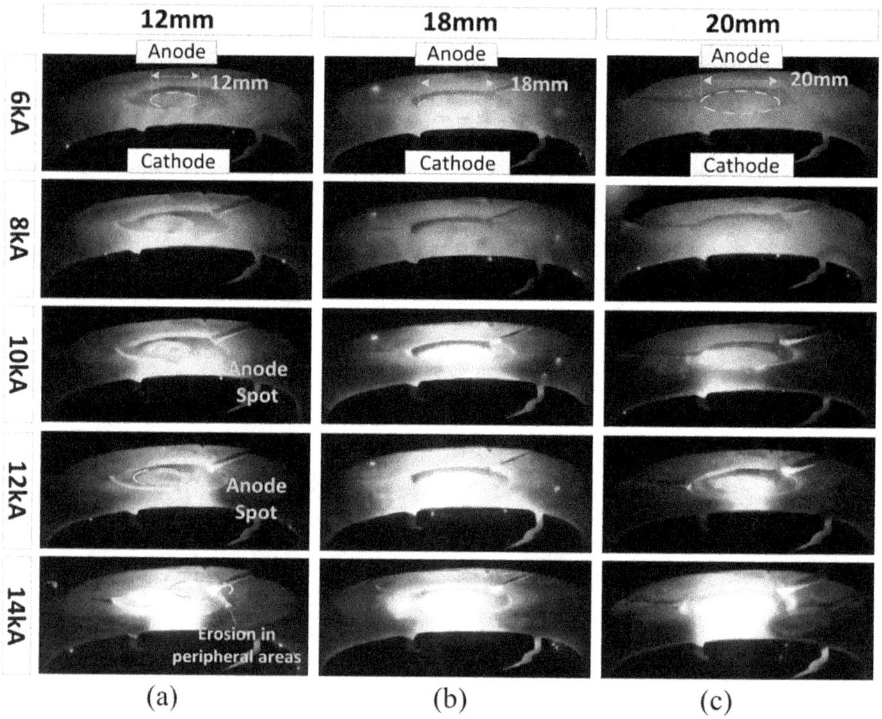

Fig. 1.34 Anode mode photos at an arcing instant of 6 ms under various arc currents at three diameters of Area I. **a** 12, **b** 18, and **c** 20 mm. Contact configurations see Fig. 1.28. Both contact diameters are 60 mm. Arcing time is 10 ms. The average opening velocity is 2.4 m/s. The contact separation length as arc extinguishing is 24 mm. The full contact gap is 35 mm. The arc current ranged between 6 to 14 kA with a step of 2 kA

between the three diameters of Area I. The highest average current density of Area I was 15.9 A/mm^2 with the diameter of Area I of 12 mm. At arc current of 10 kA, the anode spot appeared in all of the three diameters of Area I. But the anode spot erosion area did not cover the entire Area I so far, therefore the average current density at arc current of 10 kA was not regarded as the anode spot current density. When the arc current increased to 12 kA, it was found that the expansion of anode spot erosion region just covered the whole Area I with the diameter of 12 mm. In such condition, the average current density of Area I of 31.9 A/mm^2 was regarded as the current density of the anode spot. When the arc current further increased to 14

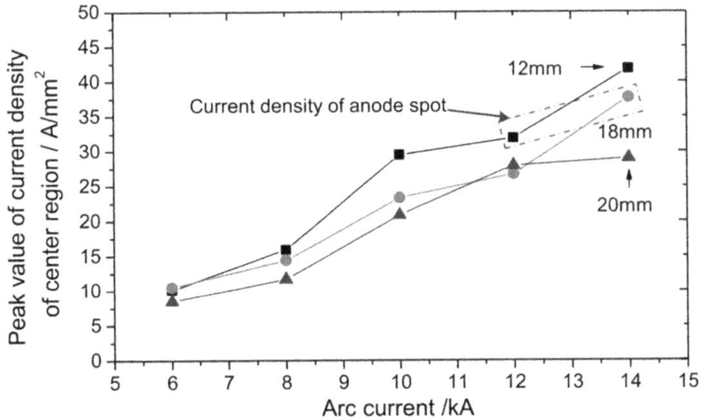

Fig. 1.35 Dependence of average current density of Area I on arc current. The average current density was measured at current peaks. The diameters of the Area I was 12, 18, and 20 mm, respectively. Contact configurations see Fig. 1.28. Both contact diameters are 60 mm. Arcing time is 10 ms. The average opening velocity is 2.4 m/s. The contact separation length as arc extinguishing is 24 mm. The full contact gap is 35 mm. The arc current ranged between 6 and 14 kA with a step of 2 kA

kA, the anode spot expanded over the Area I with the diameter of 12 mm. It reached the peripheral regions even though there was a gap. The anode spot current density of Area I reached 41.8 A/mm^2. This value reflected a part of current density inside an anode spot, which implied an uneven distribution of current density in an anode spot. Meanwhile, when the arc current increased to 14 kA, the anode spot erosion region just expanded to fully occupy the Area I with a diameter of 18 mm. In this case the average current density of Area I of 37.7 A/mm^2, which was regarded as the current density of the anode spot. In the case of diameter of Area I of 20 mm, the expansion of anode spot erosion region cannot fully occupy the Area I until arc current of 14 kA. As a summary, the current density of anode spot was estimated in a range of 31.9–37.7 A/mm^2.

In the experiments, it was found that the anode current concentrated in the central region. Figure 1.36 shows the average current density of the central and peripheral regions as a function of arc current. The average current density values corresponded to arc current peak. It shows two cases with opening velocities of 1.8 and 2.4 m/s, respectively. It shows that the current density of the central area increased rapidly with an increase of arc current. However, the average current density of the three peripheral regions increased slowly with the increasing of arc current. As the current increased, current concentrated more in the central region than in the peripheral regions, especially when the opening velocity was higher. Mitchell [17] suggested the arc current in diffused arcs constricted to the central region of the anode. However, Schellekens [48] experimentally showed that the current density of anode was quite uniform in diffused arcs. The anode current density distribution should be related to the axial magnetic field. Schellekens used an axial magnetic field of 200 mT,

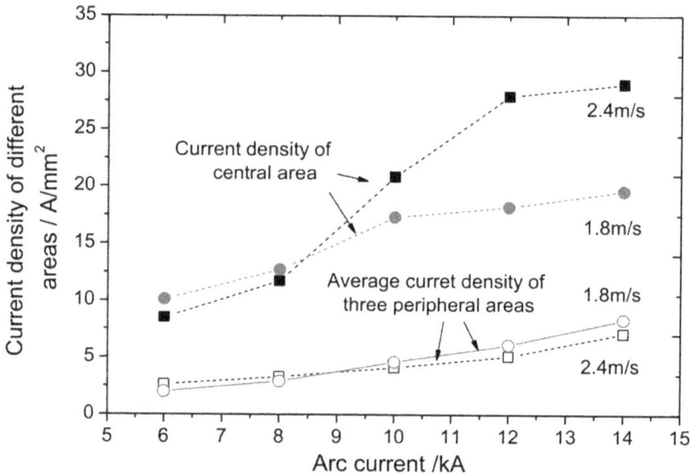

Fig. 1.36 Average current density of the central and peripheral regions as a function of arc current. The measured data corresponds to the current peak. The opening velocities are 1.8 and 2.4 m/s, respectively

which was much stronger than that in our experiment (several tens of mT). In our experiment, the cup-type axial magnetic field cathode generated an axial magnetic field of 3.5 and 2.8 mT/kA at the anode surface center point with the opening velocity of 1.8 and 2.4 m/s, respectively, as shown in Fig. 1.30. The weak axial magnetic field seemed to be responsible to the anode current concentration in the central region.

The anode spot current density is a significant influencing factor for a successful current interruption because the anode spot density can indicate the anode heat flux density distribution and the energy input to the anode. The anode energy input is determined by anode current density and anode sheath voltage. Therefore, the anode current density is a significant influencing factor contributing to the thermal processes of the anode in vacuum arcs. Moreover, it is well known that the anode thermal processes significantly influence the current interruption capability of vacuum interrupters [57–59]. Thus, the measured anode spot current density might be important for the determination of current interruption capacity of axial magnetic field vacuum interrupters. We estimated the anode spot current density by the average current density of Area I in the condition that the anode spot erosion region covers the Area I entirely. In the case of 12 mm diameter of Area I, it determined the lower limit of the anode spot current density, which was 31.9 A/mm^2 under an arc current of 12 kA. In the case of 18 mm diameter of Area I, it determined the upper limit of the anode spot current density, which was 37.7 A/mm^2 under an arc current of 14 kA. It should be noted that the estimated anode spot current density are related to the experimental conditions, especially arc current. Moreover, it is still difficult to understand the cause–effect relationship between the anode current density and the anode arc modes. It is possible that the constricted anode current density leads to the anode erosion, which may activate the anode to be a new source for the plasma.

In this process, the vacuum arc may evolve into an anode spot from the diffuse arc mode.

1.4 Thermal Process of Anode Surface Driven by High-Current Anode Mode Vacuum Arcs

In a current interruption by a vacuum interrupter, as the current of vacuum arc increases from a low current to a high current, the interelectrode plasma evolves from the diffuse arc mode to the high-current anode mode. In the high-current anode mode, the arc energy concentrated to heat a local contact region, raising the temperature of the local contact surfaces. Therefore, the temperature of a local anode can exceed the melting temperature, and often approaches the boiling temperature of the anode material. In such cases, the overheated local anode region becomes a new source of arc plasma. After current zero, the anode in arcing process becomes an active post-arc cathode. Therefore, in a high-current interruption in vacuum, the anode arcing conditions are critical.

There are two concerned effects of high-current anode mode vacuum arcs on the anode surface, which are anode erosion and anode temperature. The anode erosion effect is discussed in Sects. 1.4.1 and 1.4.2. Section 1.4.1 discusses the anode erosion caused by a blowing effect in high-current anode mode vacuum arcs under an axial magnetic field. Section 1.4.2 model how anode erosion forms by suffering the high-current anode mode vacuum arcs. The anode temperature after current zero is discussed in Sects. 1.4.3 and 1.4.4. Section 1.4.3 experimentally determines the decay modes of anode surface temperature after current zero. Section 1.4.4 investigates the impact of anode surface temperature after current zero on dielectric recovery strength.

1.4.1 Anode Erosion Caused by Blowing Effect in High-Current Anode Mode Vacuum Arcs Subjected to Axial Magnetic Field

Thermal processes play a significant role in determining the anode activity in high-current anode mode vacuum arcs, which are strongly correlated with contact material properties. In the high-current anode mode there is a blowing effect of anode materials caused by the arc, in which the arc plasma pressure can reach $\sim 1 \times 10^4$ Pa, and the anode surface melts and deforms by the blowing effect. Many researchers investigated the thermal and solidification processes of the anode in high-current vacuum arcs. Schellekens and Schulman investigated the effect of arc heating on the contact temperature and erosion in drawn vacuum arcs for several industrial designs of the axial magnetic field contacts [60]. They found that the flow of liquid metal off the cathode and anode surfaces contributed to the overall contact erosion. They showed that the melt depth on the contact surface of the Cu–Cr contacts was about 150 μm under a high-current vacuum arc. Frey et al. showed the micro-structural details of the contact materials after arcing [61]. Wei et al. revealed that liquid phase separation was involved in the microstructure evolution of Cu–Cr alloys during a series of vacuum breakdowns and that a Cr phase, comprising spheres and sheets, formed in the melt layer [62]. Jia et al. reported a swirl flow of liquid metal on the anode surface and explained this behavior because of the interaction between ions from cathode plasma jets and the anode melting pool [63]. In addition, several modeling and simulation works described the melting of anode surfaces [64–67]. This section focuses on anode erosion caused by the blowing effect in the high-current anode mode vacuum arcs subjected to axial magnetic field. The results would provide evidence for the influence of the high-current anode mode vacuum arcs on contact surfaces in vacuum interrupters.

Figure 1.37 shows the contacts used in the experiments and its axial magnetic flux

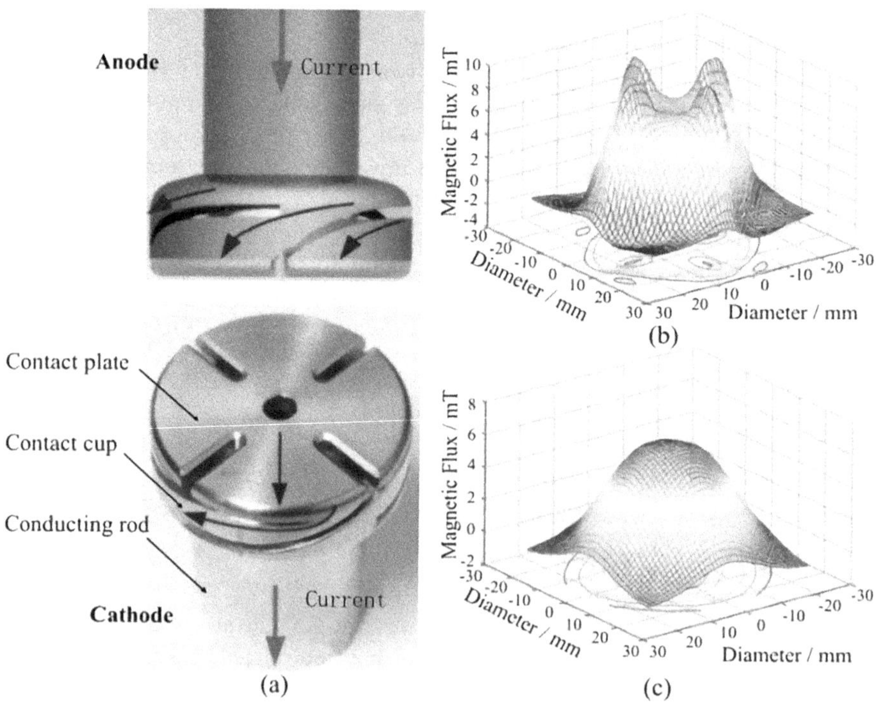

Fig. 1.37 The contacts used in the experiments and the distribution of the axial magnetic flux density. **a** The cup-type axial magnetic field contacts. **b** Distribution of axial magnetic flux density on the contact surface at current peak. **c** Distribution of axial magnetic flux density on the intermediate plane of the contact gap at current peak. The contact material is CuCr25 (25% Cr). The contact diameter is 42 mm. The contact stroke at current zero is 24 mm and the contact stroke at current peak is 12 mm. The arc is 1 kA (r.m.s.)

density. As shown in Fig. 1.37a, a pair of cup-type axial magnetic field contacts were selected. The material used for the contacts was CuCr25 (25% Cr). The diameter of the contacts was 42 mm. The average opening velocity was set as 2.4 m/s. Thus, the contact stroke at current zero was 24 mm with an arcing time of 10 ms. The arc current was 1 kA (rms) and the frequency was 50 Hz. The arc plasma was substituted by a conductor with the same diameter as the contacts. The arc conductivity was set as 1579 S/m, which was determined by the arc voltage and the arc current in the experiment. Figure 1.37b shows the axial magnetic flux density distribution on the contact surface at current peak at which the contact separation distance was 12 mm. The maximum axial magnetic field on this plane was 7.8 mT and the axial magnetic field at the central point of the contact surface was 6.3 mT. Figure 1.37c shows the distribution of the axial magnetic flux density on the intermediate plane of the contact gap at current peak at which the contact separation distance was 12 mm. The maximum axial magnetic field value was 5.5 mT, which located at the center of the intermediate plane.

The arc current was supplied by a *L–C* discharging circuit. The current frequency was 50 Hz. The applied arc current reached 20 kA (rms). The arcing time was controlled as ~10 ms. The average opening velocity was set as 2.4 m/s. Thus, the contact separation at current zero was 24 mm. The contacts were installed inside a demountable vacuum chamber, as shown in Fig. 1.38. The demountable vacuum chamber was evacuated and sustained at $10^{-3}-10^{-4}$ Pa. The chamber had two observation windows. A high-speed charge-coupled device (CCD) video camera recorded the vacuum arc evolution through one window. A recording speed of 10,000 frames/s was used with the aperture fixed at 4 and an exposure time of 2 μs. A two-color pyrometer measured the anode surface temperature through another window, which can measure temperatures in a range from 1073 to 1873 K. The two-color pyrometer was designed by Japan CHINO Corporation with a Model IR-FB. The pyrometer's measurement wavelengths were 850 and 1000 nm, respectively. The time response

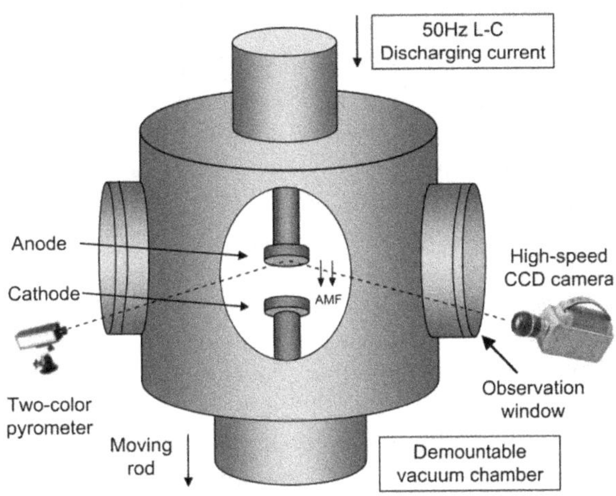

Fig. 1.38 Demountable vacuum chamber for vacuum arc evolution observation and anode surface temperature measurement

from 10 to 90% of the total range was 0.5 ms. The distance from the receiver of the pyrometer to the anode surface was fixed at 50 cm. At this distance, the view field of the pyrometer on the anode surface was approximately 100 mm^2. Experiment showed that the central region of the anode surface was seriously eroded. Thus, the pyrometer was directed toward the center of the anode surface to obtain a peak surface temperature after current zero.

Figure 1.39 shows a thermal effect of high-current anode mode vacuum arcs on the anode surface. The arc current was 10.5, 12.5, 14.1, 15.8, 16.7, and 17.8 kA (r.m.s.), respectively. Figure 1.39a shows the anode erosion photographs at the arcing instant of 8 and 9 ms. It shows that at 10.5 and 12.5 kA, there was no obvious anode surface erosion. At 14.1 kA, a bright blue region can be observed at the center of the anode surface. This bright blue region expanded when the current increased to 15.8 kA. When the arc current further increased to 16.7 and 17.8 kA, the bright blue region expanded more, and it became ring-shaped. Figure 1.39b shows a decay of the anode surface temperature after current zero. Because the time responses of the two-color pyrometer that arise from 10 to 90% of the total range was 0.5 ms, a decay from current zero to 0.3 ms seemed too rapid to be detected by the two-color pyrometer. Therefore, in this experiment a decay time was defined as a period starting from 0.3 ms after current zero to the instant when the anode surface temperature declined to the lower limit of the pyrometer measurement range (1073 K). At 10.5 and 12.5 kA, the decay time were 0.22 and 0.38 ms, respectively. The decay was fast because the arc mode was diffuse. When the arc current increased to 14.1 kA, the decay became slower with a decay time of 0.70 ms. At arc current of 15.8, 16.7, and 17.8 kA, the decay became even slower with a decay time of 2.26, 2.38, and 3.54 ms, respectively. Figure 1.39c shows a dependence of the arc energy and the decay time on the arc current. The arc energy was obtained by a calculation from the arc current and arc voltage by integrating the arcing time. The arc energy was 3.52 kJ at arc current of 10.5 kA. It increased linearly with an increasing of arc current, reaching 7.49 kJ at 17.8 kA. Moreover, the anode temperature decay time increased rapidly with arc current. The decay time increased from 0.22 to 3.54 ms as the arc current increased from 10.5 to 17.8 kA.

Figure 1.40 shows a blowing effect of high-current anode mode vacuum arcs on the anode surface. The experiments were carried out 20 times in total, in which 10 times with a low arc current (4 kA) for conditioning, another 10 times with the arc current increasing from 10.5 to 17.8 kA in a step of approximately 2 kA. Figure 1.40a shows the surface erosion photo of the anode and the cathode. It shows the erosion region nearly covered the entire anode surface and there were clear indications of liquid flowing. There was also a serious erosion of the cathode contact because of reignition arcing processes. Figure 1.40b shows a pattern of anode surface erosion. The erosion in the anode center region was labeled as Erosion Region I. The erosion in the peripheral region was labeled as Erosion Region II. A distinction between the Erosion Region I and the Erosion Region II is that the Erosion Region II was smoother, and it was exposed to a liquid metal flowing over the anode surface. Two samples were taken for observations of the microstructure of the melt layer from the Erosion Region I and the Erosion Region II. The sample taken from the Erosion

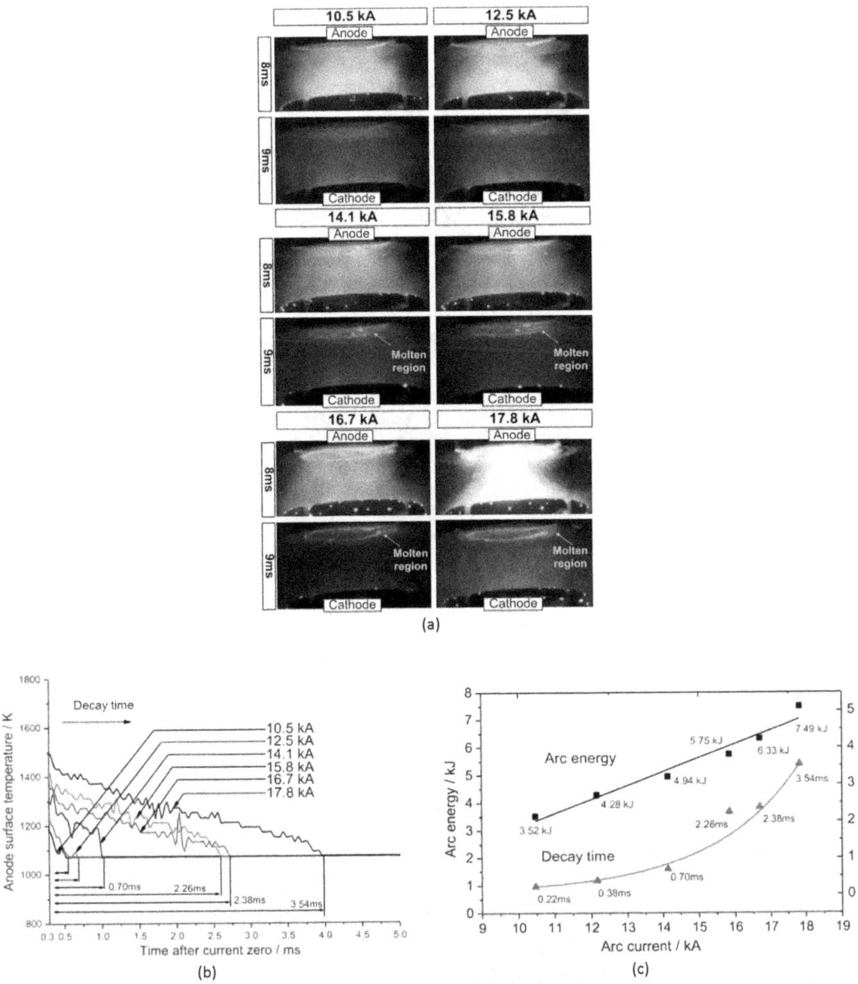

Fig. 1.39 Thermal effect of high-current anode mode vacuum arcs on the anode surface. **a** Vacuum arc photographs show the anode erosion at arcing time of 8 and 9 ms; **b** Decay of the anode surface temperature after current zero. **c** Dependence of the arc energy and the decay time on the arc current. Arc current is 10.5, 12.5, 14.1, 15.8, 16.7, and 17.8 kA r.m.s.; Arcing time is 10 ms; Contact separation distance as arc extinguishing is 24 mm; contact diameter is 42 mm; contact material is CuCr25 (25% Cr)

Region I was labeled as Sampling Area I and the sample taken from the Erosion Region I was labeled as Sampling Area II, as shown in Fig. 1.40b. Figure 1.40c shows the cross-sectional SEM photo of the Sampling Area I. The depth of the melt layer was <100 μm. It featured with a mixture of big Cr-rich spheres and many extremely small Cr-rich spheres. The big Cr-rich spheres in the surface melt layer were typically 30 μm in diameter. The small Cr-rich spheres can be seen more clearly in the higher magnification photograph. The energy dispersion spectroscopy

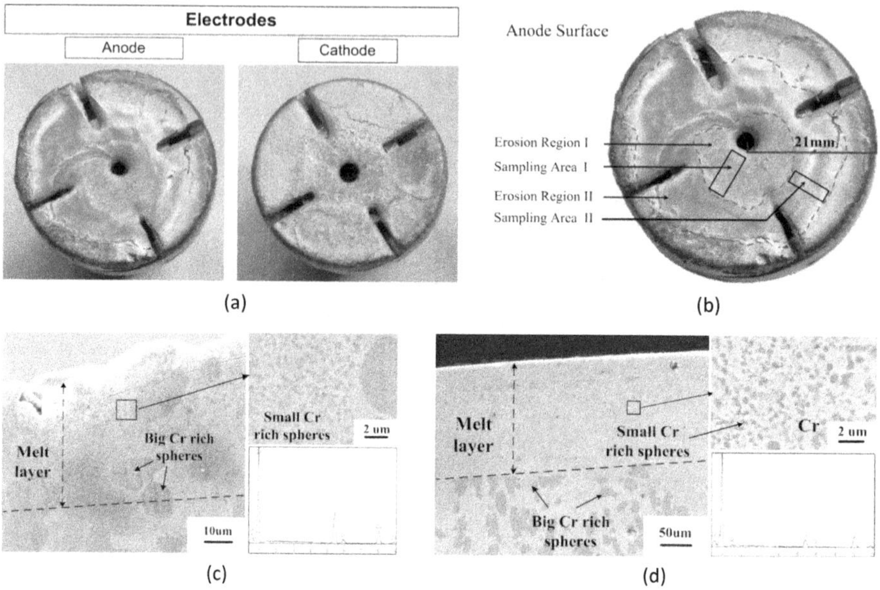

Fig. 1.40 Blowing effect of high-current anode mode vacuum arcs on the anode surface. **a** Contact surfaces erosion of the anode and cathode; **b** Anode surface erosion pattern and Sampling Areas I and II; **c** Cross-sectional SEM photo of the Sampling Area I; **d** Cross-sectional SEM photo of the Sampling Area II. Arcing time is 10 ms; Contact separation distance as arc extinguishing is 24 mm; contact diameter is 42 mm; contact material is CuCr25 (25% Cr)

(EDS) results were shown in the lower right panel of Fig. 1.40c. It indicated that the composition of Erosion Region I was to be 69% Cu and 31% Cr. Figure 1.40d shows the cross-sectional SEM photo of the Sampling Area II. Its microstructure of the surface melt layer was significantly different from that of the Sampling Area I. The depth of the surface melt layer was between 100 and 400 μm. The surface melt layer also contained many extremely small Cr-rich spheres, which can be clearly seen in the higher magnification photograph. But there were no big Cr-rich spheres can be observed in the surface melt layer. Furthermore, an interface between the surface melt layer and the bulk can be clearly seen. The extremely small Cr-rich spheres were <500 nm in diameter. The composition of Erosion Region II, determined from EDS analysis, as indicated at the lower right panel of Fig. 1.40d, was 82% Cu and 18% Cr. Table 1.4 shows the Cr content in Erosion Regions I and II as determined by EDS. Comparing to the original Cr content of 25%, the influence of high-current

Table 1.4 Composition of the melt layer on the anode surface in the different erosion regions

Material	Erosion Region I (%)	Erosion Region II (%)	Before Experiment (%)
Cr	31	18	25
Cu	69	82	75

anode mode vacuum arcs was that, the Cr content in the Erosion Region I (central region) increased to 31% and the Cr content in the Erosion Region II (peripheral region) decreased to 18%.

Blowing effect may contribute on the decay process of the anode surface temperature after current zero. As shown in Fig. 1.39a, with an increasing of the arc current, the anode erosion region expanded and became ring-shaped. This phenomenon could be explained by a blowing effect of the high-current anode mode vacuum arc. The blowing of liquid metal on the anode surface would contribute to an overall contact erosion and reduce heat concentration in the anode central region, therefore the blowing effect would significantly influence the thermal processes taking place on the anode surface and accelerate the decay process of the anode surface temperature. That may explain why the measured anode center temperature around current zero shown in Fig. 1.39b was lower than the calculated temperatures shown in [64–66], which did not consider the blowing effect. The decay of the anode surface temperature will significantly affect the dielectric recovery in the interruption of the high current. An overheated contact surface will still evaporate metal vapor after current zero because the surface will not cool down, even when the vacuum arc has extinguished. Therefore, the blowing effect may have an impact on the dielectric recovery process.

When comparing the melt layer's microstructure and depth between Erosion Region I and Erosion Region II, as shown in Fig. 1.40c with Fig. 1.40d, respectively, one can easily find the difference. The melt layer depth of the Erosion Region I was less than 100 μm. It included not only extremely small Cr-rich spheres, but also large Cr-rich spheres. The diameter of the large Cr-rich spheres was typically 30 μm, which was much smaller than the size of the Cr-rich spheres in the bulk material. The melt layer in Erosion Region II was much thicker, reaching 400 μm. The microstructure of Erosion Region II was with many extremely small Cr-rich spheres present. It is interesting to reason for the difference of the two erosion regions with different microstructure characteristics in the melt layer. Large Cr-rich spheres (~30 μm in size) in the surface layer of Erosion Region I seemed not melted completely during arcing, as remnants of the original larger Cr-rich spheres. Another explanation for the larger Cr-rich spheres in the center of the anode may be that the cooling rate of the melt pool in this region is low. Considering the arcing photographs in Fig. 1.39a, the formation of the two kinds melt layer microstructures may be attributed to the motion of the molten materials on the anode surface, which was blown from Erosion Region I to Erosion Region II in the high-arc anode mode vacuum arc.

As mentioned above, the two erosion patterns appeared on the anode surface may be explained by the motion of the molten anode material driven by the blowing effect of the high-current anode mode vacuum arc plasma pressure. Therefore, the anode erosion process would be not only a purely thermal process, but also a material transport of the anode surface. Figure 1.41 shows a schematic of the formation of the two patterns of melt layer microstructure. With an increasing of arc current, the arc enters high-current anode mode from the diffuse arc mode. The anode contact surface would be melted by the concentrated arc plasma heat flux, which may produce a molten pool. In such case, the molten material at the anode surface center might

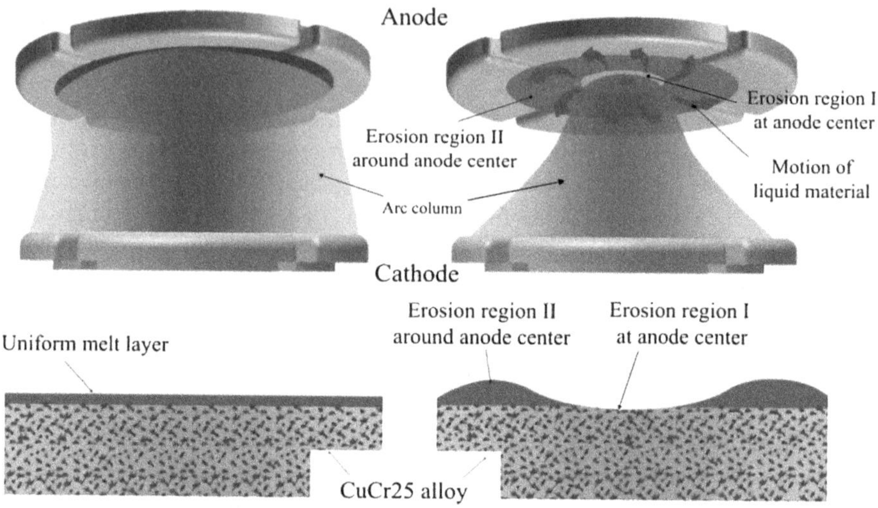

Fig. 1.41 Schematic of the formation of the two patterns of melt layer microstructure generated by diffuse arc mode vacuum arcs and by a blowing effect in high-current anode mode vacuum arcs

be blown to the peripheral region during arcing. The bright blue region expansion shown in Fig. 1.39a confirmed a local anode material transferred from the center to the periphery of the anode surface when the arc current increased to 15.8 kA and above. In Sect. 1.4.2, a modeling work will explain the blowing effect in detail.

1.4.2 Modeling of Anode Erosion Formation by High-Current Anode Mode Vacuum Arcs

Section 1.4.1 shows experimental evidences of anode erosion in high-current anode mode vacuum arcs. The anode erosion implies an interaction between a blowing effect of the vacuum arc and the molten local anode area. An anode erosion (see Fig. 1.40b) formed in arcing is solidifying in the post-arc period after current zero. The anode in arcing period becomes cathode in the post-arc period as the applied transient recovery voltage is in the reversed polarity. The solidifying anode erosion area would enhance a local electric field of the post-arc cathode surface, which probably increases the breakdown probability in the current interruption. Because the anode erosion formation implies a significant impact of the thermal process that lasts from the arcing to the dielectric recovery process, including anode melting, surface deformation, and solidification, the anode erosion formation is critical in understanding the current interruption process. Such understanding may push forward in understanding the high-current interruption in vacuum.

In this section, a model is developed to understand the anode erosion formation caused by high-current anode mode vacuum arcs, considering the heat transfer from

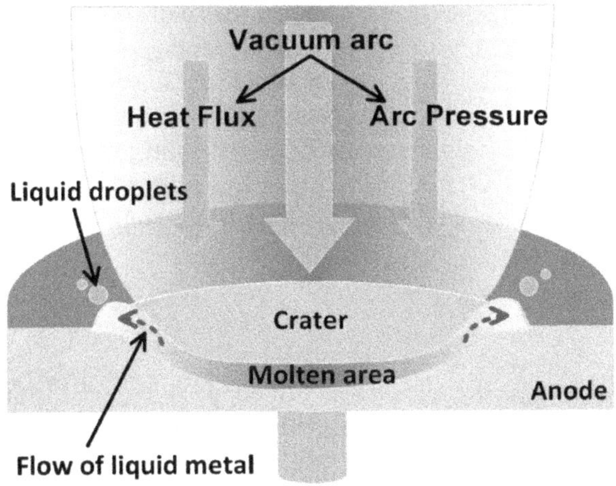

Fig. 1.42 A physical model in the arc-anode interaction in high-current anode mode vacuum arcs

the arc plasma to the anode surface, the phase transition of the anode material, and the effect of the arc pressure on the free surface of the molten area. Furthermore, the solidification process after current zero is also included in the model. The developed model would provide detailed information of the liquid metal flow, anode erosion formation, and anode temperature distribution. Moreover, the possibilities of post-arc breakdowns initiated by high anode surface temperature, anode surface deformation, and liquid droplets are discussed.

Figure 1.42 shows a physical model of the arc-anode interaction in high-current anode mode vacuum arcs. The arc plasma ejected from the cathode spots moves towards the anode. The anode surface acts as an absorber and it collects the heat and momentum transported from the arc column. The heat causes the melting, and consequently, a local molten area is formed on the anode surface. Moreover, the momentum from the arc forces the liquid material to flow away from the original position, splashing out droplets. The liquid flow changes the appearance of the anode surface and the temperature distribution. Finally, the arc extinguishes and the anode surface cools down after current zero, which leaves a visible erosion on the anode surface. In accordance to this physical description, two major physical processes are considered in the model. One is plasma flow in the arc column and the second is the anode thermal process including the melting, the surface deformation, the liquid metal flow, and the solidification. Accordingly, vacuum arc plasma flow is calculated by using a magnetohydrodynamics (MHD) model in order to obtain the heat and pressure arriving in the anode surface and then these parameters are adopted as the boundary conditions to calculate the heat transfer, the phase transition, the surface deformation, the liquid metal flow, and the solidification of the anode surface. Next it introduces the mathematical models for the vacuum arc column region and the anode region, respectively.

In the modeling of vacuum arc column region, a 2D axisymmetric magnetohydrodynamics simulation model was adopted. The equations of mass, momentum, energy, and the equation of magnetic induction were solved. The plasma was considered to be quasi-neutral, with an average ion charge of 1.9 [68, 69]. The electron energy conservation equation was not solved because the electron temperature was reported to be a uniform distribution with a value of around 3 eV both by experiments [70] and by simulations [50]. Hence, the electron temperature was set at 3 eV. The magnetohydrodynamics model was consisted by Eqs. (1.13)–(1.19).

The mass continuity equation is shown in Eq. (1.13).

$$\frac{\partial \rho}{\partial t} = \nabla \cdot (\rho \boldsymbol{u}) = 0 \tag{1.13}$$

where ρ is the mass density of copper ions and u is the ion drift velocity.

The momentum equation is shown in Eq. (1.14).

$$\frac{\partial \rho \boldsymbol{u}}{\partial t} = \boldsymbol{u} \cdot \nabla(\rho \boldsymbol{u}) = -\nabla(P_i + P_e) + \boldsymbol{J} \times \boldsymbol{B} \tag{1.14}$$

where p_{i} and p_{e} are the ion pressure and the electron pressure, respectively; J is the current density and B is the magnetic field.

The ion energy equation is shown in Eq. (1.15).

$$\frac{\partial \rho e}{\partial t} + \nabla \cdot (\phi e) + \frac{\partial \rho K}{\partial t} + \nabla \cdot (\phi K) = 3\frac{m_e}{m_i} n_e v_{ei} k_B (T_e - T_i) \tag{1.15}$$

where e is the ion thermal energy; K is the ion kinetic energy; m_{i} and m_{e} are the ion mass and the electron mass, respectively; n_{e} is the electron number density; v_{ei} is the electron-ion collision frequency; T_{i} and T_{e} are the ion temperature and the electron temperature, respectively, and k_{B} is the Boltzmann constant.

The magnetic induction equation is shown in (1.16).

$$\frac{\partial \boldsymbol{B}}{\partial t} - \nabla \times (\boldsymbol{v} \times \boldsymbol{B}) - \frac{1}{\mu_0 \sigma} \nabla^2 \boldsymbol{B} = 0 \tag{1.16}$$

where v is the electron drift velocity; μ_0 is the vacuum permeability and σ is the conductivity.

The current density J, the electron drift velocity v and the electron number density n_{e} were calculated by the Eqs. (1.17)–(1.19).

$$\nabla \times \mathbf{B} = \mu_0 \boldsymbol{J} \tag{1.17}$$

$$\boldsymbol{J} = e n_e (\boldsymbol{u} - \boldsymbol{v}) \tag{1.18}$$

$$n_{\mathrm{e}} = Z n_i \tag{1.19}$$

where Z is the average ion charge.

The boundary conditions of both the flow and the magnetic fields were set for the cathode boundary, the anode boundary, and the lateral boundary. The cathode boundary was treated as an inlet of arc plasma. Therefore, the mass flow rate of ion flow was set according to Kimblin's experimental measurements of ion erosion from cathode spots [71]. The ion temperature at the cathode boundary was set as 3 eV. Under high-current anode mode vacuum arcs, the observed cathode spot distribution showed that the current density distribution on the cathode surface was not uniform [72, 73]. In this model, the current density distribution was assumed to be highest at the center of the cathode surface and it decreased linearly to zero toward the edge of the cathode. The azimuthal component of the magnetic field was set according to the current density distribution. The axial magnetic field was assumed to be a homogeneous distribution; therefore, the axial magnetic flux density was fixed at the cathode boundary. The anode boundary was assumed to be a perfect sink for the flow field in the arc column simulation. So, an outlet boundary condition was given for the plasma flow. For the magnetic field on the anode surface, the generalized Ohm's law was applied to maintain the anode surface as an equipotential surface. Equation (1.20) described the magnetic field boundary condition on the anode surface. For the lateral boundary, a free slip boundary condition was specified for the flow field so that the normal component of the ion drift velocity was zero. Ampere's law was used to determine the value of the azimuthal magnetic field on the lateral boundary.

$$\frac{\partial \boldsymbol{B}_\theta}{\partial z} = -\mu_0 \sigma \left(-v_z B_\theta - \frac{1}{n_e e} j_\theta B_z + \frac{1}{n_e e} \frac{\partial P_e}{\partial r} \right) \tag{1.20}$$

The model equations were solved numerically using an open-source computational fluid dynamic (CFD) software OpenFOAM. As this code is generally used to solve fluid dynamics problems, some modifications are required to simulate vacuum arcs. The simulation method used here was similar to that employed by Schade and Shmelev [50], Wang et al. [54] and Langlois et al. [74]. The calculation domain was initialized with the plasma parameters at the cathode side boundary. Moreover, dependency of the results on initial conditions has been checked which showed that the variation of the initial conditions only affects the time cost to obtain a stationary solution. Benchmark simulations were carried at lower current and compared to the measurements of Schellekens [75]. It showed that the simulations results were close to that of the experiments.

In the modeling of anode region, a heat conduction and a local anode liquid material flow model was established to describe the arc-anode heat transfer, the anode melting, the surface deformation, and the liquid metal flow on the anode surfaces. The conservation equations of mass (1.13), momentum (1.14) and energy (1.15) for liquid and solid phases are re-written as Eqs. (1.21)–(1.23).

$$\nabla \cdot \boldsymbol{u} = 0 \tag{1.21}$$

$$\rho\left[\frac{\partial \boldsymbol{u}}{\partial t}+\boldsymbol{u}\cdot(\nabla\cdot\boldsymbol{u})\right]=\nabla\cdot\left[-p\boldsymbol{I}+\mu\left(\nabla\boldsymbol{u}+(\nabla\boldsymbol{u})^T\right)\right]+K\boldsymbol{u}+\gamma\boldsymbol{n}\kappa\delta(\phi) \quad (1.22)$$

$$\rho c_p\left[\frac{\partial T}{\partial t}+\nabla\cdot(\boldsymbol{u}T)\right]=\nabla\cdot(\lambda\nabla T)+S_{arc}+Q_{vap} \quad (1.23)$$

where T is the temperature. t is the time. ρ is the density. c_p is the specific heat capacity. and λ is the thermal conductivity. Since the arc-anode interface is within the simulation domain, additional terms, S_{arc} and Q_{vap}, are introduced to describe the heat sources in Eq. (1.23), which represent the arc energy and the energy loss by evaporation at the anode surface, respectively. The arc-anode interface is traced by a Level-set method [76], which is described below. The volumetric heat source term is obtained by multiplying the actual heat flux density and the delta function of the Level-Set method. $\boldsymbol{u}$ is the velocity vector. p is the pressure. $\boldsymbol{I}$ is the identity matrix. $\mu = 3.2$ mPa · s [77] is the dynamic viscosity. $\boldsymbol{Ku}$ is the Darcy damping force to cancel the velocity if the temperature is lower than the melting temperature. $\gamma\boldsymbol{n}\kappa\delta(\phi)$ is the delta function source term which acts only at the interface. Φ is the Level-Set function. γ is the surface tension coefficient. $\boldsymbol{n}$ is the surface normal vector. κ is the curvature and $\delta(\phi)$ is the delta function in the Level-Set method. The effect of vacuum arc pressure is introduced by setting the boundary condition for the pressure p in Eq. (1.22), which is according to the result of the magnetohydrodynamics simulation of the vacuum arc column. The buoyancy force is neglected here, since the arcing time is short, and no differences have been observed when an anode is the upper contact or is the lower contact.

The Level-Set method can describe the free surface flow of the liquid metal because it effectively deals with the rapid change of the interface and it can easily add energy source terms at the arc-anode interface. The Level-Set equation is given as Eq. (1.24).

$$\frac{\partial \phi}{\partial t}+\boldsymbol{u}\cdot\nabla\phi=\gamma_{1s}\nabla\cdot\left(\varepsilon_{1s}\nabla\phi-\phi(1-\phi)\frac{\nabla\phi}{|\nabla\phi|}\right) \quad (1.24)$$

where ε_{1s} is the transition distance and Υ_{1s} is the speed of re-initialization. The φ function changes from 0 to 1 at the interface with a step function representing the arc phase and the liquid (including solid) phase, respectively.

The heat transfer from an arc column to an anode surface is introduced by adding an extra source term S_{arc} in Eq. (1.23). The heat flux density value of S_{arc} is obtained from the magnetohydrodynamics simulation of vacuum arc column by considering the kinetic and thermal energies of the impinging ions and electrons, ionization and evaporation energies, and the energy is associated with the electron emission work function. The φ variable is used to track the interface. The melting/solidification process is calculated by including the latent heat of fusion in the specific heat capacity function (1.25).

$$c_p = c_{p,s} + L_f e^{-\left((T - T_{\text{melting}})^2/(T_1 - T_S)^2\right)} / \sqrt{\pi (T_1 - T_S)^2} \tag{1.25}$$

where T_1 and T_s represent the liquid and solid temperatures, respectively. The liquid and solid temperature were set near the melting point of copper to avoid a singularity at the melting temperature. The latent heat of fusion L_f is included as an apparent heat capacity by adding a Gaussian function centered at the melting temperature T_{melting} and with a width of $(T_1 - T_s)$.

The energy loss by evaporation is given by Eq. (1.26).

$$Q_{vap} = -L_v \dot{m} \delta(\phi) \tag{1.26}$$

where L_v is the latent heat of evaporation and $\dot{m}$ is the mass flux of evaporation, as shown in Eq. (1.27).

$$\dot{m} = \sqrt{\frac{m_i}{2\pi k_B}} \frac{p_{sat}(T)}{\sqrt{T}} \tag{1.27}$$

Here m_i is the atomic weight of copper. k_B is the Boltzmann constant and $p_{sat}(T)$ is the saturated vapor pressure.

Generally, the Darcy force is used to describe flow in a porous medium. Here, as both liquid and solid copper was treated as fluid, the Darcy force term in Eq. (1.22) was used to set an extremely high viscosity for the solid phase so that it does not flow. A thin mushy zone with a temperature between the liquid and solid temperature was also set at the interface of the liquid and solid phases to avoid a numerical singularity. The permeability K is given in Eq. (1.28).

$$K = -C\left(\frac{(1 - f_1)^2}{f_1^3 + b}\right) \tag{1.28}$$

where C is a constant representing the mushy zone morphology. b is a constant to avoid a division by zero and f_1 is the liquid fraction, as shown in Eq. (1.29).

$$f_1 = \begin{cases} 0, & if \quad T < T_S \\ \frac{T - T_1}{T_1 - T_s}, & if \quad T_S < T < T_1 \\ 1, & if \quad T > T_1 \end{cases} \tag{1.29}$$

where T_1 and T_s represent the liquid and solid temperatures, respectively.

The modeling parameters were assumed as following. The arc current peak were 15, 20, and 25 kA, respectively. The frequency of arc current was 50 Hz. The arcing time was 10 ms. The contact material was oxygen-free high-conductivity copper. Contact diameter and contact gap were set as 40 mm and 10 mm, respectively. An axial magnetic flux density of 10 mT/kA was applied. The vacuum arc plasma

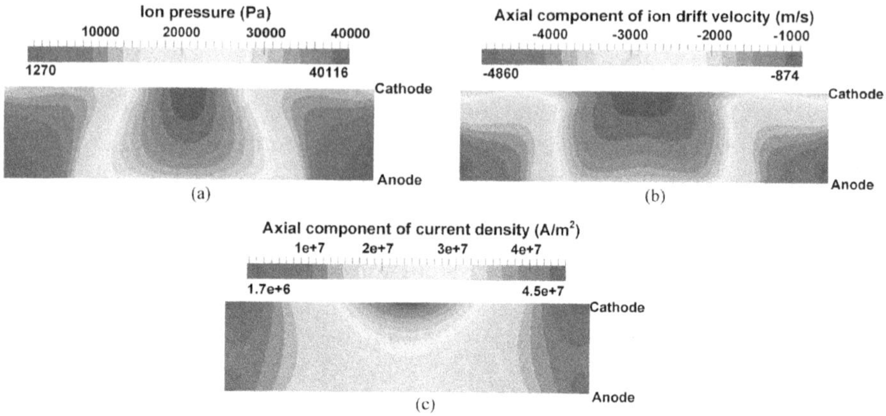

Fig. 1.43 Vacuum arc column physical parameters distribution at arc current of 20 kA peak. **a** The ion pressure; **b** the axial component of ion drift velocity and **c** the axial component of current density

parameters at an anode surface are determined by the heat flux inputting from vacuum arc column to the anode. Moreover, the plasma pressure distribution at the anode surface is also obtained, which acts as a boundary condition for the liquid metal flow on the anode.

Figure 1.43 shows a magnetohydrodynamics simulation of vacuum arc column with a current of 20 kA peak. Figure 1.43a–c show the distribution of ion pressure, the axial component of current density, and the axial component of ion drift velocity, respectively. Figure 1.43a shows that the maximum ion pressure of 40.1 kPa appeared at the center of the cathode surface. The ion pressure dropped in the axial direction from the cathode to the anode and in the radial direction from the gap center to the lateral boundary. The ion pressure located at the anode center was approximately 30 kPa, while the minimum ion pressure located at the edge of the anode surface and amounted approximately 1.3 kPa. The electron and ion pressure both contributed to the total pressure exerted on the local molten anode surface. Hence, the electron pressure was also required as a boundary condition for the liquid metal flow in the anode region. Since the ion temperature and the electron temperature in the arc were close, the electron pressure was approximately 1.9 times of the ion pressure. Figure 1.43b shows the axial component of ion drift velocity varies from 874 to 4861 m/s over the entire simulation domain. The ion drift velocity in the central region with a diameter of around 20 mm ranged between 874 to 2200 m/s, while the ion drift velocity in the side region was higher than in the central region. As a mass flow rate boundary was chosen for the velocity field at the cathode side, the total ion mass flux was set. The velocity was calculated based on the boundary condition and it was influenced by the physical conditions inside the simulation domain. Here, as the pressure was higher in the center near the cathode surface, the ion velocity is consequently slower than the periphery at the cathode. Figure 1.43c shows the current density in the cathode center was approximately 4.5×10^7 A/m^2 and was linearly reducing to near zero to

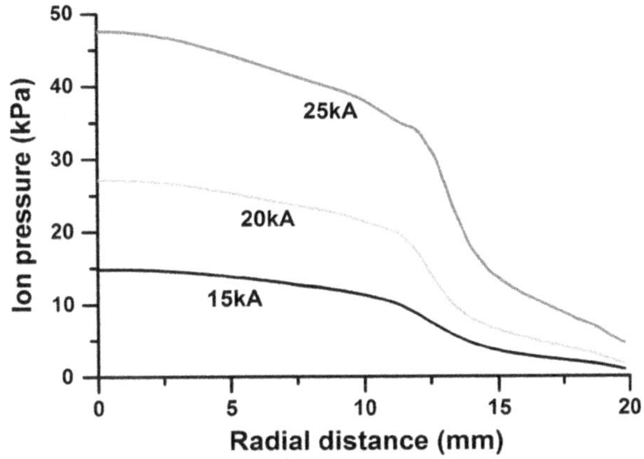

Fig. 1.44 Ion pressure distribution on the anode surface at three arc current

the edge of the cathode surface. It also shows that the current density at the center of the anode surface was lower than that of the cathode center. The axial component of the current density in the anode center was approximately 2.7×10^7 A/m^2.

Figure 1.44 shows the ion pressure distributions on anode surfaces at the three kinds of arc current of 15, 20, and 25 kA peak. The corresponding peak pressures were 14.8, 27, and 47 kPa, respectively. Since the Lorentz force is proportional to the square of the arc current, the peak pressure also followed a similar proportionality. The pressure at the anode surface center was the maximum. It slowly decreased in the radial direction initially, and then it steeply decreased after 10 mm. At the distances >15 mm, the pressure was relatively low. Such non-uniform distribution of ion pressure is the result of a combined effect of non-uniform current density distribution on the cathode surface and the pinch effect of the Lorentz force.

The heat flux transported from the arc column to the anode surface consists of the kinetic energy carried by electrons and ions. The kinetic energy includes drift velocity, the thermal energy of ions and electrons, the ionization energy, the evaporation energy, and the work function of electrons. They are summed up as shown in the Eqs. (1.30) and (1.31).

$$E_e = n_e v_{ez}\left(2kT_e + 0.5m_e v_e^2 + e\phi_w + e\phi_s\right) \tag{1.30}$$

$$E_i = n_i v_{iz}\left(2kT_i + 0.5m_i v_i^2 - eZ\phi_w - eZ\phi_s + \phi_v + \phi_Z\right) \tag{1.31}$$

where ϕ_w is the electron emission work function. ϕ_s is the anode sheath voltage. ϕ_v is the vaporization energy. ϕ_Z is the energy required to ionize an atom to the Z_{th} degree of ionization. All energies are given in units of eV.

Figure 1.45 shows the heat flux densities of the three kinds of arc current of 15, 20 and 25 kA peak. They are calculated by using Eqs. (1.30) and (1.31) by summing the electron energy flux E_{e} and the ion energy flux E_{i}. The heat flux densities were nearly proportional to the particle number density and the square of the arcing current. It

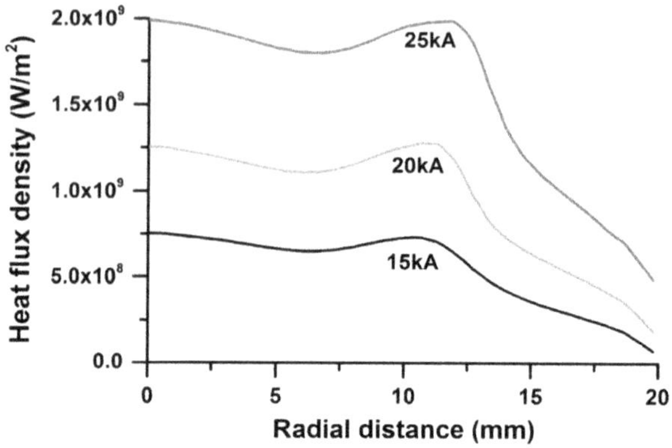

Fig. 1.45 Heat flux density distributions on anode surface at three arc current of 15, 20 and 25 kA peak

can be seen in Fig. 1.45 that the spatial distribution of the heat flux density has a two-peaks feature in the radial direction. The first peak was at the center of the anode surface. The secondary peak appeared between 10 and 15 mm, where the axial component of the electron and ion drift velocities increased dramatically. The heat flux densities of the two peaks were close to each other for the three kinds of arc current, but beyond the second peak the values decreased rapidly.

The heat flux density and the vacuum arc plasma pressure mentioned above are obtained by MHD simulations. They act as boundary conditions at the arc-anode interface, by adding extra source terms in the momentum equation Eq. (1.22) and the energy equation Eq. (1.23). The following shows the simulation results of anode region.

Figure 1.46 shows the anode temperature distribution during an arcing period of 10 ms. Figure 1.46a–c show the results of arc current of 15, 20, and 25 kA peak, respectively. The initial geometrical edge of the anode surface is marked with a red box in the first frame. The dark band at the top of each figure is outside the anode area. Figure 1.46a shows that the maximum temperature located at the center of anode surface, which was approximately 1540 K, appearing between 7 and 8 ms. As the heat flux density in the central region of the anode surface was more uniform than in the edge region, as shown in Fig. 1.45, the surface temperature in the anode central region changed little. Since the melting of the anode surface was limited during the arcing under the arc current of 15 kA peak, the flow of liquid metal and the erosion of the anode surface were not obvious. The peak anode surface temperature at current zero was 1380 K, which was close to the melting point of copper (1357.77 K). Figure 1.46b shows that the anode surface started to melt at about 4 ms during the arcing period. Then the molten radius expanded, while the molten depth increased during 4–7 ms. The molten depth remained approximately 0.5 mm after 7 ms. During the period of molten area expansion, the liquid metal was being driven towards the edge of the anode surface by the arc pressure. As a result, the anode surface started to deform, and anode erosion formed. A rim appeared at

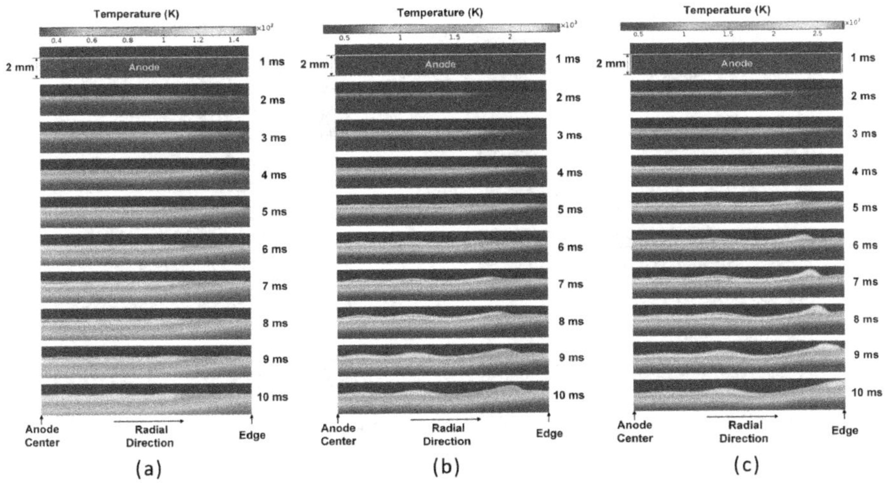

Fig. 1.46 Anode surface temperature distribution during arcing. **a** arc current of 15 kA peak **b** arc current of 20 kA peak **c** arc current of 25 kA peak

the edge of the molten area after 7 ms. The height difference between the rim and the original anode surface was approximately 1 mm. The flow of liquid metal also affected the temperature distribution on the anode surface. In this case, the maximum temperature on the anode surface was approximately 2400 K at 7 ms. The location of the maximum temperature was not at the center of the anode surface, which was unlike that shown in Fig. 1.46a. This is because the hot liquid metal was being moved away by the arc plasma pressure. There was also an obvious temperature decrease after 8 ms. As a result, at current zero the highest temperature was about 1950 K. Figure 1.46c shows that the heat flux density and the plasma pressure at arc current of 25 kA peak were much higher than that of 15 and 20 kA peak; therefore, the melting and the erosion of the anode surface was more significant. The melting of anode surface started from 3 ms in the arcing. From 6 ms and then on, the melting area expanded across the entire anode surface. The rim at the edge of the anode surface was about 1 mm in height. The rim moved towards the edge of the anode surface during the arcing, and it reached the lateral boundary of the simulation domain at current zero. The anode surface temperature was also much higher at arc current of 25 kA. The maximum temperature was approximately 2800 K on the anode surface, which appeared between 6 and 7 ms. While the highest temperature at current zero was approximately 2320 K.

Figure 1.47 shows the molten liquid metal flowing in the radial direction towards the edge of the anode surface, blown by the arc pressure during a high-current anode mode arc. Figure 1.47a and b show the results of arc current of 20, and 25 kA peak, respectively. Figure 1.47a shows the liquid metal flow started when part of the anode surface melted. At 5 ms, the highest flow velocity located at approximately 12 mm from the anode surface center. The highest flow velocity location located at the edge of the molten area, where the plasma pressure started to drop dramatically, as shown

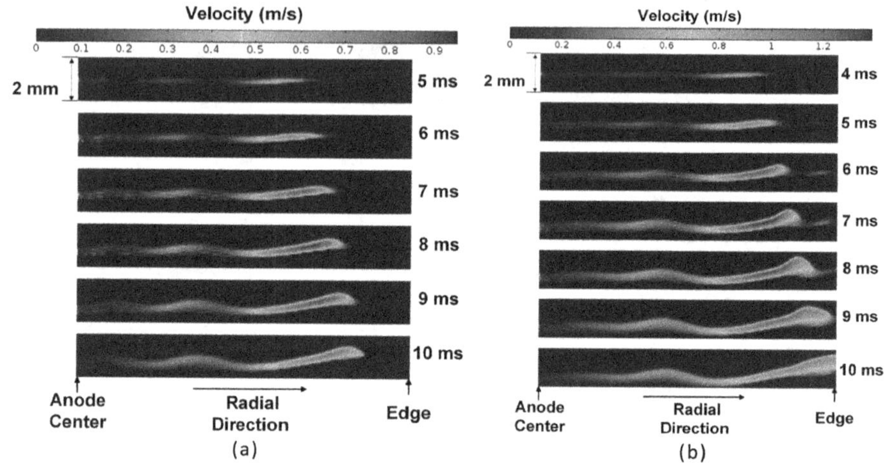

Fig. 1.47 Flow velocity of anode molten liquid metal towards the edge of the anode surface. **a** Arc current of 20 kA peak. **b** Arc current of 25 kA peak

in Fig. 1.44. The arc pressure drove the liquid metal, which led to form a rim at the edge of the molten area. The maximum liquid flow velocity appeared at the inner side of the rim. The maximum velocity was approximately 0.95 m/s at 8 ms. At current zero, the maximum velocity dropped to approximately 0.8 m/s as the arc pressure decreased to zero. Figure 1.47b shows the flow of liquid metal started from about 4 ms in the arcing. The maximum flow velocity was approximately 1.39 m/s at roughly 7 ms during the arcing. The rim of the molten area was higher and in a sharp shape, comparing to that shown in Fig. 1.47a. The liquid metal reached the lateral boundary at current zero and flowed out of the simulation domain.

The temperature distribution on the anode surface is influenced by the liquid metal flow driven by the arc pressure in the high-current anode mode vacuum arcs. As the hot liquid metal is blown away from the anode central region, the maximum temperature does not remain at the center of the anode surface. Figure 1.48 shows

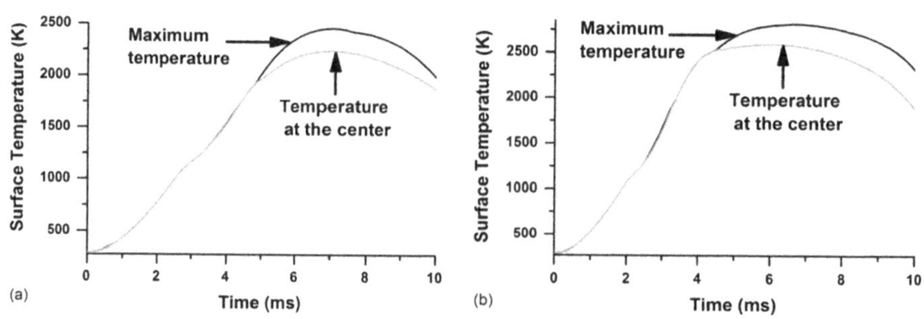

Fig. 1.48 The time evolution of the maximum anode surface temperature and the surface temperature at the anode center. **a** Arc current of 20 kA peak. **b** Arc current of 25 kA peak

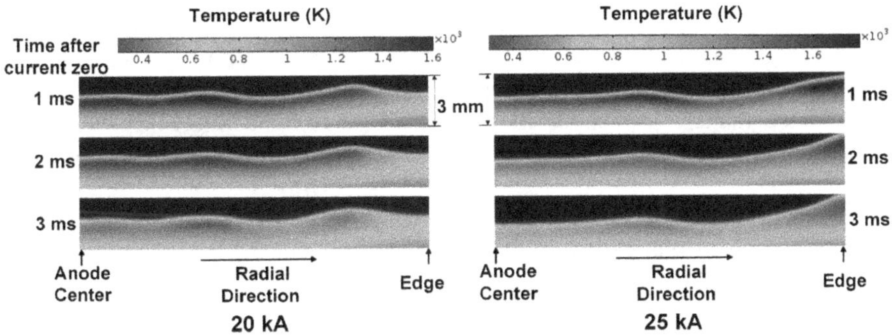

Fig. 1.49 Anode temperature distribution after current zero at arc current of 20 and 25 kA peak

the time evolution of the maximum temperature on the anode surface and the surface temperature at the anode center in the arcing time of 10 ms for the arc current of 20 and 25 kA peak. The maximum temperature maintained at the anode center for the first 4–5 ms, when the liquid metal flow was not obvious. However, the maximum temperature on the anode surface became higher than the surface temperature at the center of the anode surface, once the liquid metal flow started. The maximum temperature was approximately 200 K higher than the surface temperature at the anode center at 7 ms, for both arc current.

After current zero, the anode temperature decreases, and the molten liquid metal solidifies. Figure 1.49 shows the solidification process after current zero for the arc current of 20 and 25 kA peak. It illustrates the anode temperature distribution. Figure 1.49a shows that the anode surface topography did not change notably. The anode surface temperature decreased gradually with time. At about 2 ms after current zero, the maximum temperature on the anode surface was approximately 1330 K, which was lower than the melting temperature of copper (1357.77 K). Hence, the solidification duration in this case was about 2 ms. Figure 1.49b shows the liquid metal continued to flow after current zero. Therefore, the anode surface topography slightly changed, as the height of the ridge at the edge of the anode surface increased. The maximum temperature on the anode surface at 3 ms after current zero was almost 1600 K and the solidification duration was about 4 ms.

The above results show that the liquid metal flow redistributed the thermal energy on the anode surface and it influenced the temperature distribution. Because of the liquid metal flow, the maximum temperature on the anode surfaces appeared in a ring-shaped area rather than at the anode center. The liquid metal flow on the anode surface was supported by the experiments shown in Sect. 1.4.1. The high-speed photos showed a bright ring-shaped area appearing on the anode surfaces. In addition, two kinds of erosion regions were identified by the microstructure of the anode surface molten layer. The thickness of the molten layer in the central region of the anode surface was thinner than in the peripheral region. Alterations of Cr content on the anode surface after arcing implied that the central anode region surface temperature was lower than that in the peripheral region. The anode erosion pattern supported well to this modeling work. The area of Erosion Region I in the experimental work appeared to be smaller than the modeling work. This was probably due to the difference of contact materials. In the experiment, it used Cu–Cr (75/25 wt%) contact material, rather than the pure Cu used in the modeling. Since the melting point of Cr is higher than Cu, the liquid metal flow can be easily impeded by the Cr particles.

Schellekens and Schulman [60] observed a spatially homogeneous contact temperatures at current zero in high-current diffuse vacuum arcs. They suggested that flowing liquid metal was a significant mechanism for redistributing the arcing energy into a more uniform coverage of the surfaces. Furthermore, as the current increases the contact erosion became visibly dominated by liquid metal flowing off from the contact faces during arcing. This provided a further evidence that the liquid metal flow may influence the anode surface temperature distribution.

Anode melting and erosion formation is believed to strongly influence the interruption capacity of vacuum interrupters, because post-arc breakdown of a vacuum interrupter is related to the following three aspects: the anode surface temperature and hence the metal vapor evaporation, the anode surface temperature, anode surface deformation under high electric fields, and liquid droplets formation.

Firstly, the anode surface temperature is important for the recovery of dielectric strength. If the anode surface temperature is too high, a high density of evaporated metal vapor would lead to a post-arc Townsend breakdown. The critical surface temperature corresponding to the interruption limit was reported to be about 2000 K, according to an estimation by Schade and Dullni [7]. Figure 1.46a showed that the anode surface temperature at current zero was about 1380 K in a 10 ms arcing duration under an arc current of 15 kA peak. Hence, the probability of a post-arc Townsend breakdown would be relatively low in this case. However, the anode surface temperatures at current zero under the arc current of 20 and 25 kA peak reached around 1950 K and 2320 K, respectively, which suggested that a failure of current interruption would likely occur in the vacuum interrupters with the same conditions as in the simulation.

Secondly, the anode surface deformation and melting time may influence the dielectric recovery. Figure 1.49 shows that the anode surface was still molten at current zero, and anode surface deformed. The presence of a strong and fast rising electric field after current zero such as a transient recovery voltage would enhance the deformation, further enhancing the local electric field, and it may lead to a Taylor cone [78, 79] (see Fig. 1.4). A formation time of a Taylor cone is about 1–4 ms [80]. Figure 1.49 showed the solidification duration was close to the formation time of the Taylor cone. Therefore, the probability of initiating a Taylor cone is high during the solidification duration, which may lead to a current interruption failure with a high-current anode mode vacuum arc. A detail introduction of the Taylor cone will be given in Sect. 2.3.

Thirdly, liquid droplets emitted from the anode surfaces are possibly related to the post-arc breakdown of vacuum interrupters. Tian et al. pointed that it met the criterion for the formation of liquid droplets on the anode surface in high-current anode mode vacuum arcs [81]. When the metal vapor alone is not able to cause breakdown due to a low density, the bombardment of the liquid droplets may be responsible for initiating a post-arc breakdown [7]. Moreover, the hot liquid droplets in the interelectrode gap could act as a vapor source in the post-arc recovery process. Therefore, evaporation from the liquid droplets surfaces is important and can retard the decay of the vapor density after current zero. Thus, the probability of a Townsend breakdown is higher in presence of liquid droplets.

1.4.3 Decay Modes of Anode Surface Temperature After Current Zero in Vacuum Arcs—Experimental Study

In vacuum interrupters, anode surface conditions after current zero have a significant impact for a successful interruption. Because an anode in arcing period becomes a post-arc cathode when a transient recovery voltage is applied. A hot anode surface will evaporate an amount of metal vapor whose density dominates the dielectric recovery strength of the vacuum interrupters [7, 21]. In addition, metal vapor is expected to be positively correlated with the anode surface temperature. Thus, a dielectric recovery process will be retarded significantly if the post-arc cathode surface temperature is still high after current zero, especially when the surface remains in a molten state. That is why axial magnetic field and transverse magnetic field (also known as radial magnetic field) technology are adopted in vacuum interrupters, which controls arc energy input into contact surfaces. The technologies aim that there is no (local) overheating on the contact surfaces. Thus, a successful current interruption can be achieved.

The significance of anode surface temperature has drawn a lot of researchers' interests. Dullni et al. [82] measured the decay of anode surface temperature by adopting a pyrometer and a thermionic current method, respectively. They suggested that the boiling temperature of the anode material could be reached if an anode spot appears during the arcing period. Schellekens and Schulman [60] also measured the anode surface temperature after extinguishing a drawn vacuum arc using an infrared pyrometer. Their results showed that the peak value of anode surface temperature at current zero increased linearly with the peak value of the arc current. In addition, the temperature distribution on the surface was relatively homogeneous after extinguishing a high-current vacuum arc. Watanabe et al. [57] measured the melting time of the anode surface after current interruptions and they calculated the decay process of the anode surface temperature. They found that there was an interruption limit for CuCr50 (50% of Cr) contact material when anode surface temperature reached 1750 K at current zero. Niwa et al. [66] in a further investigation showed that for Cu–Cr contacts with an axial magnetic field, the current interruption limit was reached when the anode temperature reached 2000–2500 °C. They also revealed that anode surface temperature at current zero increased as the proportion of Cr content increased for CuCr contact material. Comparing different contact materials, Ide et al.

[83] found that anode surface melting time after current zero were associated with contact material and increased in the order of Cu, CuCr, and AgWC.

However, anode surface temperature characteristics are not well understood after extinguishing high-current anode mode vacuum arcs in transmission voltage level vacuum circuit breakers , whose contact gaps are typically high. For instance, the full contact gap of a 126 kV single break VCB reaches 60 mm [22], which is much larger than that of a medium voltage VCB (typically <20 mm). In addition, the vacuum arcs are easily approaching unstable in high contact gaps. A comparative strong magnetic field is needed to confine the arcs well. This section experimentally understands the decay modes of anode surface temperature after extinguishing vacuum arcs with high contact gaps. The decay of anode surface temperature can be classified into two modes which corresponds to diffuse arc mode and high-current anode mode, respectively. The two decay modes of anode surface temperature after current zero have different impacts on the dielectric recovery process (see Sect. 1.4.4).

Figure 1.50 shows a sketch of the contacts used in the experiments. The experiments were designed to measure the anode temperature after current zero, especially after high-current anode mode vacuum arc extinction. A pair of asymmetrical contacts with different contact diameter was used. The asymmetrical contacts make it less difficult to achieve the experiment's objective by taking account of a limited capacity of the power source supply. In Fig. 1.50 the fixed upper contact was assigned as anode, which was convenient to observe the activities of the surfaces. The lower contacts which were able to move were assigned as cathode. Table 1.5 shows the

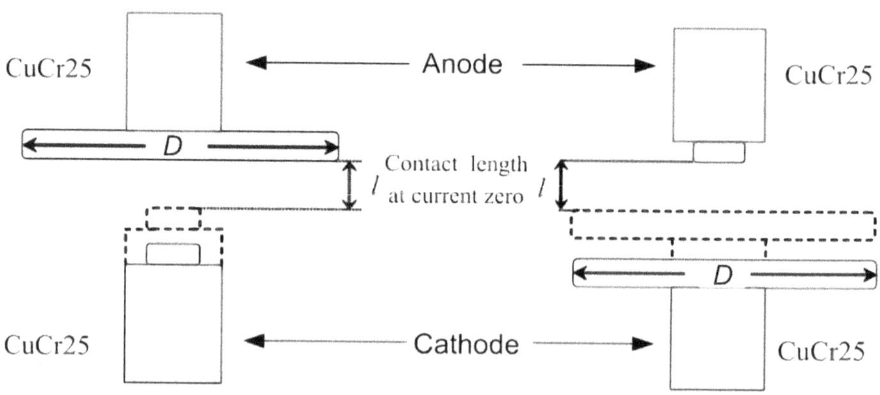

Fig. 1.50 Sketch of a pair of asymmetrical contacts arranged in the experiment. The upper contact was fixed as an anode and the lower one was movable as a cathode

Table 1.5 Arrangement of experiment

Experimental arrangement	Contact diameter (mm)	
	Anode	Cathode
Case I	60	12
Case II	12	60

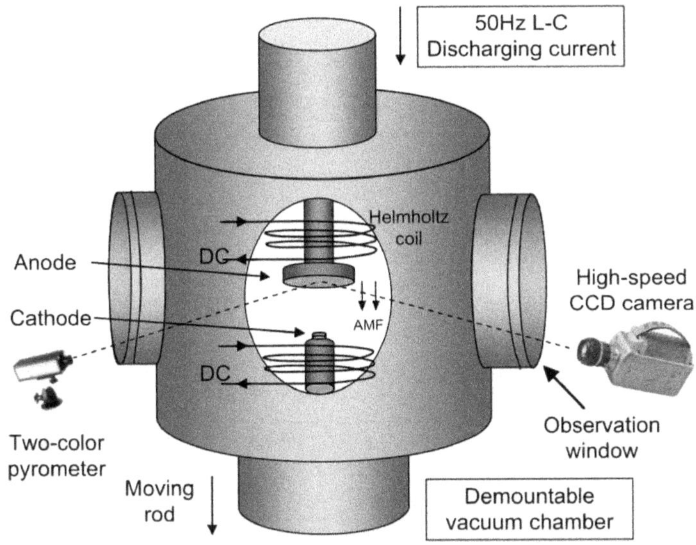

Fig. 1.51 Experimental setup for anode surface temperature measurement

arrangement of the experiments. There are two cases were investigated in the experiment. In case I, the anode diameter was 60 mm and the cathode diameter were 12 mm. In case II, the anode diameter was 12 mm and the cathode diameter was 60 mm. The thickness of the contacts was 5 mm. The contact material of both anode and cathode was CuCr25 (25% of Cr).

Figure 1.51 shows an experimental setup for measuring anode surface temperature. The asymmetrical contacts were installed inside a demountable vacuum chamber which was evacuated and sustained to 10^{-3}–10^{-4} Pa. As shown in Fig. 1.51, the contacts were surrounded by an external Helmholtz coil, which generated a uniform axial magnetic field. The Helmholtz coil has a radius of 16 cm, which was installed coaxially with the asymmetrical contacts in the demountable chamber. There was a 380-A dc current passing through the Helmholtz coil, which induced an axial magnetic flux density of 110 mT. The axial magnetic field was with a direction from the upper contact to the lower contact in the chamber. There were two observation windows on the demountable vacuum chamber. A high-speed CCD video camera was used to record the vacuum arc evolution through one window and a two-color pyrometer measured anode surface temperature through the other one. The CCD video camera had a recording speed of 10 000 frames/s. Its aperture was fixed at four and its exposure time was set as 2 μs.

A L–C discharging circuit was adopted to provide arc currents. The current frequency was 50 Hz and the arc current reached 10 kA rms. The applied arc current can be controlled by adjusting the voltage of the capacitor banks. The movable lower contact can be controlled electronically to realize an arcing time of ~10 ms. A contact separation length l, which was equal to the length of contact separation at current

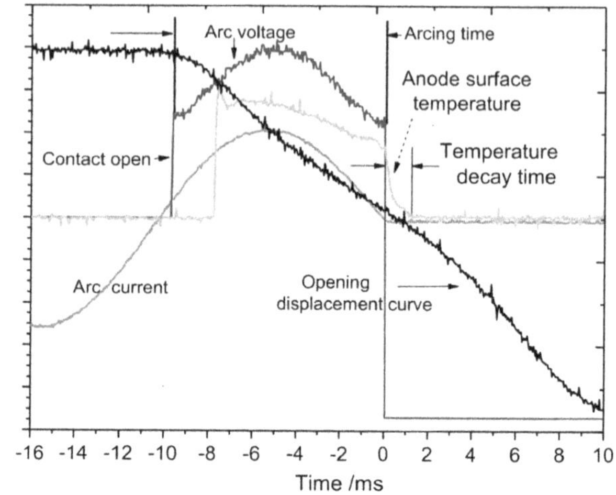

Fig. 1.52 Typical results of the measured arc current, arc voltage, post-arc anode surface temperature, and opening displacement

zero, was set as 24 mm. The average opening velocity used in the experiments was set as 2.4 m/s which were referred as an average velocity during the arcing period. This was chosen according to the 126-kV vacuum interrupters [22]. Figure 1.52 shows typical results of the measured arc current, arc voltage, post-arc anode surface temperature, and opening displacement. It should be noted that only the anode temperature measured by the pyrometer after current zero can be processed as valid data. This is because the glaring arc emission during arcing period overwhelmed the emission from contact surface, so it made the data before current zero unreasonable.

A two-color pyrometer designed by Japan CHINO Corporation with a Model IR-FB was chosen to measure the anode surface temperature. The two-color pyrometer can measure temperature in a range from 1073 to 1873 K. The measurement wavelengths were 850 and 1000 nm, respectively. Its time response that arise from 10 to 90% of the total range was 0.5 ms. A distance from the receiver of the two-color pyrometer to the measured anode surface was fixed at 50 cm in the experiments. To aim at the peak value of the surface temperature, one must figure out the region of anode surface that was eroded by arcs seriously. Usually, this region was associated with anode spots, which can be determined by observations using the CCD camera. Thus, the pyrometer aimed at the area where an anode spot appeared. It was found that anode spots resided at the same region for the same contacts if an anode spot formed on the surface in the previous experiment. Therefore, the region at which the pyrometer aimed every time did not need to adjust. In addition, it should be noted that only the temperature measured by the pyrometer after current zero can be processed as valid data. This is because the glaring arc emission during arcing period overwhelmed the emission from contact surface, so it made the data unreasonable. Therefore, the data before current zero should be neglected.

The two-color pyrometer measures anode surface temperatures by obtaining a ratio of the spectral radiation energy at one wavelength to that at another wavelength based on Planck's law. In addition, it has the advantage that the measurement errors can be limited by choosing feasible wavelengths, without knowing the emissivity of the measured surfaces. Sometimes the temperature measured by the two-color pyrometer is called color temperature. The relation between color temperature T_R and true temperature T can be expressed as Eq. (1.32) [84].

$$T_R = \left(\frac{1}{T} + \frac{\ln(\varepsilon_1/\varepsilon_2)}{C_2\left(\lambda_2^{-1} - \lambda_1^{-1}\right)}\right)^{-1} \tag{1.32}$$

where $C_2 = 1.4388 \times 10^{-2}$ m · K. λ_1 and λ_2 are the wavelengths and ε_1 and ε_2 are the emissivities at the two wavelengths. The temperature error E_T can be defined as Eq. (1.33).

$$E_T = \frac{|T_R - T|}{T} \times 100(\%) \tag{1.33}$$

To estimate the error, the emissivity needs to be obtained in advance. In addition, the emissivity varies with the wavelength. The emissivity ε_λ at wavelength λ can be expressed empirically as Eq. (1.34) [85].

$$\varepsilon_\lambda = \frac{1}{1 + a_0\lambda^2} \tag{1.34}$$

where a_0 is a constant which can be deduced from experimental data. Schellekens and Schulman [60] determined the emissivity of CuCr contact surface as 0.63 at a wavelength of 1000 nm. Therefore, the constant a_0 was 0.58 by substituting the emissivity of 0.63 into (1.34). In this experiment, the contact materials were CuCr and the wavelengths λ_1 and λ_2 were 850 and 1000 nm, respectively. Thus, the emissivity at the wavelength 850 nm was 0.7 calculating from Eq. (1.34) with $a_0 = 0.58$. According to (1.32) and (1.33), the measured temperature error was determined as ~5% at a temperature of 1300 K.

For case I shown in Table 1.5, Fig. 1.53 shows the arc evolution, arc voltage, post-test anode photo and anode surface temperature decay after current zero under three kinds of arc current. The arc current (rms) was 3.1, 5.8, and 8.5 kA, respectively. The anode was 60 mm in diameter and the cathode was 12 mm in diameter. The applied axial magnetic flux density was 110 mT. The arcing period was 10 ms and the contacts opened with a velocity of 2.4 m/s. The contact separation length was 24 mm at current zero. Figure 1.53a shows the arc evolution. At the arc current of 3.1 kA, the arc appeared as a diffuse column that emitted a luminous light. The anode seems quiet. When the arc current increased to 5.8 kA, both cathode and anode became more active with a bright light. The arc column became more luminous too, and an anode spot formed. At the arc current of 8.5 kA, it was observed that the anode was quite active, with an anode spot formed again. The anode was observed to project particles into the contact gap. The vacuum arc evolves into a high-current

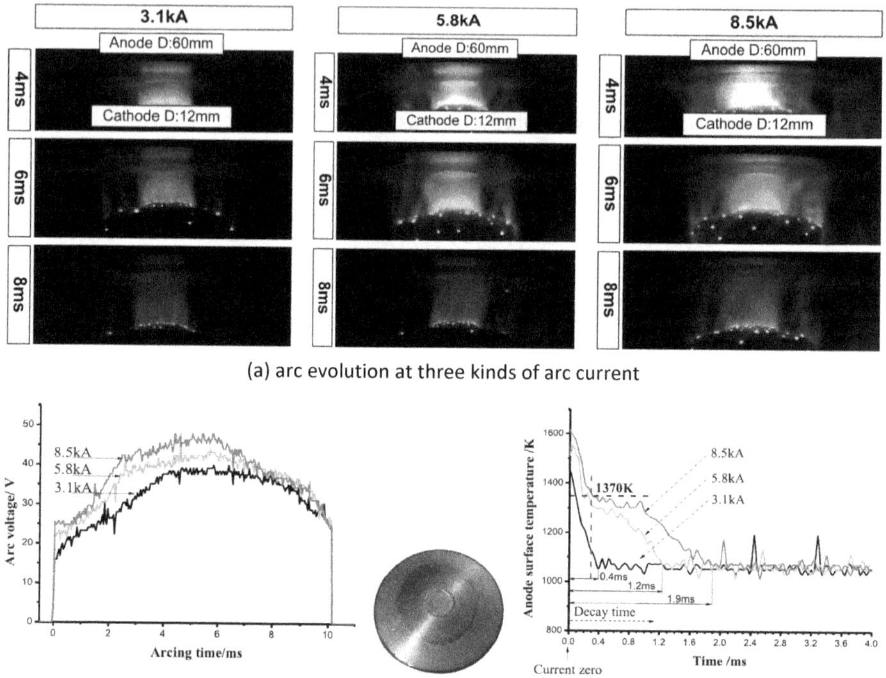

(a) arc evolution at three kinds of arc current

(b) Arc voltage and anode surface

(c) anode surface temperature decay after current zero

Fig. 1.53 The arc evolution, arc voltages, post-test anode photo and anode surface temperature decay after current zero for case I. Anode diameter: 60 mm; cathode diameter: 12 mm

anode mode. Figure 1.53b shows the arc voltage and post-test anode erosion photo. The arc voltage remained in a range of 15–50 V during the arcing period. The arc voltage looked smooth at the arc current of 3.1 kA. With an increase of arc current to 5.8 kA, there were several arc voltage spikes appearing. The anode spot formation at the arc current of 5.8 kA can be confirmed by the arc photos, the arc voltage spikes and a severe local anode erosion by the post-test examination. As the arc current further increased to 8.5 kA the arc voltage spikes disappeared. The further formation of anode spot was possible to ease the arc voltage spikes appearing. Figure 1.53c shows the anode surface temperature decay after current zero. At the arc current of 3.1 kA, the anode surface temperature decayed exponentially after current zero. However, the decay curve of the arc current of 8.5 kA were quite different, which showed three stages in the decay processes. In the first stage, the temperature decayed exponentially from the initial temperature at current zero to a value around 1370 K. In the second stage, the surface temperature remained as a plateau, which was quite close to the melting point of copper (1357.77 K). It implied the anode surface remained in a molten state after extinguishing the 8.5 kA current. In the third stage, the anode temperature continued decaying exponentially to the lower measuring limit of the pyrometer (1073 K). The decay curve of the arc current of 5.8 kA looked like that of 8.5 kA. However, the three stages were not obvious. A decay time can be defined as a period from current zero to the instant when anode surface temperature dropped to the lower measuring limit of the pyrometer. The decay time was 0.4, 1.2, and 1.9 ms for the arc current of 3.1, 5.8, and 8.5 kA, respectively.

For case II shown in Table 1.5, Fig. 1.54 shows the arc evolution, arc voltage, post-test anode photo and anode surface temperature decay after current zero under four kinds of arc current. The arc current (rms) was 3.2, 7.8, 8.8, and 9.5 kA, respectively. The anode was 12 mm in diameter. The cathode was 60 mm in diameter. Both the anode material and the cathode material were CuCr25 (25% Cr). Figure 1.54a shows the arc evolution. At the arc current of 3.2 kA, the arc remained in a diffuse arc mode. The anode seemed quiet. After the arc current increased to 7.8 kA and up to 9.5 kA, the arc column constricted obviously with a luminous light. The anode became quite bright and an anode spot formed, which can be confirmed by an assistance of arc voltage and a post-test anode erosion examination. Figure 1.54b shows the arc voltage during arcing period and post-test anode erosion photo. The arc voltage remained in a range of 15–50 V. The arc voltage was low, and it appeared smooth at the arc current of 3.2 kA. The arc voltage became higher with the arc current increased to 7.8 and 8.8 kA and a few spikes appeared on the arc voltage curve. When the arc

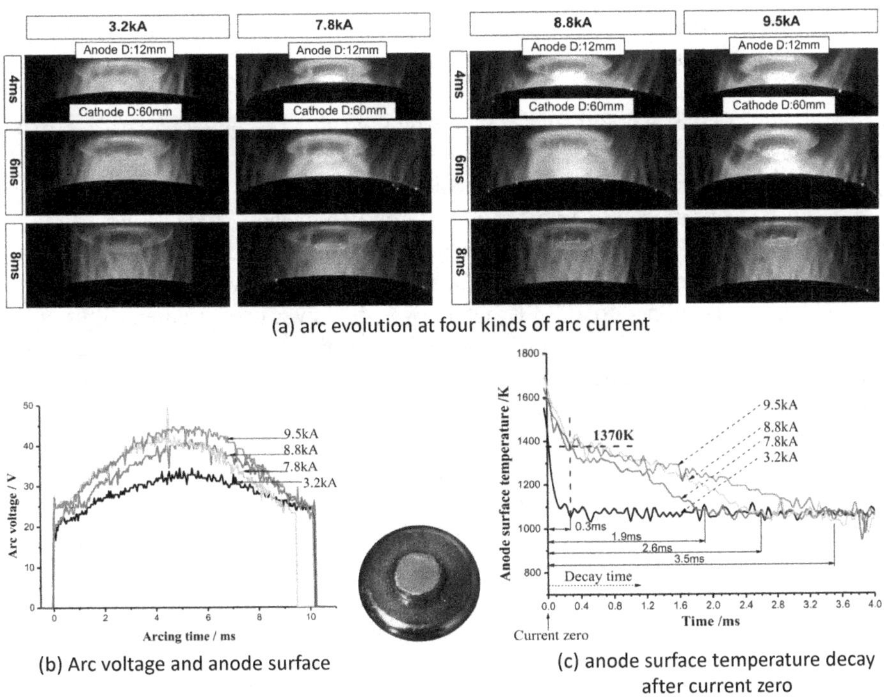

Fig. 1.54 The arc evolution, arc voltages, post-test anode photo and anode surface temperature decay after current zero for case II. Anode diameter: 12 mm; cathode diameter: 60 mm

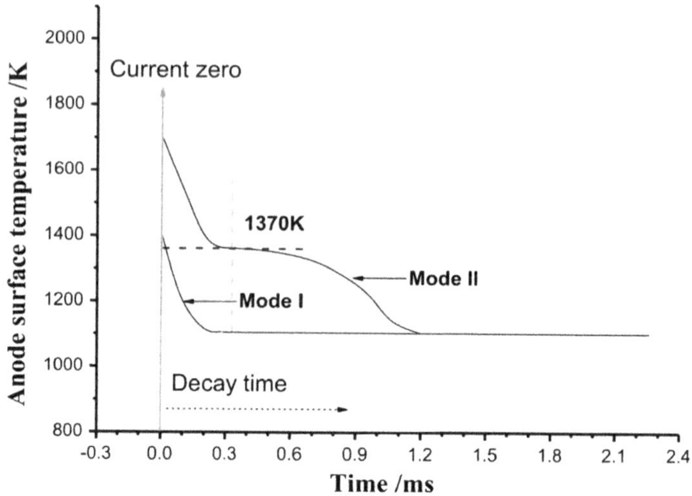

Fig. 1.55 Two decay modes of anode surface temperature after current zero

current further increased to 9.5 kA, the spikes disappeared by a further formation of the anode spot. Figure 1.54c shows the anode surface temperature decay after current zero. At the arc current of 3.2 kA, the anode surface temperature decayed exponentially to 1073 K. As the arc current increased to 7.8, 8.8, and 9.5 kA, the anode surface temperature also showed the three stages mode mentioned above. The decay time were 0.3, 1.9, 2.6, and 3.5 ms when the arc current was 3.2, 7.8, 8.8 and 9.5 kA, respectively.

As shown in Figs. 1.53c and 1.54c, the decay process of the anode surface temperature can be classified into two basic patterns, which are Mode I and Mode II, respectively. Figure 1.55 showed the two decay modes of anode surface temperature after current zero. In Mode I, anode surface temperature decayed exponentially after current zero and dropped to 1073 K, which was the lower measurement limit of the two-color pyrometer. It is reasonable to infer that the surface temperature finally dropped to the room temperature far below 1073 K. In addition, the whole decay process depended on the initial temperature at current zero. With a decrease of the initial anode surface temperature at current zero, the duration of the decay processes decreased as well. In Mode II, the anode temperature decay curve can be divided into three stages. In the first stage, the temperature dropped from the initial temperature at current zero to 1370 K. In the second stage, the temperature remained as a plateau, which was close to the melting point of copper. In the third stage, the anode temperature exponentially dropped to the lower measuring limit of the pyrometer (1073 K). The duration of the decay process was also influenced by the initial temperature at current zero.

After classifying the decay process of anode surface into the two modes, it is natural to relate the decay modes to their corresponding arc modes, which are believed to have an impact on the interruption capacity of vacuum interrupters. Mode I correspond to a thermal conduction on anode surface with a diffuse arc mode. As shown in Fig. 1.53a, the vacuum arc remained in a diffuse arc mode at the current of 3.1 kA. Figure 1.53c shows the corresponding anode surface temperature decay curve with the arc current of 3.1 kA can be classified as the Mode I. Mode II corresponds to the anode surface decay process after extinguishing a current with the high-current anode mode. As shown in Fig. 1.53a, the vacuum arc was in a high-current anode mode at the arc current of 5.8 and 8.5 kA. The corresponding temperature in Fig. 1.53c can be classified as the Mode II. This correlation can also be verified by Fig. 1.54. One may find that it is not easy to judge a boundary between the diffuse are mode and the high-current anode mode. In such case, the anode surface temperature decay curves after current zero provide further information to assist the judgment. Therefore, a clear information with a physical meaning can make the judgment more reasonable. The post-test anode surfaces, typically the one shown in Fig. 1.53b, are consistent with the experimental results and the theoretical discussion shown in Sects. 1.4.1 and 1.4.2. The anode surface was locally eroded by a blowing effect in a high-current anode mode arc. As the anode surface temperature in arcing period dominates the generation of metal vapor in post-arc period, the anode surface temperature measurement and the classification of the decay modes after current zero has a definite physical meaning to the interruption capacity of vacuum interrupters.

So far it reveals that four elements determine the current interruption capability of vacuum interrupters. The four elements are the arc mode, the anode erosion, the anode temperature and the decay mode, respectively. A successful current interruption is related to a diffuse arc mode, a low anode erosion, a low anode temperature at current zero (less than ~2000 K) and the Mode I of the anode temperature decay. On the other hand, a failure current interruption is related to a high-current anode mode, a severe anode erosion, a high anode temperature at current zero (higher than 2000–2500 °C) and the Mode II of the anode temperature decay.

1.4.4 Decay Modes of Anode Surface Temperature After Current Zero in Vacuum Arcs-Impact on Dielectric Recovery Strength

In Sect. 1.4.3, the decay processes of the anode surface temperature after current zero exhibited two modes. The two decay modes are correlated with the arc modes. Furthermore, Watanabe et al. [57] and Niwa et al. [66] introduced melting time as an indicator to describe the decay of the anode surface temperature after current interruption, and they proposed that the interruption capacity of a vacuum interrupter was associated with the duration of melting time. The melting time results shown in Sect. 1.4.3 agreed with them. However, one still does not explicitly know that how anode surface temperature determines current interruptions. There are many calculations on thermal transfer process on an anode [86–88] and there are also

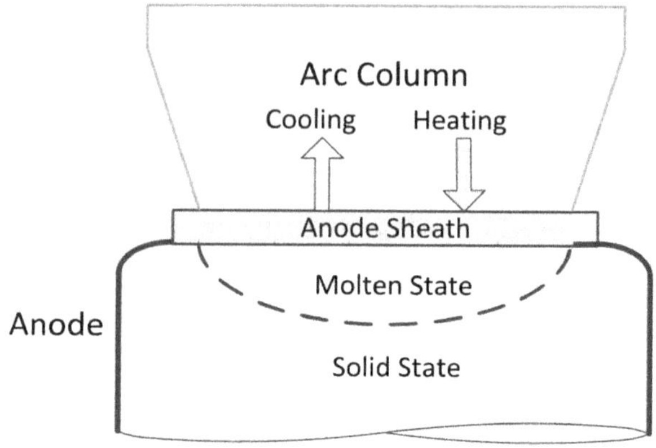

Fig. 1.56 Physical model for anode surface temperature calculation

many calculations on breakdown in metal vapor with a low pressure [89, 90], but a relationship between anode surface temperatures and breakdowns in the post-arc period, which is critical to understand the physical processes of current interruption in a vacuum interrupter, is still unclear.

This section theoretically determines a relationship between breakdowns and anode surface temperature after current zero. Thus, a heat conduction model was used to calculate the anode surface temperature after interrupting a vacuum arc. This was followed using the Particle-in-Cell/Monte Carlo collision (PIC-MCC) method to simulate an evolution of post-arc breakdowns in metal vapor via the temperature calculations. The physical meaning of the two decay modes can thus be interpreted by the temperature calculations. Finally, the relationship was found as a Paschen curve, whose metal vapor density was determined by the anode surface temperature.

A thermal model is established to describe a heat conduction process of an anode surface after interrupting a vacuum arc, taking the phase transition process into account, as shown in Fig. 1.56. There are four regions in the model, which includes an arc column, an anode sheath, a molten region, and a solid region. The arc is considered as axisymmetric, which corresponds to the arcs under an axial magnetic field control. The energy input provided by the arc column heats the contact surface, resulting in melting. Thus, the contact can be divided into a solid part and a molten part. The boundary can move according to the input and output of the energy. The energy dissipates by thermal conduction and evaporation, while the effects of radiation and convection are considered playing a minor role. In addition, copper was selected as the anode material.

A 3-D transient model of the anode surface temperature is adopted to describe the heat conduction by the Eq. (1.35).

$$\frac{\partial}{\partial t}(\rho H) = \nabla \cdot (k\nabla \mathrm{T}) \tag{1.35}$$

where T is the anode temperature. ρ is the density of the contact material copper. k is the thermal conductivity, and H is the enthalpy. The enthalpy of the material is calculated by the Eq. (1.36).

$$H = h + \beta L \tag{1.36}$$

where h is the sensible enthalpy. β is the liquid fraction, and L is the latent heat of the material. The liquid fraction β is defined as Eq. (1.37).

$$\begin{cases} \beta = 0, & if \quad T < T_S \\ \beta = 1, & if \quad T > T_l \\ \beta = (T - T_s)/(T_l - T_s), & if \quad T_S < T < T_l \end{cases} \tag{1.37}$$

where T_s is the solidus temperature of the contact material and T_l is the liquidus temperature.

The distribution of heat flux input as a function of time can be empirically described by the Eq. (1.38) according to Wang et al. [86].

$$H = H_P \times e^{-5000r^2} \times \sin(100\pi t) \tag{1.38}$$

where r is the radial position at the anode surface. t is the simulation time, and H_p is the peak value of heat flux density. The value of H_p is related to the arc current, which was obtained from a magnetohydrodynamic simulation of the vacuum arc [50]. Four values of 5×10^8, 8×10^8, 1.2×10^9, and 1.5×10^9 W/m^2 are chosen, which are based on a study by Zhang et al. [91]. They showed that the peak values of heat flux density when anode spot formats fell in a range of (6.1–10.5) $\times 10^8$ W/m^2, which was obtained by three-dimensional magneto-hydrodynamic model simulation results referring to the experimental results for anode spot formation (see Sect. 1.2).

Evaporation can vaporize anode material as well as take some energy away from the anode surface. The energy carried away by evaporation per unit area can be expressed as the Eq. (1.39) [64].

$$L_{evap} = cV\frac{P_s(T_s)}{\sqrt{T_s m}} \tag{1.39}$$

where $c = 5.76 \times 10^9$ is a constant. $V = 3.19$ eV is the vaporization enthalpy per atom. T_s is the surface temperature. P_s is the saturated vapor pressure, and m is the atomic mass of the anode material (copper). The saturated vapor pressure for copper can be expressed by the Clausius–Clapeyron Eq. (1.40)

$$P_s = P_{atm}\exp\left[-\frac{\Delta_{vap}H}{R}\left(\frac{1}{T_s} - \frac{1}{T_{boil}}\right)\right] \tag{1.40}$$

where $P_{atm} = 10^5$ Pa represents the atmospheric pressure. R is the gas constant. $\Delta_{vap}H = 300.4$ kJ/mol is the latent heat of vaporization, and $T_{boil} = 2868$ K is the boiling temperature of copper. Figure 1.57 shows the calculated curves for saturated pressure and evaporated energy as functions of surface temperature T_s.

The Particle-in-Cell/Monte Carlo collision (PIC-MCC) method is adopted to simulate the evolution of breakdowns in metal vapor. PIC method is a simulation method for studying plasma physics, based on the evolution of plasma distribution function in phase space [92]. In a discharge, a short-range collision must be considered. Thus, MCC method is introduced to PIC's scheme. A detailed discussion of the PIC–MCC can be found in Ref. [93]. The simulations were implemented by the VORPAL code [94].

A model with two dimensions in physical space and three dimensions in velocity space (2d3v) was developed, as shown in Fig. 1.58. The cathode was designed with a rounded end for avoiding a significant concentrating electric field near the edge, and a negative direct voltage was connected to it. In addition, the anode was connected to the ground.

Only the monovalent ions were considered because the initial stage of breakdowns were focused on, in which Cu^+ ions were mostly generated. Thus, there were three types of particles in the model including Cu vapor, Cu^+, and electrons, which followed a relationship described by the Eq. (1.41). The particles were absorbed by the boundaries when they reached the boundaries or contact surfaces of the model. Elastic scatter, ionization, and excitation were considered, and the cross-sectional data were obtained from evaluated electron data library [95].

$$\mathrm{Cu} + \mathrm{e}^- \rightarrow \mathrm{Cu} + \mathrm{e}^-$$

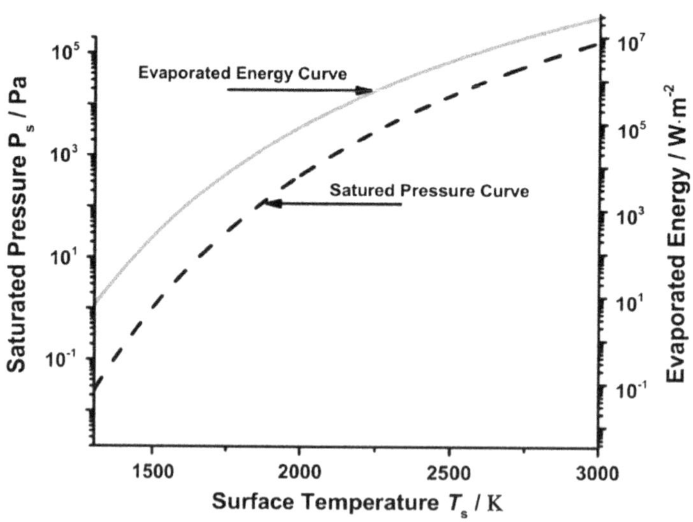

Fig. 1.57 Saturated pressure and evaporated energy as functions of surface temperature T_s

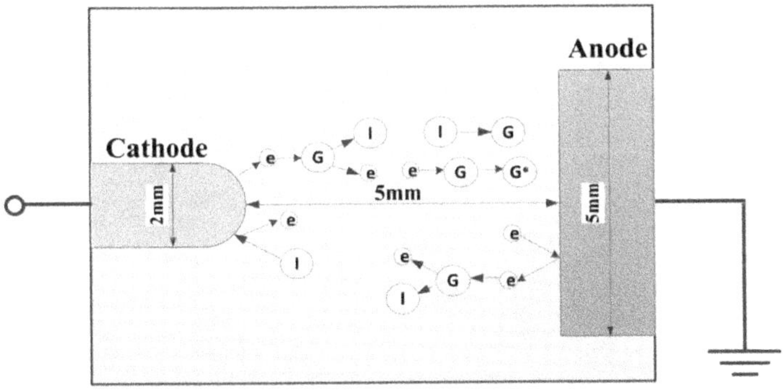

Fig. 1.58 Schematic of a Particle-in-Cell/Monte Carlo collision breakdown model

$$\begin{aligned}&\mathrm{Cu} + \mathrm{e}^- \rightarrow \mathrm{Cu}^+ + 2\mathrm{e}^- \\ &\mathrm{Cu} + \mathrm{e}^- \rightarrow \mathrm{Cu}^* + \mathrm{e}^-\end{aligned} \tag{1.41}$$

Assuming that the background metal vapor has a uniform unchanged temperature of T = 2000 K, the density N can be obtained by substituting the Eq. (1.40) into the ideal gas law $N = P_s/kT$. Moreover, it was assumed that a breakdown was initialized by seed electrons emitted from the cathode surface. The Richardson–Dushman Equation (1.42) was adopted to describe the emission of the electrons.

$$J = M\frac{4\pi me}{h^3}T^2 exp\left(-\frac{W}{T}\right) \tag{1.42}$$

where h is the Planck constant. T is the anode surface temperature. M is a constant parameter. m is the mass of an electron, and e is the charge of an electron. W can be expressed as

$$W = W_0 - \frac{e}{4\pi\varepsilon_0}|fE| \tag{1.43}$$

where W_0 is the work function of the contact material. ε_0 is the vacuum permittivity. E is the electric field strength, and f is the field enhancement factor.

Secondary electron emission from contact surfaces is another important physical effect that must be considered. Vaughan [96] gave an empirical formula to estimate that secondary electron emission yield and can be expressed as Eq. (1.44).

$$\delta(E_{inc}, \theta) = \delta_{max}\left(1 + k_s\frac{\theta^2}{2\pi}\right) \times f(w, k) \tag{1.44}$$

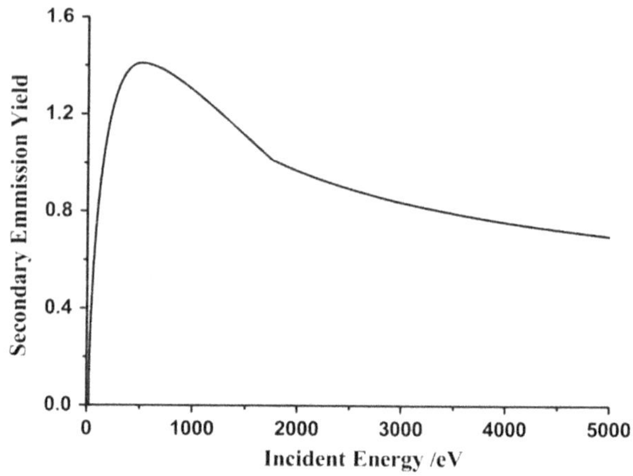

Fig. 1.59 Secondary electron emission yield for copper

where $E_{\rm inc}$ is the energy of an incident particle. $\delta_{\max}$ is the maximum value of the secondary electron emission yield. θ is the angle between the incident particle and contact surface, and k_s is the surface coefficient. Moreover

$$f(w,k) = \begin{cases} \left(we^{(1-w)}\right)^k, & k = \begin{cases} 0.56, & w \le 1 \\ 0.25, & 1 < w \le 3.6 \end{cases} \\ 1.125w^{-0.35}, & w > 3.6 \end{cases}$$

And $w = (E_{inc} - E_{min})/(E_{max} - E_{min})$, where E_{min} is the minimum energy that can induce secondary electron emission from contact surface and $E_{\rm max}$ is the maximum energy corresponding to the maximum value of the secondary electron emission yield. According to the data for copper [90], the secondary electron emission yield for the copper is shown in Fig. 1.59.

Therefore, a 3-D transient model of anode surface temperature was established by the above-mentioned equations. It was used to determine the evolution and distribution of an anode surface temperature. The model considered heat conduction, phase transition, and evaporation. In the model the anode was considered as copper. It was a butt type electrode with a diameter of 60 mm. The input and the output of the energy on the anode surface were described in Eqs. (1.38) and (1.39), respectively. From the anode temperature simulations results, the evaporated metal vapor density can be estimated quantitatively. The decay characteristics of the metal vapor, which are correlated with the dielectric recovery process of the vacuum interrupter, will be obtained.

Figure 1.60 shows the anode temperature distribution and evolution from the views of the cross section and anode surface, respectively. Figure 1.61 shows a radial distribution of the anode temperature. The heat flux flowing from the arc into the anode was a 50 Hz sine wave with a peak value of 1.2×10^9 W/m^2. The heat energy caused melting on the anode. The melting lasted well after 10 ms arcing time when the arc extinguished. The maximum temperature, following the distribution of energy

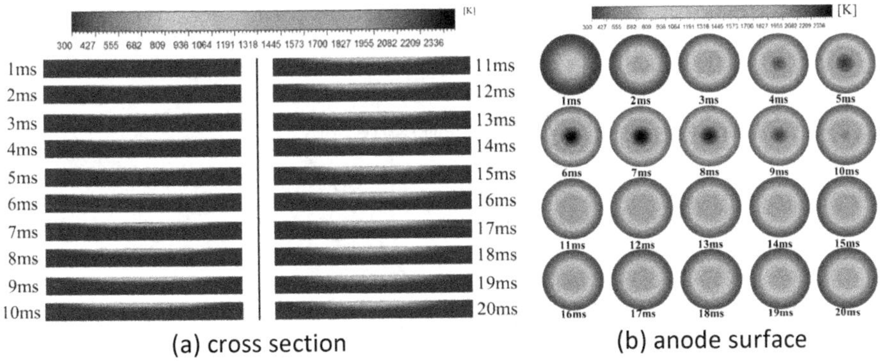

Fig. 1.60 Distribution and evolution of the anode temperature during the arcing period

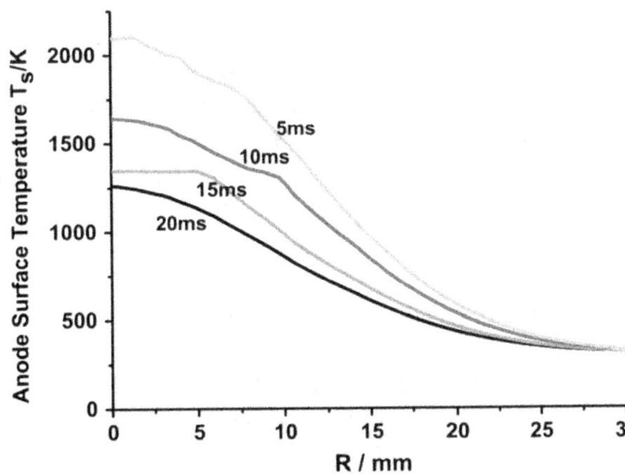

Fig. 1.61 Radial distribution of surface temperature

input, appeared at the anode surface center. The temperature decreased with the radial direction. The energy was conducted into the inside of the anode, causing a depth of melting area. The whole melting area extended to half of the anode surface. It remained as a large area after current zero with a temperature well above the melting point of the anode material.

Figure 1.62 shows the evolution of the anode surface center temperature. The heat flux flowing into the anode were 50 Hz sine waves with peak values of 5×10^8, 8×10^8, 1.2×10^9, and 1.5×10^9 W/m^2, respectively. A higher energy input resulted in a higher temperature curve. The anode surface temperatures curve varied in accordance with a sine wave, as the wave of the arc current. The temperatures curves followed different frequencies. Furthermore, the temperature curves achieved their peaks after 5 ms at which the arc current was at its peak value. The delay was because of the thermal capacity of the contact material, which impeded the change of the temperature rapidly. For the heat flux density of 5×10^8 W/m^2, the peak value during the arcing was about 1250 K, which was lower than the melting point of

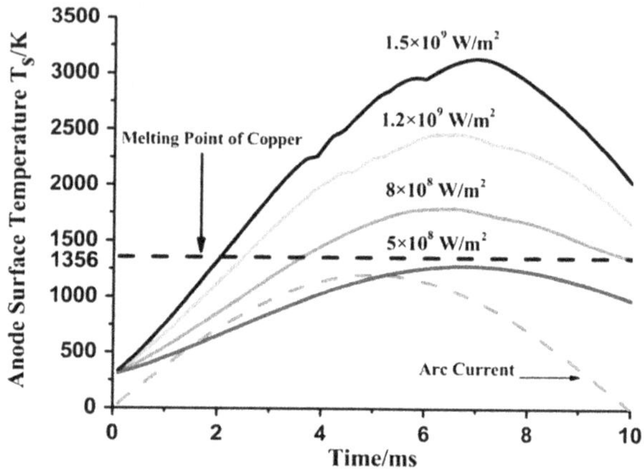

Fig. 1.62 Evolution of the anode surface center temperature during arcing period of a half cycle

copper, the anode material, 1356 K. For the heat flux of 8 × 10^8, 1.2 × 10^9, and 1.5 × 10^9 W/m^2, however, the peak values were higher than the melting point of copper, so melting occurred. Furthermore, the temperature at current zero was still higher than the melting temperature for the case of the heat flux density of 1.2 × 10^9 and 1.5 × 10^9 W/m^2, which led to a liquid state of the anode surface at current zero.

Figure 1.63 shows the decay of the anode surface center temperature after current zero. It shows the anode temperature exponentially decayed after current zero for the heat flux density input of 5 × 10^8 and 8 × 10^8 W/m^2. At current zero, there was no more heat flux density inputting into the anode. From then on, the anode temperature declined to the ambient temperature according to the thermal capacity of the anode. For the heat flux density of 1.2 × 10^9 and 1.5 × 10^9 W/m^2, the anode surface temperature curve showed two stages in the decaying periods. In the first stage, the temperature decayed from an initial value at current zero to the melting point of the contact material and it held for a while. In the second stage, the temperature

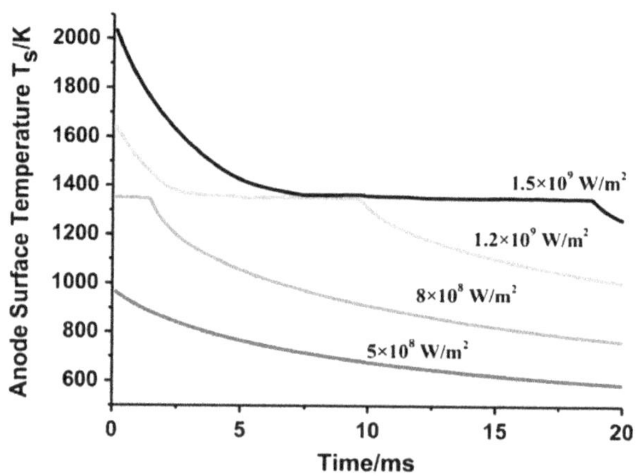

Fig. 1.63 Decay of anode surface center temperature after current zero

exponentially decayed to the ambient temperature. The stagnation was caused by the material recovering from a liquid to a solid state, in which it released excess energy. The released energy also impeded the decay of the anode temperature, so the temperature remained at the melting point for several milliseconds.

In the model shown in Fig. 1.58, there were three types of particles including Cu atoms, Cu^{+} ions, and electrons. The model accounted for elastic scatter, ionization, and excitation. The detailed process of a breakdown can be obtained by the model. The model can also obtain a relationship between the background metal vapor densities and breakdown voltages. Consequently, a relationship between the anode surface temperature and the breakdown voltage can be derived, which assumed that the background metal vapor density depends on the anode surface temperature at an equilibrium state. In the following simulation results, the background neutral metal vapor was copper with a density of $10^{22}/m^3$. The applied voltage was 100 V, and the gap distance was 10 mm.

Figure 1.64 shows the multiplication of electrons and ions evolved with time during a breakdown process. The neutral background vapor was ionized, and the number of the particles increased exponentially due to electron avalanche. The ions curve was smooth comparing with the electrons curve, because copper ion mass is much higher than electron mass. Thus, the ions remained in the calculation region, but the electrons were absorbed by the boundaries of the calculation region. For the same reason, the ions quantity was higher than that of the electrons.

Figure 1.65 shows the spatial evolution of ions and electrons during a breakdown. The evolutions of ions and electrons were different during a breakdown. In the initial stage of the breakdown, the neutral metal vapor near the cathode was ionized by the emitted electrons from the cathode surface. Then the gap was gradually filled with ions, and the ion density increased. Since the electrons moved faster than the ions,

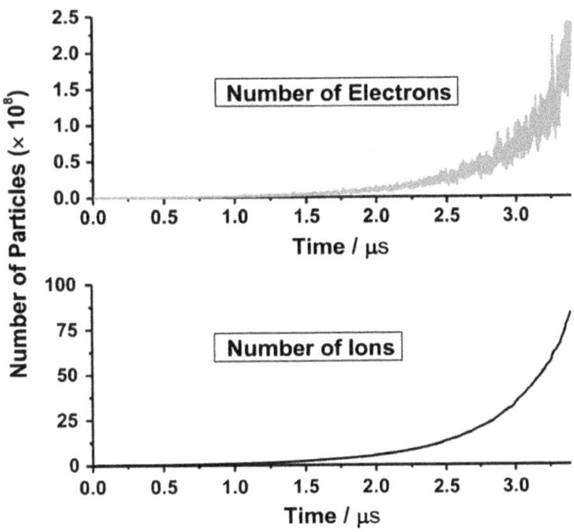

Fig. 1.64 Multiplication of electrons and ions during a breakdown

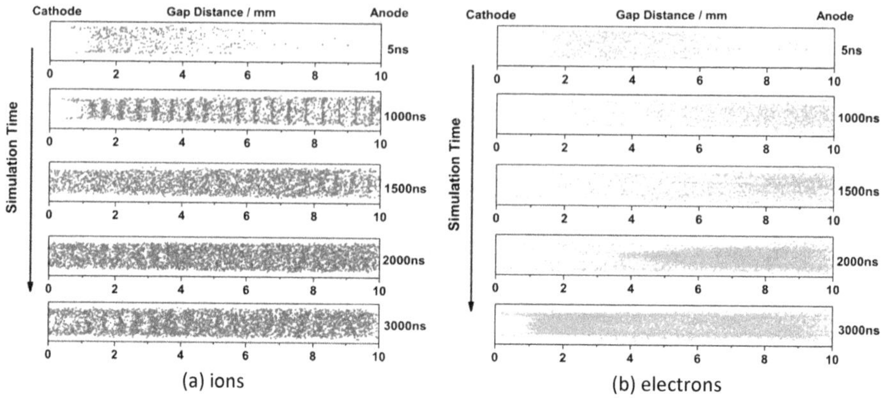

Fig. 1.65 Spatial evolution of ions and electrons during a breakdown

they initially moved to the anode by the electric field force and concentrated in the anode region. Then they expanded from the anode to the cathode. The evolutions combined with the multiplication shown in Fig. 1.64 can help to judge the breakdown instant. If the electrons were uniformly distributed across the gap while multiplying, this meant that a discharge channel formed.

Figure 1.66 shows the evolution of electric potential distribution across the gap during a breakdown. In the initial stage, the electric potential was uniformly distributed across the gap since the number of the particles was too low to affect the distribution. However, the electric field strength adjacent to the cathode began to increase after 1500 ns, which was corresponding to that the electron distribution expanded from the anode to the cathode, as shown in Fig. 1.65b. Eventually, a sheath formed adjacent to the cathode, which resulted in a strong electric field that accelerated the ions to bomb the cathode. The high energy bombarding ions led

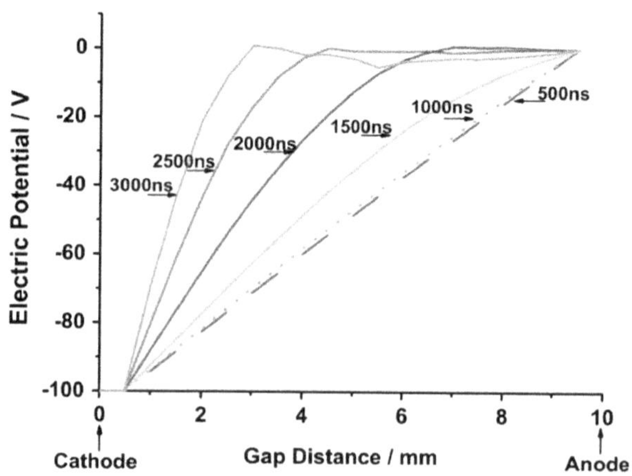

Fig. 1.66 Evolution of electric potential distribution across the gap during a breakdown

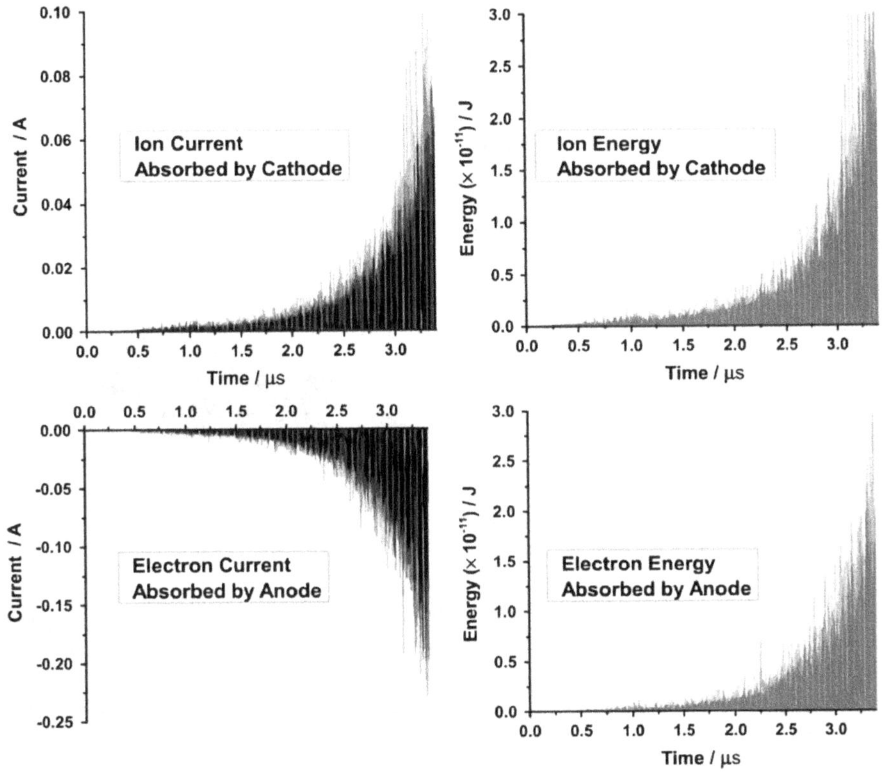

Fig. 1.67 Energy and current absorbed by the anode and cathode surfaces

to secondary electron emission, as shown in Fig. 1.59. Therefore, secondary electron emission would act as a major electron emission source instead of the primary electron emission.

Figure 1.67 shows the energy and current absorbed by the anode and cathode surfaces, respectively. The electron current absorbed by the anode was higher than that of ions absorbed by the cathode, but it was still in the same order of magnitude. Both the electron energy absorbed by the anode and the ion energy absorbed by cathode surfaces exponentially increased. However, the ion energy absorbed by the cathode was slightly higher than that of electron because the sheath formed adjacent to the cathode giving more energy to ions by the electric field.

Figure 1.68 shows a typical multiplication processes of ions and electrons without a breakdown, in which the applied voltage was 40 V. The applied voltage was lower than the previous 100 V. Other parameters kept as same as the previous ones. The electron and ion quantity linearly increased in the initial stage, but soon they saturated, which was an equilibrium between the generation and dissipation of the particles.

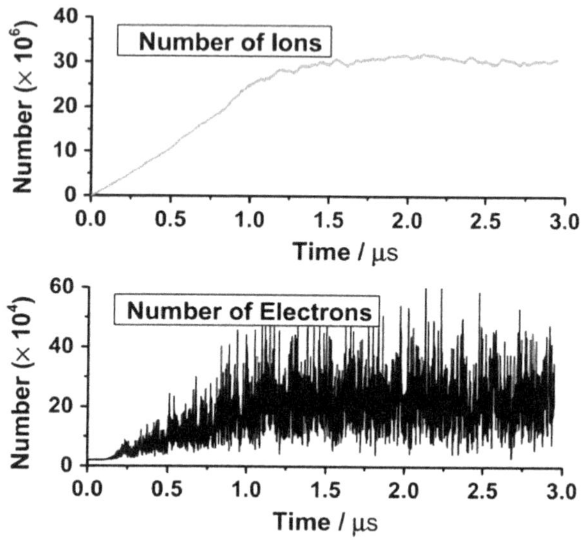

Fig. 1.68 A typical curve of electron and ion quantity without a breakdown

This equilibrium could be achieved after several microseconds or more. The equilibrium can help to judge a breakdown did not occur. If a breakdown occurs, the ions and electrons quantity would exponentially increase, as shown in Fig. 1.64.

Figure 1.69 shows a comparison between breakdown occurrence for 1 and 10 mm gaps. The applied voltages were 50, 100, 200, 300, and 400 V, respectively. It took a longer time for a breakdown of a larger gap than a smaller one. Their difference was ~5 μs when a low voltage was applied. However, their difference reduced to about 1 μs when a higher voltage was applied. A larger gap distance required more time for ions and electrons to be absorbed by the contact surfaces, during which secondary

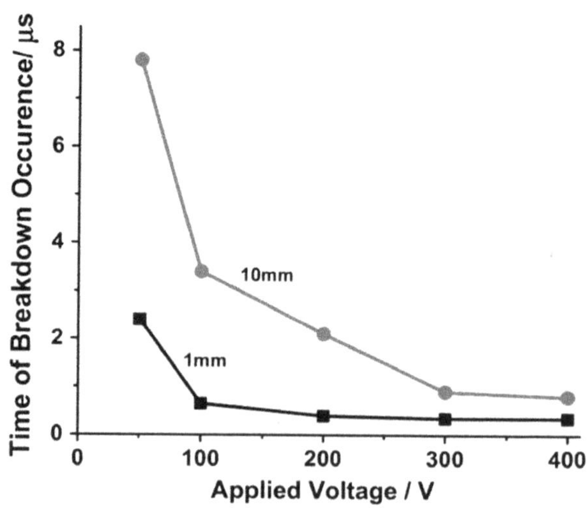

Fig. 1.69 A comparison between breakdown occurrence for 1 mm and 10 mm gaps varied with applied voltages

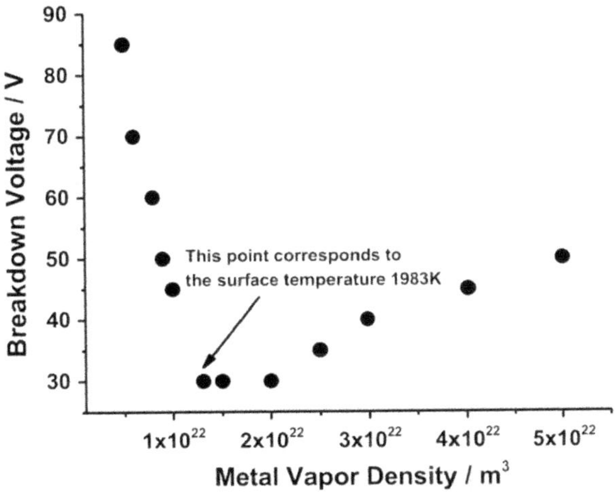

Fig. 1.70 Paschen curve of the copper metal vapor

electron emission became an important mechanism to induce a discharge. However, increasing applied voltage reduced the difference, because it accelerated the ions and electrons bombarding the contact surfaces.

Figure 1.70 shows the calculated the Paschen curve of the metal vapor. In the low-density region, the breakdown voltages declined with the increase in the metal vapor densities. The minimum breakdown voltage of 30 V occurred at a density about $(1.3–2.0) \times 10^{22}/m^3$. Then the breakdown voltage began to increase with the increase of the metal vapor density. Corresponding to the minimum breakdown voltage, the anode surface temperature was calculated as 1983 K. The temperature was derived from Eq. (1.40) and the ideal gas law, by assuming a constant metal vapor temperature of 2000 K. The anode surface temperature of 1983 K was considered as the critical temperature for breakdowns.

The dielectric recovery after current zero plays a decisive role in the current interruption of vacuum interrupters. There is a transient recovery voltage applied upon the switching gap after current zero. At the same time, the plasma, metal vapor, liquid droplets generated during the arcing period remain in the gap and decay with time. Usually, there are many factors that affect the interruption process. [97]. Here the metal vapor density is focused on. The dielectric recovery is considered complete when the metal vapor density drops below a certain level, at which the electron mean free path in the vapor is of the order of the gap length [98].

The Rich–Farrall model [99] is widely used to describe the dissipation of metal vapor after current zero. The model calculates the decay of the metal vapor. The metal vapor density n at the center of the switching gap volume at time t after current zero can be described as Eq. (1.45).

$$n\left(0, \frac{L}{2}, t\right) = n_0\left[1 - e^{-\frac{1}{\alpha^2}}\right] erf \frac{1}{2}\left(\frac{L}{R}\right)\left(\frac{1}{\alpha}\right) \tag{1.45}$$

where $\alpha = t/R(2k_B T/m)^{1/2}$. R is the radius of the contact. L is the gap length. m is the mass of a metal atom. k_B is the Boltzmann constant, and the error function $erf(x) = 2/\sqrt{\pi}\int_0^x e^{-\tau^2/2}d\tau$. However, Takahashi et al. [100] measured the metal vapor density after interrupting a vacuum arc, and their experimental results showed that the measured dissipation of metal vapor was slower than that predicted by the Rich–Farrall model. There seem other factors impeding the dissipation of the metal vapor predicted by the Rich–Farrall model.

Surface temperature is obviously an important influencing factor that postpones the dissipation of metal vapor. Figure 1.63 suggests that the anode surface temperature decays with a time constant of the order of milliseconds. That means the anode surface temperature stays nearly constant in several microseconds. A hot contact surface impedes the dissipation of metal vapor because of evaporation. Therefore, it is reasonable to take account of the effect of anode surface temperature to estimate a dielectric recovery process, especially after interrupting a high-current anode mode vacuum arc with which the anode surface temperature is high.

There is a lack of the experimental data for copper metal vapor breakdowns. To confirm the simulation results, it is better to compare with other theoretical models. Burm [101] gave a set of equations that describe the Paschen curves in metal vapors and he also verified the theory by comparing the theoretical results with experiments for Hg and Zn [102]. According to his theory, the breakdown voltages can be expressed as Eq. (1.46).

$$V_{breakdown} = \frac{B_0 pd}{ln(A_0 pd) - ln\left(ln\left[\gamma + 1/\gamma\right]\right)} \tag{1.46}$$

where A_0 and B_0 are constants for breakdowns depending on the type of vapor. d is the gap distance. p is the vapor pressure in the gap, and $\gamma = 0.1$ [103] is the second Townsend coefficient. The constants A_0 and B_0 can be written as Eqs. (1.47) and (1.48).

$$A_0 = 0.224\frac{e^4}{16\pi\varepsilon_0^2 E_{12}^2}\frac{3}{k_B T_h} \tag{1.47}$$

$$B_0 = \frac{5}{4}\frac{\pi a_r^2}{k_B T_h}\frac{E_{12}}{e} \tag{1.48}$$

where e is the electron charge. ε_0 is the dielectric constant of vacuum, $E_{12} =$ 7.72 eV [104] is the first step ionization energy. k_B is the Boltzmann constant. T_h is the vapor temperature, and a_r is the atomic radius. For the Paschen curve, the minimum breakdown voltage and corresponding pressure multiplied by the gap distance ($p \times d$) can be derived as Eqs. (1.49) and (1.50).

$$V_{min} = exp(1)\frac{B_0}{A_0}ln\left(\frac{\gamma + 1}{\gamma}\right) \tag{1.49}$$

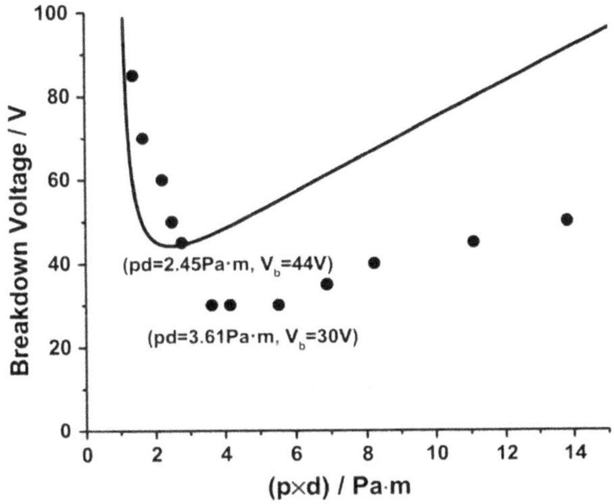

Fig. 1.71 Paschen curve for copper vapor. The solid line follows Burm's equations and the dots come from PIC–MCC simulations

$$(p \times d)_{min} = \frac{exp(1)}{A_0} ln\left(\frac{\gamma + 1}{\gamma}\right) \quad (1.50)$$

According to Burm's equations and parameters described above, the Paschen curve for copper vapor is shown in Fig. 1.71. The minimum point of corresponded to $p \times d = 2.45$ Pa · m and 44 V. With an assumption of copper vapor temperature of 2000 K, the minimum point obtained from the PIC–MCC method corresponded to $p \times d = 3.61$ Pa · m and 30 V. Therefore, the minimum point obtained from the PIC–MCC method was lower than that of Burm's equations. This is partly because only the first step ionization was considered in Burm's equations, which reduced the possibility of ionizations in metal vapor by neglecting processes such as excitations and elastic scatters. Moreover, the PIC–MCC method used the secondary emission yield as shown in Fig. 1.59 in contrast with a fixed value of 0.1 used in Burm's equations. Furthermore, the metal vapor can be removed or absorbed by the boundary in the PIC model, which reduced the metal vapor density, and then the minimum point of the PIC-MCC method may shift right resulting in a higher $p \times d$ value. In addition, Schade and Dullni [7] proposed a critical breakdown value of metal vapor density was 3×10^{21}/m^3 at a gap distance of 10 mm. The value corresponds to 0.83 Pa · m if the vapor temperature is 2000 K. In summary, the above-mentioned results suggests a Paschen breakdown mechanism of the metal vapor in vacuum interrupters.

A liquid contact surface caused by arcing can also increase the probability of breakdowns after current zero even if the contact temperature drops below a critical value. The liquid local contact surface is easily deformed by an electric field caused by a transient recovery voltage, enhancing the local electric field. The polarity of the voltage applied upon a switching gap is reversed after current zero, and an anode in the arcing period becomes a post-arc cathode. Sandolache and Rowe [9] observed a deformation process of a liquid surface under the influence of an applied voltage and they found that a Taylor cone appeared on the cathode surface. Therefore, it believes that a liquid local post-arc cathode surface will reduce the interruption ability of a vacuum interrupter significantly. This will be discussed in Chap. 2.

References

1. Slade P G, The Vacuum Interrupter: Theory, Design, and Application, Boca Raton, FL: CRC Press, 2008.
2. Smeets R, Sluis L, Kapetanovic M, Peeko D, Janssen A, Switching in Electrical Transmission and Distribution Systems, John Wiley & Sons, Ltd, 2015, pp. 190–194.
3. Wang H, Wang Z, Liu J, Zhou Z, Wang J, Geng Y, Liu Z, Optical absorption spectroscopy of metallic (Cr) vapor in a vacuum arc, Journal of Physics D: Applied Physics, 51(2018) 035203.
4. Jüttner B 2001 J. Phys. D: Appl. Phys. 34, 103–23.
5. Daalder J E 1981 Physica B + C 104, 91–106.
6. Miller H C 1983 IEEE Trans. Plasma Science 11, pp. 122–127.
7. Schade E and Dullni E, "Recovery of breakdown strength of a vacuum interrupter after extinction of high currents," IEEE Trans. Dielectr. Electr. Insul. 2002, Vol. 9, No. 2, pp. 207–215.
8. Yu, L, Wang J, Geng Y, Kong G, Liu Z, High-Current Vacuum Arc Phenomena of Nanocrystalline CuCr25 Contact Material, IEEE Trans. Plasma Science, Vol. 39, No. 6, 2011, pp. 1418–1426.
9. Sandolache G, Rowe S W. Vacuum breakdown between molten metal electrodes. 22nd International Symposium on Discharges and Electrical Insulation in Vacuum (ISDEIV), Matsue, Japan: IEEE, 2006: 13–16.
10. Wang Z, Wang H, Zhou Z, Tian Y, Geng Y, Wang J and Liu Z 2016 J. Appl. Phys. 120, 083301.
11. Miller H C, Anode phenomena, in Handbook of Vacuum Arc Science and Technology, Fundamentals and Applications, R. L. Boxman, P. J. Martin, and D. M. Sanders, Eds. Park Ridge, NJ, USA: Noyes, 1995, pp. 308–364.
12. Miller H C, Anode modes in vacuum arcs, IEEE Trans. Dielectr. Electr. Insul., vol. 4, no. 4, pp. 382–388, Aug. 1997.
13. Miller H C, Anode modes in vacuum arcs: update, IEEE Trans. Plasma Science, Vol.45, No.8, 2017, pp. 2366–2374.
14. Miller H C, Discharge modes at the anode of a vacuum arc, IEEE Trans. Plasma Sci., vol. PS-11, no. 3, pp. 122–127, Sep. 1983.
15. Batrakov A V, Vacuum-arc anode phenomena: new findings and new applications, Proceedings of 28th International symposium on discharges and electrical insulation in vacuum, September, 2018, Greifswald, Germany, pp. 163–168.
16. Heberlein J V R and Gorman J G, The high-current metal vapor arc column between separating electrodes, IEEE Trans. Plasma Sci., vol. 8, no. 4, pp. 283–288, Dec. 1980.
17. Mitchell G R, "High-current vacuum arcs. Part 1: An experimental study," Proc. Inst. Electr. Eng., vol. 117, no. 12, pp. 2315–2326, Dec. 1970.
18. Rich J A, Prescott L E, and Cobine J D, "Anode phenomena in metal-vapor arcs at high current," J. Appl. Phys., vol. 42, no. 2, pp. 587–601, Feb. 1971.
19. Boxman R L, Harris J H, and Bless A, "Time-dependence of anode spot formation threshold current in vacuum arcs," IEEE Trans. Plasma Sci., vol. 6, no. 3, pp. 233–237, Sep. 1978.
20. Watanabe K, Kaneko E, and Yanabu S, Technological progress of axial magnetic field vacuum interrupters, IEEE Trans. Plasma Sci., vol. 25, no. 4, pp. 609–616, Aug. 1997.
21. Schade E, Physics of high-current interruption of vacuum circuit breakers, IEEE Trans. Plasma Sci., vol. 33, no. 5, pp. 1564–1575, Oct. 2005.
22. Zhang Y, Yao X, Liu Z, Geng Y, and Liu P, Axial Magnetic Field Strength Needed for a 126-kV Single-Break Vacuum Circuit Breaker During Asymmetrical Current Switching, IEEE Transactions on Plasma Science, Vol. 41, No. 6, June 2013, pp. 1670–1678.

23. Kong G, Liu Z, Geng Y, Ma H, Xue X, Anode Spot Formation Threshold Current Dependent on Dynamic Solid Angle in Vacuum Subjected to Axial Magnetic Fields, IEEE Transactions on Plasma Science, Vol. 41, No. 8, August 2013, pp. 2051–2060.
24. Liu Z, Kong G, Geng Y, Wang J, Estimation of critical axial magnetic field to prevent anode spot in vacuum interrupters, IEEE Transactions on Plasma Science, Vol. 42, No. 9, September 2014, pp. 2277–2283.
25. Drouet M G, Current distribution at anode and current flow in the interelectrode region of the vacuum arc, in Proc. 12th Int. Symp. Discharges Electr. Insul. Vacuum, Shoresh, Israel, 1986, pp. 120–124.
26. Shmelev D L, MHD model of plasma column of high current, in Proc. 19th Int. Symp. Discharges Electr. Insul. Vacuum, Xi'an, China, 2000, pp. 214–217.
27. Liu Z, Cheng S, Zheng Y, Rong M, and Wang J, Comparison of vacuum arc behaviors between axial-magnetic-field contacts," IEEE Trans. Plasma Sci., vol. 36, no. 1, pp. 200–207, Feb. 2008.
28. Zalucki Z, and Janiszewski J, Transition from constricted to diffuse vacuum arc modes during high AC current interruption," IEEE Trans. Plasma Sci., vol. 27, no. 4, pp. 991–1000, Aug. 1999.
29. Kimblin C W, Arcing and interruption phenomena in AC vacuum switchgear and in DC switches subjected to magnetic-fields, IEEE Trans. Plasma Sci., vol. 11, no. 3, pp. 173–181, Sep. 1983.
30. Kaneko E, Tamagawa T, Okumura H, and Yanabu S, Basic characteristics of vacuum arcs subjected to a magnetic-field parallel to their positive columns, IEEE Trans. Plasma Sci., vol. 11, no. 3, pp. 169–172, Sep. 1983.
31. Yanabu S, Satoh Y, Tamagawa T, Kaneko E, and Sohma S, 10 years' experience in axial magnetic field-type vacuum interrupters, IEEE Trans. Power Del., vol. 1, no. 4, pp. 202–208, Oct. 1986.
32. Schulman M B and Bindas J A, Evaluation of AC axial magnetic fields needed to prevent anode spots in vacuum arcs between opening contacts, IEEE Trans. Compon. Packag. Technol. A, vol.17, no. 1, pp. 53–57, Mar. 1994.
33. Shkol'nik S M, Anode phenomena in arc discharges: A review, Plasma Sources Sci. Technol., vol. 20, no.1, 013001, Feb. 2011.
34. Londer Y I and Ulyanov K N, Mathematical model of the vacuum arc in an external axial magnetic field, IEEE Trans. Plasma Sci., vol. 35, no. 4, pp. 897–904, Aug. 2007.
35. Dyuzhev G A, Lyubimov G A, and Shkolnik S M, Conditions of the anode spot formation in a vacuum-arc," IEEE Trans. Plasma Sci., vol. 11, no. 1, pp. 36–45, Mar. 1983.
36. Frind G., Carroll J J, and Tuohy E J, Recovery times of vacuum interrupters which have stationary anode spots, IEEE Trans. Power Appl. Syst., vol. 101, no. 4, pp. 775–781, Apr. 1982.
37. Chaly A M, Magnetic control of high current vacuum arcs with the aid of an axial magnetic field: A review, IEEE Trans. Plasma Sci., vol. 33, no. 5, pp. 1497–1503, Oct. 2005.
38. Gorman J G, Kimblin C W, Voshall R E, Wien R E, and Slade P G, The interaction of vacuum arcs with magnetic fields and applications, IEEE Trans. Power App. Syst., vol. PAS-102, no. 2, pp. 257–266, Feb. 1983.
39. Heberlein J V R, Gorman J G, and Bhasavanich A D, Magnitude of magnetic fields required to prevent anode spots in vacuum arcs, in Proc. Rec. IEEE Conf. Plasma Sci., May 1981, p. 97.
40. Fenski B, Heimbach M, and Shang W, The influence of unipolar axial magnetic field on the behaviour of vacuum arcs, IEEE Trans. Plasma Sci., vol. 31, no. 2, pp. 299–302, Apr. 2003.
41. Stoving P N and Bestel E F, Finite element analysis of AMF vacuum contacts, in Proc. 18th Int. Symp. Discharges Elect. Insul. Vac., Eindhoven, The Netherlands, Aug. 1998, pp. 522–529.
42. Schellekens H and Schram D C, IEEE Trans. Plasma Sci. 13, 261 (1985).
43. Boxman R L and Goldsmith S, J. Appl. Phys. 54, 592 (1983).
44. Keidar M, Beilis I, Boxman R L, and Goldsmith S, J. Phys. D: Appl. Phys. 29, 1973 (1996).
45. Rondeel W G J, J. Phys. D: Appl. Phys. 8, 934 (1975).
46. Schellekens H, J. Appl. Phys. 54, 144 (1983).
47. Sherman J C, Webster R, Jenkins J E, and Holmes R, J. Phys. D: Appl. Phys. 11, 379 (1978).

48. Schellekens H, A current constriction phenomenon in high current vacuum arcs, Phys. B + C, vol. 104, no. 1–2, pp. 130–136, 1981.
49. Drouet M G, Poissard P, and Meunier J L, IEEE Trans. Plasma Sci. 15, 506 (1987).
50. Schade E and Shmelev D L, IEEE Trans. Plasma Sci. 31, 890 (2003) and Erratum IEEE Trans. Plasma Sci. 32, 829 (2004).
51. Chaly A M, Logatchev A A, Zabello K K, and Shkol'nik S M, IEEE Trans. Plasma Sci. 31, 884 (2003).
52. Nagashima M, Yogi K, Tsuji T, Kaneko E, Ide N, and Yanabu S, in Proceedings of the XXIInd International Symposium on Discharges and Electrical Insulation in Vacuum (2006), p. 415.
53. Izraeli I, Boxman R L, and Goldsmith S, IEEE Trans. Plasma Sci. 15, 502 (1987).
54. Wang L, Jia S, Shi Z, and Rong M, J. Phys. D: Appl. Phys. 38, 1034 (2005).
55. Wang L, Jia S, Liu Y, Chen B, Yang D, and Shi Z, J. Appl. Phys. 107, 113306 (2010).
56. Zhang Z, Ma H, Liu Z, Geng Y, Wang J, Cathode-constriction and column-constriction in high current vacuum arcs subjected to an axial magnetic field, J. Phys. D: Appl. Phys., 2018, 51, 145203.
57. Watanabe K, Sato J, Kagenaga K, Somei H, Homma M, Kaneko E, Takahashi H, The anode surface temperature of CuCr contacts at the limit of current interruption, IEEE Trans. Plasma Sci., vol. 25, no. 4, pp. 637–641, Aug. 1997.
58. Pieniak T, Kurrat M, Gentsch D, Surface Temperature Analysis of Transversal Magnetic Field Contacts Using a Thermography Camera, IEEE Transactions on Plasma Science, 2017, Volume: 45, Issue: 8, pp. 2157–2163.
59. Poluyanova I N, Zabello K K, Logatchev A A, Yakovlev V V, Shkol'nik S M, Measurements of Thermal Radiation Brightness of Anode Surface After Current Zero for a Range of Current Levels, IEEE Transactions on Plasma Science, 2017, Volume: 45, Issue: 8, pp. 2119–2125.
60. Schellekens H and Schulman M B, Contact temperature and erosion in high-current diffuse vacuum arcs on axial magnetic field contacts, IEEE Trans. Plasma Sci., vol. 29, no. 3, pp. 452–461, Jun. 2001.
61. Frey P, Klink N, Michal R, and Saeger K E, Metallurgical aspects of contact materials for vacuum switching devices, IEEE Trans. Plasma Sci., vol. 17, no. 5, pp. 734–740, Oct. 1989.
62. Wei X, et al., Liquid phase separation of Cu–Cr alloys during the vacuum breakdown, J. Alloys Compounds, vol. 509, pp. 7116–7120, Jun. 2011.
63. Jia S, Yang D, Wang L, and Shi Z, Investigation of the swirl flow on anode surface in high-current vacuum arcs, J. Appl. Phys., vol. 111, no. 4, pp. 043301–043306, 2012.
64. Gellert B, Egli W, Melting of copper by an intense and pulsed heat source, J. Phys. D, Appl. Phys., vol. 21, no. 12, pp. 1721–1726, Dec. 1988.
65. Wang L, et al., Anode activity in a high-current vacuum arc: Three-dimensional modeling and simulation, IEEE Trans. Plasma Sci., vol. 40, no. 9, pp. 2237–2246, Sep. 2012.
66. Niwa Y, et al., The effect of contact material on temperature and melting of anode surface in the vacuum interrupter," in Proc. 19th International Symposium on Discharge and Electrical Insulation in Vacuum (ISDEIV), Xi'an, China, 2000, pp. 524–527.
67. Goldsmith S, Boxman R L, Sapir E, Cohen Y, Yaloz H, and Brosh N, Distribution of peak temperature and energy flux on the surface of the anode in a multi-cathode spot pulsed vacuum arc," IEEE Trans. Plasma Sci., vol. 15, no. 5, pp. 510–514, Oct. 1987.
68. Miller H C and Kutzner J, Contrib. Plasma Phys. 31, 261 (1991).
69. Kutzner J and Miller H C, J. Phys. D: Appl. Phys. 25, 686 (1992).
70. Arai K, Takahashi S, Morimiya O, and Niwa Y, IEEE Trans. Plasma Sci. 31, 929 (2003).
71. Kimblin C W, J. Appl. Phys. 44, 3074 (1973).
72. Shkol'nik S M, Afanas'ev V P, Barinov Y A, Chaly A M, Logatchev A A, Malakhovsky S I, Poluyanova I N, and Zabello K K, IEEE Trans. Plasma Sci. 33, 1511 (2005).
73. Jia S, Zhang L, Wang L, Chen B, Shi Z, and Sun W, IEEE Trans. Plasma Sci. 39, 3233 (2011).
74. Langlois Y, Chapelle P, Jardy A, and Gentils F, J. Appl. Phys. 109, 113306 (2011).
75. Schellekens H, Ph.D. thesis, Eindhoven University of Technology, 1983.
76. Olsson E and Kreiss G, J. Comput. Phys. 210, 225 (2005).

77. Assael M J, Kalyva A E, Antoniadis K D, Michael Banish R, Egry I, Wu J, Kaschnitz E, and Wakeham W A, J. Phys. Chem. Ref. Data 39, 033105 (2010).
78. Taylor G I and McEwan A D, J. Fluid Mech. 22, 1 (1965).
79. Elele E O, Shen Y, Pettit D R, and Khusid B, Phys. Rev. Lett. 114, 054501 (2015).
80. Wang H, Wang Z, Zhou Z, Jiang Y, Wang J, Geng Y, and Liu Z, J. Appl. Phys. 120, 053303 (2016).
81. Tian Y, Wang Z, Jiang Y, Ma H, Liu Z, Geng Y, Wang J, Nordlund K, and Djurabekova F, J. Appl. Phys. 120, 183302 (2016).
82. Dullni E, Gellert B, and Schade E, Electrical and pyrometric measurements of the decay of the anode temperature after interruption of high-current vacuum arcs and comparison with computations, IEEE Trans. Plasma Sci., vol. 17, no. 5, pp. 644–648, Oct. 1989.
83. Ide N, Sakuma R, Kaneko E, and Yanabu S, The electrode surface state after current interruption in vacuum circuit breaker, in Proc. 22nd Int. Symp. Discharges Electr. Insul. Vac., Matsue, Japan, 2006, pp. 396–399.
84. Muller B and Renz U, Development of a fast fiber-optic two-color pyrometer for the temperature measurement of surfaces with varying emissivities," Rev. Sci. Instrum., vol. 72, no. 8, pp. 3366–3374, 2001.
85. Duvaut T, Comparison between multiwavelength infrared and visible pyrometry: Application to metals, Infr. Phys. Technol., vol. 51, no. 4, pp. 292–299, 2008.
86. Wang L, Jia S, Yang D, Liu K, Su G, and Shi Z Q, "Modelling and simulation of anode activity in high-current vacuum arc," J. Phys. D, Appl. Phys., vol. 42, no. 14, p. 145203, Jul. 2009.
87. Ul'yanov K N, "A physical model of anode spot formation in a high-current vacuum arc," High Temperature, vol. 41, no. 2, pp. 135–140, Mar. 2003.
88. Shashurin A, Beilis I I, and Boxman R L, "Heat flux to an asymmetric anode in a hot refractory anode vacuum arc," Plasma Sour. Sci. Technol., vol. 19, no. 1, p. 015002, Feb. 2010.
89. Loffhagen D, Uhrlandt D, and Hencken K, "Monte Carlo simulation of the breakdown in copper vapour at low pressure," in Proc. 24th Int. Symp. Discharges Elect. Insul. Vacuum, Braunschweig, German, Aug./Sep. 2010, pp. 7–10.
90. Sarrailh P, Garrigues L, Boeuf J P, and Hagelaar G J M, "Modeling of breakdown during the post-arc phase of a vacuum circuit breaker," Plasma Sour. Sci. Technol., vol. 19, no. 6, 065020, Dec. 2010.
91. Zhang Z, Ma H, Yi X, Liu Z, Geng Y, Wang J, Study on the heat flux density delivered to the anode at the transition to anode spot formation in high current vacuum arcs, in Proceedings of 4th international conference on electric power equipment—switching technology, xi'an, China, 2017, pp. 590–593.
92. Joy D C, A Database of Electron-Solid Interactions, EM Facility Knoxville, TN, USA: University of Tennessee, 2008.
93. Vahedi V and Surendra M, "A Monte Carlo collision model for the particle-in-cell method: Applications to argon and oxygen discharges," Comput. Phys. Commun., vol. 87, nos. 1–2, pp. 179–198, May 1995.
94. Nieter C and Cary J R, "VORPAL: A versatile plasma simulation code," J. Comput. Phys., vol. 196, no. 2, pp. 448–473, May 2004.
95. Perkins S T, Cullen D E, and Seltzer S M, "Tables and graphs of electron-interaction cross sections from 10 eV to 100 GeV derived from the LLNL evaluated electron data library (EEDL), Z = 1–100," Lawrence Livermore Nat. Lab, Livermore, CA, USA, Tech. Rep. UCRL-50400, 1991.
96. Vaughan J. R. M., "A new formula for secondary emission yield," IEEE Trans. Electron Devices, vol. 36, no. 9, pp. 1963–1967, Sep. 1989, and "Secondary emission formulas," IEEE Trans. Electron Devices, vol. 40, no. 4, p. 830, Apr. 1993.
97. Dullni E and Schade E., "Investigation of high-current interruption of vacuum circuit breakers," IEEE Trans. Electr. Insul., vol. 28, no. 4, pp. 607–620, Aug. 1993.
98. Farrall G A. "Recovery of dielectric strength after current interruption in vacuum," IEEE Trans. Plasma Sci., vol. 6, no. 4, pp. 360–369, Dec. 1978.

99. Rich J A and Farrall G A, "Vacuum arc recovery phenomena," Proc. IEEE, vol. 52, no. 11, pp. 1293–1301, Nov. 1964.
100. Takahashi S, Arai K, Morimiya O, Hayashi K, and Noda E, "Laser measurement of copper vapor density after a high-current vacuum arc discharge in an axial magnetic field," IEEE Trans. Plasma Sci., vol. 33, no. 5, pp. 1519–1526, Oct. 2005.
101. Burm K T A L, "Calculation of the Townsend discharge coefficients and the Paschen curve coefficients," Contrib. Plasma Phys., vol. 47, no. 3, pp. 177–182, May 2007.
102. Burm K T A L, "Paschen curves for metal plasmas," J. Plasma Phys., vol. 78, no. 2, pp. 199–202, Apr. 2012.
103. Kutzner J and Załucki Z, "Secondary effects on solid surfaces stimulated by the cathode plasma flux in vacuum arcs," Czechoslovak J. Phys., vol. 45, no. 12, pp. 1075–1082, Dec. 1995.
104. Lide D R, CRC Handbook of Chemistry and Physics, 89th ed. New York, NY, USA: Taylor & Francis, 2008.

Chapter 2
Dielectric Recovery Properties After Current Interruption in Vacuum

In Chap. 1, we discussed the high-current vacuum arcs phenomena. We especially discussed the high-current anode mode vacuum arc because of its negative influences on current interruptions. In this chapter, we focus on the dielectric recovery properties after the arcing period. The importance of the dielectric recovery properties is that it determines the performance of the current interruption in vacuum. The dielectric recovery process is a complex physical process. There are many impact factors on the dielectric recovery process, such as metal vapor density, particles, droplets, residual ions, residual magnetic field etc. The metal vapor density around the current zero region and the particles, play decisive roles in the dielectric recovery process. First, this chapter discusses a free recovery processes after diffuse vacuum arcs extinction, in which no transient recovery voltage is applied. Then it discusses the metal vapor density in current zero region, including Cu and Cr vapor density. After that it discusses a phenomena of Taylor cone generated on the post arc cathode after high-current interruptions, which is pulled by a strong electric field under a high transient recovery voltage, and how the dielectric recovery properties are influenced by the Taylor cone. Finally, this chapter discusses the dielectric breakdowns in current interruptions in vacuum, which includes instantaneous breakdown in cathode sheath expansion, late breakdowns induced by particles, and how mechanical shocks influence the late breakdowns.

2.1 Free Recovery Processes After Diffused Vacuum Arcs Extinction

Dielectric recovery behavior after current zero is especially important for vacuum circuit breakers because it ultimately determines the interruption capacity of the VCBs [1, 2]. As a power frequency AC current in a vacuum interrupter approaches sinusoidal zeros, cathode spots, which are the sources of emission, gradually turn off until the final cathode spot ceases to function. However, the current interruption

Z. Liu et al., *Switching Arc Phenomena in Transmission Voltage Level Vacuum Circuit Breakers*, https://doi.org/10.1007/978-981-16-1398-2_2

process does not come to an end at that time because many particles produced in the arcing period remain in the switching gap including metal vapor [1, 3], ions [4], and droplets [5]. Therefore, the ability of the switching gap to withstand a high voltage is finally reached until the residual particles dissipate from the switching gap. A transient recovery voltage is immediately applied upon the switching gap after current zero. If the dielectric strength of the gap cannot withstand the rapidly rising voltage, a re-ignition may occur. This physical process of dielectric recovery after current zero develops very quickly, and the time scales for the residual particles to disperse are different. The decay of metal vapor density may last for several milliseconds [5, 6], while ions may take only a few microseconds to dissipate from the gap [5, 7]. The density of these residuals depends on the contact material, the arcing time, the amplitude of arc current, and the magnetic field of contacts.

Although the dielectric recovery after high current interruption are more concerned, a study on low current interruption may reveal some fundamental information about current interruption processes. Because the vacuum arc is a diffuse arc under a low current, with which the condition is relatively simple. For example, when a stationary anode spot forms, the contact surfaces melt where they are attached to the arc roots and continue to evaporate metal vapor after current zero. In addition, the contact surfaces are roughened because of high energy input from the vacuum arcs, as shown in Sect 1.4.1. So, the low current experiments help to discover the nature of the dielectric recovery processes. It is fundamental to understand the recovery processes after vacuum arcs extinguishing under a "free" condition i.e., without an influence of a transient recovery voltage. Moreover, the knowledge of free recovery behaviors after diffuse vacuum arcs extinguishing forms a basis for further understanding the dielectric recovery process of a vacuum interrupter under a high current interruption. This section is to understand a free recovery process in vacuum interrupters after diffuse vacuum arcs extinguishing.

Figure 2.1 shows a schematic of a test vacuum interrupter. The contacts were butt-type with a diameter of 12 and 25 mm, respectively. The thickness of the contact plate was 4 mm. Three kinds of contact materials were used in the experiments: oxygen-free high-conductivity Cu (OFHC), CuCr25 (25% of Cr) and CuCr50 (50% of Cr). All of the test vacuum interrupters were conditioned by low current before the experiments. In order to confirm that the vacuum arcs remained in a diffuse arc mode, the arc modes were observed before measuring the recovery processes. Thus, several vacuum interrupters were specifically designed for the vacuum arc observation. The specially designed vacuum interrupters had an observation window (40 mm $\times$ 40 mm) on its stainless-steel shield. The insulation envelopes of the vacuum interrupters were made of glass for optical observation. After confirming the vacuum arc mode as diffuse arc mode, normal vacuum interrupters were used for the free dielectric recovery experiments. The normal vacuum interrupters were with the same structure as shown in Fig. 2.1 except that the shield did not have a window.

The experiments were conducted using a circuit shown schematically in Fig. 2.2. An arc current was provided by a L-C discharging circuit. A drawn arc was initiated between a pair of butt contacts in a test vacuum interrupter. At beginning the main switch (SW1) was opened, and the auxiliary switch (SW2) was closed. A test

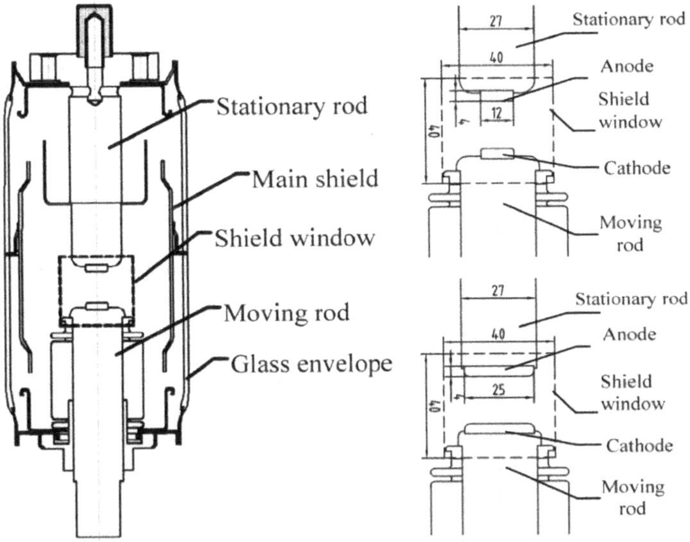

Fig. 2.1 Schematic of a tested vacuum interrupter. The upper electrode was fixed in the position; the lower one was able to move. Contact diameters were 12 mm and 25 mm, respectively

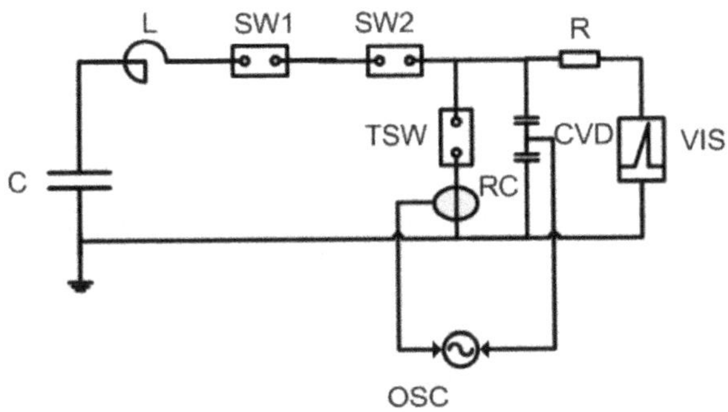

Fig. 2.2 Sketch of experimental circuit. C: Capacitor banks; L: Reactors; SW1: Main switch; SW2: Auxiliary switch; TSW: Test switch with a tested vacuum interrupter involved; R: Resistor (50 Ω); CVD: Capacitive voltage divider (3 pF, 4.2:1); VIS: Voltage impulse source; RC: Rogowski coil; OSC: Oscilloscope

switch (TSW), in which a test vacuum interrupter was assembled, was kept closed. In addition, SW1, SW2, and TSW were driven by permanent magnetic actuators to efficiently control the close and open time accurately. Before the experiments, the capacitor banks were charged to an appropriate voltage that controlled a discharging arc current. First, SW1 was closed, which initiated a power frequency current of 50 Hz passing through the reactors L. After a predetermined interval to control the arcing time, TSW and SW2 were controlled to open at the same time by the permanent magnetic actuators. In this way, the first half-wave of the initiated discharging

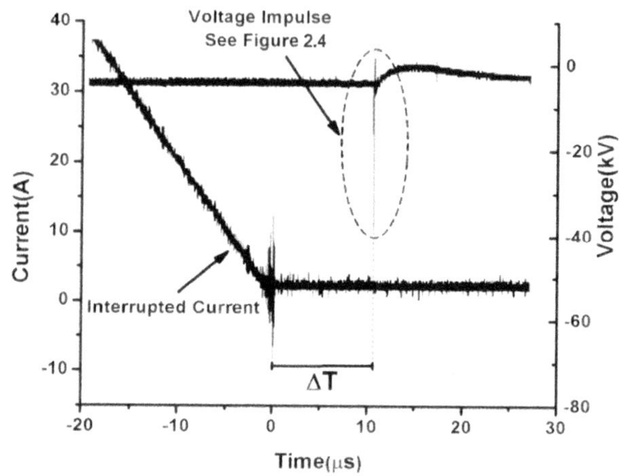

Fig. 2.3 Current and voltage waveforms around current zero. A negative voltage impulse was applied to a pair of contacts of a tested vacuum interrupter after a predetermined interval ΔT. The detailed waveform of the impulse is shown in Fig. 2.4.

current passed through the test vacuum interrupter and a vacuum arc was drawn. The predetermined open time of TSW and SW2 can adjust the arcing time from 1 to 9 ms. In the experiment, the arcing time was set to 9 ms if the vacuum arc was extinguished at the first current zero point. Then, a negative voltage impulse was applied to a pair of contacts of the test vacuum interrupter through a current-limiting resistor R (50 Ω) after a predetermined interval ΔT. ΔT can be adjusted from 0 μs to 15 μs as shown in Fig. 2.3. After repeating the application of the voltage impulses, one can obtain the free recovery behaviors of a test vacuum interrupter from 0 μs to 15 μs after current zero.

Figure 2.3 shows a waveform of a voltage impulse that was generated by a voltage impulse source (VIS) and it can be electronically controlled to apply to the test vacuum interrupter. Figure 2.4 shows the waveform of the voltage impulse shown in Fig. 2.3. The impulse peaked at 90 kV with a rising time of 150 ns (rate of rise 480 kV/μs). The voltage impulse was measured by a Tektronix high-voltage probe P6015A (1000:1) through a capacitive voltage divider (CVD) (3 pF, 4.2:1). The arc current was measured by a Rogowski coil.

Figure 2.5 shows the waveforms of arc current, arc voltage, and displacement curve of moving contact. For the permanent magnet operating mechanism, the average opening velocity remained quite stable during the experiments. The opening velocity for the test vacuum interrupters was 1.2 m/s, and the scatter of the arcing time was less than 1 ms. The gap length of the test vacuum interrupter was measured as ~10 mm when the voltage impulses were applied. To ensure the measurement accuracy of gap length at current zero, a KTC-100 type linear displacement transducer was used to measure the displacement of the actuator during the experiments.

A high-speed charge-coupled device (CCD) video camera Phantom V10 was used to record the evolution of the vacuum arc modes in the experiments. The recording velocity was set to 4000 frames per second. The camera aperture was fixed at 4, and the exposure time was set at 2 μs.

Figure 2.6 shows the arc voltage and vacuum arc photos with 3 kinds of contact

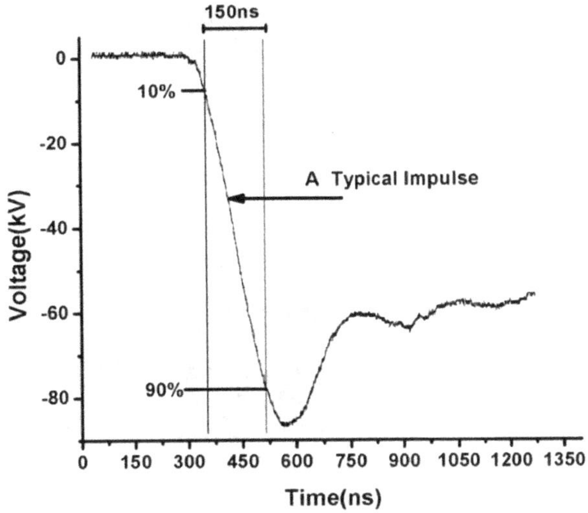

Fig. 2.4 Waveform of a voltage impulse. The impulse peaked at 90 kV with a rising time of 150 ns (rate of rise ~480 kV/μs)

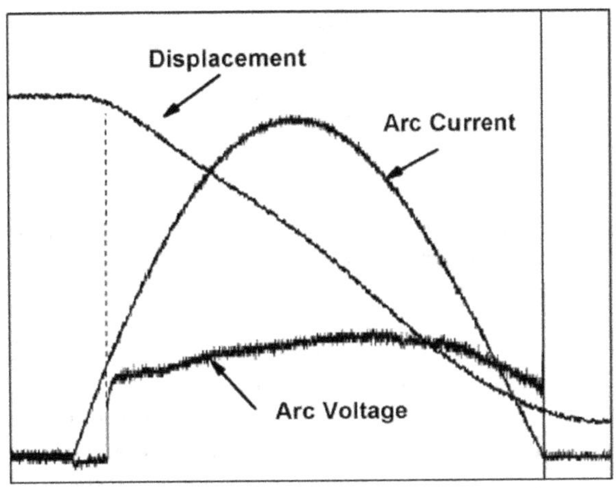

Fig. 2.5 Waveform of arc current, arc voltage, and displacement curve of moving contact

materials and 2 kinds of contact diameters. The contact materials were Cu, CuCr25 (25% of Cr), and CuCr50 (50% of Cr), respectively. The contact diameters were 12 and 25 mm, respectively. The current peak was 2.1 kA. As shown in Fig. 2.6, the arc voltages for the 3 kinds of contact materials and 2 kinds of contact diameters remained in a range of 20–35 V and they appeared smooth during the arcing period. Figure 2.6 also shows the vacuum arcs for the 3 kinds of contact materials and 2 kinds of contact diameters remained in a diffuse arc mode during the arcing period. The anode in the arcing period acted as a passive collector of electrons which were emitted from cathode spots without appearing anode spots or footpoints.

Figure 2.7 shows the free recovery of dielectric behaviors of 3 kinds of contact materials and 2 kinds of contact diameters. The contact materials were Cu, CuCr25

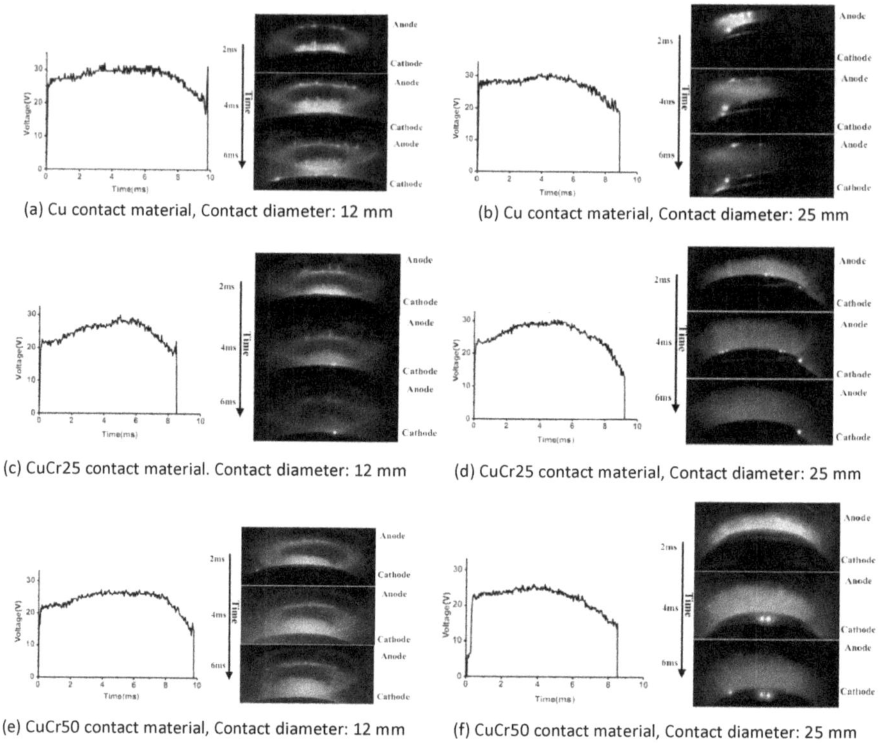

Fig. 2.6 Arc voltage and vacuum arc photos with 3 kinds of contact materials and 2 kinds of contact diameters. Current peak: 2.1 kA, Frequency: 50 Hz

(25% of Cr), and CuCr50 (50% of Cr), respectively. The contact diameters were 12 and 25 mm, respectively. The current peak was 2.1 kA. By repeating the applications of the voltage impulses shown in Fig. 2.4, the free recovery of dielectric behaviors of the tested vacuum interrupters were obtained. The crosses indicate that the gap could not withstand the voltage impulse, and the circles indicate that no breakdown occurred. As Fig. 2.7 shows, the breakdown voltage did not increase continuously after current zero. However, there was a stepwise behavior in the free recovery process. The voltage stepped from ~28 to ~70 kV at ~4 μs after current zero. The stepwise free recovery behavior dominated all the investigated contact materials and contact diameters. The measured data can be divided into three zones: a low voltage zone, a high voltage zone, and a transition zone. In the low voltage zone, which began from 0 to ~4 μs after current zero, the breakdown voltage fluctuated between 15 and 45 kV, and the mean breakdown voltage for the 3 contact materials and 2 contact diameters was 28 kV. In the transition zone, which ranged from 3 to 6 μs after arc extinction and the medium value for the stepping instant was ~4 μs, the breakdown voltage rose rapidly from ~28 to ~70 kV. In the high voltage zone, which began from ~4 to ~15 μs after current zero, the mean breakdown voltage for the 3

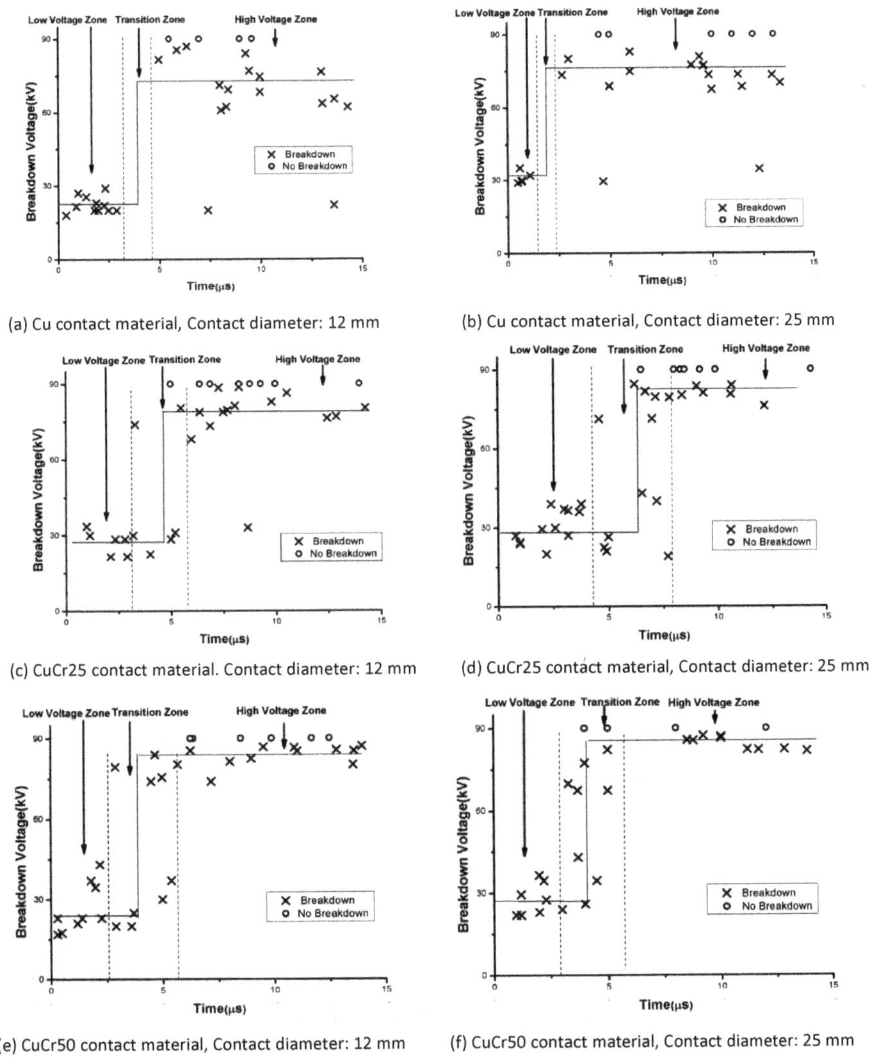

Fig. 2.7 Free recovery of dielectric behaviors of 3 kinds of contact materials and 2 kinds of contact diameters. Cross: breakdown voltage; Circle: no breakdown occurred

kinds of materials and 2 contact diameters was ~70 kV. The voltage pulses applied across the contact gaps did not lead to a breakdown every time. Moreover, the contact diameters (12 and 25 mm) have little influence on the final breakdown voltage for the same contact material.

Figure 2.7 indicated a fast recovery process (~4 μs after current zero) following diffused vacuum arcs extinction. The stepwise free recovery processes may be caused by a fast decay of residual particles such as metal vapor, ions, and droplets. Such particles that are produced during the arcing time remain in the gap after current extinction and decay over time. Now we will discuss the roles of metal vapor and ions during the free recovery process, thus the stepwise behavior may be interpreted.

First, the role of metal vapor in a free recovery process was considered. It is difficult to find an accurate model to describe the metal vapor recovery process because of its complexity. However, the model of Rich and Farrall [8] provides a quantitative way to calculate metal vapor density after vacuum arc extinguishing. The formula of Lins [9] was adopted to calculate the initial density of the metal vapor n after current extinction, as shown in Eq. (2.1).

$$n(t) = \frac{S_m}{\omega^2 + \beta^2}(\beta \sin \omega t - \omega \cos \omega t + \omega e^{-\beta t}) \tag{2.1}$$

where S_m is the rate at which the metal atom is evaporated from contact surfaces. β is the particle loss per second at the bounding surfaces of the gap volumes, and ω is the angular frequency of the current. Moreover, S_m is shown in Eq. (2.2)

$$S_m = \frac{K I_m E}{m V} \tag{2.2}$$

where E is the erosion rate of the contact materials. I_m is the peak value of currents. m is the mass of a metal atom, and K is a parameter to adjust metal vapor generation rates. $V = \pi R^2 L$ is the gap volume where R is the radius of the contact and L is the gap length. In addition, if one assumes that the condensation coefficients C_i for the contact surfaces are equal to 1, the particle loss rate can be rewritten as Eq. (2.3).

$$\beta = \frac{\overline{v} \sum A_i C_i}{4V} = \sqrt{\frac{8k_B T}{\pi m}} \frac{2\pi RL + \pi R^2}{4\pi R^2 L} \tag{2.3}$$

where $\overline{v}$ is mean thermal velocity. A_i is the surface area of the i_{th} surface. C_i is the condensation coefficient of i_{th} surface. m is the mass of a metal atom. k_B is Boltzmann's constant, and T is the temperature of metal vapor. We assumed 2000 K as the metal vapor temperature [10–12]. We employed 115 μg/C [13], 33.2 μg/C, and 31 μg/C [14] as the erosion rates for Cu, CuCr25 and CuCr50 contact materials, respectively. In addition, we set K at 8% as Lins did [9]. The expression for the metal vapor density at the center of the gap volume at time t after current extinction can be described by Eq. (2.4).

$$n\left(0, \frac{L}{2}, t\right) = n_0\left[1 - e^{-\frac{1}{\alpha^2}}\right] erf\left[\frac{1}{2}\left(\frac{L}{R}\right)\left(\frac{1}{\alpha}\right)\right] \tag{2.4}$$

where $\alpha = \frac{t}{R}\sqrt{\frac{2k_B T}{m}}$ and $erf(x) = \frac{2}{\sqrt{\pi}} \int_0^x e^{-t^2} dt$ is an error function.

Figure 2.8 shows the decay of the metal vapor density. The contact diameter was 12 mm and 25 mm, respectively. The initial densities of the metal vapor are

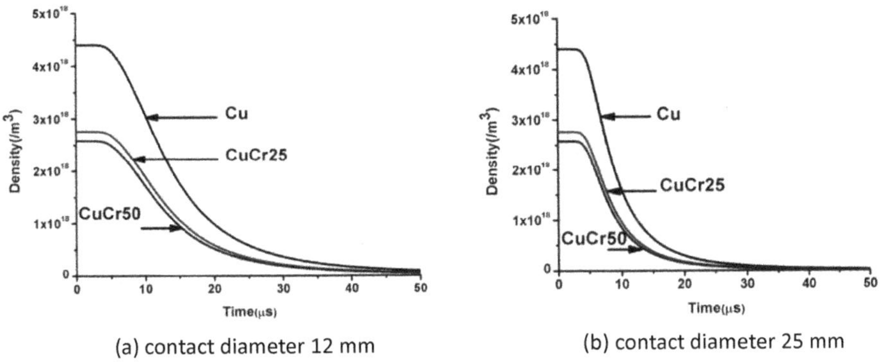

Fig. 2.8 Decay of the metal vapor density after current zero

of the same order of $10^{18}/\mathrm{m}^{-3}$ and the metal vapor decayed exponentially over the recovery time. Furthermore, the calculated results are similar to the experimental results obtained by Lins [6], Gellert et al. [11], and Takahashi et al. [15]. Rich and Farrall [8] concluded that the decay of metal vapor dominates the recovery process after a diffuse vacuum arc extinction and that the recovery process will end when the metal vapor density drops below the level where the electron's mean free path in the vapor is similar to the gap length. Frind et al. [16] confirmed this conclusion by investigating the recovery process after interrupting a high current that was above the threshold to form an anode spot.

However, the stepwise free recovery characteristics may not be attribute to the metal vapor. It is shown in Fig. 2.8 that the fast decay of metal vapor density ranged on the order of ~10 μs rather than ~ μs, as shown in Fig. 2.7. So, the metal vapor dissipation cannot explain the stepwise behaviors in the sense of time scale. Schade and Dullni [2] proposed a critical value of $n \times d = 3 \times 10^{19}/\mathrm{m}^2$ to determine an instantaneous breakdown. If the product of metal vapor density and contact gap is below the critical value, breakdown may be attributed to a type of late breakdown. In the case of Fig. 2.7, the maximum value of $n \times d$ is $5 \times 10^{16}/\mathrm{m}^2$, which was three order lower than the critical value. Since the arc current in the tests was so low and the arc mode was in a diffuse mode, metal vapor density decayed while the arc was still burning. Thus, only a small quantity of metal vapor remained after arc extinction, in contrast to a high current interruption. Therefore, the reason for the free stepwise dielectric recovery behavior after current zero cannot be due to metal vapor alone.

Second, the role of ions in a free recovery process was considered. Residual plasma, which is composed of ions and electrons, also has an influence on dielectric recovery processes. There are many researchers [4, 17, 18] who focused on sheath development when a transient recovery voltage is applied upon the residual plasma, which results in a post-arc current after current zero. After current zero, the voltage across the inter-electrode plasma changes its polarity. Under the influence of the transient recovery voltage, the electrons quickly reverse their moving direction toward the post-arc anode, and the heavier positive ions first remain there. Thus, a sheath forms in front of the post-arc cathode and develops toward the post-arc anode associated with a post-arc current. This sheath takes the whole transient recovery voltage. When the sheath can not withstand the transient recovery voltage, a breakdown occurs.

The free dielectric recovery process is simpler than a real current interruption. There is no transient recovery voltage applied in the free dielectric recovery process of a vacuum interrupter. In such a situation, the post-arc current is not notable. In this situation, one is easier to focus on a "free" dissipation process of the inter-electrode particles. The understanding of the free dielectric recovery process is helpful to get an insight of a dielectric recovery process in a vacuum interrupter under a transient recovery voltage. In the free recovery process, the decay rate of ions is much slower than that of electrons. Although the ions may dissipate from the inter-electrode gap in ~1 μs based on the calculation of the Maxwellian velocity distribution [1], some experimental results [4, 17, 19] indicated that the decay time for the ions may be longer than the time calculated by the Maxwellian velocity distribution function. In other words, some ions that have very low velocities may remain after vacuum arc extinguishing, and Dullni et al. called them "slow ions" [3]. However, the Rich-Farrall model is not applicable in the "slow ion" case because the effect of collisions is not considered. To calculate the decay time of the ions, A Particle in Cell—Monte Carlo Collision (PIC-MCC) method was adopted to describe an ion transport process after vacuum arc extinction.

Figure 2.9 shows the computing sequence for Particle in Cell—Monte Carlo Collision method. It consists of four modules including movement module, collision module, weighting module and electric field calculation module. In the particle's movement module, one can obtain the motion of charged particles by solving Newton's Eq. (2.5).

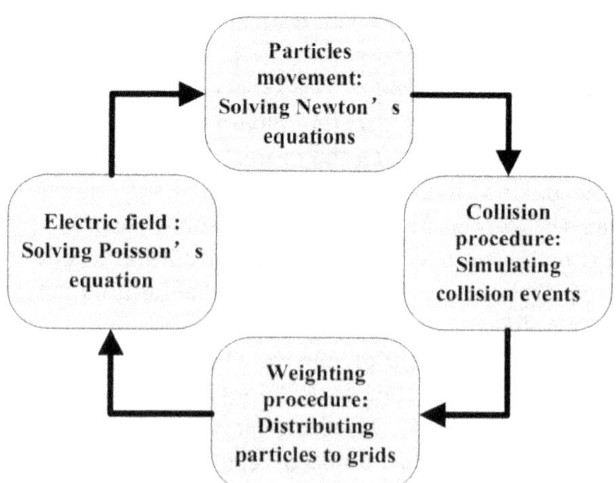

Fig. 2.9 computing sequence for Particle in Cell—Monte Carlo Collision method

$$\frac{d\vec{v}}{dt} = \frac{q\vec{E}}{m} \tag{2.5}$$

where v is the velocity of the charged particles. q is the electric charge. m is the mass of the charged particles, and E is the electric field that forces charged particles to move. The collision module is used to model collisions between particles and background neutrals by the Monte Carlo method based on atomic cross-sections. The weighting module obtains the density of the charged particles and distributes them to cell grids. The electric field module calculates the electric field at cell grids by solving Poisson's Eq. (2.6).

$$\nabla^2\varphi = -\frac{e}{\varepsilon_0}(n_i - n_e) \tag{2.6}$$

where φ is the electric potential of the field. ε_0 is the vacuum permittivity. e is electron charge. n_i is ion density, and n_e is electron density. Furthermore, the electric field force is provided to the movement module, which continues to simulate the motion of the charged particles. A review of the Particle in Cell—Monte Carlo Collision method can be found in [20].

With the Particle in Cell—Monte Carlo Collision method, one can calculate the decay rate of Cu ions in the free recovery processes considering the collisions between the charged particles and background neutrals. A 2D cylindrical model is developed with a gap length of 10 mm between the contacts. The contact diameters are 12 and 25 mm, respectively. Lins [7] measured a plasma density of $10^{17}/\mathrm{m}^3$ after an artificial interruption of a sinusoidal current of 200 A, which was adopted in the simulation. A background neutral metal vapor density of $10^{18}/\mathrm{m}^3$ was set (See Fig. 2.8). A Maxwellian velocity distribution was assumed for the ions and electrons at a temperature between 0.2 and 1 eV [21–24]. The influence of ion-neutral and electron-neutral collisions was considered. The cross-section curve for Cu atoms can be found in reference [25]. Secondary emission from the contact surfaces was not considered during the simulation. The charged particles were absorbed by the boundaries when the particles achieve the boundaries of the model. The cell grid spacing was equal to one Debye length and the time step was equal to a tenth of plasma oscillation period.

Figure 2.10 shows the ion density decay over time. The ion density was $10^{17}/\mathrm{m}^3$ at current zero and it dropped to $10^{14}/\mathrm{m}^3$ in ~4 μs, considering the collision effect. However, if the ions leave the gap freely without the collisions, the ion density dropped from $10^{17}/\mathrm{m}^3$ to $10^{14}/\mathrm{m}^3$ in ~0.5 μs. The decay time for the ions without the collisions is much faster than it is with the collisions. The decay rates calculated in the Particle in Cell—Monte Carlo Collision model are like the results measured by Lins [7]. Moreover, the ion density falls by three orders of magnitude in ~4 μs.

The decay of the ions may have a strong impact on the free recovery behaviors after current zero. Compared to the decay time of the metal vapor ~10 μs (see Fig. 2.8), the decay time of the ions is ~4 μs, which is much closer to the stepwise behavior shown in Fig. 2.7. The ion density falls by three orders of magnitude while the metal vapor

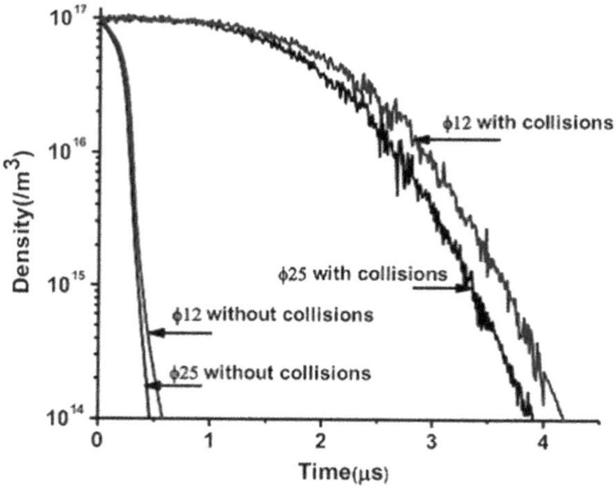

Fig. 2.10 Decay of the ion density over time. Contact diameter of 12/25 mm. The initial ion density is assumed to be a Maxwellian velocity distribution. Contact material: Cu; Ion temperature: 0.2 eV; Electron temperature: 1 eV

density remains almost constant. Thus, the stepwise behavior of the free recovery is more likely to be explained by the influence of the ions rather than that of the metal vapor. Nevertheless, metal vapor, which serves as background neutrals, still be necessary for breakdown to occur. According to a calculation by Wang et al. [26], the residual charge after extinction has a significant influence on the distribution of the electric field in vacuum interrupters. They pointed out that if the residual charge reached a certain value, the electric field intensity on the surface of the contacts and shields may rise steeply and this has a significant influence on the breakdowns of the vacuum interrupters.

Based on the above analysis, we intended to attribute the stepwise breakdown behavior in a free dielectric recovery process after interrupting a diffuse vacuum arc to an ion-assisted breakdown. The possibility of a breakdown drops with the decay of the ion density, and the voltage impulse cannot initiate a breakdown if the ion density drops to a certain level. Moreover, the metal vapor which plays an effective role in the collision processes also has an influence on the decay of ions.

2.2 Metal Vapor Density in Current Zero Region

After current zero in a vacuum interrupter, an amount of metallic vapor is left in the inter-electrode gap. Schade and Dullni proposed that metallic vapor played a vital role in the post-arc breakdown, which could directly lead to a failure of interruption [2]. Therefore, it is meaningful to measure the density of metallic vapor around current zero and to determine its evolution process for better understanding the dielectric recovery properties in current interruption of vacuum interrupters.

Measurements of metallic vapor density in the vacuum arc in both arcing phase and post-arc phase have been performed by various authors, who adopted the interferometric [27, 28], optical emission spectroscopy [29], and optical absorption spectroscopy [15, 30]. Laser-induced fluorescence (LIF) is also adopted in vacuum arc diagnostics, by which the local information of plasma can be obtained. Both Lins [6, 31] and Gellert [11, 32] used laser-induced fluorescence to measure the local density of metallic vapor near current zero for dielectric recovery research. In vacuum circuit breakers, CuCr contact material is mostly adopted. Therefore, it is interesting to measure the Cu and Cr vapor density, especially in the current zero region.

2.2.1 Cu Vapor Density

In this section, a technique of planar laser-induced fluorescence (PLIF) is adopted for measuring a 2-D copper vapor distribution when anode spots occur during the arcing period. When anode spots occur in the arcing phase, there could be a non-uniform distribution of metallic vapor in the contact gap near current zero, which has a significant impact on the dielectric recovery process.

Figure 2.11 shows an experimental setup for the Cu vapor density measurement by planar laser-induced fluorescence. As shown in Fig. 2.11a, triggered vacuum arcs are generated inside a cylindrical demountable vacuum chamber. The chamber was equipped with silicon windows to enable both vacuum arc observation and laser probing. A vacuum system ensures that the pressure within the chamber was at ~5×10^{-4} Pa. In the vacuum chamber, a pair of copper butt type contacts with a diameter of 40 mm were adopted. The gap between the contacts was fixed to 15 mm. The arc current was generated by a *L-C* discharging circuit, which consisted of pre-charged capacitors, inductors, and a main switch. With a combination of inductors (0.1 mH) and the capacitor (100 mF), a power frequency current of 50 Hz can be generated. The discharging arc current can be controlled by the pre-charge voltage of the capacitors. A triggered vacuum arc was initiated by an external copper electrode, which connected to a high-voltage impulse generator.

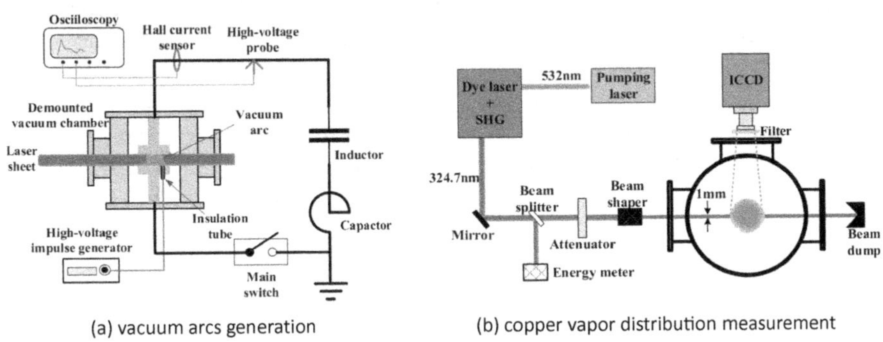

Fig. 2.11 Experimental setup for the Cu vapor density measurement by planar laser-induced fluorescence

Figure 2.11b shows a pumping laser (Quantel, Qsmart 850-SLM) generated a laser beam at 532 nm with a pulse duration of 5 ns and a linewidth below 0.005 cm^{-1}. This laser beam was used to pump a tunable dye laser (Quantel, TDL 90). With a combination of second harmonic generator, a dye laser generated a laser beam at 324.7 nm, which was used to excite the copper atoms in the vacuum arc. Then the energy of the probing laser beam was monitored by an energy meter (Ophir, NOVA II) via a beam splitter (9:1). The laser energy was attenuated to about 500 μJ ensuring the planar laser-induced fluorescence in the linear regime [33]. A beam shaper consisted of a series of lens, which re-shaped the laser beam to a laser sheet with a size of 20 mm × 1 mm. The laser sheet passed the vacuum arc and it excited the copper atoms on its path. On its perpendicular direction, the fluorescence emitted by excited copper atoms (510.6 nm) was recorded by an intensified-CCD (i-CCD) camera via a band-pass filter (center wavelength 511.32 nm, 8.84 nm FWHM). According to laser-induced fluorescence theory [33, 34], the intensity of fluorescence is proportional to the target particle density. Therefore, the 2-D fluorescence signal distribution in this experiment exactly represented the 2-D mapping of copper vapor distribution. Besides, in the experiment, optical absorption spectroscopy technique was adopted for absolute density calibration.

Figure 2.12 shows the evolution of copper vapor density distribution in current zero region for a high-current anode mode vacuum arc. The peak current was 5 kA. Figure 2.12a shows it was a high-current anode mode vacuum arc with a significant anode spot appearing on the anode. Figure 2.12b shows the Cu vapor density distribution. In the figure, $t = 0$ μs was the current zero instant. When $t = -70$ μs, the arc was still burning with a few cathode spots. However, an obvious copper vapor source can be found near anode, indicating an intense evaporation of a local anodic molten region. In the center of anodic vapor source, the copper vapor density was 7×10^{19} m^{-3}. In the remaining gap, the copper vapor diverged. The Cu vapor density decreased sharply along the gap. At the intermediate region of the gap, the copper vapor density decreased to about 5×10^{18} m^{-3}. When $t = -20$ μs, the copper vapor density near anode decreased about 1.5×10^{19} m^{-3}. When $t = 0$ μs, which is the current zero instant, the vacuum arc extinguished. However, a high-concentration copper vapor still existed, whose density was about 6×10^{18} m^{-3}. Subsequently, at $t = 100$ μs after current zero, with a cooling of the anodic molten region, the copper vapor density in the center of anodic vapor source decreased to a value of about 3×10^{18} m^{-3}. Then at $t = 200$ μs, the anodic vapor source disappeared at this moment, and the copper vapor density have an almost uniform distribution in the gap with a density of about 8×10^{17} m^{-3}. After that, the copper vapor started to dissipate from the gap freely. At $t = 300$ μs, the copper vapor density decreased to a value of about 1.5×10^{17} m^{-3}.

It is surprising to find that the metal vapor density was so low even in the case of the high-current anode mode vacuum arc. It found that the maximum metal vapor density at current zero was lower than 10^{19} m^{-3} even with an anode spot. However, if a breakdown is to be initiated based on Townsend discharge mechanism, the metal vapor density of about 10^{21} ~10^{22} m^{-3} is needed, as discussed in Sect. 1.4.4. The measured density of 10^{19} m^{-3} has little chance causing breakdowns, because the density is 2 ~3

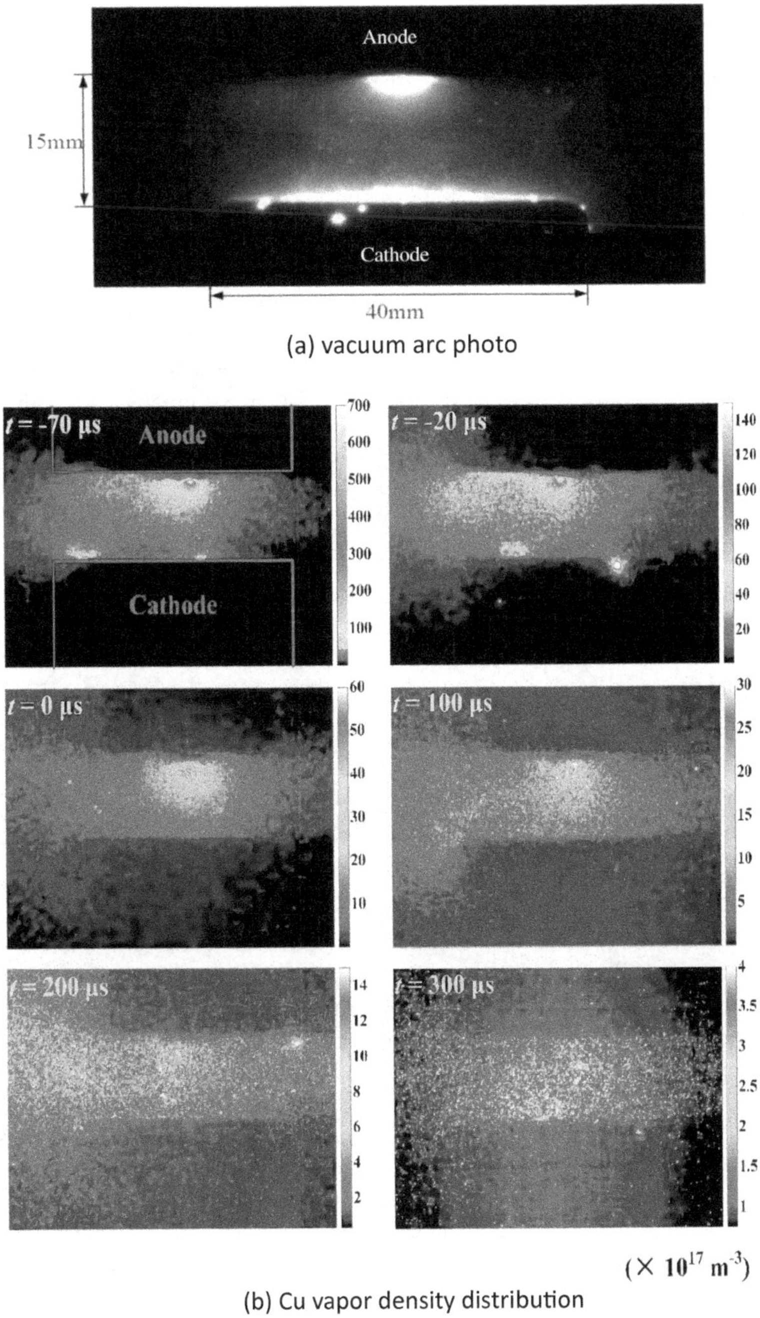

Fig. 2.12 Evolution of 2-D metallic vapor distribution in current zero region for a high-current anode mode vacuum arc. The arc current is 5kA peak. $t = 0$ μs is the current zero

orders lower than the threshold density for Townsend discharge. Therefore, it implies that, besides residual metallic vapor-initiated Townsend discharge mechanism, other mechanisms are also contributing to the post-arc breakdown phenomenon. The possible candidates may include liquid metal on the molten region (see Sect. 2.3), cathode sheath (see Sect. 2.4.1), particles in the post-arc gap (see Sect. 2.4.2), and mechanical shocks (see Sect. 2.4.3), which may play an important role as triggering the breakdown phenomenon. Schade and Dullni [2] also proposed that metallic vapor might initiate the breakdown in the post-arc phase and re-ignite the vacuum arc, but additional triggering processes were required. Liquid protrusions, which are produced in the molten region at anode in arcing period, may trigger the post-arc breakdowns in the dielectric recovery process [35]. In this process, an electric field which is as high as 10^6~10^7 V/m is necessary to deform the liquid metal into a Talyor cone. Such an extremely high electric field can be created in a cathode ion sheath. After the vacuum arc extinction, some residual plasma remains in the gap. Due to a transient recovery voltage, which begins to increase at current zero, an ion sheath forms near the cathode. The cathode ion sheath withstands the whole voltage drop between the contacts. Hence a high electric field can be generated between the contacts, which deforms the liquid metal [36]. In addition, the metallic vapor can also efficiently impede the expansion of ion sheath, which led to a higher electric field near the anode [36]. Therefore, metallic vapor, with a lower density within the post-arc gap, may be capable to re-ignite the vacuum interrupters assisted by the above candidates.

2.2.2 *Cr Vapor Density*

In this section, the metal vapor density in vacuum arcs of pure Cr contact material is measured by using spatial optical absorption spectroscopy with a broadband light source. The technique helps to obtain the distribution of the absolute metallic vapor density of both diffuse arc and high-current anode mode vacuum arcs during the arcing period. Figure 2.13 shows the optical absorption spectroscopy experimental setup for the measurement of Cr vapor density in the Cr vacuum arc. The Cr vacuum arcs were generated inside a cylindrical demountable vacuum chamber with a diameter of 0.4 m and a height of 0.5 m. The demountable chamber was equipped with two glass windows (BK7) to enable both vacuum arc observation and spectroscopy. A pump ensured that the pressure within the chamber was ~5×10^{-4} Pa. Figure 2.14 shows the configuration of the contacts. A pair of pure chromium contacts were used. The contacts were butt type with a diameter of 40 mm. The contacts were fixed. The contact gap was 10 mm. The arcs were initiated by a trigger electrode located in the cathode. The trigger electrode connected to a high voltage impulse generator. An external axial magnetic field of 90 mT was applied in order to ensure the vacuum arc stable. A stable arc is desired in the measurements of the optical absorption spectroscopy.

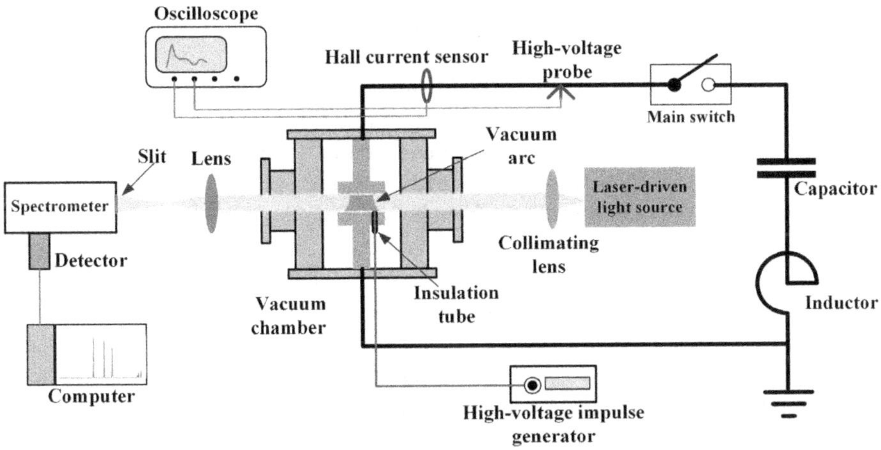

Fig. 2.13 The schematic of experimental setup

Fig. 2.14 The configuration of a pair of contacts

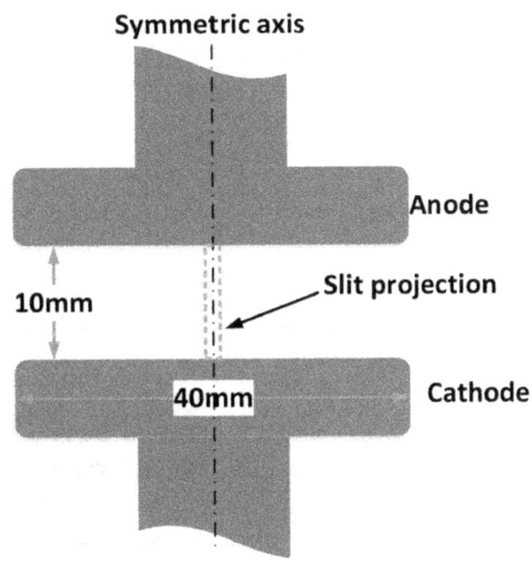

The arc current was generated by a *L–C* discharging circuit, shown in Fig. 2.13. Capacitors of 100 mF and inductors of 0.1 mH were used to generate a 50 Hz AC waveform. The capacitor was pre-charged to 1 kV to discharge a 20-kA peak-current. The vacuum arc voltage and current were measured using a Tektronix high-voltage probe P6015A (1000:1) and a Hall-current sensor, respectively. The data were recorded with a Tektronix four-channel oscilloscope. The arcing period from 5 ms (current peak) to 10 ms (current zero) were specially studied. An optical recording of the vacuum arc was done with a high-speed camera (Photron UV50), with a speed of 10 000 fps and an exposure time of 4 μs.

In the optical setup, a broadband laser driven light source (LDLS, EQ-99) was used in the optical absorption spectroscopy. By using a collimating lens, a dispersive light from the light source was converted to a parallel light-beam with a diameter of about 12 mm. After passing the vacuum arc region, the parallel light beam reached the entrance slit of a spectrometer via a lens (see Fig. 2.13). To perform an axially, spatially-resolved optical absorption spectroscopy, the entrance slit (100 μm width and 12 mm height) only permitted the light beam along the axis of the contact gap to enter the spectrometer (see Fig. 2.14). The light was then spectrally dispersed by a spectrometer (Andor Shamrock 500i, 0.5 m focal length) equipped with an intensified charge-couple device (iCCD) camera (Andor iStar DH334T, 1024 × 1024 pixels). Considering the dynamics of the vacuum arcs, a 50 μs exposure time for the iCCD camera was adopted. In addition, to obtain a higher spectral resolution, a grating of 2400 l/mm was used during the acquisition of the spectrum. The whole system was calibrated using a low-pressure Hg lamp.

Optical absorption spectroscopy is known as a reliable tool to measure absolute density and it is commonly used for plasma diagnostics. Mitchell and Zemansky [37] and other authors [38–40] discussed its theory in detail. To measure the absorption spectrum, four spectra were recorded: the light source passing through the plasma ($I_{\text{light+plasma}}$), the light source only (I_{light}), the plasma emission only (I_{plasma}), and the background light ($I_{\text{background}}$). Then, the fractional absorption $A(\lambda)$ could be calculated as Eq. (2.7):

$$A(\lambda) = 1 - I_t(\lambda)/I_o(\lambda) = 1 - \frac{I_{light+plasma}(\lambda) - I_{plasma}(\lambda)}{I_{light}(\lambda) - I_{background}(\lambda)} \tag{2.7}$$

where λ is the wavelength. $I_t(\lambda)$ is the transmitted spectral intensity, and $I_0(\lambda)$ is the initial spectral intensity.

Figure 2.15 shows a typical optical absorption spectroscopy result of the Cr vacuum arc at $t = 5$ ms (current peak). The arc current was 4 kA peak, and the axial magnetic field flux density was 90 mT. Figure 2.15a shows a 2D spectrum of the plasma with the light source. Figure 2.15b shows four spectra including the light source passing through the plasma ($I_{\text{light+plasma}}$), the light source only (I_{light}), the plasma emission only (I_{plasma}), and the background ($I_{\text{background}}$), respectively. Figure 2.15c shows a fractional absorption near cathode in the range between 425 and 429.5 nm. Here, three Cr resonance lines at 425.43, 427.48, and 428.97 nm can be used to calculate the density of Cr atoms at the ground state.

The absorption profile of Cr atoms in vacuum arc involves several broadening mechanisms, such as Stark broadening, Doppler broadening, and instrumental broadening. The instrumental profile should be de-convoluted to obtain the true-line absorption profile for accurate determination of the Cr vapor density. By using one of the lines of a low-pressure Hg calibration lamp (at 435.8 nm, which is near the Cr lines), the instrumental profile suggested a Gaussian profile with a full-width at half-maximum (FWHM) of about 0.0461 nm. However, the FWHM of the absorption profile shown in Fig. 2.15c, was about 0.0615 nm. Therefore, instrumental broadening proved to be stronger than other broadening mechanism, which will cause

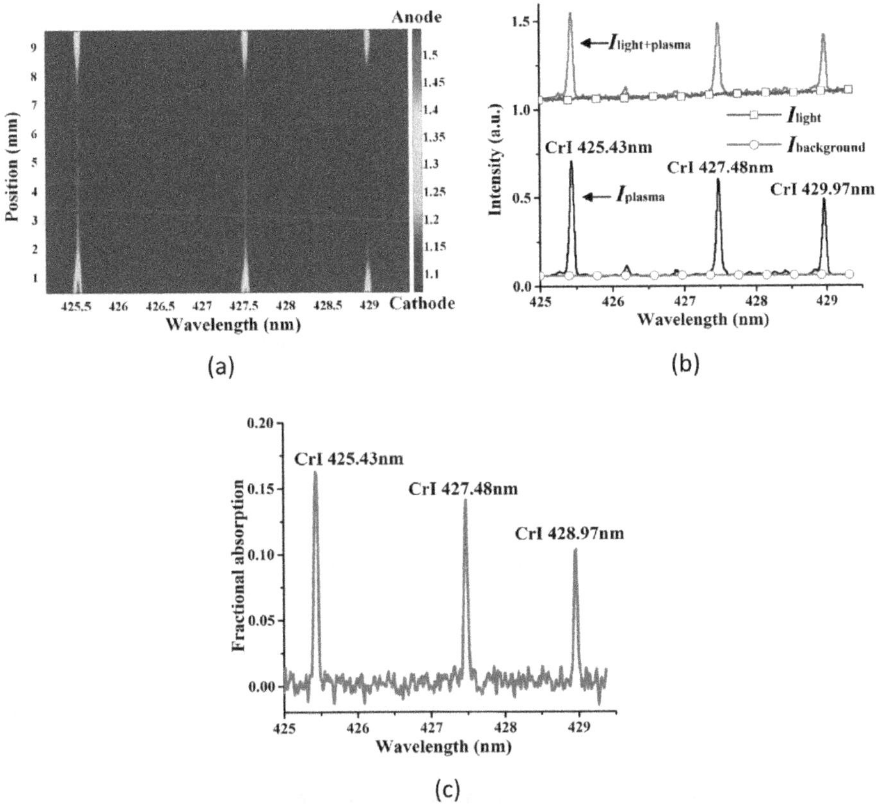

Fig. 2.15 Typical optical absorption spectroscopy results of the Cr vacuum arc at t = 5 ms. The arc current was 4 kA peak, and the flux intensity of the axial magnetic field was 90 mT. **a** A 2D spectrum of the plasma with light source; **b** Original spectra near cathode: the light source passing through the plasma ($I_{light+plasma}$), the light source only (I_{light}), the plasma emission only (I_{plasma}), and the background ($I_{background}$); c fractional absorption of the chromium atoms

an unneglectable and unacceptable error when determining the ground-state density by de-convoluting the spectral profile [39]. A simple method, known as the line equivalent width method, can solve this problem [41]. The area below the fractional absorption shown in Fig. 2.15c is independent of the line profile [42], as shown in Eq. (2.8).

$$W = \int A(\lambda)d\lambda = \int \left(1 - \frac{I_t(\lambda)}{I_0(\lambda)}\right)d\lambda \tag{2.8}$$

Therefore, the species density can be calculated precisely, even for strongly absorbing lines, and regardless of the instrumental function of the spectrometer [42]. After combining Eq. (2.8) with the Beer–Lambert law, one obtains Eq. (2.9).

$$W = \int \left[1 - \exp(-\alpha(\lambda)L)\right] d\lambda \tag{2.9}$$

where $\alpha(\lambda)$ is the absorption coefficient. L is the absorption length. Here L is 0.04 m, equals to the diameter of the contacts. Because the light generating from the source is not strongly absorbed, the plasma is optically thin and $\alpha(\lambda)L$ <<1 is satisfied. Hence, an approximation can be formulated as Eq. (2.10) for a simpler calculation:

$$\exp(-\alpha(\lambda)L) \approx 1 - \alpha(\lambda)L \tag{2.10}$$

The optical depth, $\alpha(\lambda)L$, in the experiment can be as high as 0.4. The approximation in Eq. (2.10) is still valid for an error limited within 10%. Thus, Eq. (2.9) can be transformed into Eq. (2.11).

$$W = L \int \alpha(\lambda) d\lambda \tag{2.11}$$

According to the theoretical framework given in [37], a linear relationship between the ground-state density and the integral of the absorption coefficient is given as Eq. (2.12).

$$\int \alpha(\lambda) d\lambda = \frac{\lambda_0^4 A_{21}}{8\pi c} \frac{g_2}{g_1} N_g \tag{2.12}$$

where A_{21} is the spontaneously emission probability between the high energy level (2) to low energy level (1). g_1 and g_2 are the statistical weights of the levels. N_g is the ground-state density. These spectroscopic parameters of Cr resonance lines could be found in the NIST database [43]. Consequently, the ground-state density can be obtained via the Eq. (2.13).

$$N_g = \frac{8\pi c}{\lambda_0^4 A_{21}} \frac{g_1}{g_2} \frac{W}{L} \tag{2.13}$$

The fluctuation of both the light source and the spontaneous radiation of the vacuum arc, as well as the detector noise decrease the signal-to-noise ratio substantially. As shown in Fig. 2.15c, the minimum fractional absorption coefficient, which can be efficiently distinguished from noise signal, was about 0.02. This led to a minimum detection limit of about 1.5×10^{17} m^{-3} in the measurement.

The measurement results of the Cr vapor density were shown under two cases, which is diffuse vacuum arc and high-current anode mode vacuum arc, respectively. Firstly, the measurement results of the diffuse vacuum arc are shown. Figure 2.16 shows a diffuse Cr vacuum arc with a current peak of 4 kA. Figure 2.16a shows the waveforms of arc current and the arc voltage during the arcing period. The arc voltage was stable at about 25 V. The vacuum arcs were measured by optical

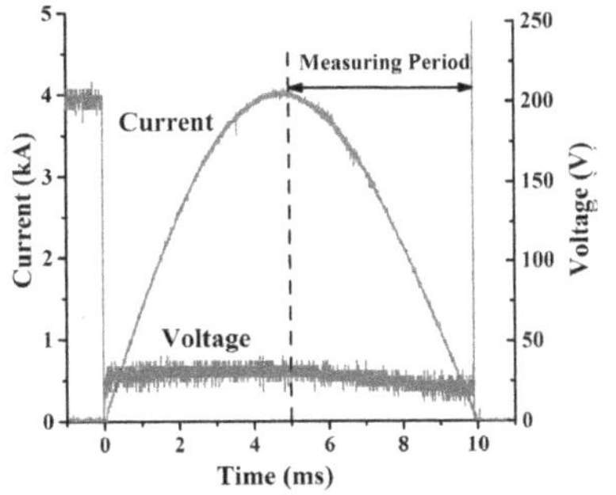

(a) Waveforms of arc current and arc voltage

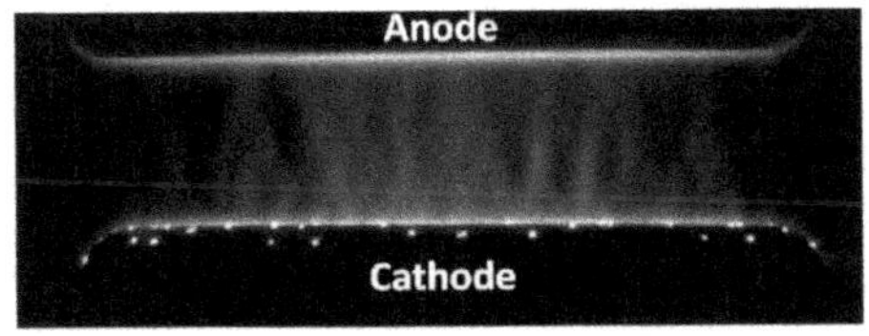

(b) high-speed photograph of the vacuum arc at current peak t = 5 ms

Fig. 2.16 A diffuse Cr vacuum arc with a current peak of 4 kA. An axial magnetic field of 90 mT was applied

absorption spectroscopy during a period from 5 ms (current peak) to 10 ms (current zero) with a time interval of 1 ms. This is a period approaching the extinction process of the vacuum arc. Figure 2.16b shows a high-speed photograph of the vacuum arc. This was a typical diffuse arc mode. An axial magnetic flux intensity of 90 mT was applied. The cathode jets separated each other in the gap. Cathode spots, as the main source of the vacuum arc, continuously provide massive neutral plasma and metallic vapor when moving over the cathode. As a result, a bright light can be seen near the cathode. However, a bright glowing region near the anode can also be seen in Fig. 2.16b. The glowing near the anode may be explained by a presence of a zone with high-density metallic atoms [29].

Figure 2.17 shows an axial Cr vapor density distribution along the contact gap. The axial position was from 0.5 mm from cathode to 9.5 mm from the cathode. To apply an axial magnetic field of 90 mT upon the arc proved to maintain a relative stable arc. The data in the figure were obtained by repeating many times of measurement. In this plasma diagnosis, the line equivalent width method was adopted for the Cr vapor density determination, in which the de-convolution operation was avoided. Hence, the error induced by an instrumental profile and its deconvolution can be neglected. In addition, the error caused by Hg lamp calibration can be neglected because the spectral resolution has the same deviation. Figure 2.17 shows that the axial distribution of Cr vapor density in the contact gap can be divided into three regions: the cathode region (0.5–3 mm), the arc column region (3–7 mm), and the anode region (7–9.5 mm). During the whole arcing period, except when $t = 10$ ms (current zero), Cr vapor has a higher density near the cathode, indicating the cathodic vapor source. Besides, in the anode region, Cr vapor density was also high. While in the arc column region, Cr vapor density was much lower than that in the cathode and the anode region. Therefore, it showed a bathtub curve. The Cr density in the anode and the cathode region gradually decreased as the current approaching zero. However, the Cr vapor density in the arc column region kept almost unchanged. At current zero, the Cr vapor density distribution was almost uniform in the contact gap,

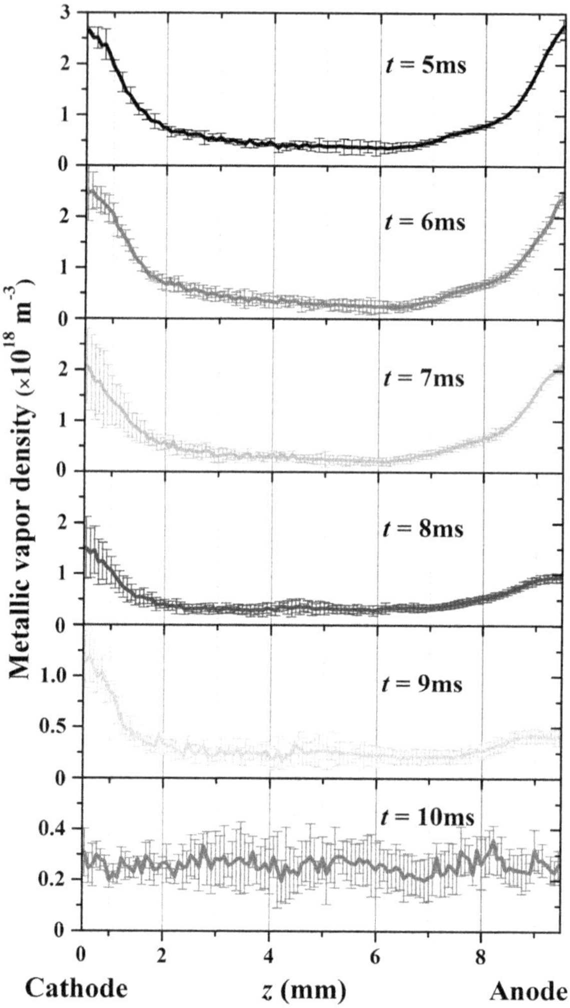

Fig. 2.17 The axial Cr vapor density distribution along the contact gap at different moments during an arcing period from 5 to 10 ms. The peak arc current was 4 kA, and the applied axial magnetic flux intensity was 90 mT

with a value of about 3×10^{17} m^{-3}. In addition, Fig. 2.17 confirmed the presence of high metallic density near the anode even when the vacuum arc was in the diffuse mode. At t = 5 ms, the Cr vapor density showed a maximum of about 2.5×10^{18} m^{-3} near the anode, which was much higher than that in the middle region of the gap. Miller suggested that atoms entered the contact gap from the anode with a help of energetic ions from the cathode plasma jet [44]. As a result, the sputtered atoms can be excited by electrons in their vicinity, thus intense light was produced.

Figure 2.18 shows a dependence of Cr vapor density on time in the cathode, anode and arc column regions. It shows the Cr vapor density evolution at positions 0.5, 5, and 9.5 mm from cathode. The arc current waveform was also shown. It shows that two curves almost coincided each other, which were the arc current curve and the Cr vapor density curve at the position 0.5 mm from the cathode, respectively. The coincidence

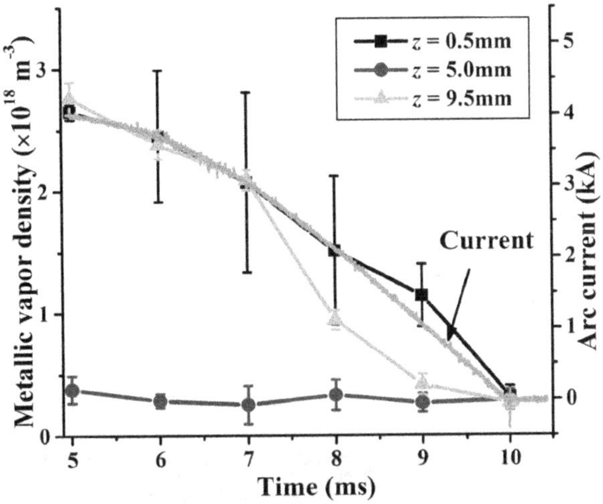

Fig. 2.18 Dependence of Cr vapor density on time in the cathode, anode and arc column regions. The peak arc current was 4 kA, and the applied axial magnetic flux intensity was 90 mT

could be explained as follows. The cathode spots maintained an average current per spot of 30–50 A for Cr [45]. The number of cathode spots changed following the current. Moreover, the cathode spots repelled each other with a 'retrograde motion' [46], which led to a homogeneous distribution of the cathode spots. If each cathode spot supplies a certain amount of metallic vapor, one can deduce that the metallic-vapor density near the cathode should be approximately proportional to both the quantity of cathode spots and the instantaneous arc current. In the anode region at a position 9.5 mm from the cathode, Cr vapor density decreased faster than that of cathode region after t = 7 ms. The high-density atoms near anode may be sputtered by the ion bombardment from cathode jets [44]. Therefore, it implied that, with a decrease of arc current that defined the ion flux flowing toward the anode, the ion bombardment decreased as well. Furthermore, a cooling of the anode, which was affected by both evaporation rate [47] and sputtering yield [48], may also cause a faster decrease of the metallic vapor density near the anode. In the arc column region at a position 5 mm from cathode, the density of Cr vapor was almost unchanged, with a low value of about 3×10^{17} m^{-3} during the measured arcing period. This implied a high ionization of the vacuum arc in the arc column region.

Secondly, the measurement results of the high-current anode mode vacuum arc are shown. Figure 2.19 shows a high-current anode mode Cr vacuum arc with a current peak of 14 kA. A uniform axial magnetic field of 90mT was applied. Figure 2.19a shows the waveforms of arc current and the arc voltage during the arcing period. The arc voltage has a peak value of 45 V, which was much higher than the 25 V for the diffuse vacuum arc shown in Fig. 2.16a. Figure 2.19b shows high-speed photographs of the vacuum arc. It noticed that from $t = 5$ ms to $t = 9$ ms, a thin bright region appeared at the anode, which expanded into the contact gap. It marked an evaporation process from an anodic molten region.

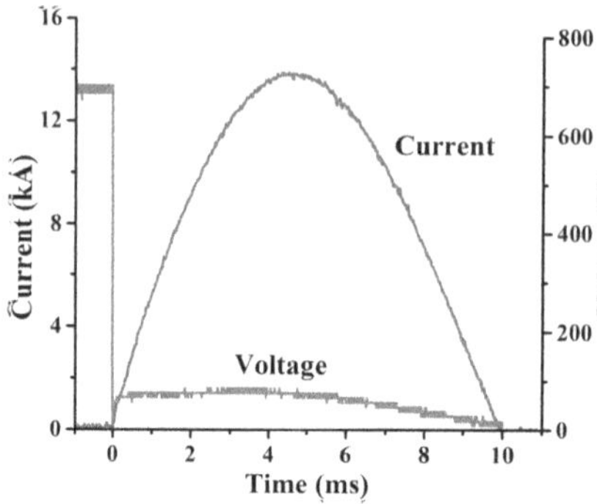

(a)Waveforms of arc current and arc voltage

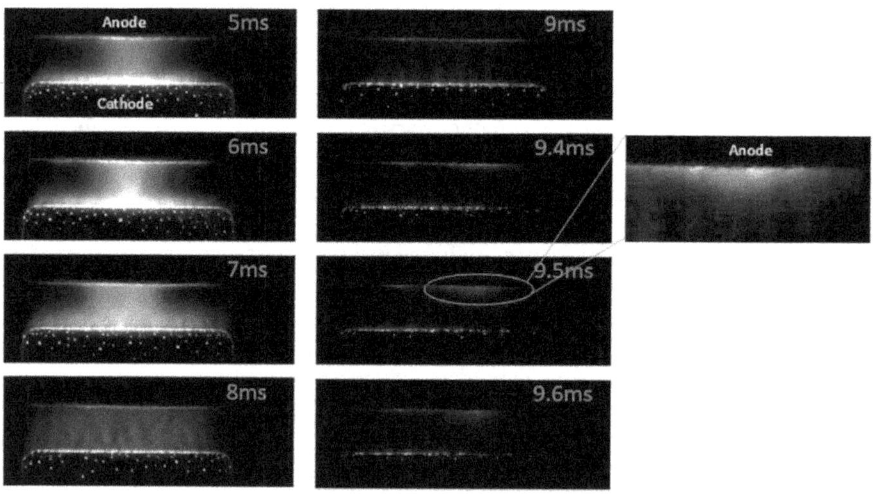

(b) high-speed photograph of the vacuum arc

Fig. 2.19 A high-current anode mode Cr vacuum arc with a current peak of 14 kA. An axial magnetic field of 90 mT was applied

Figure 2.20 shows an axial Cr vapor density distribution along the contact gap. The axial position was from 0.5 mm from cathode to 9.5 mm from the cathode. An axial magnetic field of 90 mT was applied. The errors were mainly caused by the instability of vacuum arc radiation. In contrast to the diffuse arc mode, the high-current anode mode vacuum arc was more difficult to be controlled. Figure 2.20 also shows that the axial distribution of Cr vapor density in the contact gap can be divided into three regions: the cathode region (0.5–3 mm), the arc column region (3–7 mm), and the anode region (7–9.5 mm). During $t = 5$–8 ms, the Cr vapor has a high density in the cathode region, which indicated a cathodic vapor source. At the same time, the anode had been severely heated by the arc column. It was considered that the anodic

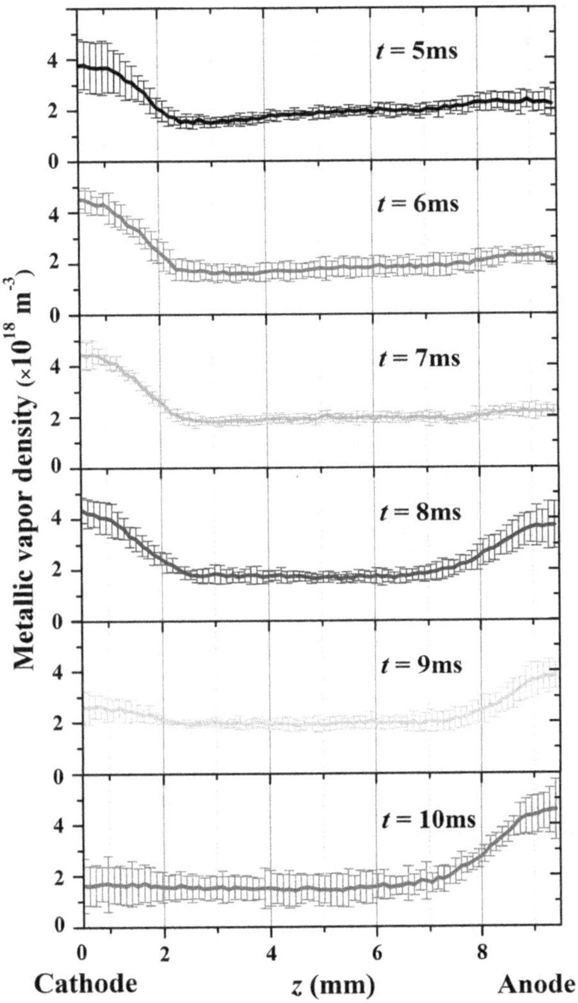

Fig. 2.20 The axial Cr vapor density distribution along the contact gap at different moments during an arcing period from 5 to 10 ms. The peak arc current was 14 kA, and the applied axial magnetic flux intensity was 90 mT

evaporated or sputtered atoms were ionized near the anode by the dense plasma in the arc column [49]. At $t = 8$ ms, the Cr vapor density in the anode region increased, reaching about 4×10^{18} m^{-3}. At $t = 9$–10 ms, the Cr vapor density in the cathode region decreased as the cathode spots gradually extinguished, while the Cr vapor density kept high in the anode region. It implied that the anodic evaporation process of a local molten anode region played an important role. It also implied that the dielectric recovery process is substantially slow under this condition, and the risk of a breakdown increased.

Figure 2.21 shows a dependence of Cr vapor density on time in the cathode, anode and arc column regions. It shows the Cr vapor density evolution at positions 0.5, 5, and 9.5 mm from cathode. The Cr vapor density near the cathode maintains a high

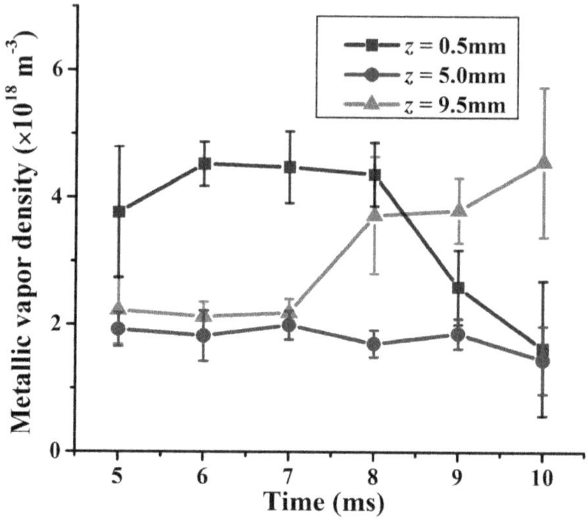

Fig. 2.21 Dependence of Cr vapor density on time in the cathode, anode and arc column regions. The peak arc current was 14 kA, and the applied axial magnetic flux intensity was 90 mT

value (about 4×10^{18} m^{-3}) until $t = 8$ ms. The Cr vapor density near the anode started to increase from $t = 7$ ms. This implied that atoms ejected from the anodic molten region to be ionized by the incoming cathodic plasma. This vapor ejecting phenomenon continued to current zero. In the arc column region, the Cr-vapor density kept almost unchanged at about 2×10^{18} m^{-3} during the measured arcing period.

Again, as discussed in last Sect. 2.2.1, the metallic vapor density at current zero was much lower than expected. Even in high-current anode mode, the metallic vapor density rarely exceeds 10^{19} m^{-3}. Under such condition, the mean free path of electrons is considerably longer than the contact gap. Therefore, it is quite difficult for metallic vapor to cause a breakdown in the post-arc phase via Townsend avalanches. The liquid protrusions, which are produced in the molten region of the arcing anode, may trigger the breakdown, which is discussed in Sect. 2.3. Aided by a strong electric field applied by a transient recovery voltage, a liquid protrusion developed during arcing period on a local anodic molten region can serve as a trigger for a post-arc breakdown. Therefore, metallic vapor, which has a low density in the contact gap, can be re-ignited in the post-arc phase.

2.3 Taylor Cone Initiated on Contact Surface After High-Current Interruptions

In current interruptions in vacuum, after vacuum arcs extinguish at current zero, there is a transient recovery voltage imposed on vacuum interrupters. If the applied voltage exceeds the dielectric strength of the vacuum interrupter, breakdowns occur, leading to a failure of current interruptions. After current zero, the breakdowns depend on not only the density of metal vapor in the contact gap (see Sect. 2.2), but also the contact conditions. After a high current interruption, the contact surfaces might be severely damaged, resulting in a (local) molten area (see Sect. 1.4). Furthermore, the molten area might be deformed by the transient recovery voltage, leaving protrusions on the contact surfaces, which enhance a local electric field and make the breakdowns occur easier. Moreover, the molten surfaces last for several milliseconds and continue to influence on the dielectric strength. Dullni and Schade [2, 5] proposed that breakdowns can be initiated by the protrusions in a relative long time after current zero and referred this kind of breakdowns as late breakdowns. Sandolache and Rowe [50] and Proskurovsky [51] experimentally confirmed a protrusion was pulled out by an electric field on a liquid–metal cathode. A bright spot at the tip of the protrusion was clearly observed in their experiments. A "Taylor cone" [52] forms under a strong electric field. Orlov et al. observed that breakdowns occurred when the electric field near the tip of Taylor cone was enhanced above a critical value [53].

Although there have been many studies on Taylor cones in various research fields due to its special features, e.g., Electrohydrodynamic atomization [54, 55] and electrospinning [56, 57], the deformation process of the liquid metal on the contact after interrupting a high-current vacuum arc has not been investigated thoroughly. A relationship between Taylor cones and breakdowns is not clear. In addition, it is difficult to understand the deformation process by experiments under transient recovery voltages. Simulation work helps to get insight into the phenomena. In this section, the formation process of the Taylor cones under a high electric field was simulated by adopting the finite element method. A time-dependent Electrohydrodynamic model was established, and the liquid–gas interface was tracked by the level set method. The results show the entire formation process of a Taylor cone as well as the local field enhancement. Moreover, the critical electric field strengths above which a Taylor cone could be initiated were obtained, whose buildup time obeyed a power law. In addition, a relationship between the molten contact surface deformation and the breakdown is given, which has an impact on the current interruption of vacuum interrupters.

A physical model of a Taylor cone formation is built up for a molten cathode surface under a high electric field. As shown in Sect. 1.4, after suffering a high-current anode mode vacuum arc, a crater with a crown-shape rims as well as droplets can be observed on the arcing anode surfaces. After current zero, the voltage polarity is reversed. Figure 2.22 shows the physical model of a Taylor cone formation. A tiny hemispheric liquid metal was adopted in the model to represent a local molten post arc cathode surface. This was a time dependent axisymmetric model. The grey region is the external region, and the blue region is the liquid metal with an initial hemispheric shape on a grounded plane. By a perfect conductor assumption, the electric field in the liquid metal was set to zero with the same potential to the grounded plane. An average electric field E_0 (4.5–17 MV/m), which obtained from the applied potential divided by the gap length, was uniform in the most region except at the tip on the liquid metal, where the local field was enhanced. As a result, a combination effect of the electric stress $\boldsymbol{F}_{es}$ and the surface tension $\boldsymbol{F}_{st}$ led to a further deformation of the liquid metal. A free interface was created between two regions to present the deformation process of liquid metal. In addition, the initial radii of the hemispheric metal liquid were set

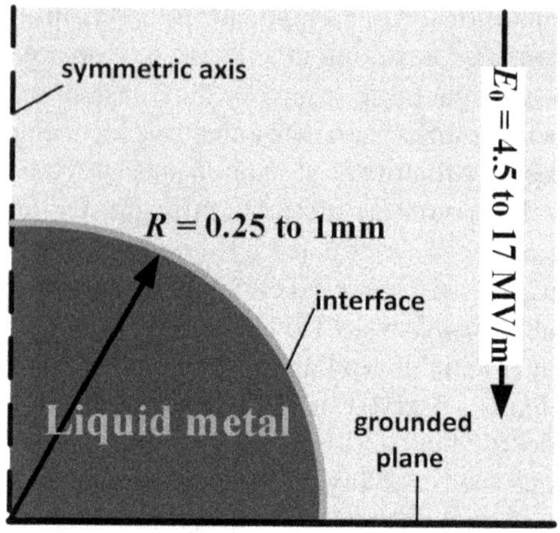

Fig. 2.22 A physical model of a Taylor cone formation

as 0.25 mm, 0.5 mm, and 1 mm, respectively. For a sake of simplicity, a constant temperature was assumed, and the solidification was neglected. The material of liquid metal in this model was copper, and the temperatures were set as 1356 K (melting temperature of copper), 2000 K, and 2300 K, respectively. The property of liquid copper varied with temperature is listed in Table 2.1 [58]. Additionally, the influence of temperature on dynamic viscosity was neglected compared to surface tension. Therefore, the dynamic viscosity was set as a constant value of 4.34 mPa s in the liquid metal region. While in the external region, the density was set as 0.01 kg/m^3, and the dynamic viscosity was set as 0.001 mPa s.

Based on the physical model, an Electrohydrodynamic model of the Taylor cone formation is built up. In this model, two kinds of fluid phases were taken into consideration: the liquid phase and the gas phase. The liquid phase represents the molten metal on the contact surface, and the gas phase represents the external region above the contact by assigning a quite low density and low dynamic viscosity. The fluid flow was governed by the mass conservation and Navier–Stokes equations. The Navier–Stokes equations took account of the surface tension force ($\boldsymbol{F}_{st}$) on the liquid–gas interface and the electrical stress ($\boldsymbol{F}_{es}$) deduced from the applied electric field. The gravitational Bond number is defined as Eq. (2.14), which reflects the relative importance of the gravity and surface tension.

Table 2.1 The property of liquid copper varied with temperature [58]

Temperature (K)	Density (kg/m^3)	Surface tension coefficient (mN/m)
1356	8000	1303
2000	7484.80	1154.88
2300	7244.80	1085.88

$$B_{Og} = \frac{\rho L^2 g}{\sigma} \tag{2.14}$$

where ρ is the density of the fluid; L is the characteristic length in the model; g is the gravity coefficient; σ is the surface tension coefficient. Based on the parameters shown in Table 2.1, the bond number is calculated between 0.003 and 0.06, which indicates that the gravity effects on the whole domain can be neglected. Assuming that the fluid was incompressible, the Electrohydrodynamic equations can be expressed as Eqs. (2.15) and (2.16)

$$\nabla \cdot \boldsymbol{u} = 0 \tag{2.15}$$

$$\rho\frac{\partial \boldsymbol{u}}{\partial t} + \rho(\boldsymbol{u} \cdot \nabla)\boldsymbol{u} = \nabla \cdot \left[-p\boldsymbol{I} + \mu(\nabla\boldsymbol{u} + (\nabla\boldsymbol{u})^T)\right] + \boldsymbol{F}_{st} + \boldsymbol{F}_{es} \tag{2.16}$$

where $\boldsymbol{u}$ is the velocity of the fluid. p is the pressure of the fluid. $\boldsymbol{I}$ is the identity tensor. Since the liquid metal was considered as a perfect conductor, only the electric field in the vacuum region was calculated. The Laplace's equation in gas phase was shown in Eq. (2.17)

$$\nabla^2\varphi = 0 \tag{2.17}$$

where φ is the electric potential. Thus, the electric field was calculated as Eq. (2.18)

$$\boldsymbol{E} = -\nabla\varphi \tag{2.18}$$

where $\boldsymbol{E}$ is the electric field. Because the perfect conductor assumption, the electric field near the metal was perpendicular to its interface pointing to the external region, and the electrical stress ($\boldsymbol{F}_{es}$) in Eq. (2.16) can be deduced as Eq. (2.19)

$$\boldsymbol{F}_{es} = -\frac{1}{2}\varepsilon_0 \boldsymbol{E}_s^2 \delta(\phi)\boldsymbol{n} \tag{2.19}$$

where $\boldsymbol{E}_s$ is the electric field adjacent to the liquid surface. $\delta(\Phi)$ is the delta function used in the level set method. $\boldsymbol{n}$ is the surface normal vector.

The liquid–gas interface can be tracked by the level-set method, which has been widely adopted to model incompressible two-phase fluid flow. The details can be found in [59]. The level-set method adopts a level-set variable Φ to describe two phase fluids based on a fixed mesh grid. The variable Φ is equal to 1 in the liquid phase and Φ is equal to 0 in the gas phase. The movement of the liquid–gas interface was realized by solving the Φ variable in the transport equation, which is shown in Eq. (2.20).

$$\frac{d\phi}{dt} + \boldsymbol{u} \cdot \nabla\phi = \gamma\nabla \cdot \left(\varepsilon\nabla\phi - \phi(1-\phi)\frac{\nabla\phi}{|\nabla\phi|}\right) \tag{2.20}$$

where ε is the thickness of transition region and γ is the speed of re-initialization of the level set function. Here γ is set as 2.5 m/s. Then, the surface tension force ($\boldsymbol{F}_{st}$) acting on the liquid–gas interface in Eq. (2.16) can be calculated by Eq. (2.21)

$$\boldsymbol{F}_{st} = \sigma\boldsymbol{n}k\delta(\phi) \tag{2.21}$$

where κ is the curvature of liquid–gas interface. The model was solved by the COMSOL Multiphysics. The governing equations of the model are strongly coupled. In addition, uniform and fixed mesh grids were adopted in the model. Different grid lengths, 0.018, 0.013, and 0.006 mm, were chosen to discretize the model, and a comparison shows that the results are independent on the grid lengths.

Figure 2.23 shows a simulated deformation process of the liquid copper and how it distorted the electric field strength profiles. The initial conditions were as following. The average electric field applied upon the gap was $E_0 = 7$ MV/m. The radius of the hemispheric liquid metal was 1 mm. The liquid metal temperature was 1356 K. At the beginning $t = 0$ ms, the hemispheric liquid metal was attached to a grounded plane. The electric field was almost uniform except the field, to some extent, concentrated at the top of the hemisphere. At $t = 1$ ms, the hemisphere was pulled out to be a hemi-ellipsoidal shape. Meanwhile, the electric potential profile was further distorted, with which the electric field strength kept growing at the top. At $t = 1.8$ ms, the original hemispheric shape was stretched to be a cone with a sharp tip at which the electric field strength was significantly concentrated. Because of the restriction of the grid length in the simulations, it is not easy to obtain an unlimited sharp tip. So, the simulations terminated when the radius of the tip was as similar size as the grid length.

Figure 2.24 shows the maximum electric field strength during the deformation process. The initial conditions were as following. The liquid metal temperature was 1356 K. The radius of the initial hemisphere was 1 mm. The average applied electric field E_0 was 7 MV/m. The maximum electric field strength located at the tip. In the level-set method, the interface between two phases in the model is not an ideal layer but with a certain thickness. So that the interface could fluctuate with the flow or phase change, especially at the region where the interface changes sharply, which slightly fluctuated the maximum electric field. The evolution of the maximum electric field strength as a function of time showed an exponential form of $E_{max} = ae^{bt}$, where a and b are constant. The maximum electric field fast increased from $t = 1.5$ ms. At $t = 1.8$ ms, the maximum electric field exceeded 200 MV/m, which was about 28.5 times of E_0. The deformation process accompanied by an electric field enhancement. The enhanced electric field, in turn, accelerated the deformation, and finally, a Taylor cone formed.

Figure 2.25 shows a velocity field of the liquid metal and the electric field strength on the interface at $t = 1$ ms. In the velocity field, the gas–liquid interface is the black

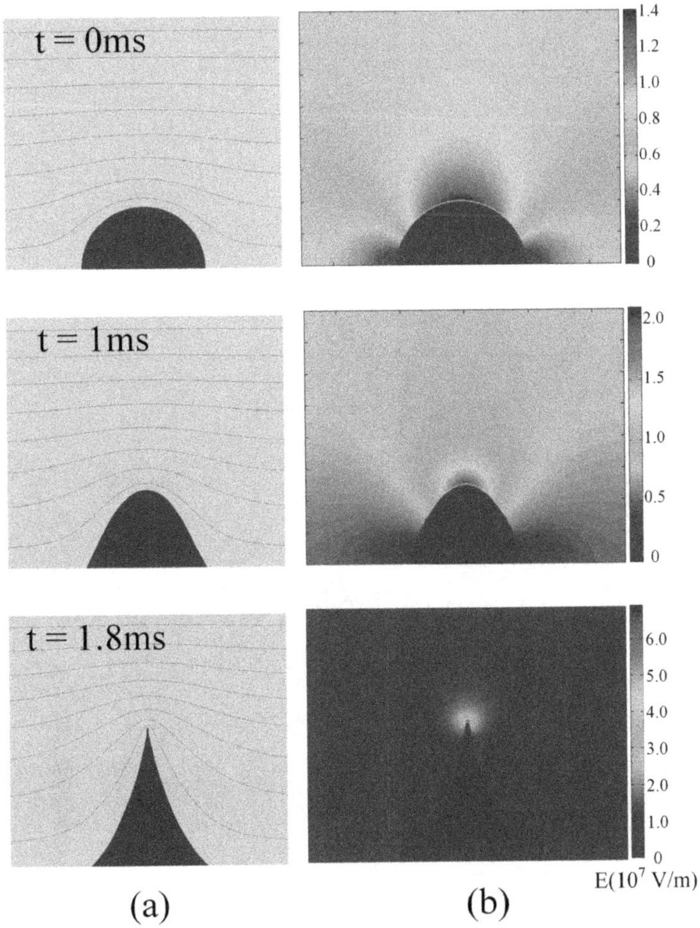

Fig. 2.23 A simulated deformation process of the liquid copper. **a** The electric potential profiles; **b** The electric field strength profiles

line. There was a clear vortex in the external region while there was no vortex in the internal of the liquid metal. It suggested that the velocity of the external liquid had a higher speed comparing to the internal fluid and peaked near the tip. Moreover, in the electric field, the electric stress $\boldsymbol{F}_{es}$ was completely imposed on the liquid–gas interface. Since the maximum electric field strength located at the tip, the electric stress $\boldsymbol{F}_{es}$ reached the maximum at the tip but was almost zero at the bottom. Again, following the perfect conductor assumption, the electric stress was perpendicular to the metal surface pointing to the external region. Meanwhile, according to Eq. (2.21), the surface tension force $\boldsymbol{F}_{st}$ was also imposed on the interface, but the force had an opposite direction to electric stress $\boldsymbol{F}_{es}$, pointing from the external to the internal. Once the electric field stress force was higher than the local surface tension force, a

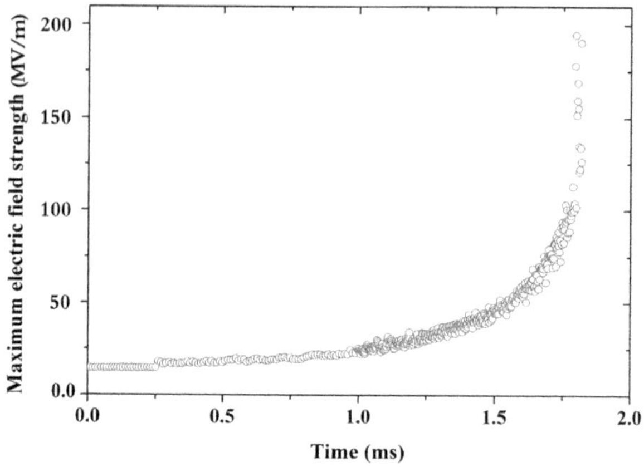

Fig. 2.24 The maximum electric field strength during the deformation process of the liquid copper. The initial temperature: $T = 1356$ K; the radius of the initial hemisphere: $R = 1$ mm; the average applied electric field: $E_0 = 7$ MV/m

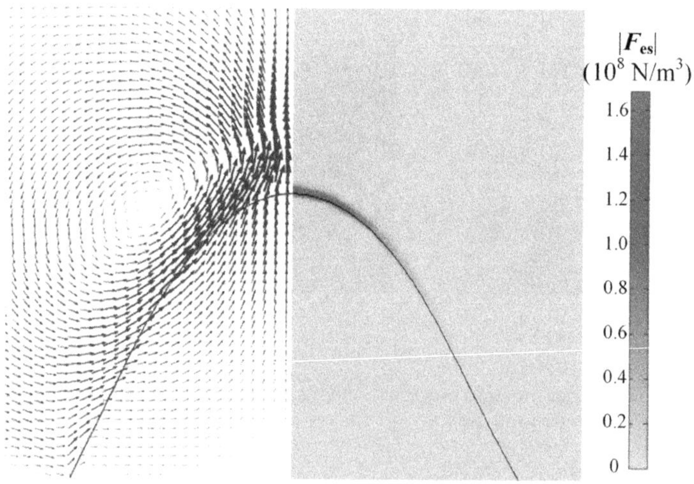

Fig. 2.25 The velocity field of fluids and the electric stress profile (|$\boldsymbol{F}$es|) at $t = 1$ ms. The velocity field is shown on the left. The electric stress is shown on the right. The black line is the liquid–gas interface. The initial temperature: $T = 1356$ K; the radius of the initial hemisphere: $R = 1$ mm; the average applied electric field $E_0 = 7$ MV/m

net non-uniform force applied on the interface pulled the liquid metal upwards, thus the initial hemisphere became a cone with a sharp tip.

However, if an initial electric field is low, the electric stress force cannot overcome the surface tension force, the hemisphere will not form a Taylor cone. Figure 2.26 shows a failure of a Taylor cone formation from a liquid copper. The initial conditions were as following. An average electric field applied upon the gap was $E_0 = 5.1$ MV/m.

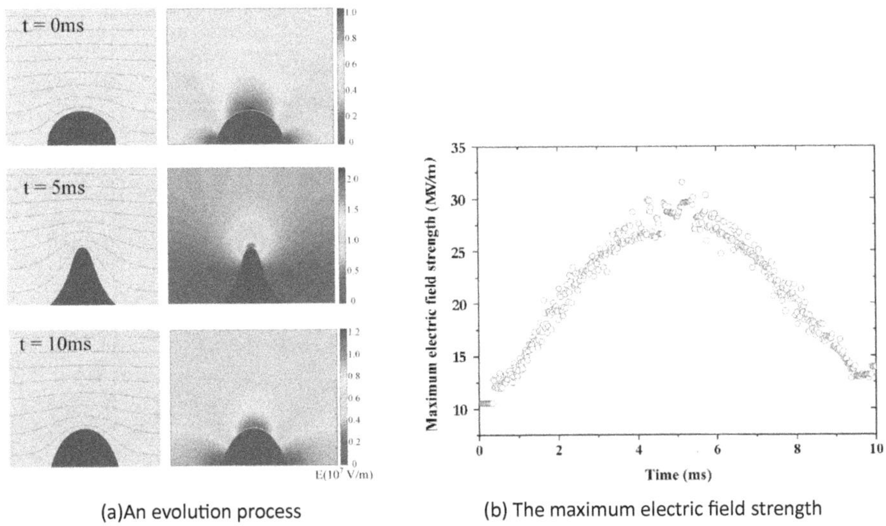

Fig. 2.26 A failure of a Taylor cone formation from a liquid copper

The radius of the hemispheric liquid metal was 1 mm. The temperature was 1356 K. The hemispheric liquid copper was attached to a grounded plane. Figure 2.26a show an evolution process. At $t = 5$ ms, the hemisphere was pulled out and the electric field enhanced near the top. However, at $t = 10$ ms, the liquid copper ceased stretching and contracted into a hemisphere again, while the electric field strength near the top recovered to a value similar to the initial one. The final shape of the liquid copper slightly deformed comparing to the initial shape. Figure 2.26b shows the maximum electric field strength at the top. It reflected a failure formation of a Taylor cone. Before $t = 5$ ms, the electric field strength gradually increased to about 30 MV/m. After that, the electric field strength deceased gradually because the electric stress force cannot win the inward direction surface tension force caused by a highly curved tip. Therefore, the liquid metal contracted and the formation of a Taylor cone failed.

Figure 2.27 shows a buildup time of Taylor cones formation as a function of the average electric fields E_0. It shows that the buildup time was inversely proportional to the applied electric field strength. For the three radii of 1, 0.5, and 0.25 mm, there existed a critical electric field strength, above which Taylor cones can form. The critical electric fields were 5.1, 7.28, and 10.25 MV/m corresponding to the hemisphere radius of 1, 0.5, and 0.25 mm, respectively. It suggested that the critical value significantly increased as the radius decreased. An electric Bond number Bo_e can represent a relative effect of electric stress force versus surface tension force, which is given in Eq. (2.22) [60, 61].

$$B_{Oe} = \frac{E_0^2 L}{\sigma} \tag{2.22}$$

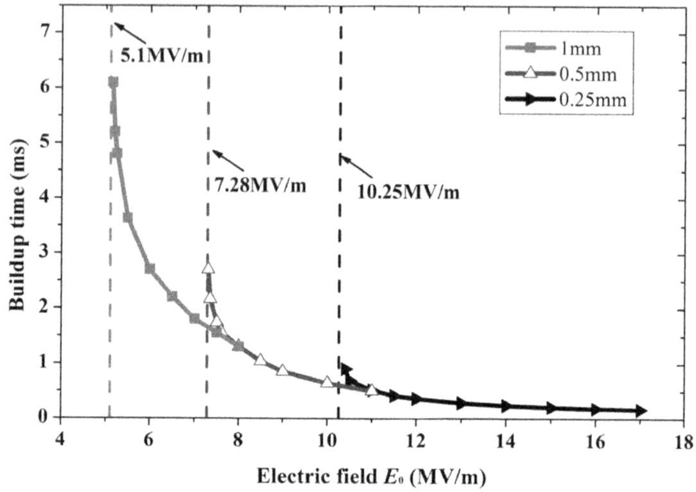

Fig. 2.27 The buildup time of Taylor cones versus applied electric fields E_0. The dashed lines indicate the crtitical values.The initial temperature: T = 1356 K; radii: R = 1, 0.5, and 0.25 mm

Only if the electric Bond number exceeds a critical level, a Taylor cone can be formed. Otherwise, the electric stress force cannot overcome the surface tension force, under which the liquid copper returned to the initial shape after a fluctuation (see Fig. 2.26).

Figure 2.28 shows the buildup time of a Taylor cone as a function of the initial electric field E_0. The initial temperature was set as 1356 K (the melting point of copper), 2000 K, and 2300 K, respectively. It shows that the buildup time exponentially decreased with an increasing of the initial electric field. It also shows that the Taylor cone formation critical electric field value decreased with an increasing of the initial temperature. The critical value was 5.1, 4.8 and 4.65 MV/m when the

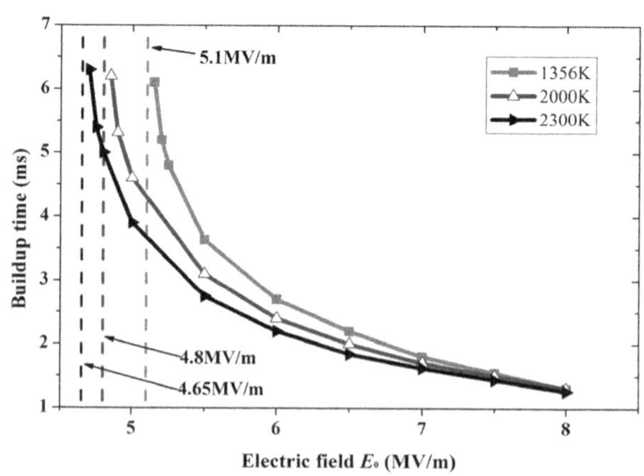

Fig. 2.28 The buildup time of a Taylor cone as a function of the initial electric field E_0. The dash lines indicate Taylor cone formation critical electric field value. The initial temperature: T = 1356 K, 2000 K, and 2300 K; hemisphere radius R = 1 mm

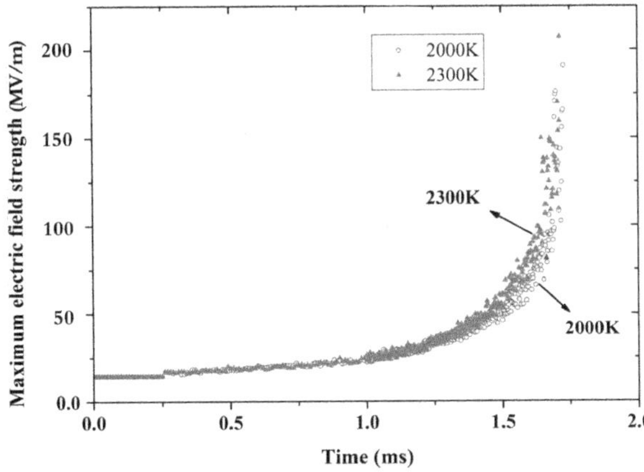

Fig. 2.29 The evolution of the maximum electric field strength during the deformation processes. The initial temperature: $T =$ 2000 K and 2300 K; hemisphere radius $R =$ 1 mm; the average applied electric field: $E_0 = 7$ MV/m

initial temperature was 1356, 2000 and 2300 K, respectively. It may be related to the temperature dependence of the material. After the electric field exceeded $E_0 =$ 7 MV/m, the influence of the temperature can be neglected.

Figure 2.29 shows the maximum electric field at the tip during the deformation process of the liquid copper. The evolution of the maximum electric field was almost identical for the two kinds of initial temperature of 2000 and 2300 K. It shows that the initial temperature did not play a significant role on maximum electric field strength of a Taylor cone.

As shown in Fig. 2.24, the development of a Taylor cone is accompanied by an enhancement of the local electric field strength. Therefore, it is expected that more electrons are emitted according to the field emission theory, which is crucial to initiate breakdowns. The density of the electron emission current, calculated by the Murphy and Good theory, [62] can be found in Eq. (2.23).

$$j_{em}(T_w, E_w, \phi_w) = e\int_{-\infty}^{\infty} N(T_w, E_w, \phi_w)D(E_w, W)dw \tag{2.23}$$

where T_{w} is the temperature of the emitter; E_{w} is the electric field at the emitter; Φ_{w} is the work function of the material. Equation (2.23) can also be simplified as Eq. (2.24) [63].

$$j = K_1\left(120T_w^2 + 406E_w^{9/8}\right) \times \exp\left[-\left(\frac{T_w^2}{2.727 \times 10^9} + \frac{E_w^2}{4.252 \times 10^{17}}\right)^{-1/2}\right] \tag{2.24}$$

where

$$K_1 = 1.45 \times 10^4 \frac{8100T_w^3 + 0.35E_w^2}{9100T_w^3 - 5300E_wT_w + 0.75E_w^2} \tag{2.25}$$

Fig. 2.30 The thermo-field electron emission current density during the deformation. The initial temperature: $T = 2000$ K and 2300 K; hemisphere radius $R = 1$ mm; the average applied electric field: $E_0 = 7$ MV/m

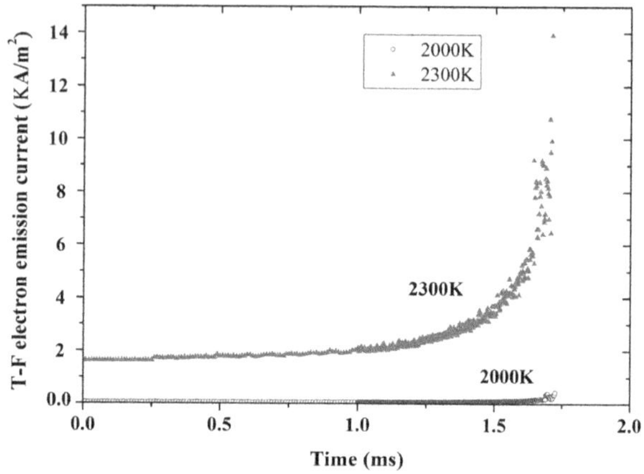

Figure 2.30 shows the thermo-field electron emission at the tip with two kinds of initial temperatures, based on the above-mentioned theory and the data shown in Fig. 2.29. The emission electron current density at the tip started to exponentially increase from 1.5 ms when the initial temperature was 2300 K. This curve coincides with the curve shown in Fig. 2.29. This suggests that the electron growth originated from the enhancement of the local electric field. However, the electron emission curve with the initial temperature of 2000 K did not follow the same behavior. This curve appeared fairly low and flat, suggesting that the emission remained at a low level during the deformation process, with an emission density several orders lower than that with the temperature 2300 K. In contrast to the corresponding curve in Fig. 2.29, the emission current density did not increase as the electric field increased. Therefore, the influence of temperature has a significant impact on the electron emission. The possible reason might be that the electron emission depends on both the electric field and the thermal field. The high temperature could initiate a breakdown rapidly rather than waiting for a sharp tip formation. At the same time the high temperature leads to more vapor evaporation, resulting in a higher probability of breakdowns. In addition, a Taylor cone will not be initiated if the surface temperature is cooled down quickly after current zero. Thus, it is important to control the contact surface temperature in a low level in order to get a successful current interruption.

In summary, the local electric field can be enhanced by the Taylor cone, and, in turn, the Taylor cone will be accelerated, which may trigger a restrike. The breakdown did not occur immediately when the voltage is applied but is postponed to several milliseconds later when the Taylor cone forms on the liquid metal, which could interpret some late breakdowns phenomena in vacuum interrupters. It is worth to note that late breakdowns can also be triggered by particles (see Sects. 2.4.2 and 2.4.3).

2.4 Dielectric Breakdowns

2.4.1 Breakdown in Cathode Sheath: Metal Vapor and Residual Plasma

Current interruption in vacuum interrupters leaves considerable residual plasma and metal vapor after vacuum arcs extinguish. The residual plasma and metal vapor affect the post-arc physical process of the vacuum interrupters. When a transient recovery voltage applies, which begins to rise around current zero, the residual plasma forms a sheath expanding from the cathode to the anode. During this sheath expansion period, the residual plasma and the metal vapor may induce a sheath breakdown by suffering the transient recovery voltage, hence an instantaneous breakdown occurs. Therefore, it is meaningful to understand the breakdowns during the sheath expansion.

Sheath expansion were modeled by various researchers. Andrews and Varey described the sheath by a continuous-transition model [64], which accounted for Poisson's equation, continuity equation, and the momentum equations. This model worked well to predict sheath expansion after the current zero [65]. Childs and Greenwoods subsequently improved this model by taking the following factors into consideration, which includes the post-arc current density, the electric field at the post-arc cathode surface, the power input to the surface, and the sheath thickness [66, 67]. The model was verified by their experiments on direct current interruptions. After that, sheath expansion was further modeled with the particle-in-cell method, which considered some microscopic behaviors inside the plasmas [68, 69]. By using the particle-in-cell method, Mo et al. studied how the ion and electron temperatures affected sheath expansion [70], and Changho solved the ion distributions in a collision-free sheath [71]. Sarrailh et al. developed a one-dimensional hybrid Maxwell–Boltzmann model, which accounted for the collisions between particles [72, 73]. They demonstrated a breakdown in sheath expansion and discussed an impact of contact surface temperature, which was related to the metal vapor density [74].

Despite the above-mentioned achievements, an impact of the residual plasma on the sheath breakdown was not considered, such as a contribution of the high-energy electrons in the discharges [75], which may lead to an overestimation of the role of the metal vapor density in current interruptions in vacuum. Moreover, a non-equilibrium nature of a dynamic sheath expansion and the breakdown process should be considered. Therefore, a fully kinetic model is needed to estimate how the residual plasma affects the breakdown process in the sheath expansion.

In this section, a fully kinetic particle-in-cell Monte Carlo collision model is developed. The model was with one spatial dimension and three velocity dimensions (1D3V). It also included an external circuit to ensure the calculations were self-consistent. By using the model, it found an impact of the residual plasma and the metal vapor density on the sheath expansion process, thus a breakdown probability can be obtained. This is proposed as a mechanism for instantaneous breakdowns during sheath expansion.

Figure 2.31 shows a fully kinetic particle-in-cell Monte Carlo collision model with one spatial dimension and three velocity dimensions. In the model, the electrode gap was 10 mm, and the contact material was copper. At the beginning of the post-arc phase, the gap was filled with a uniform neutral plasma composed of electrons and Cu^{2+} ions [76]. The particles were assumed to be a Maxwell velocity distribution. The electron temperatures was set as $T_e = 3$ eV as a typical value and the copper ion temperature was set as $T_i = 2$ eV [77]. Five sets of initial values of plasma density and metal vapor density are shown in Table 2.2.

The physical description of the model is as following. In the sheath expansion, the post-arc contact gap consists of two regions, which includes a neutral plasma region and a cathode sheath region, respectively. Driven by the transient recovery voltage, the sheath starts from the cathode and expands to occupy the whole gap. The transient recovery voltage is the driving force of the sheath expansion. The whole transient recovery voltage applies on the sheath, which forms a high electric field in front of the cathode and causes an intense electron emission. The electron emission was determined by the Fowler–Nordheim (F–N) theory [78]. The post-arc cathode is the arcing anode, which may be severely damaged by the arcs (see

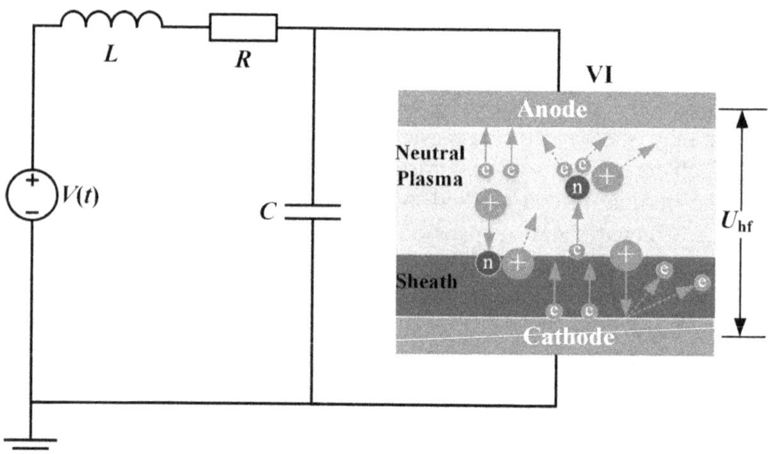

Fig. 2.31 A schematic of the sheath expansion model. V(t) is a voltage source; R, L, and C are the lumped resistance, inductance, and capacitance, respectively; VI is the vacuum interrupter

Table 2.2 Five sets of initial values of plasma density and metal vapor density

No	Initial plasma density (m^{-3})	Metal vapor density (m^{-3})
1	2×10^{17}	0
2	2×10^{17}	10^{21}
3	2×10^{17}	5×10^{21}
4	2×10^{17}	10^{22}
5	4×10^{17}	5×10^{21}

Sect. 1.4). To model the surface deformation, a field enhancement factor of $\beta = 50$ was set in the F–N equation. An influence of metal vapor on the sheath expansion was taken into consideration because there are collisions between the metal vapor and the charged particles. The model accounted for charge-exchange and momentum-exchange collisions between ions (Cu^{+}, Cu^{2+}) and neutral atoms (Cu), through elastic and excitation collisions. The model also accounted ionization collisions between electrons (e) and neutral atoms (Cu). Because the model focused on an origin of a restrike, only single-charged copper ions (Cu^{+}) generation was considered in the ionization collisions [79, 80], while at the beginning of the simulations, the residual plasma consisted of $Cu2^{+}$ ions and electrons. The cross-sectional data for the relevant collisions came from reference [81–83]. Furthermore, the sheath expansion took several tens of microseconds, while the metal vapor kept a much longer time of several milliseconds because a local contact surface was still in a hot molten liquid state (see Sect. 1.4.4). Therefore, the model considered the metal vapor density to be a constant in the sheath expansion period. In addition, the model took accounted for the secondary electron emission (SEE) from the cathode surface, which is caused by ion bombardment (Cu^{+}, $Cu2^{+} \rightarrow$ e). The yield rate was estimated from an empirical formula given by Vaughan [84].

Particle-in-cell simulations have certain restrictions. To obtain a stable and reliable solution, the grid length Δx and the time step Δt in the simulations should be chosen based on the Debye length λ_D and the plasma frequency ω_{pe}. The rules can be expressed as $\Delta x \leq \lambda_D$ and $\Delta t \leq 0.2/\omega_{pe}$ [68]. Based on the initial plasma conditions, the model used a grid length of $\Delta x = 2 \times 10^{-5}$ m and a time step of $\Delta t = 10^{-12}$ s.

Figure 2.32 shows a scheme for the particle-in-cell Monte Carlo collision simulation. The motion of the particles is described by Newton's laws as Eq. (2.26) [68].

$$\frac{d\boldsymbol{v}}{dt} = \frac{\boldsymbol{F}}{m} = \frac{q\boldsymbol{E}}{m} \tag{2.26}$$

where v is the velocity of the particles. $\boldsymbol{F}$ is the force acting on the particles. m is the mass of the particles. $\boldsymbol{E}$ is the total electric field to advance the charged particle

Fig. 2.32 Scheme for the particle-in-cell Monte Carlo collision simulation

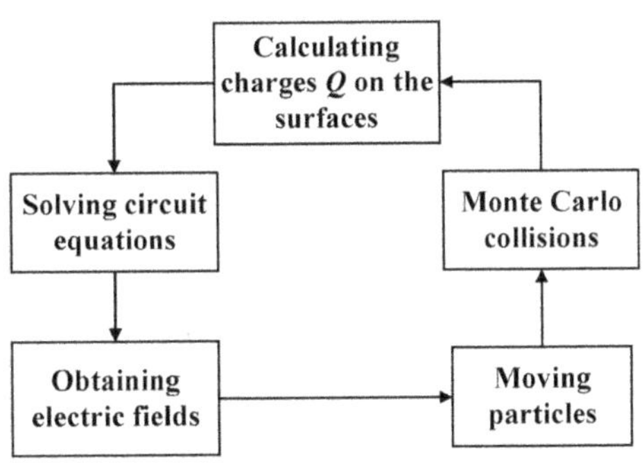

dynamics. The electric field $\boldsymbol{E}$ and the corresponding electric potential u are obtained by solving Poisson's Eqs. (2.27) and (2.28).

$$\nabla^2\varphi = \frac{-e(n_i - n_e)}{\varepsilon_0} \tag{2.27}$$

$$\boldsymbol{E} = -\nabla\varphi \tag{2.28}$$

where n_i is the ion density. n_e is the electron density. e is the electron charge. u is the electric potential. ε_0 is the vacuum permittivity. To consider short-range particle collisions, such as excitation and ionization, the particle-in-cell model was combined with the Monte Carlo collision model [20].

The sheath expansion is also influenced by the external circuit, which applies a voltage drop across the contact gap. The model provided an external circuit to simulate a transient recovery voltage U_{hf}. The external circuit is shown in Fig. 2.31, where R, L, and C are the lumped resistance, inductance, and capacitance, respectively. A DC voltage source $V(t)$ was set because the sheath expansion only lasts for a few microseconds, while the voltage source remains as a constant during this period. Additionally, Q was defined as the total electrical charges absorbing by the electrodes of the vacuum interrupter. Therefore, the circuit equation is shown as Eq. (2.29).

$$LC\frac{d^2U_{hf}}{dt^2} + \frac{L}{R}\frac{dU_{hf}}{dt} + U_{hf} = V(t) - L\frac{d^2Q}{dt^2} \tag{2.29}$$

Figure 2.33 shows a transient recovery voltage. It was calculated by the following circuit parameters: $V(t) = -40.5$ kV, $L = 3.5$ mH, $C = 0.4$ μF, and $R = 20$ Ω. The voltage curve followed the international standard IEC 62,271–100. A coupling between the external circuit and the plasma was implemented inside the model. The electric field $\boldsymbol{E}$ in the gap can be obtained from the source terms and the boundary conditions, as shown in Eq. (2.30) [85].

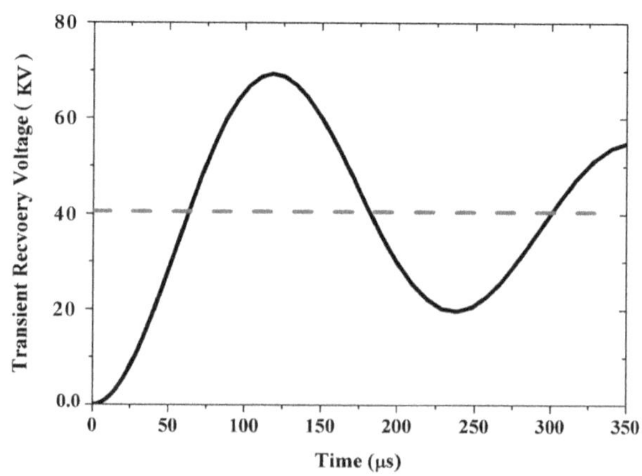

Fig. 2.33 Model calculated transient recovery voltage

$$\boldsymbol{E} = \boldsymbol{E}_p + U_{hf}\boldsymbol{E}_0 \tag{2.30}$$

where E_p is the electric field solved by Poisson's equation with zero boundary conditions to account for the space charge effects, assuming that the gap is filled with plasma and that the electrodes are perfect conductors grounded to earth; E_0 is a Laplace electric field only accounting for the vacuum field effects, which is obtained by imposing a voltage drop of 1 V across the sheath without any charged particles; U_{hf} is the voltage drop across the electrodes. At each time step Δt, the particles that are absorbed or emitted by the electrodes are calculated to obtain a convection current I_{tot}. Therefore, the total charge on the electrodes Q can be derived as Eq. (2.31).

$$Q = U_{hf} \times Q_0 - I_{tot} \times \Delta t \tag{2.31}$$

where Q_0 is the electric charge absorbed or emitted by the electrodes with an applied voltage drop of 1 V. Subsequently, Q was substituted into Eq. (2.29) to calculate U_{hf} for the next time step. This method is described in details in [86].

The simulation results show that the metal vapor has an impact on the sheath expansion. Figure 2.34 shows the sheath expanded over time under three kinds of metal vapor densities of 0, 10^{21} m^{-3}, and 5×10^{21} m^{-3}. The initial plasma density was 2×10^{17} m^{-3}. It shows that comparing with the case without the metal vapor, the metal vapor retarded the sheath expanding from the cathode (at 0 mm) to the anode (at 10 mm). As the metal vapor density increased from 10^{21} m^{-3} to 5×10^{21} m^{-3}, the sheath expansion duration increased from 10 to 23.5 μs.

Figure 2.35 shows the evolution of particles quantities over time at two kinds of background metal vapor densities of 10^{21} m^{-3}, and 5×10^{21} m^{-3}, respectively. It shows that all the ionized particles decayed to zero approximately at the same time under each background metal vapor density. The Cu^{2+} ions decreased exponentially from the beginning. The electrons and Cu^{+} ions initially increased and then decreased

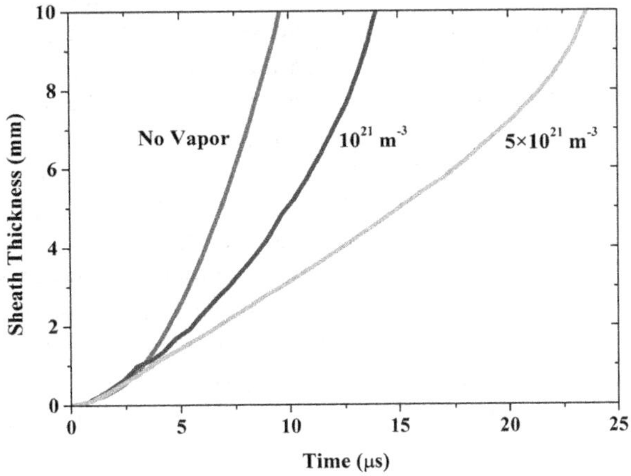

Fig. 2.34 Sheath thickness expands with time. The initial plasma density: 2×10^{17} m^{-3}; the initial plasma temperature: $T_e = 3$ eV for electrons and $T_i = 2$ eV for Cu^{2+} ions; The metal vapor densities: 0, 10^{21} m^{-3}, and 5×10^{21} m^{-3}

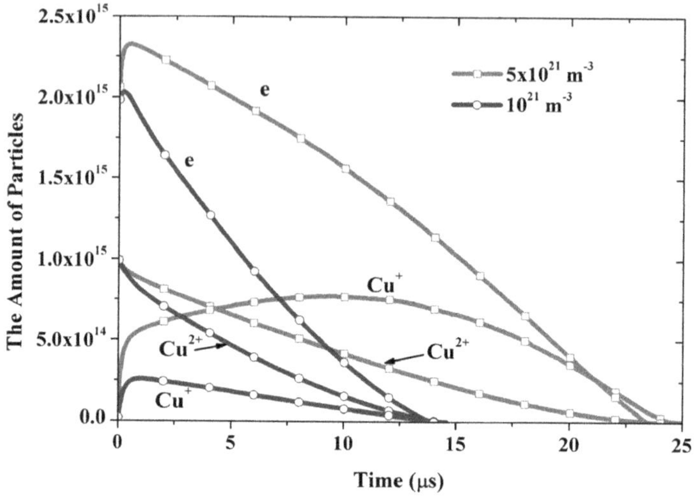

Fig. 2.35 Evolution of particles number with time. The initial plasma density: 2×10^{17} m^{-3}; the initial plasma temperature: $T_e = 3$ eV for electrons and $T_i = 2$ eV for Cu^{2+} ions; The metal vapor densities: 10^{21} m^{-3}, and 5×10^{21} m^{-3}

to zero, which resulted from an ionization of the neutral atoms into electrons and Cu^+ ions.

Sheath breakdown occurs if a high metal vapor density exists. Figure 2.36 shows an evolution of a sheath breakdown in an current interruption of vacuum arcs. Figure 2.36a, b, c and d shows the spatial–temporal evolution of Cu^+ ions density, electrons density, charged particles numbers and the sheath electric field, respectively. The initial plasma density was 2×10^{17} m^{-3}. The initial plasma temperature was $T_e = 3$ eV for the electrons and $T_i = 2$ eV for the Cu^{2+} ions. The metal vapor density was 10^{22} m^{-3}. Figure 2.36a and b shows that the sheath boundary gradually expanded from cathode to anode from 0 to 2 μs. However, the sheath boundary stagnated at 1.5 mm away from the cathode surface at 6 μs. Thereafter, the ion and electron densities further increased from 6 to 8 μs. After 8 μs, the sheath boundary stopped expanding and even returning to the cathode surface, which suggested a breakdown. Figure 2.36c also indicated that the numbers of residual Cu^+ ions and electrons increased exponentially, especially fast after 8 μs, which was driven by an enhanced electric field applied on the sheath, as shown in Fig. 2.36d. The number of Cu^{2+} ions decreased slightly over time because a high metal vapor density prevented the Cu^{2+} ions from dissipating from the gap. There were strong collisions between the particles. An electron avalanche ionization process was triggered.

The breakdowns in sheath expansion are influenced not only by the metal vapor density but also by the residual plasma density. Figure 2.37 shows a sheath breakdown at a high residual plasma density of 4×10^{17} m^{-3}. Figure 2.37a and b shows the evolution of charged particles quantities and the electric field across the sheath, respectively. The initial temperature was $T_e = 3$ eV for the electrons and $T_i =$

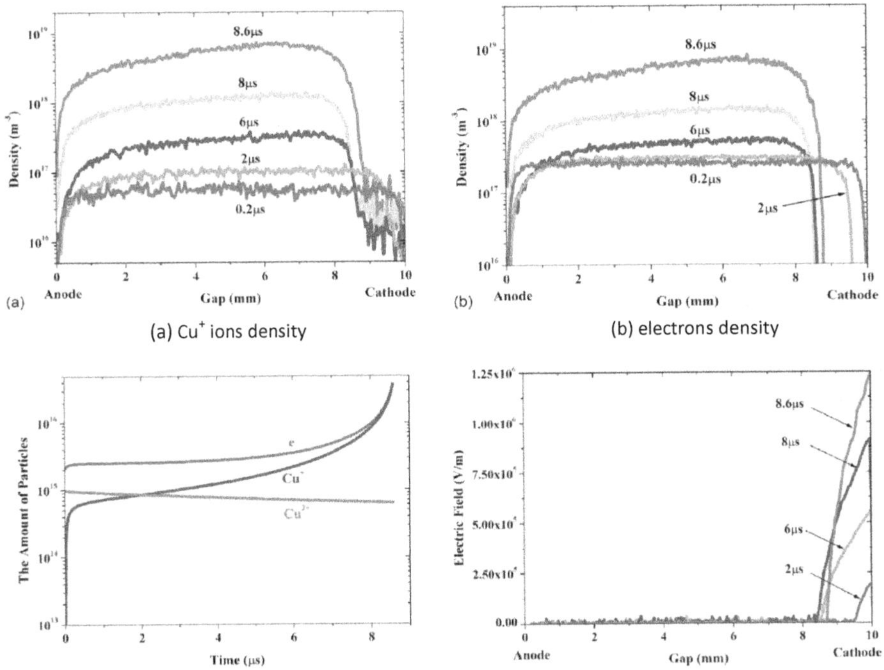

(a) Cu^{+} ions density (b) electrons density

(c) Charged particles number (d) Sheath electric field

Fig. 2.36 Evolution of sheath breakdown after interruption of vacuum arcs. The initial plasma density: 2×10^{17} m^{-3}; the initial plasma temperature: $T_e = 3$ eV for electrons and $T_i = 2$ eV for Cu^{2+} ions; The metal vapor densities: 10^{22} m^{-3}

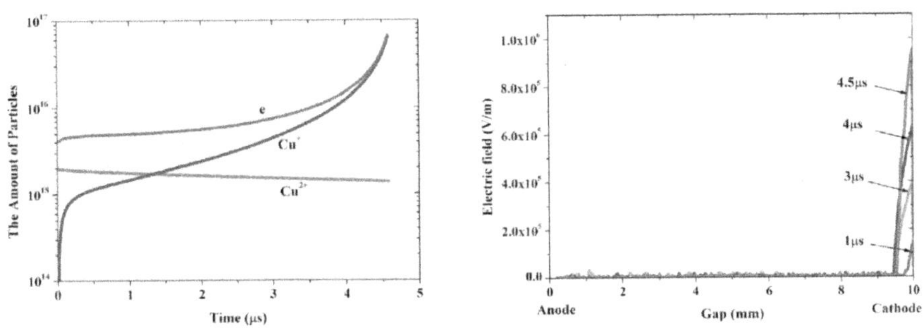

(a) Charged particles number (b) Sheath electric field

Fig. 2.37 Sheath breakdown at a high residual plasma density. Initial plasma density: 4×10^{17} m^{-3}; electron temperature: 3 eV; ion temperature: 2 eV; and metal vapor density: 5×10^{21} m^{-3}

2 eV for the Cu^{2+} ions. The metal vapor density was 5×10^{21} m^{-3}. It shows that a breakdown occurred in the sheath expansion. The breakdown followed a similar way to that shown in Fig. 2.36, even though the metal vapor density was as low as 5×10^{21} m^{-3}. At the metal vapor density of 5×10^{21} m^{-3}, the threshold residual plasma density for breakdowns was found to be 2×10^{17} m^{-3}. Therefore, breakdowns in sheath are dependent not only on the metal vapor density but also on the initial residual plasma density.

The influence of metal vapor density and the initial residual plasma density on the sheath breakdown can also be found by a comparison among Figs. 2.34, 2.36 and 2.37. With a metal vapor density of 5×10^{21} m^{-3} and an initial plasma density of 2×10^{17} m^{-3}, the post-arc plasma was dissipated from the gap in 23 μs (see Fig. 2.34). When the metal vapor density increased to 10^{22} m^{-3}, the sheath stopped expanding at 8.6 μs (see Fig. 2.36). When the initial plasma density increased to 4×10^{17} m^{-3}, the sheath stopped expanding at 4 μs (see Fig. 2.37).

Figure 2.38 shows how the electric field strength across the sheath initiates breakdowns. The electric field strength increased as the sheath expanded from the cathode to the anode. When the initial plasma density was 2×10^{17} m^{-3} and the metal vapor density was 5×10^{21} m^{-3}, all of the plasma dissipated from the contact gap during 23 μs. The electric field strength increased much quickly when the initial plasma density increased to 4×10^{17} m^{-3} or the metal vapor density increased to 10^{22} m^{-3}. A strong electric field causes intense electron emissions, depending on the *F–N* equation, and the existence of the cathode sheath enhanced this effect.

Figure 2.39 proposes a breakdown mechanism during the sheath expansion. When the metal vapor density in the post-arc gap increases, the ionization probability increases. This results in more charged particles entering the post-arc gap. In addition,

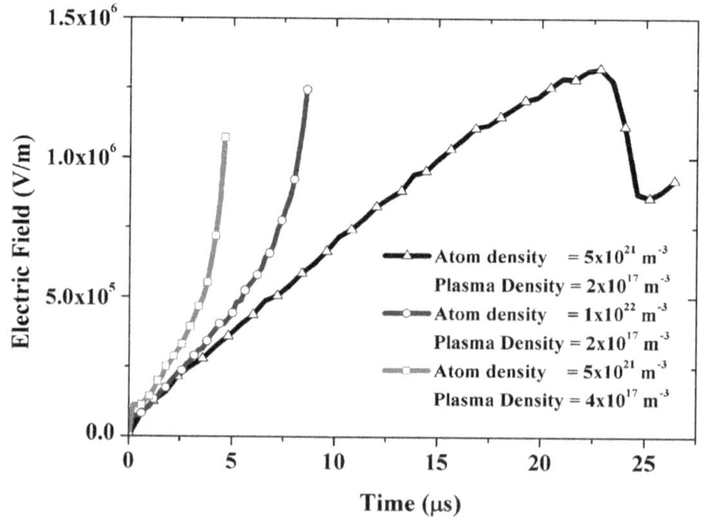

Fig. 2.38 The electric field strength across the sheath

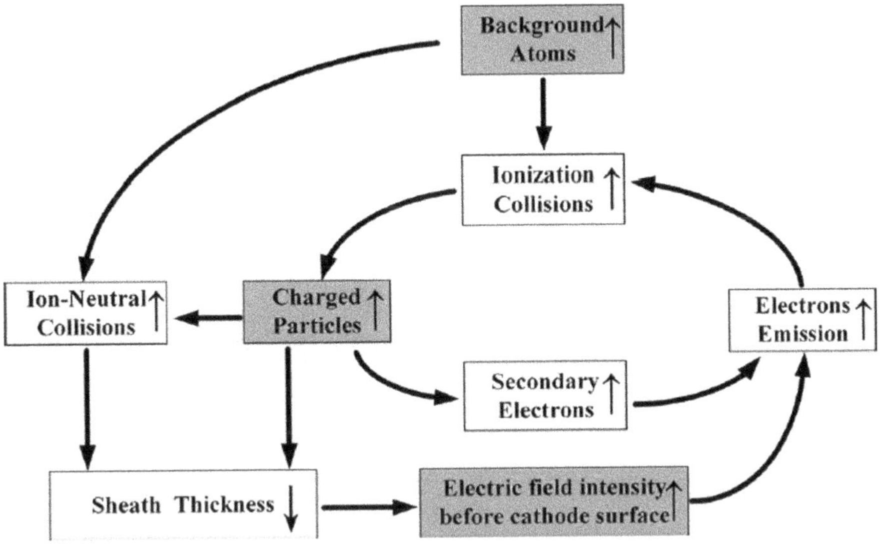

Fig. 2.39 A breakdown mechanism during the sheath expansion

the ions-neutral particles collisions slow down the decay of the ions and it impedes the sheath growth. Consequently, the sheath ceases to expand and even returns to the cathode. At the same time, the voltage applied on the gap continues to rise, thus the electric field strength across the cathode sheath increases. According to the *F–N* equation, a high electric field strength causes a high electron emission rate and results in a high ionization probability. The charged particles generated from the ionization process bombards the cathode surface, which generates secondary electron yields. This again increases the probability of ionization collision. All these factors cause an avalanche of charged particles generation, which leads to a breakdown in the cathode sheath.

It is interesting to obtain a breakdown boundary in the cathode sheath expansion under the effects of both metal vapor and residual plasma. Figure 2.40 shows the boundary between breakdown and no-breakdown as a function of metal vapor density and residual plasma density. There was also an uncertain region in the figure, because the boundary was determined through numerous simulation cases. When the residual plasma density was 10^{15} m^{-3}, the metal vapor density for a breakdown initiation was about 10^{22} m^{-3}. Moreover, the boundary was insensitive to the residual plasma density until it reaches 10^{17} m^{-3}. When the residual plasma density exceeded 10^{17} m^{-3}, the critical metal vapor density for a breakdown initiation steeply decreased. When the plasma density reached 10^{18} m^{-3}, the metal vapor density for a breakdown initiation was about 5×10^{20} m^{-3}. If the residual plasma was not considered, one would overestimate the role of the metal vapor density in the cathode sheath breakdowns in vacuum interrupters.

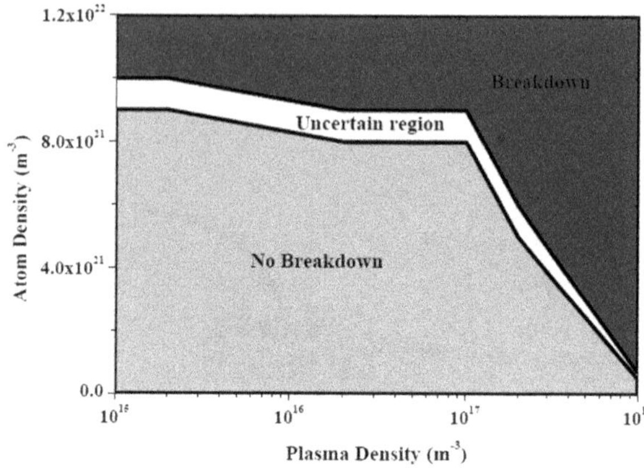

Fig. 2.40 The boundary between breakdown and no-breakdown as a function of metal vapor density and residual plasma density in the cathode sheath expansion

2.4.2 Particles Induced Late Breakdown

The dielectric recovery after vacuum arc extinguishing is important because it determines the current interruption capability of vacuum circuit breakers. The previous sections discussed several influencing factors on the dielectric recovery process, which included metal vapor (see Sect. 2.2), Taylor cone (see Sect. 2.3) and residual plasma (see Sect. 2.4.1). Such influencing factors may explain instantaneous breakdowns. However, late breakdowns were observed occurring hundreds of milliseconds after arc extinguishes [87, 88]. This phenomenon cannot be attributed to the influencing factors mentioned above. Locally molten contacts that evaporate neutral metallic vapor, solidify with a time constant of ms (see Sect. 1.4.3). Tayor cone forms with a time constant of ms (see Sect. 2.3). The residual plasma dissipates with a time constant of several microseconds (see Sect. 2.4.1) after current zero. Such influencing factors cannot trigger the late breakdowns. Therefore, researchers speculated that the late breakdowns are triggered by particles [89–92].

The effects of particles on electrical breakdown have been investigated by many researchers. Ejiri et al. injected artificial particles in vacuum gaps to simulate an impact of particles caused by arcs or mechanical shock [93]. They observed particles to bounce up and down between the contacts. They found that some charged particles with a low impact velocity of 50 ms^{-1} induced breakdown. Dascalescu et al. found that the particles usually exhibited oscillating motion near the contact surface [94]. This behavior could be enhanced by increasing the applied voltage, thus triggering breakdown. Sakai et al. investigated particle-triggered breakdown [95, 96]. They found that micro-discharge, which could be observed between particles and the contacts, would electrify the particles and decrease the threshold breakdown voltage. They also found that the motion of particles was affected by the Coulomb force. Chatterton studied breakdown triggered by low-velocity particles and he found that insensitive field emission could occur if particles moved toward the cathode, leading to discharge between the particle and the cathode [97]. Chakrabarti and Chatterton investigated impact damage caused by particles with 5–10 μm radius [98]. They found that particle-induced discharge occurred when particle bombardment exceeded a critical limit. Gellert et al. used a laser-shadow technique and laser-induced fluorescence to observe the particles in vacuum arcs [11]. Meanwhile, theoretical analysis was done to understand the influencing factors on particles behaviors, including particle size, specific mass, electrode structure, and electric field strength [99, 100]. A smoothed particle hydrodynamics method simulated the particle projectile impact process with impact velocities ranging from 105 ms^{-1} to 631 ms^{-1} [101]. The results showed that enormous secondary particles and contact damage was generated when particles collided with the electrode.

Some researchers provided an opposing opinion regarding the effects of particles on late breakdowns in vacuum. Xu et al. determined the influence of size, speed, charge quantity, and field enhancement factor of particles by establishing a synthetic experimental system [102]. They speculated that the late breakdown might be due to Townsend discharge resulting from high-temperature electrode surface and protrusions because particles cannot directly cause the observed discharging current. Meanwhile, Anders et al. argued that the late breakdowns was a consequence of mechanical shock or vapor deposition instead of the presence of particles [103]. Dullni et al. thought that field electron emission was responsible for the restrike while switching a capacitive current rather than the particles [104]. Hurly et al. considered that a charged particle itself could be a source of field emission, which greatly affected its behavior and possibly initiated a field emission induced breakdown [105].

The above-mentioned studies implied that particles would have a significant influence on the dielectric recovery process in current interruptions in vacuum. Vacuum arcs generate particles from contact materials. Particles can be generated from cathode spots or from the anode in the case of high current interruptions. Particles coming from anode are usually in a presence of liquid metal with large size and low velocity. Late breakdowns after arc extinction in vacuum may be related to these particles. Therefore, it is interesting if one can directly observe the particles induced breakdowns during the dielectric recovery period after vacuum arcs extinguish. An optical diagnostic platform based on a laser-shadow technique was built up to achieve the goal, which filters out light emission from vacuum arcs and allows a direct observation of particles motion. The results showed a direct evidence that particles triggered a late breakdown 132 ms after vacuum arcs extinguish.

Figure 2.41 shows the experimental setup for the particles induced late breakdown observation. Figure 2.41a shows an experimental setup for vacuum arc initiation. It included a current source that was an L–C discharging circuit, where $L = 0.1$ mH and $C = 100$ mF. The discharge current frequency was 50 Hz. The discharging current peak was in a range between 2 and 16 kA, which was controlled by the charging voltage of C. The arcing time was controlled at ~10 ms. A pair of butt type contacts was vertically installed in a demountable vacuum chamber. The chamber was pumped at 1.3×10^{-5} Pa by a molecular pump. Two kinds of contact materials were chosen, which were CuCr10 (10% of Cr) and CuCr50 (50% of Cr), respectively. The contact diameter was 40 mm. The contact gap was fixed, which was set as 13 mm. A triggered vacuum arc was initiated. When the arc extinguished, a residual voltage remained on C applied on the contacts, which was usually below 100 V.

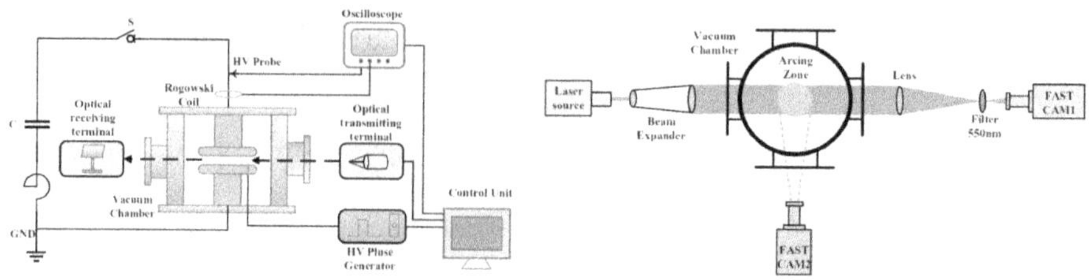

(a)Vacuum arc initiation setup (b) Optical setup for particles observation

Fig. 2.41 Diagram of the experimental setup for the particles induced late breakdown observation

Figure 2.41b shows an optical setup for particles observation. The optical observation is based on a laser-shadow technique. It captures silhouettes of flying droplets and particles in the contact gap around current zero. A laser beam (550 nm, 100 mW) with Gaussian profile was adopted as a light source. The laser beam passed through a beam expander to generate a circular beam spot with a diameter of 50 mm. This circular beam spot size was large enough to cover the contact gap. This circular beam was homogeneous and it collimated through the contact gap. Then it focused with a convex lens with a diameter of 100 mm and a focal length of 1000 mm. Next, the beam passed through a narrow-band filter with a center wavelength of 550 nm. This filter blocked the majority of light emission from the vacuum arcs. Theoretically, only laser light with the wavelength of 550 nm can pass through the filter. The silhouettes of particles can be recorded with a high-speed camera, CAM1. Another high-speed camera, CAM2, placed vertically with respect to the main optical path and it recorded the behaviors of the vacuum arcs. The particles induced late breakdowns can be recorded by the two cameras.

A control unit synchronized the signals between the electrical and optical unit. When the capacitor bank *C* was charged, switch S was closed. Then a high voltage pulse generator initiated a vacuum arc in the fixed contact gap via a trigger electrode, which resulted in a discharging current passing through the circuit. Meanwhile, the optical unit tracked the particles. An oscilloscope recorded the arc voltage and arc current waveforms. The arc voltage was measured by a Tektronix high-voltage probe P6015A (1000:1) and the arc current was measured by a Rogowski Coil.

Figure 2.42 shows particles observed during arcing. Figure 2.42a shows a case of particles observed with contact material of CuCr10 (10% of Cr). Figure 2.42b,c shows two cases of particles observed with contact material of CuCr50 (50% of Cr). Figure 2.42d shows the anode surface erosion after arcing. The arc current was 16 kA (peak). The time zero was current zero. Figure 2.42a shows that there was a liquid metal pool forming on a local anode surface, which developed into a protrusion on the local anode surface at −6.95 ms. Thus, the local anode surface temperature should exceed the melting point of the contact material. Then the protrusion stretched and subsequently broke into particles, rushing towards the cathode. The particle velocities were in a range from 10.8 ms^{-1} to 21.3 ms^{-1}. Figure 2.42b shows that the observed particles became much less with the contact material of CuCr50. There

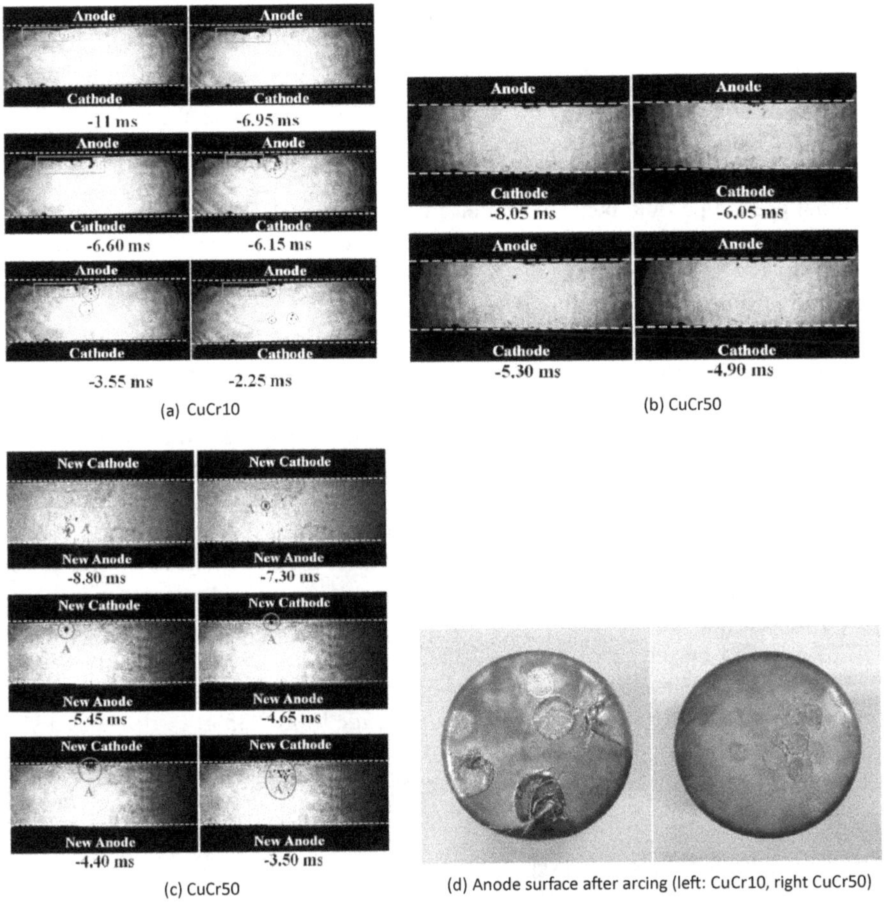

Fig. 2.42 Shadowgraphs of particles observed during arcing. Arc current is 16 kA peak

was also a local melting region on the anode. But the area of the melting region was much less than that of CuCr10. An instability appeared on the melting region of the anode surface at −6.05 ms. Some particles erupted from a tiny spot on the anode, which moved towards the cathode, but returned back toward the anode before reaching the cathode. In addition, some particles attached on the anode from escaping. Figure 2.42c shows another observed pattern of particles with contact material of CuCr50. In this case, the vacuum arc did not extinguish at first current zero. Then in the second half wave of arc current, the electrode polarity reversed. The upper electrode became new cathode and the lower electrode became new anode. A large particle ejected from the new anode. The large particle moved toward the new cathode and it was accompanied by a group of small particles. The large particle collided with the new cathode at a velocity of 65 ms^{-1}. The collision broke the large particle into many small debris particles. The debris conically rushed to the new anode.

Figure 2.42d shows the anode erosion photos after arcing with the contact materials of CuCr10 and CuCr50, respectively. The anode of CuCr50 eroded slightly. However, the anode of CuCr10 eroded heavily. There were several significant erosion pools and liquid flowing tracks appearing on the anode of CuCr10. Some sharp tips can also be found. The anode erosion patterns of the two kinds of contact materials coincided with the particle observations shown in Fig. 2.42a, b and c.

What are the particle behaviors after arc extinguishes? Two examples are given in the following, in both of which the arc current was 16 kA (peak) and the contact material was CuCr10. Figure 2.43 shows a particle behavior after arc extinguishes. The time zero was current zero and the electrode polarity reversed after current zero. Figure 2.43a shows the particles' consecutive shadowgraphs, and Fig. 2.43b shows a graph of the particle trajectory in space. The particle was observed near the new anode 2.50 ms after current zero. It subsequently approached the new cathode with a velocity of 11.28 m s^{-1}. Instead of adhering to the surface when it collided to the new cathode, it rebounded from the new cathode to the new anode with a velocity of 0.43 m s^{-1}. Before reaching the new anode, the particle slowly returned back towards the new cathode. Eventually it adhered to the new cathode.

Figure 2.44 shows another particle behavior after arc extinguishes. Figure 2.44a shows the particles' consecutive shadowgraphs, and Fig. 2.44b shows a graph of the particle trajectory in space. The particle was observed near the new cathode 132.25 ms after current zero. During 132.25 to 203.25 ms, it bounced three times between the new cathode and the new anode. The particle eventually adhered at the new anode surface. Its flight velocity was in a range from 2.35 m s^{-1} to 0.32 m s^{-1}. The low velocity may be related that the recovery voltage much decreased 100 ms after current zero in the experiment. One may expect that a high recovery voltage

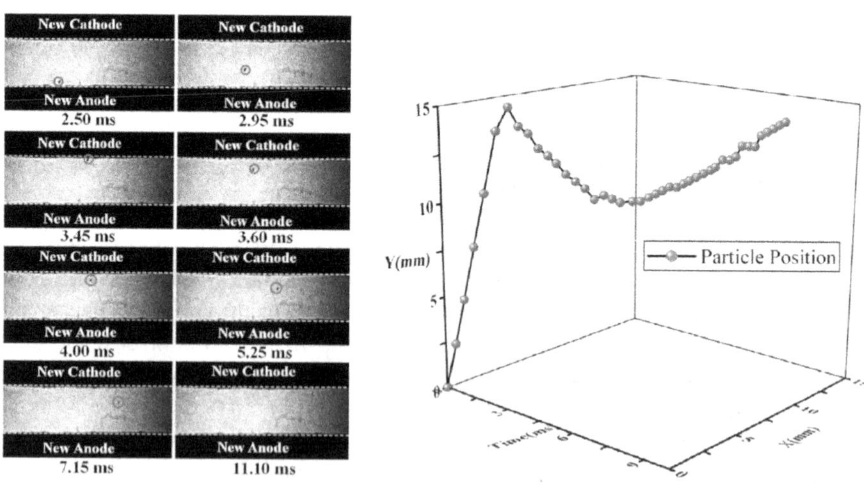

(a) Shadowgraphs showing particle motion

(b) 3D plot of the particle's position

Fig. 2.43 A particle behaviors after arc extinguish

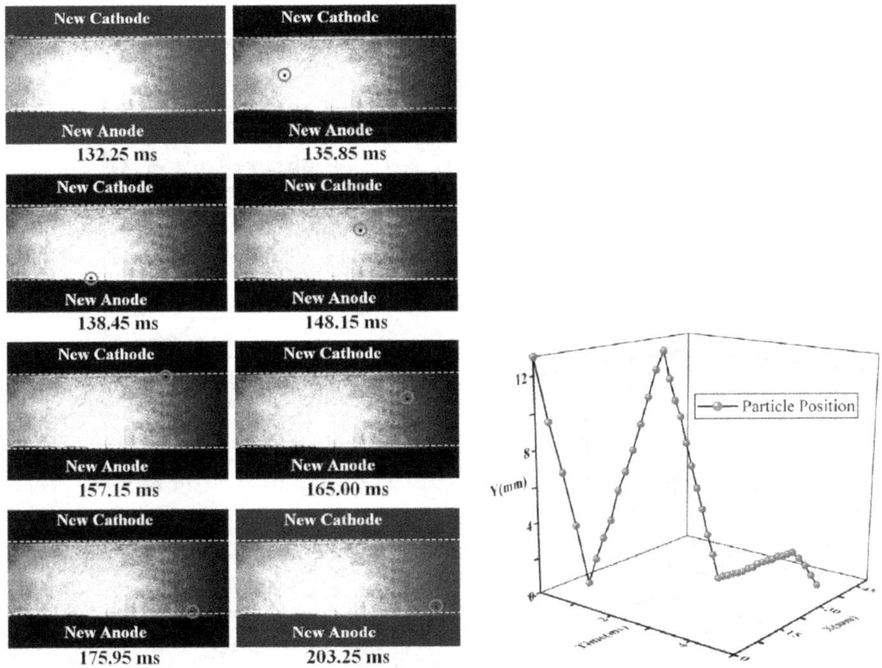

(a) Shadowgraphs showing particle motion

(b) 3D plot of the particle's position

Fig. 2.44 Another particle behavior after arc extinguish

leads to a high particle velocity. In addition, particle did appear in the post-arc gap 132.25 ms after current zero without any sign of pre-existence. This was an evidence that residual particles suddenly appeared in the post-arc gap a long time after current zero. It implied that some particles loosely adhered to the contact surfaces or other inner surfaces of vacuum interrupters. The particles may return back to the post-arc gap, either due to mechanical shocks or due to the electric field force. The late appeared particles would significantly increase the randomness of late breakdowns.

Figure 2.45 shows a protrusion generation after current zero. The arc current was 16 kA (peak) and the contact material was CuCr10. A particle appeared at current zero. Its velocity was quite low and it suspended in the post-arc gap for tens of milliseconds. During the suspension, its volume gradually increased, which implied that the particle was not solid. It should be a liquid droplet since a solid one cannot expand in vacuum. Eventually, it adhered on the new anode. Then it became a protrusion on the new anode. This new born protrusion will enhance the local electric field and increase the possibility of breakdowns (see Sect. 2.3).

Next there are two examples showing that particles directly induced post-arc breakdowns, in which one was an instantaneous breakdown and the other was a late breakdown. Figure 2.46 shows an instantaneous breakdown after current zero. The arc current peak was 12 kA and the contact material was CuCr10 (10% of Cr).

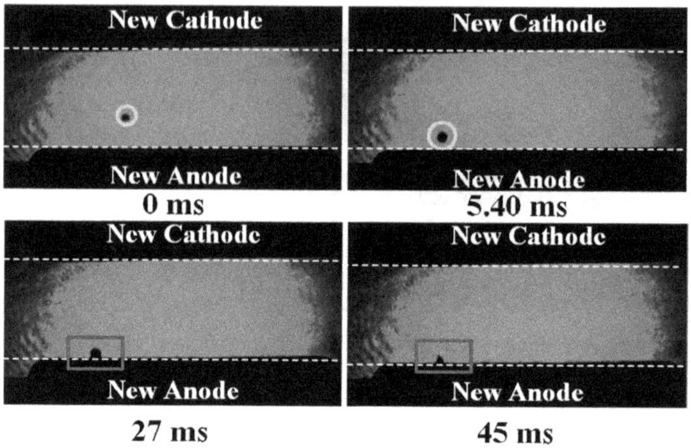

Fig. 2.45 Shadowgraphs showing a protrusion generation caused by a post-arc particle

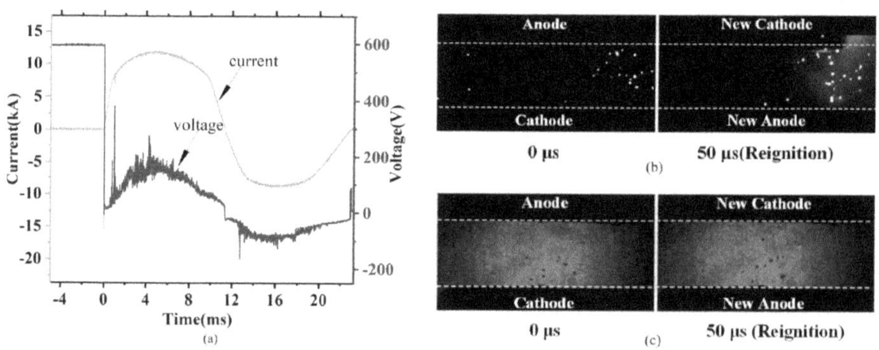

Fig. 2.46 Instantaneous breakdown phenomenon at current zero. **a** Arc current and arc voltage waveforms **b** High-speed camera arc photographs and **c** Shadowgraphs. Contact material: CuCr10; Peak arc current: 12 kA

Figure 2.46a shows the arc voltage and arc current waveforms. Figure 2.46b shows arc images captured by the high-speed cameras. Figure 2.46c shows the shadowgraphs at the corresponding instants. As shown in Fig. 2.46a, the current and arc voltage waveforms continued until the second half wave, indicating an instantaneous breakdown at the first current zero. As shown in Fig. 2.46b, many luminous particles with high temperature were observed in the contact gap at the current zero region. The particles scattered in the contact gap. An instantaneous breakdown was initiated at 50 μs. Figure 2.46c confirmed that numerous particles existed in the contact gap. There were fewer particles presented in the Fig. 2.46c compared to that in Fig. 2.46b. That was related to the observation methods. The high-speed camera images are captured directly with a CCD, which requires the targets emit light by their own. In comparison, the shadowgraph images are generated by using the radiance of a background light source. Moreover, a filter was used to remove interference from self-luminescence. Therefore, some particles or droplets were not seen clearly in Fig. 2.46c. In any way, there were many particles with high temperature existing in the contact gap at the current zero.

Figure 2.47 shows a late breakdown caused by particles. The arc current was 12 kA peak and the contact material was CuCr10 (10% of Cr). Figure 2.47a shows the arc voltage and arc current waveforms. Figure 2.47b shows arc images captured by the high-speed camera. Figure 2.47c shows the corresponding shadowgraphs. As shown in Fig. 2.47a, the vacuum arc extinguished at the first current zero. After that a breakdown occurred 4.8 ms later. Figure 2.47b shows that an incandescent particle appeared at the middle position of the contact gap at the current zero. It approached the new cathode with a velocity of 1.23 m s^{-1}. The particle's radiance implied its temperature was quite high. At 4.70 ms, the particle flew to a position close to the new cathode, and a micro-discharge was observed between the particle and the new cathode. At 4.80 ms, the particle collided with the new cathode and emitted an intense light. Figure 2.47c shows the shadowgraph trajectory of the particle, which verified the observation shown in Fig. 2.47b. The particle's flight duration was 4.80 ms, which coincided with the late breakdown interval shown in Fig. 2.47a. This proved that the late breakdown was triggered by the particle.

Contact materials have a significant influence on particles generation, which is related to the contact erosion pattern. Figure 2.42c shows CuCr10 (10% of Cr) contact

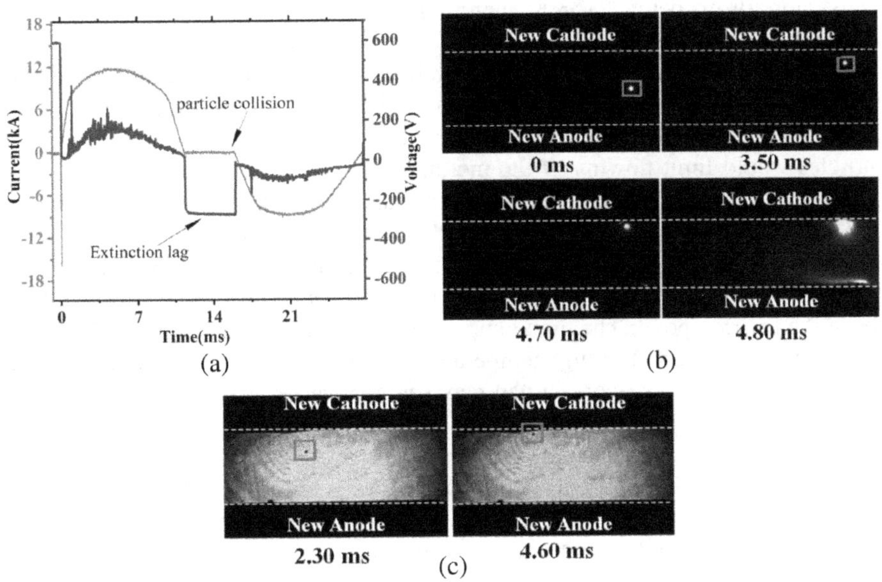

Fig. 2.47 Late breakdown phenomenon induced by a particle after arc extinction. **a** Arc current and arc voltage waveforms **b** High-speed camera arc photographs **c** Shadowgraphs. Contact material: CuCr10; Peak arc current: 12 kA

material was much more eroded by the vacuum arcs than CuCr50 (50% of Cr) contact material. For the arcing with CuCr10 contact material, there was a melting pool forming on the anode. Then protrusions formed on the rim of the melting pool, which may be caused by the blowing effect of the arcs (see Sect. 1.4). After that the rim of the melting pool broke into many particles. A Rayleigh–Plateau instability may explain the particles generation. On the contrary, for the arcing with CuCr50 contact material, there was much less particles than that of CuCr10 contact material. The particles ejected from a tiny spot on the anode, and the momentum of the particles seemed to be insufficient to overcome the pressure of the vacuum arc. Thus, the released particles were pushed back. This implied that the momentum of ejected particles was correlated with the flowing velocity of liquid metal on the anode surface. A large and active melting pool may be able to generate more particles with high energy. Low chromium content contact material tended to be damaged more seriously by high-current anode mode vacuum arcs, forming a broad melting pool on the contact surface. In contrast, high chromium content contact material can effectively impede the flow of metal liquid, which may suppress particles generation.

Moreover, the metallurgical properties of the contact material also affect particles generation. Based on a Cu–Cr binary phase diagram [106], Cu and Cr are totally miscible when the temperature is greater than 2173 K. A solidification process begins as the temperature decreases. Because Cr has a much higher melting point than Cu, the fraction of Cr in the alloy will solidify into particles, mixing with liquid Cu. If the weight fraction of chromium is in a range between 40 and 94.5%, the original uniformly mixed alloy will first form a Cu-rich phase and a Cr-rich phase as the temperature drops below 2173 K, then Cr particles emerge in the Cu liquid as the temperature drops below 2023 K. The size of the Cr-rich particles is related to the cooling rate of the liquid [107, 108]. Generally, a slower cooling rate results in larger particles. In case of power frequency current, the cooling rate is in the order of the power frequency cycle, which is slow enough for solidification. Therefore, large Cr particles would limit flowing of the melting pool. Therefore, the contact material with more Cr content exhibited better performance in reducing particle generation.

The temperature of the particles would be quite high. Figures 2.46b and 2.47b shows the observed particles in the contact gap was quite luminous. Light emission indicated the temperature of the particles. The temperature of the particles would be above the melting point. The high temperature of the particles suggested that some of them were droplets. The high temperature also related to vapor evaporation. The particles scattered in the contact gap may serve as vapor evaporating sources, and the evaporated metal vapor would form a conductive channel. Lins [9] believed that an existing of droplets was a major contributor to the metal vapor density after arc extinguished, even the amplitude of the current was as low as 500 A. The existence of high temperature particles would retard the decay of metal vapor and it may increase the possibility of breakdowns. Moreover, solid particles have temperatures below the melting point. Thus, the solid particles would have a much longer lifetime. When the liquid droplets and solid particles eventually adhered to the contact surfaces or the shields, they would be released by mechanical shocks or by high electric fields.

Therefore, it is interesting to understand how temperature affects the behavior of particles and post-arc breakdowns.

Assuming a particle is floating in vacuum. The particle's temperature is responsible for evaporation and surface radiation. The two physical processes can be expressed as Eq. (2.32) [109].

$$\frac{dT}{dt} = -\left(\frac{3}{\rho_p r C_p}\right)\left[G\left(L + \frac{2k_B T}{m}\right) + \sigma_s\left(T^4 - T_0^4\right)\right] \tag{2.32}$$

where t is time; T is the particle temperature; r is the radius of the particles; C_p is the specific heat; ρ_p is the particle density; m is the atomic mass; T_0 is the ambient temperature, which is set as 298 K; L is the latent heat of evaporation; k_B is the Boltzmann constant; σ_s is the Stefan–Boltzmann constant.

Based on the evaporation principle, a relationship between saturated vapor pressure P and temperature follows Eq. (2.33) [73].

$$P = 10^5 \times \exp\left[-\frac{L}{R}\left(\frac{1}{T} - \frac{1}{T_{boil}}\right)\right] \tag{2.33}$$

where R is the gas constant and T_boil is the boiling temperature. Considering the mass loss due to evaporation, the radius of the particle follows Eq. (2.34).

$$\frac{dr}{dt} = -\frac{G}{\rho} \tag{2.34}$$

where G is the rate of mass loss per unit area. According to the theory of gas molecular motion, the atomic evaporation rate J, which states how many atoms evaporated from the surface per second per unit area, varies with temperature T as Eq. (2.35).

$$J = \frac{P}{\sqrt{2\pi m k_B T}} \tag{2.35}$$

The rate of mass evaporation G is related to pressure P and temperature T, which follows Eq. (2.36).

$$G = m \times J = \sqrt{\frac{m}{2\pi k_B T}} \times P = 4.37 \times 10^{-3} \times \sqrt{\frac{M}{T}} \times P \tag{2.36}$$

where M is the relative atomic mass. The initial temperature of a Cu particle was set to 2835 K (the boiling point) and 1357 K (the melting point), respectively. The initial temperature of a Cr particle was set to 2944 K (the boiling point) and 2180 K (the melting point), respectively. In addition, the particle radii were set to 1 μm, 10 μm, and 100 μm, respectively.

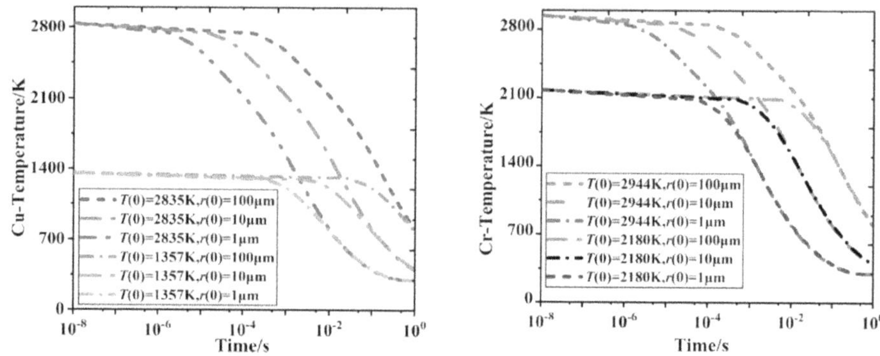

(a)Cu particles (b) Cr particles

Fig. 2.48 Particle temperature decay with various initial particle radii and particle temperatures

Figure 2.48 shows Cu and Cr particles temperature decay. It shows that the particle size of both Cu and Cr has a significant impact on the temperature decay. The temperature of small particles decayed faster than that of larger ones. It is interesting that the particles temperature decay converged together after a short time if the particles had same size. The convergence time depends on the size of the particle. The convergence time was several milliseconds for particles radius of 1 μm, and the convergence time rose to dozens of milliseconds for particles radius of 100 μm. This indicated that the particle size rather than the initial temperature was a more significant influencing factor on the particle temperature decay. The experiment also found particles existing several or dozens of milliseconds after current zero.

Particles have two effects on breakdowns during the post-arc period. The first effect could induce a micro-discharge when a particle approaches a contact surface. Sarrailh et al. estimated the metal vapor density could achieve 10^{22} m^{-3} when the contact surface temperature reached 2000 K [73]. If the temperature of a particle is 2000 K and the particle approaches close enough to a contact surface, the metal vapor density in this small gap will reached 10^{22} m^{-3}. Such high metal vapor density will reach a threshold density of breakdown (see Sect. 1.4.4). Moreover, when the particle approaches quite close a contact surface, the electric field between the particle and the contact surface would be quite high. As a result, a micro-discharge would easily form. The micro-discharge may lead to a full discharge across the contact gap, as shown in Fig. 2.47. It shows a light emitted from a region between the particle and the new cathode at 4.80 ms, then a full breakdown occurred.

The second effect could change contact surface conditions by the molten droplets adhering effect. As shown in Fig. 2.45, a particle fell on the surface of new anode. It gradually solidified, then it merged with the surface as a tip. The tip enhances the local electric field. Thus, breakdowns became much easier with the tip than that on a smooth surface. A Taylor cone forms from a molten contact surface under a transient recovery voltage within several milliseconds (see Sect. 2.3). It is notable that Taylor cone formation was much less than 45 ms indicated in Fig. 2.45. This implies that the contact surface conditions may change over a long time after arc extinguishes, except the Taylor cone effect. Consequently, breakdowns could be triggered by the contact surface conditions changing long after arc extinguishes.

The experimental results revealed two kinds of particle motions after arc extinguishes, which was shown in Figs. 2.43 and 2.44, respectively. Figure 2.44 showed that the particle bounced between the contacts with collision. In this case it is most probably a charged particle, whose bounce was driven by the electric field. Its initial polarity would be correlated with the contact surface from which it originated. A particle ejected from an anode will carry positive charge, and a particle ejected from a cathode will carry negative charge. The particle would change its charge polarity in two ways. One way is that it directly contacts the opposite electrode. The other way is via a field emission when it closely approaches the opposite electrode. The charging process may reduce or accumulate charge on the particle, depending on it approaches to anode or cathode and itself charge polarity. For instance, when a negatively charged particle is ejected from the cathode, field emission may occur when it approaches the anode, resulting in a strong local electric field. Field emission results in electron loss and, under some circumstances, the net charge could reverse polarity. As a result, the particle would rebound from the anode and flies back to the cathode. In the same way, it flies back to the anode again. That explains the particles bouncing between the contacts, as shown in Fig. 2.44. Figure 2.43 showed that the particle flied from the new anode to the new cathode and it finally suspended close to the new cathode. The particle probably carries net positive charge when it leaves the new anode. After colliding with the cathode, it takes some negative charge from the cathode. But the collected negative charge may be insufficient to reverse its polarity. Thus, it may suspend near the cathode due to a balance of gravity and the Coulomb force.

Figure 2.49 shows a simulation model for particle induced local electrical field enhancement effect. It shows a particle reached to a close position to the lower contact with a distance of d. The upper contact is anode, on which a voltage of U_{upper}

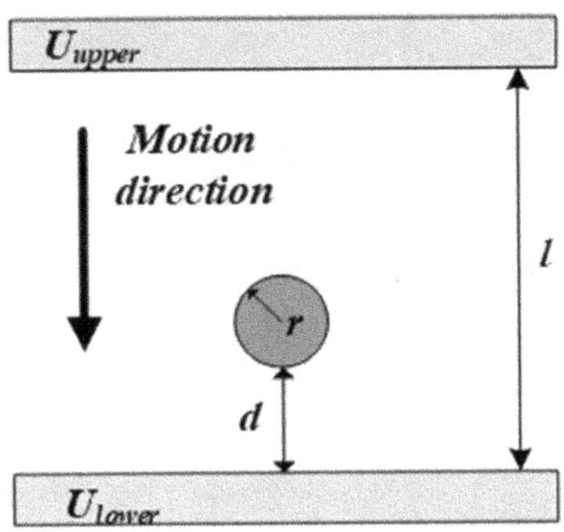

Fig. 2.49 A simulation model for particle induced local electrical field enhancement effect

is applied. The lower contact is cathode, which is grounded. The radius of the contact edge R is set to 5 mm. The contact gap l is set to 10 mm. The radius of particle r is set to 150 μm, which is much less than the contact gap length and electrode radius, thus the fringe effects can be neglected. Charges carried by a spherical particle in a uniform electrostatic field follows Eq. (2.37) [110, 111].

$$Q = -\frac{2\pi^3 \varepsilon_0 r^2 U_{upper}}{3l} \tag{2.37}$$

where the permittivity in vacuum ε_0 is 8.85×10^{-12} F m^{-1}. The voltage across the contact gap is normalized to 1 V. If there are no particles in the contact gap, the electric field distribution would be uniform except at the edges. The uniform electric field is E_0. Therefore, calculated electric field strength can be taken as per unit value normalized to the E_0.

Figure 2.50 shows the electric field distribution and electric strength when a particle approaches the cathode. It shows the maximum electric field strength E_{max} located at the bottom of the particle. Figure 2.50a, b and c shows three cases of d/r as 3/5, 2/5 and 1/5, respectively. The E_{max} increased as d decreased, which was 5.90, 6.79 and 8.99, respectively. Figure 2.50d shows that when d/r was less than 0.5, E_{max} increased rapidly and tended towards infinity. It suggests that E_{max} would exceed the breakdown electric field strength threshold value when the particle approaches close enough to the cathode, which triggers a micro-discharge. Such a micro-discharge may cause intense metal vapor evaporation and ionization, eventually it possibly led to breakdowns across the whole contact gap, either an instantaneous breakdown shown in Fig. 2.46 or a late breakdown shown in Fig. 2.47. In such cases, breakdowns may occur at low voltages.

Apart from the local electric field enhancement effect, the particles bouncing is also meaningful to breakdowns. Latham [112] proposed that a particle can gain energy continuously during the period of oscillation, and its velocity increased until it reached a critical value to trigger breakdowns. He also provided a criterion for acceleration, as shown in Eq. (2.38).

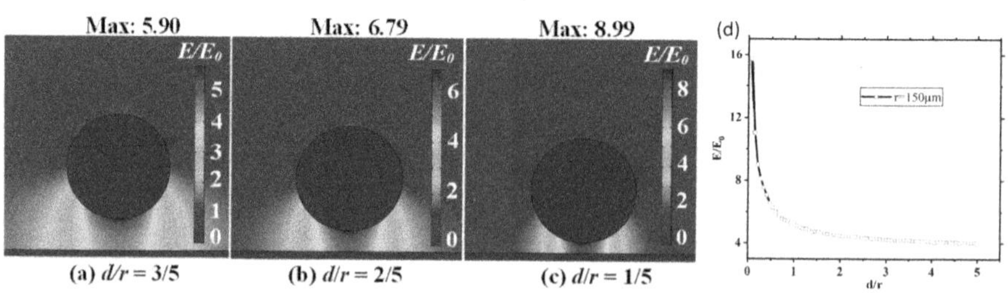

Fig. 2.50 Electric field distribution and electric strength when a particle approaches the cathode

$$c^2 + \left| \frac{Q_{p2}}{Q_{p1}} \right| > 1 \tag{2.38}$$

where c is the collision coefficient between the particle and the target contacts; Q_{p1} is the charge before collision, and Q_{p2} is the charge after collision. The proposed mechanism can be used to interpret breakdown under relatively low voltages. The ability to gain energy primarily depended on how much charge the particle can take after collision, which was strongly related to local processes, e.g. micro-discharges. Moreover, the contact surface conditions combined with other impact factors made particles bouncing difficult to predict and to control. This unpredictable bouncing behavior implies that there is a certain probability for an electric field to provide a sufficient high energy to a particle, eventually leading to breakdowns. If the applied voltage is not adequate to trigger breakdowns directly, particles bouncing would assist to induce breakdowns at low voltages.

The experimental results suggested that the particles generation seemed to be inevitable, especially in the case of high current. Particles' lifetime spans up to hundreds of milliseconds. The existence of particles is crucial for late breakdowns. Metal vapor, residual plasma, or hot droplets cannot fully explain the late breakdowns hundreds of milliseconds after current zero. Their time scale is much less than that of late breakdowns. Breakdowns in vacuum are strongly related with the contact surface conditions. The micro-protrusions on contact surfaces determine the local electric fields and electron emissions, which may assist particles induced breakdowns. Therefore, particles seem to be a good candidate that is responsible for late breakdowns. Particles may appear quite late after arc extinguishes. In Fig. 2.44 it suddenly appeared about 132 ms after current zero. Particles generated during arcing may loosely adhere to the inner surfaces of vacuum interrupters' parts. Some external factors, such as mechanical shocks, may trigger the particles detaching from the surfaces. It is reasonable to expect that the detaching time scale is much longer than that of the above-mentioned physical processes. This may explain the time scale of late breakdowns.

2.4.3 Mechanical Shocks and Late Breakdowns

This section reveals an influence the mechanical shocks on late breakdowns in current interruptions in vacuum. The phenomena were investigated under capacitive switching. Many works proved that capacitive switching was an appropriate field for investigation of late breakdowns in vacuum. A special vacuum interrupter was designed for the investigation. Figure 2.51 shows a schematic of the vacuum interrupter with a fixed break that electrically shields the movable break in the full open position. In the vacuum interrupter a floating shield and a terminal shield composed a fixed break. From the point view of electric field, the fixed break shielded the movable break, which composed of the movable contact and the stationary contact. Thus, it

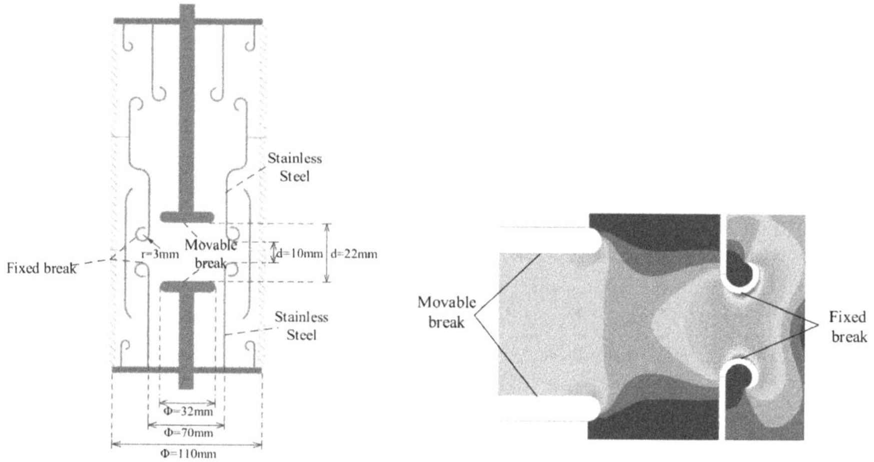

Fig. 2.51 A vacuum interrupter with a fixed break electrically shielding the movable break

weakened the electric field strength on the movable contact surface and the stationary contact surface. The shielding effect released the movable break from withstanding the recovery voltage, therefore the influence of the field emission current on the late breakdowns would be reduced. The capacitive switching test include a making operation and a breaking operation. During the making operations, an inrush current (20 kA peak, 4250 Hz) flowed through the movable break. Then, during the breaking operations, the movable break was shielded by the fixed break. The floating shield and the other terminal shield formed the second fixed break, which was electrically in series with the fixed break shielding the movable break. The double fixed breaks design aimed to decrease the possibilities of breakdowns passing through the path across the shields. The shields were made of stainless steel. The full contact gap of the movable break was 22 mm. The gap distance of the fixed break shielding the movable break was 10 mm. The diameter of the movable contact and the stationary contact were 32 mm. The contact material was made of CuCr25 (25% of Cr). The gap between the movable break and the fixed break was 19 mm. The diameter of the vacuum interrupter was 110 mm.

Figure 2.52 shows a synthetic circuit for capacitive switching. The test circuit is composed of two sub-circuits: an inrush current source circuit and a power frequency power source circuit. The inrush current source circuit provided the inrush current in making operation. The power frequency power source circuit provided a power frequency current and a recovery voltage after the power frequency current was interrupted.

The making operation and the breaking operation was done separately. The making operation was as following. C was pre-charged to a voltage. Then close SW_{inrush} and SW_{test} in sequence. In such way C is discharged through L. An inrush current up with a peak value of 20 kA at a frequency of 4250 Hz flowed through the test vacuum interrupter. Figure 2.53 shows a typical inrush current waveform in the

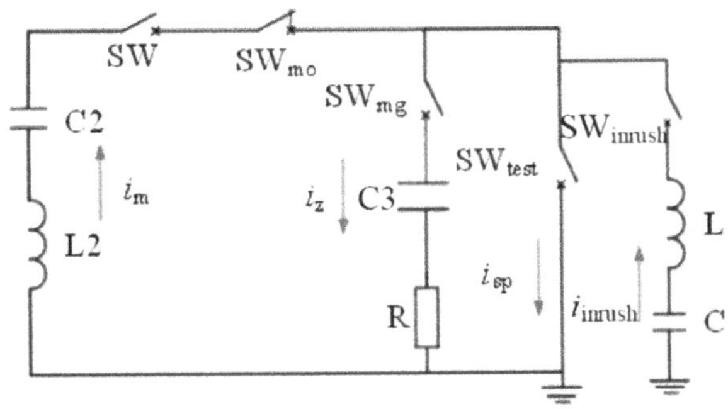

Fig. 2.52 A synthetic test circuit for capacitive current switching. SW_{test} : test vacuum interrupter; Inrush current source: C-pre-charged capacitor; L-inductance; SW_{inrush}-switch; power frequency power source: C2-pre-charged capacitor; C3-capacitor; L2-inductance; R-resistance; SW-closing switch; SW_{mo}-opening switch; SW_{mg}-isolating switch

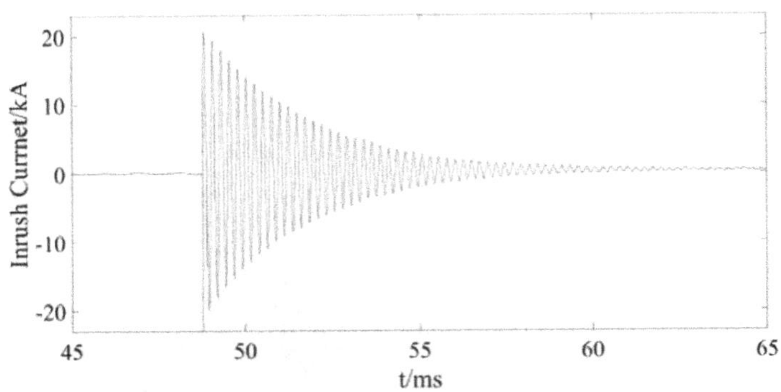

Fig. 2.53 A typical waveform of inrush current

making operation. After the making operation, SW_{inrush} is set to open position and SW_{test} is set in close position.

The breaking operation was done immediately after the making operation. C2 was pre-charged to a voltage. SW and SW_{mg} were in open position, while SW_{mo} and SW_{test} were in close position. Then close SW. In such way, C2 is discharged through L2. A power frequency current I_m(400 A r.m.s, 50 Hz) flows through the tested vacuum interrupter. SW_{test} was opened according to a pre-set arcing time. SW_{mg} was closed before current zero of the power frequency current. After current zero, C3 is charged by C2 through L2 and R with a current of I_z. Before I_z current zero, open SW_{mo}. Thus the voltage of C3 is applied across the test interrupter as a recovery voltage. Generally, a standard capacitive recovery voltage is of (1-cosωt) waveform. The recovery voltage generated in this circuit was a slowly decaying DC voltage with a rising time of 4 ms and a peak value of 82.7 kV. Figure 2.54 shows a typical waveform of power frequency current and the recovery voltage. This is a

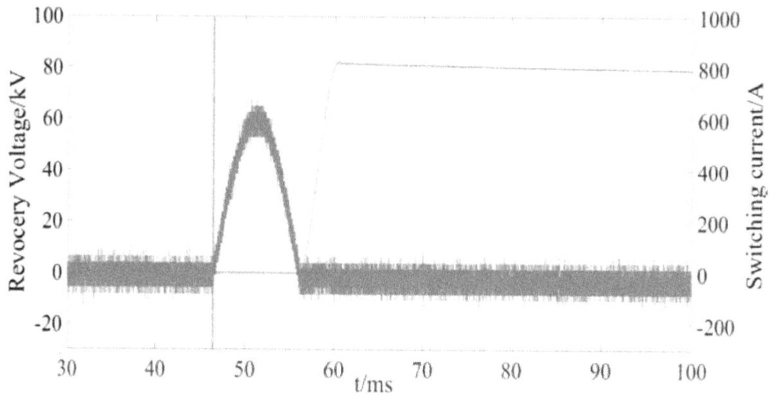

Fig. 2.54 A typical waveform of capacitive switching current and recovery voltage

more severe condition for 40.5 kV class capacitive switching than that specified in the IEC standard.

Figure 2.55 shows a photo of vacuum circuit breaker, in which the test vacuum interrupters were assembled. A spring operating mechanism was used to drive the test vacuum interrupter. The measured average closing speed was 0.9 m/s and the average opening speed was 1.9 m/s. The average closing speed was an average speed

Fig. 2.55 Photo of the test vacuum circuit breaker

Table 2.3 Three test series

Test series		I	II	III
Inrush current	Amplitude	20kA		
	Frequency	4250 Hz		
Switching current		400A(RMS)	0 A	
dt		7 ms		1 min

in the last half contact travel before the contacts mate and the average opening speed was an average speed in the first half contact travel when the contacts open. The test vacuum interrupters were immerged in oil to protect their external insulation. Pole A was the tested pole. An acceleration of the movable contact of Pole A was measured. An accelerometer with type MEMSIC CXL25GP3, was used to measure the acceleration of the movable contact, which is shown in an enlarged photo in Fig. 2.55. It was a triaxial acceleration sensor with a measuring range up to ±25 g. The output were three voltage signals representing acceleration of the three axes. The sensitivity of each axis was 0.081 V, which corresponded to 0.81 g. In order to determine the moment of contacts separation of Pole A, a battery applied a DC voltage on Pole C through a resistor in series. By measuring the voltage on the resistor, one can obtain the moment of contacts separation of Pole C. The non-synchronization between contacts separation of Pole A and Pole C was measured and the maximum one was less than 0.3 ms. So that the moment of contacts separation of Pole A can be determined.

Table 2.3 shows there are three test series. In each test series there were 24 making and breaking operations on the test vacuum interrupter. During the three test-series, the inrush current was 20 kA peak and the inrush current frequency was 4250 Hz. During the breaking operation, there set three intervals dt, which was the interval between contacts separation and the application of recovery voltage. In test series I, the breaking current was 400A r.m.s and the arcing time was 7 ms. The recovery voltage was applied at current zero, so that dt was 7 ms in test series I. When the recovery voltage reached its peak, the contact gap reached its full contact gap of 22 mm. In test series II, the breaking current was 0 A. The recovery voltage was applied 7 ms after contacts separation, thus dt was also 7 ms in test series II. Again, the contact gap reached its full contact gap at the recovery voltage peak. In test series III, the breaking current was 0 A and dt was set to 1 min, which was a long enough duration for mechanical shocks eliminating when the recovery voltage was applied.

Figure 2.56 shows a typical waveform of a breaking failure in capacitive switching. t1, t2 and t3 are moments of contacts separation, current zero and restrike, respectively. The arc current was 400 A. Before current zero, the contacts separated at t1. At current zero instant t2, a slowly decaying DC recovery voltage was applied. The recovery voltage has a rising time of 4 ms and a peak value of 82.7 kV. At t3 there was a late breakdown. So that the voltage waveform collapsed. This is an indication of breaking failure.

Table 2.4 summarized the experimental results of the making and breaking oper-

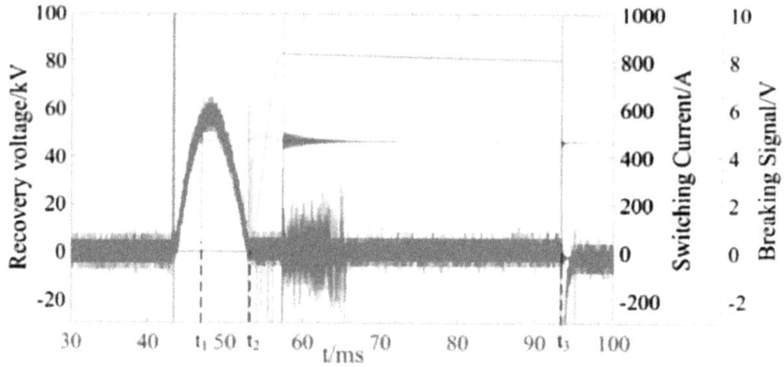

Fig. 2.56 A typical waveform of a failure breaking in capacitive switching

Table 2.4 Experimental results of the making and breaking operations

Test series	I	II	III
Total number of operations	24		
Number of late breakdowns	3	3	0
Restrike probability	12.5%	12.5%	0

ations for the three test-series. It shows that the late breakdown probability for test series I and test series II were 12.5%. The late breakdown probability for test series III was 0%.

Figure 2.57 shows a typical acceleration curve of the moving contact during the breaking operations. The acceleration measurement curves were almost same. Two dotted lines marked measuring limits of the accelerometer. The figure also marked the late breakdown events in test series I and II. The time zero was the t2 in Fig. 2.56. In test series I, three late breakdowns occurred, which were at 28.7 ms, 46.4 ms, and 72.1 ms, respectively; In test series II, there were also three late breakdowns occurring, which were at 18.4 ms, 55.1 ms, and 146.5 ms, respectively. For test

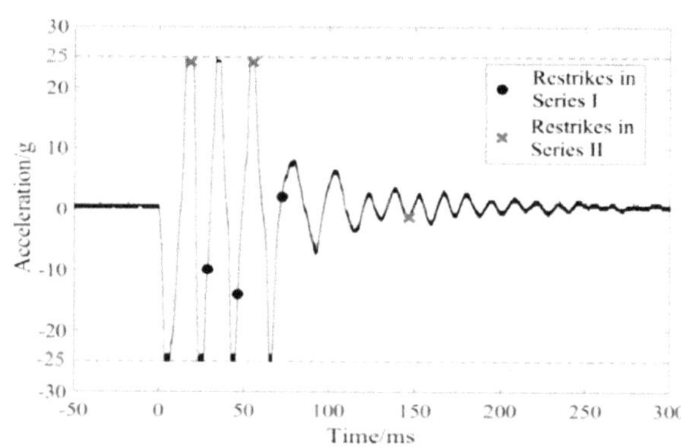

Fig. 2.57 Acceleration curve of the moving contact when late breakdowns occurred

series II, two out of the three late breakdowns occurred at the positive acceleration peaks. One late breakdown occurred when the acceleration was quite weak. For test series I, there was no obvious correlation between the accelerations and late breakdowns.

The experimental results implied that mechanical shocks had a significant impact on late breakdowns. Among the three test-series, test series I and II showed some late breakdowns, while test series III showed no late breakdowns occurring. In test series III the mechanical shocks disappeared 1 min later, so the applied recovery voltage can not induce late breakdowns. The late breakdowns occurred in test series I and II seemed to be induced by mechanical shocks. In test series II, the late breakdowns may correlate to the acceleration peaks. In test series I there seemed no such a correlation. The breaking current of 400A in test series I have some conditioning effects on the contact surfaces, which may have an impact on the correlation. During the making operation of inrush current, the movable contact surface and the stationary contact surface will be locally melted and it may cause a contact welding when the contacts mated. During the following breaking process, the locally welded area would be ruptured and then dielectric strength between the contacts would be deteriorated. By the electrically shielding effect of the fixed break, the electric field strength on the movable contact surface and the stationary contact surface was decreased. It will reduce the field emission current. In addition, it weakens the electric field force to drive the particles adhered on movable contact surface and the stationary contact surface. In such conditions, mechanical shocks appear to be a most possible cause for detaching and driving particles to induce late breakdowns in current interruptions in vacuum, especially when it suffers a continuous electric field stress.

References

1. Farrall G A, "Recovery of Dielectric Strength after Current Interruption in Vacuum", IEEE Trans. Plasma Sci. Vol. 6, No. 4, pp.360–369, 1978.
2. Schade E, and Dullni E, "Recovery of breakdown strength of a vacuum interrupter after extinction of high currents", IEEE Trans. Dielectr. Electr. Insul. Vol. 9, No. 2, pp.207–215, 2002.
3. Dullni E, Schade E, and Gellert B, "Dielectric Recovery of Vacuum Arcs after Strong Anode Spot Activity", IEEE Trans. Plasma Sci. Vol. 15, No. 5, pp. 538–544, 1987.
4. Yanabu S, Homma M, Kaneko E, and Tamagawa T, "Post Arc Current of Vacuum Interrupters", IEEE Trans. Power App. Syst. Vol. 104, No. 1, pp. 166–172, 1985.
5. Dullni E, and Schade E, "Investigation of High-Current Interruption of Vacuum Circuit-Breakers", IEEE Trans. Electr. Insul. Vol. 28, No. 4, pp. 607–620, 1993.
6. Lins G, "Measurement of the Neutral Copper Vapor Density around Current Zero of a 500-A Vacuum-Arc Using Laser-Induced Fluorescence", IEEE Trans. Plasma Sci. Vol. 13, No. 6, pp. 577–581, 1985.
7. Lins G, "Influence of Electrode Separation on Ion Density in the Vacuum-Arc", IEEE Trans. Plasma Sci. Vol. 19, No. 5, pp. 718–724, 1991.
8. Rich J A, and Farrall G A, "Vacuum Arc Recovery Phenomena", Proceedings of the IEEE, Vol. 52, No. 11, pp. 1293–1301, 1964.
9. Lins G, "Evolution of Copper Vapor from the Cathode of a Diffuse Vacuum-Arc", IEEE Trans. Plasma Sci. Vol. 15, No. 5, pp. 552–556, 1987.

10. Boxman R L, and Goldsmith S, "The Interaction between Plasma and Macroparticles in a Multi-Cathode-Spot Vacuum-Arc", J Appl Phys, Vol. 52, No. 1, pp.151–161, 1981.
11. Gellert B, Schade E, and Dullni E, "Measurement of Particles and Vapor Density after High-Current Vacuum Arcs by Laser Techniques", IEEE Trans. Plasma Sci. Vol. 15, No. 5, pp. 545–551, 1987.
12. Lins G, "Measurement of the Temperature of Evaporating Macroparticles after Current Zero of Vacuum Arcs", IEEE Trans. Plasma Sci. Vol. 16, No. 4, pp. 433–437, 1988.
13. Kimblin C W, "Erosion and Ionization in Cathode Spot Regions of Vacuum Arcs", J. Appl. Phys. Vol. 44, No. 7, pp. 3074–3081, 1973.
14. Schulman M B, Slade P G, and Bindas J A, "Effective Erosion Rates for Selected Contact Materials in Low-Voltage Contactors", IEEE Trans Compon. Pack. A, Vol. 18, No. 2, pp. 329–333, 1995.
15. Takahashi S, Arai K, Morimiya O, Hayashi K, and Noda E, "Laser measurement of copper vapor density after a high-current vacuum arc discharge in an axial magnetic field", IEEE Trans. Plasma Sci. Vol. 33, No. 5, pp. 1519–1526, 2005.
16. Frind G, Carroll J J, and Tuohy E J, "Recovery Times of Vacuum Interrupters Which Have Stationary Anode Spots", IEEE Trans. Power App. Syst. Vol. 101, No. 4, pp. 775–781, 1982.
17. Duning G, and Lindmayer M, "Plasma density decay of vacuum discharges after current zero", Proc. Xviiith International Symposium on Discharges and Electrical Insulation in Vacuum (ISDEIV), Eindhoven, The Netherlands, pp. 447–454, 1998.
18. van Lanen E P A, Popov M, van der Sluis L, and Smeets R P P, "Vacuum circuit breaker current-zero phenomena", IEEE Trans. Plasma Sci. Vol. 33, No. 5, pp. 1589–1593, 2005.
19. Gilles R, Weltmann K D, Schade E, and Claessens M, "Determination of the residual charge after current extinction-An integral approach", Proc. Xixth International Symposium on Discharges and Electrical Insulation in Vacuum (ISDEIV), Xi'an, China, pp. 481–484, 2000.
20. Birdsall C K, "Particle-in-Cell Charged-Particle Simulations, Plus Monte-Carlo Collisions with Neutral Atoms, Pic-Mcc", IEEE Trans. Plasma Sci. Vol. 19, No. 2, pp. 65–85, 1991.
21. Kutzner J, and Miller H C, "Integrated Ion Flux Emitted from the Cathode Spot Region of a Diffuse Vacuum-Arc", J Phys D Appl Phys, Vol. 25, No. 4, pp. 686–693, 1992.
22. Sarrailh P, Garrigues L, Hagelaar G J M, Sandolache G, Rowe S, Jusselin B, and Boeuf J P, "Expanding sheath in a bounded plasma in the context of the post-arc phase of a vacuum arc", J. Phys. D Appl. Phys. Vol. 41, 015203 (11 pp.), 2008.
23. Kutzner J, and Miller H C, "Ion Flux from the Cathode Region of a Vacuum-Arc", IEEE Trans. Plasma Sci. Vol. 17, No. 5, pp. 688–694, 1989.
24. Boxman R L, Sanders D M, and Martin P J, Handbook of Vacuum Arc Science and Technology: Fundamentals and Applications, Noyes Publications: New Jersey, pp. 198–199, 1995.
25. Tkachev A N, Fedenev A A, and Yakovlenko S I, "Townsend coefficient, escape curve, and fraction of runaway electrons in copper vapor", Laser Physics, Vol. 17, No. 6, pp. 775–781, 2007.
26. Z. Wang, and J. Wang, "Distribution of electric field in vacuum interrupter during interrupting", Proc.Xixth International Symposium on Discharges and Electrical Insulation in Vacuum (ISDEIV), Xi'an, China, pp. 471–474, 2000.
27. Boxman R L, "Interferometric measurement of electron and vapor densities in a high-current vacuum arc," Journal of Applied Physics, vol. 45, pp. 4835–4846, 1974.
28. Schellekens H, "The high-current vacuum arc in an axial magnetic field: An experimental investigation," Journal of Applied Physics, vol. 54, pp. 144–149, 1983.
29. Lisnyak M, Pipa A V, Gorchakov S, Iseni S, Franke S, Khapour A, Methling R and Weltmann K-D. J. Appl. Phys. 118, 123304, 2015.
30. Jenkins J E, Sherman J C, Webster R, and Holmes R, "Measurement of the neutral vapour density decay following the extinction of a high-current vacuum arc between copper electrodes," Journal of Physics D: Applied Physics, vol. 8, p. L139, 1975.
31. Lins G, "Collisional transfer and neutral copper vapour density during a diffuse vacuum arc," Journal of Physics D: Applied Physics, vol. 23, p. 784, 1990.

32. Gellert B, "Measurement of high copper vapour densities by laser induced fluorescence," Journal of Physics D: Applied Physics, vol. 21, p. 710, 1988.
33. Amorim J, Baravian G, and Jolly J, "Laser-induced resonance fluorescence as a diagnostic technique in non-thermal equilibrium plasmas," Journal of Physics D: Applied Physics, vol. 33, p. R51, 2000.
34. Britun N, Palmucci M, Konstantinidis S, and Snyders R, "Particle visualization in high-power impulse magnetron sputtering. I. 2D density mapping," Journal of Applied Physics, vol. 117, p. 163302, 2015.
35. Wang H, Wang Z, Zhou Z, Jiang Y, Wang J, Geng Y, and Liu Z, "Deformation of contact surfaces in a vacuum interrupter after high current interruptions," Journal of Applied Physics, vol. 120, p.053303, 2016.
36. Wang Z, Wang H, Zhou Z, Tian Y, Geng Y, Wang J, Liu Z, "Fully kinetic model of breakdown during sheath expansion after interruption of vacuum arcs," Journal of Applied Physics, vol. 120, p.083301, 2016.
37. Mitchell A C G and Zemansky M W, Resonance Radiation and Excited Atoms, Cambridge: Cambridge University Press, 2009.
38. Reuter S, Sousa J S, Stancu G D and Helden J-P H V, Plasma Sources Sci. Technology, **24**, 054001, 2015.
39. Britun N, Minea T, Konstantinidis S and Snyders R, J. Phys. D: Appl. Phys. **47**, 224001, 2014.
40. Ruhrmann C, Westermeier M, Höbing T, Bergner A, Denissen C, Suijker J, Awakowicz P and Mentel J, J. Phys. D: Appl. Phys. **46**, 295202, 2013.
41. Thorne A, Litzén U and Johansson S. Spectrophysics: Principles and applications (Berlin: Springer), 1999.
42. Martin K, Gilles C and Nader S, J. Phys. D: Appl. Phys. **37**, 1954, 2004.
43. Kramida A, Ralchenko Y, Reader J and NIST ASD Team. NIST Atomic Spectra Database Version 5.5.1 (Gaithersburg, MD: National Institute of Standards and Technology) (https://physics.nist.gov/asd), 2017.
44. Miller H C, IEEE Trans. Plasma Sci. 11, 122–127, 1983.
45. Slade P G, The Vacuum Interrupter: Theory, Design, and Application, Boca Raton, FL: CRC Press, 2008.
46. Burkhard J and Ingmar K, J. Phys. D: Appl. Phys. **33**, 2025, 2000.
47. Gellert B and Egli W, J. Phys. D: Appl. Phys. **21**, 1721–6, 1988.
48. Sigmund P and Szymonski M, Appl. Phys. A **33**, 141–52, 1984.
49. Shkol'nik S M, IEEE Trans. Plasma Sci. No.31, pp. 832–846, 2003.
50. Sandolache G and Rowe S W, in Proceedings of the 22nd International Symposium on Discharges and Electrical Insulation in Vacuum, pp. 13–16, 2006.
51. Proskurovsky D I, IEEE Trans. Plasma Sci. No.37, pp.1348, 2009.
52. Taylor G and McEwan A, J. Fluid Mech. 22, 1, (1965).
53. Orlov A M, Yavtushenko I O, Rynkova O G, Boronenko M P, Bodnarskii D S, and Solov'ev A A, Tech. Phys. Lett. 41, 221, 2015.
54. Xie J, Jiang J, Davoodi P, Srinivasan M P, and Wang C H, Chem. Eng. Sci. 125, 32, 2015.
55. Xu Q, Qin H, Yin Z, Hua J, Pack D W, and Wang C H, Chem. Eng. Sci. 104, 330, 2013.
56. Xia X, Wang X, Zhou H, Niu X, Xue L, Zhang X, and Wei Q, Electrochim. Acta 121, 345, 2014.
57. Wang C, Wang L, and Wang M, Mater. Lett. 124, 192, 2014.
58. Iida T and Guthrie R I, The Physical Properties of Liquid Metals, Clarendon Press, 1988.
59. Olsson E, Kreiss G, and Zahedi S, J. Comput. Phys. 225, 785, 2007.
60. Reznik S N, Yarin A L, Theron A, and Zussman E, J. Fluid Mech. 516, 349, 2004.
61. Lee B S, Cho H J, Lee J G, Huh N, Choi J W, and Kang I S, J. Colloid Interface Sci. 302, 294, 2006.
62. Murphy E L and Good R H, Phys. Rev. 102, 1464 (1956).
63. Benilov M S and Benilova L G, J. Appl. Phys. 114, 063307, 2013.
64. Andrews J G and Varey R H, Phys. Fluids 14, 339, 1971.
65. Holmes R and Yanabu S, J. Phys. D: Appl. Phys. 6, 1217, 1973.

66. Childs S E and Greenwood A N, IEEE Trans. Plasma Sci. 8, 289, 1980.
67. Childs E, Greenwood A N, and Sullivan J S, IEEE Trans. Plasma Sci. 11, 181, 1983.
68. Tskhakaya D, Matyash K, Schneider R, and Taccogna F, Contrib. Plasma Phys. **47**, 563, 2007.
69. Verboncoeur P, Plasma Phys. Controlled Fusion **47**, A231, 2005.
70. Mo Y, Shi Z, Jia S, and Wang L, Phys. Plasmas 22, 023511, 2015.
71. Changho Y, Won N, and Moohyun C, Plasma Sources Sci. Technol. 25, 025009, 2016.
72. Sarrailh P, Garrigues L, Hagelaar G J M, Sandolache G, Rowe S, Jusselin B, and Boeuf J P, J. Phys. D: Appl. Phys. **41**, 015203, 2008.
73. Sarrailh P, Garrigues L, Hagelaar G J M, Boeuf J P, Sandolache G, and Rowe S, J. Appl. Phys. 106, 053305, 2009.
74. Sarrailh P, Garrigues L, Boeuf J P, and Hagelaar G J M, Plasma Sources Sci. Technol. 19, 065020, 2010.
75. Cartwright K, Verboncoeur J, and Birdsall C, Phys. Plasmas **7**, 3252, 2000.
76. Anders A, Ion charge state distributions of vacuum arc plasmas: The origin of species, Physical Review E 55(1), pp .969, 1997.
77. Boxman R L, Sanders D M, and Martin P J, Handbook of Vacuum Arc Science and Technology: Fundamentals and Applications Noyes Publications, New Jersey), 288, pp. 89, 1995.
78. Fowler R H and Nordheim L, Proc. R. Soc. London, Ser. A **119**, pp.173, 1928.
79. Timko H, Matyash K, Schneider R, Djurabekova F, Nordlund K, Hansen A, Descoeudres A, Kovermann J, Grudiev A, and Wuensch W, Contrib. Plasma Phys. 51, 5, 2011.
80. Timko H, Ness Sjobak K, Mether L, Calatroni S, Djurabekova F, Matyash K, Nordlund K, Schneider R, and Wuensch W, Contrib. Plasma Phys. 55, 299, 2015.
81. Trajmar S, Williams W, and Srivastava S, J. Phys. B: At. Mol. Phys. 10, 3323, 1977.
82. Perkins S, Cullen D, and Seltzer S, "Tables and graphs of electron interaction cross sections from 10 eV to 100 GeV derived from the LLNL Evaluated Electron Data Library (EEDL), Z=1–100," Technical Report UCRL-50400, Vol. 31, 1991.
83. Mason E A and McDaniel E W, Transport Properties of Ions in Gases (Wiley, New York), 1988.
84. Vaughan J R M, IEEE Trans. Electron Devices 36, 1963, 1989.
85. Vahedi V and DiPeso G, Simultaneous potential and circuit solution for two-dimensional bounded plasma simulation codes, J. Comput. Phys. Vol. 131, Issue 1, pp.149–163, 1997.
86. Verboncoeur J P, Alves M V, Vahedi V, and Birdsall C K, J. Comput. Phys. **104**, 321, 1993.
87. Radic N, Sternberg Z W Long-delayed reignitions of the vacuum arc, IEEE Transactions on Plasma Science, Vol. 19, Issue: 5, pp. 894–905, 1991.
88. Smeets R P P, Thielens D W ; Kerkenaar R W P, The duration of arcing following late breakdown in vacuum circuit breakers, IEEE Transactions on Plasma Science, Vol. 33, Issue: 5, pp. 1582–1588, 2005.
89. Dullni E, IEEE Trans. Electr. Insul. Vol. 28, pp. 454–460, 1993.
90. Slattery J, Friichtenicht J and Hansen D, Appl. Phys. Lett. Vol. 7, pp. 23–25, 1965.
91. Farrall G A, IEEE Trans. Electr. Insul. Vol. 28, pp. 580–591, 1993.
92. Gebel R and Hartmann W, IEEE Trans. Electr. Insul. Vol. 28, pp. 468–472, 1993.
93. Ejiri H, Abe K, Kikuchi Y, Kumada A, Hidaka K, Donen T and Tsukima M, IEEE Trans. Dielectr. Electr. Insul. No. 24, pp. 3374–3380, 2017.
94. Dascalescu L, Tobazeon R and Atten P, Behavior of conducting particles in corona-dominated electric-fields, J. Phys. D: Appl. Phys. Vol. 28, Issue 8, pp.1611–1618, 1995.
95. Sakai K-I, Abella D L, Suehiro J and Hara M, IEEE Trans. Dielectr. Electr. Insul. Vol. 9, pp. 577–588, 2002.
96. Sakai K-I, Abella D L, Khan Y, Suehiro J and Hara M, IEEE Trans. Dielectr. Electr. Insul. Vol. 10, pp. 404–417, 2003.
97. Chatterton P, Menon M and Srivastava K, J. Appl. Phys. **43**, pp. 4536–4542, 1972.
98. Chakrabarti A and Chatterton P, J. Appl. Phys. **47**, pp. 5320–5328, 1976.
99. Dunlap D and Kenkre V, Phys. Rev. B **34**, pp. 3625, 1986.
100. Vlad S, Iuga A and Dascalescu L, J. Phys. D: Appl. Phys. **33**, pp. 127, 2000.

101. Zhang Y, Jin L, Xu X, Zhang Y, Wang J and Geng Y, IEEE Trans. Dielectr. Electr. Insul. Vol. 23, pp. 3771–3778, 2016.
102. Xu S, Hidaka K, Kaneko E, Kumada A and Ikeda H, IEEE Trans. Plasma Sci. Vol. 42, pp. 3077–3082, 2014.
103. Anders S, Juttner B, Lindmayer M, Rusteberg C, Pursch H and Unger-Weber F, IEEE Trans. Electr. Insul. Vol. 28, pp. 461–467, 1993.
104. Dullni E, Shang W K and Gentsch D, IEEE Trans. Dielectr. Electr. Insul. Vol. 13, pp. 65–71, 2006.
105. Hurley R and Parnell T , J. Phys. D: Appl. Phys. **2**, pp. 881, 1969.
106. Okamoto H, J. Phase Equilib. Diff. Vol. 33, pp. 342–343, 2012.
107. Sun Z, Song X, Hu Z, Yang S, Liang G and Sun J, J. Alloys Compd. Vol. 319, pp. 266–270, 2001.
108. Sun Z, Wang Y, Guo J, Zhu Y, Song X and Zhu R, Mater. Sci. Eng. A Vol. 452–3, pp. 411–416, 2007.
109. Tuma D, Chen C and Davies D, J. Appl. Phys. Vol. 49, pp. 3821–3831, 1978.
110. Anis H and Srivastava K, Free Conducting Particles in Compressed Gas Insulation, IEEE Transactions on Electrical Insulation, Vol: EI-16, Issue: 4, pp. 327–338, 1981.
111. Wang J, Li Q, Li B, Chen C, Liu S and Li C, Theoretical and experimental studies of air gap breakdown triggered by free spherical conducting particles in DC uniform field, IEEE Trans. Dielectr. Electr. Insul. Vol. 23, Issue: 4, pp. 1951–1958, 2016.
112. Latham R V, High Voltage Vacuum Insulation: Basic Concepts and Technological Practice (London: Academic), 1995.

Chapter 3
Vacuum Interrupters at Transmission Voltage Level

Developing transmission voltage level vacuum interrupters poses a number of challenges. As the voltage to be interrupted increases from medium voltage, large contact gaps are needed to be able to withstand the high voltages that occur at transmission voltage levels, especially the basic insulation level, power frequency withstand voltage, and any other switching transients. It becomes necessary to increase the open gap between the contacts to provide sufficient insulation, and this in turn brings its own problems. The present vacuum interrupter contacts designs used in medium voltage would lead to less control of the vacuum arcs and a dielectric failure of the gap which would cause a failure to interrupt. In addition, the increased applied voltage will definitely cause an increase of X ray emission of the vacuum interrupters. This chapter will address three main issues in transmission voltage level vacuum interrupters. The first one is vacuum insulation for transmission voltage vacuum interrupter. The second one is the contact design with large contact gaps. The third one is to determine that the X ray emission of the transmission voltage vacuum interrupters exceeds the standards or not.

3.1 Vacuum Insulation

3.1.1 Contact Design Parameters and Lightning Impulse Voltage Breakdown Characteristics Under Large Contact Gaps

Vacuum insulation in medium voltage vacuum interrupters is well understood, in which the contact gaps are within 20 mm. Shioiri et al. [1] revealed the effects of test methods, electrode area, and electrode materials on the breakdown probability distributions of uniform field gaps with contact gap of 3 mm. Sato et al. [2] found a relationship between breakdown voltage of vacuum gaps and electrode surface

Z. Liu et al., *Switching Arc Phenomena in Transmission Voltage Level Vacuum Circuit Breakers*, https://doi.org/10.1007/978-981-16-1398-2_3

roughness for plate-to-plate electrodes with contact gap of 12.1 mm. Miyazaki et al. [3] investigated the breakdown voltage between a pair of rod-plane electrodes whose materials were Cu–Cr, stainless steel and Cu, respectively. The contact gap lengths were 5 and 10 mm, respectively. They found that the order of dielectric strength of different materials is stainless steel > CuCr > Cu. Budde and Kurrat [4] studied the conditioning characteristics of the plate-to-plate electrodes from various materials including stainless steel, brass, copper and copper-chromium. The contact gap lengths were within 20 mm. They found that the lighting impulse voltage increased through the conditioning process. Okawa et al. [5] investigated the area effect on vacuum breakdown for various contact materials. The contact gap lengths were within 20 mm. Toya et al. [6] experimentally studied the statistical property of area effect on electric breakdown with the contact gap lengths 3–20 mm.

The breakdown voltage characteristics of vacuum interrupters with contact gaps larger than 20 mm seemed to be different with that of medium voltage vacuum interrupters whose contact gaps were 20 mm and bellow. Fukuoka et al. [7, 8] studied the conditioning characteristics of large contact gaps (up to 50 mm) with rod-to-plane electrodes in a vacuum chamber. Their results showed a different dependency of gap distance between a short gap range and a large gap range. Kārner et al. [9] studied the physical processes in large gaps between the broad-shaped electrodes ($\varphi = 180$ mm). The results showed that the processes in large gaps were different with that in small gaps. Spolaore et al. [10] studied the large gap effects for gaps up to ~20 cm in a cylindrical stainless-steel vacuum chamber. They found an apparent voltage saturation effect in the large gaps. Giere et al. [11] studied the dielectric strength of double and single-break vacuum interrupters. The contact gap lengths were up to 40 mm. In addition, it is meaningful to consider an influence of contact design parameters on the lightning impulse voltage breakdown characteristics of vacuum interrupters with contact gaps larger than 20 mm.

This section is to determine the influence of contact design parameters on the lightning impulse voltage breakdown characteristics of vacuum interrupters with contact gaps of 10–50 mm. The investigated contact design parameters include contact diameters of 75 and 60 mm, contact surface roughness of 1.6 and 3.2 μm, and contact radius of curvature of 6 and 2 mm. The results would be useful for high voltage vacuum interrupters insulation design.

Figure 3.1 shows an experimental setup. Figure 3.1a shows the configuration of test vacuum interrupters. The height of the test vacuum interrupter was 471.5 mm and the diameter of the vacuum interrupter was 132 mm. The vacuum interrupters' contact gaps can be adjusted in a range from 10 to 50 mm. The inner pressure was kept at the order of 10^{-5} Pa. Figure 3.1b shows the test vacuum interrupters were assembled into a porcelain envelope, in which SF_6 gas was used as an external insulation media of the test vacuum interrupters. The SF_6 gas pressure in the porcelain envelope was 0.25 MPa. The contact gap was adjusted manually through a gap spacing adjuster. A positive polarity lightning impulse voltage (1.2/50 μs) was applied to the movable terminal. The stationary terminal was grounded. The injected energy in each shot of impulse voltage application was about 3 kJ. Dielectric tests were carried out by up

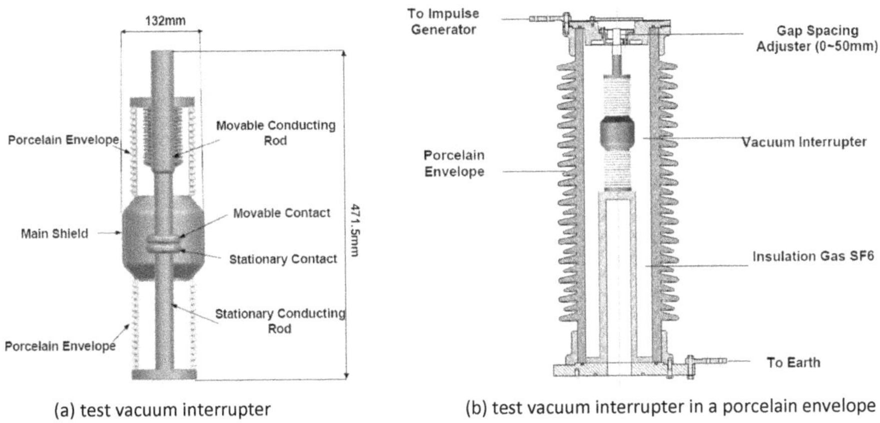

Fig. 3.1 Experimental setup

and down method [12] and ΔV was set as ~4 kV. The occurrence of a breakdown was determined by observing the voltage waveforms on a digital oscilloscope.

Figure 3.2 shows a butt type contact design in the test vacuum interrupters. The contact material was CuCr40 (40% of Cr). The thickness of the electrodes was 15 mm. There are three contact design parameters considered, which are radius of contact edge, contact surface roughness and contact diameter, respectively. Based on the three contact design parameters, four test vacuum interrupters were designed, as shown in Table 3.1. By using the No. 1 vacuum interrupter as a benchmark, one can understand the influence of the three contact design parameters upon the lightning impulse voltage breakdown characteristics. No. 2 vacuum interrupter were used to compare radius of contact edge of 2 and 6 mm. No. 3 vacuum interrupter was used

Fig. 3.2 Butt type contact design in the test vacuum interrupters

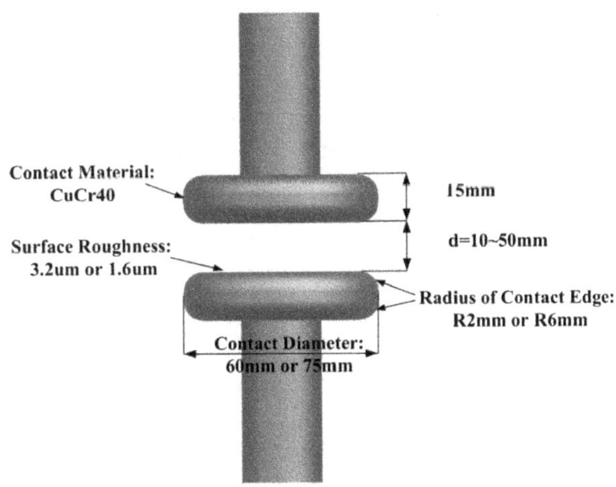

Table 3.1 Design of four test vacuum interrupter based on three contact parameters

VIs	Radius of contact edge (mm)	Surface roughness (μm)	Contact diameter (mm)
No. 1	6	1.6	60
No. 2	2	1.6	60
No. 3	6	3.2	60
No. 4	6	1.6	75

to compare contact surface roughness of 1.6 and 3.2 μm. No. 4 vacuum interrupter was used to compare contact diameter of 60 and 75 mm.

In order to analyze the breakdown probability distribution in the test vacuum interrupters with contact gaps 10–50 mm, a conditioning saturation region should be defined. It is well known that the conditioning process of the breakdown voltage followed a relationship shown in Eq. (3.1) [13].

$$U_{BN} = U_1 + U_A\left(1 - \exp\left(-\frac{n-1}{\tau}\right)\right) \tag{3.1}$$

where U_{BN} is the breakdown voltage at the n_{th} voltage application; $U_{\mathrm{A}} = U_{\mathrm{L}} - U_1$; U_1 is the initial value; U_{L} is the limit value; τ is the conditioning factor. In a conditioning process, the breakdown voltage at the n_{th} voltage application U_{BN} is increasing as the voltage application N increases. But the U_{BN} tends to a limited value U_{L} ($N \to \infty$). When the ratio of $U_{\mathrm{BN}}/U_{\mathrm{L}}$ is 0.95 and above, the breakdown voltage is considered entering saturation, which is the criterion of "saturation" in the experiments.

Figure 3.3 shows the conditioning history with up-and-down method for the four test vacuum interrupters shown in Table 3.1. The arrows in Fig. 3.3 indicated the start points of entering the saturation region. In the saturation region, the breakdown voltage probability was analyzed for the four vacuum interrupters with contact gaps 10–50 mm.

For contact gaps in medium voltage vacuum interrupters, the breakdown voltage probability distribution met the Weibull distribution [6]. Equation (3.2) shows the Weibull distribution between the accumulative breakdown probability $F(U)$ and the applied voltage U [14].

$$F(U) = 1 - \exp\left[-\left(\frac{U - U_0}{\Theta - U_0}\right)^m\right] \tag{3.2}$$

where $F(U)$ is the cumulative probability at the voltage U; U_0 is the location parameter that represents the voltage that makes breakdown probability reduce to zero; m is the Weibull shape or "shape" factor; Θ is the scale parameter.

Figure 3.4 shows the breakdown voltage probability of the four vacuum interrupters with contact gaps 10–50 mm. It is interesting to find that all of the breakdown

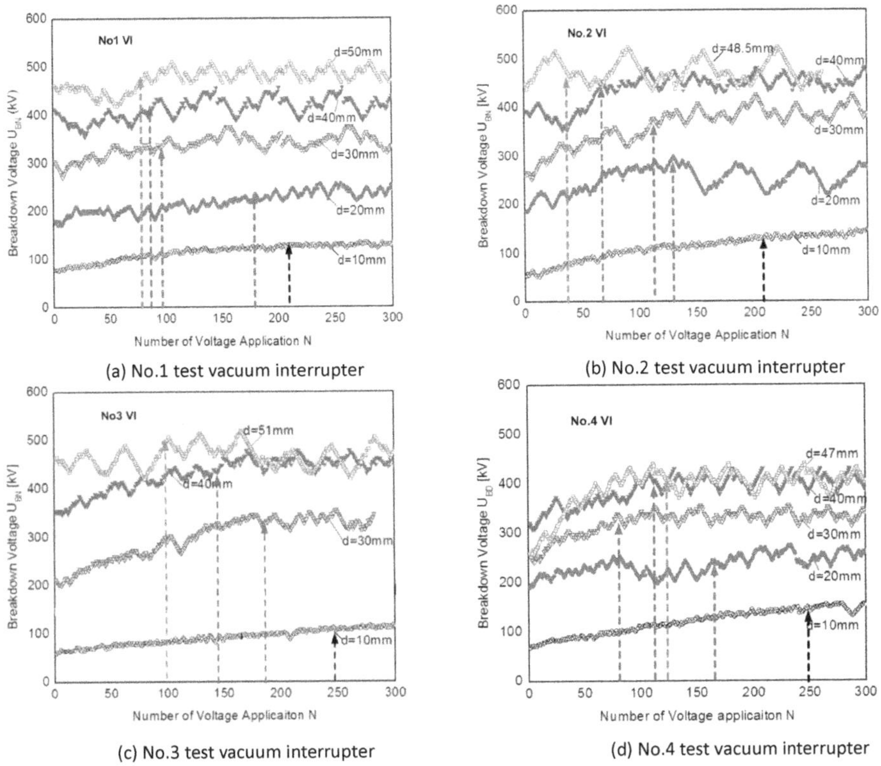

Fig. 3.3 Conditioning history with up-and-down method for the four test vacuum interrupters shown in Table 3.1

voltage probability distributions followed Weibull distribution, which was applicable for not only contact gaps in medium voltage vacuum interrupters, which is 20 mm and below, but also contact gaps up to 50 mm. Toya et al. [6] experimentally studied the statistical property of breakdown voltage of vacuum gaps using parallel plane copper electrodes in a gap length of 3–20 mm. They found that the cumulative breakdown probability followed Weibull distribution by introducing a parameter for the weak point of breakdown and the electrode surface area. Zhang et al. [15] studied an influence of contact contour on breakdown characteristics in vacuum under uniform field with a contact gap of 6 mm by applying standard lightning impulse voltage. They found that the breakdown probability distribution followed Weibull distributions when the breakdown voltage saturated. The experimental results of the four vacuum interrupters shown in Fig. 3.4 proved that, Weibull distribution fitted in not only the medium voltage contact gaps that is equal to or less than 20 mm, but also the transmission voltage contact gaps up to 50 mm.

Table 3.2 shows the values of the Weibull parameters for the four vacuum interrupters. The R-square was used to confirm the goodness of fit. The values of R-square were above 0.9 in all of the cases. The 50% breakdown voltage U_{50} can be obtained

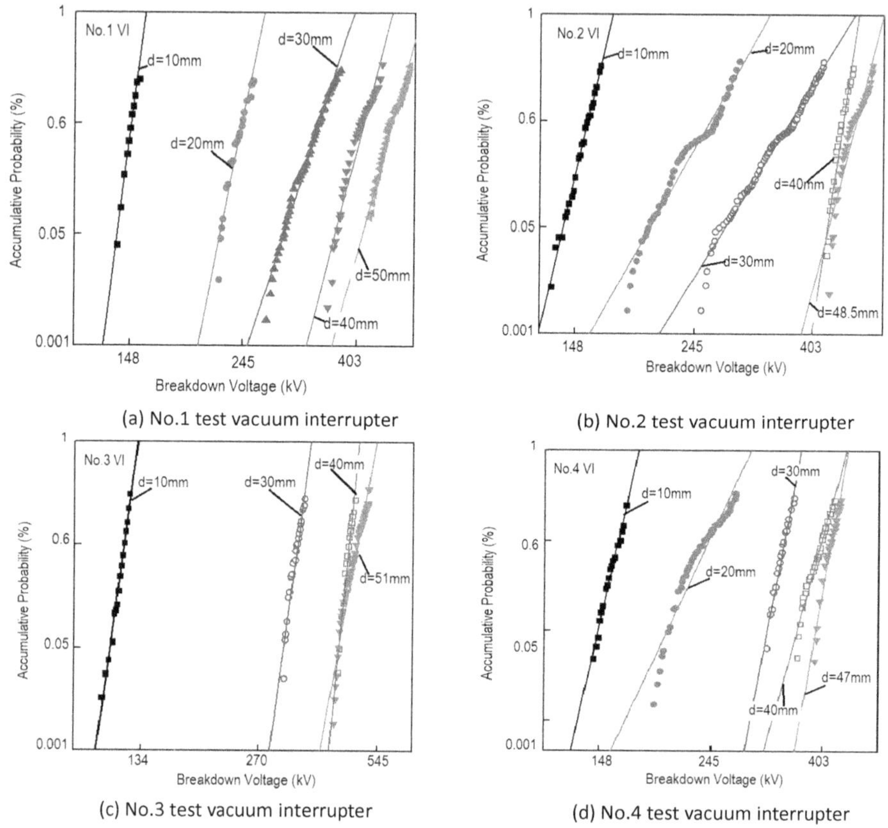

Fig. 3.4 Breakdown probability of the four vacuum interrupters plotted with Weibull probability distribution

from Eq. (3.2). The calculated U_{50} was shown in Table 3.2. It shows that U_{50} increased with increasing of contact gap d.

The relationship between U_{50} and d followed a power equation $U_{50} = k\mathrm{d}^{\alpha}$, where α and k are constants. Figure 3.5 shows the relationship between U_{50} and gap length d for the four vacuum interrupters. Figure 3.5a shows that, for No. 1 vacuum interrupter, which is a benchmark, the relationship between U_{50} and d was fitted by Eq. (3.3).

$$U_{50} = 30 \times d^{0.70}(10mm \leq d \leq 50mm) \tag{3.3}$$

The R-square, which was used to confirm the goodness of fit, was 0.99. Figure 3.5b shows that, for No. 2 vacuum interrupter with the radius of contact edge of 2 mm, the relationship between U_{50} and d was fitted by Eq. (3.4).

$$U_{50} = 40 \times d^{0.65}(10\mathrm{mm} \leq d \leq 48.50\mathrm{mm}) \tag{3.4}$$

Table 3.2 The gap length, number of voltage applications until U_{BN} saturated; 50% breakdown voltage U_{50} and R-square of the four vacuum interrupters

	Gap length (mm)	Number of voltage application till U_{BN} is saturated	50% breakdown voltage U_{50} [kV]	R-Square
No. 1 VI	10	210	149	0.940
	20	180	239	0.992
	30	97	331	0.995
	40	90	402	0.992
	50	82	457	0.997
No. 2 VI	10	211	164	0.991
	20	126	276	0.977
	30	114	380	0.991
	40	70	451	0.997
	48.5	41	473	0.980
No. 3 VI	10	257	154	0.990
	20	161	181	0.995
	30	175	331	0.996
	40	148	400	0.997
	51	97	469	0.994
No. 4 VI	10	250	176	0.992
	20	164	230	0.990
	30	79	333	0.997
	40	105	394	0.998
	47	124	412	0.996

The R-square was 0.98. Figure 3.5c shows that, for No. 3 vacuum interrupter with the surface roughness of 3.2 μm, the relationship between U_{50} and d was fitted by Eq. (3.5)

$$U_{50} = 20 \times d^{0.79} (10\text{mm} \le d \le 51\text{mm}) \quad (3.5)$$

The R-square was 0.99. Figure 3.5d shows that, for No. 4 vacuum interrupter with the contact diameter of 75 mm, the relationship between U_{50} and d was fitted by

$$U_{50} = 40 \times d^{0.60} (10\text{mm} \le d \le 47\text{mm}) \quad (3.6)$$

The R-squared was 0.98.

Figure 3.6 shows an influence of contact edge radius on breakdown characteristics within the contact gaps of 10–50 mm. Figure 3.6a shows a relationship between U_{50} and d with two kinds of contact edge radius. It shows that the U_{50} of No. 2

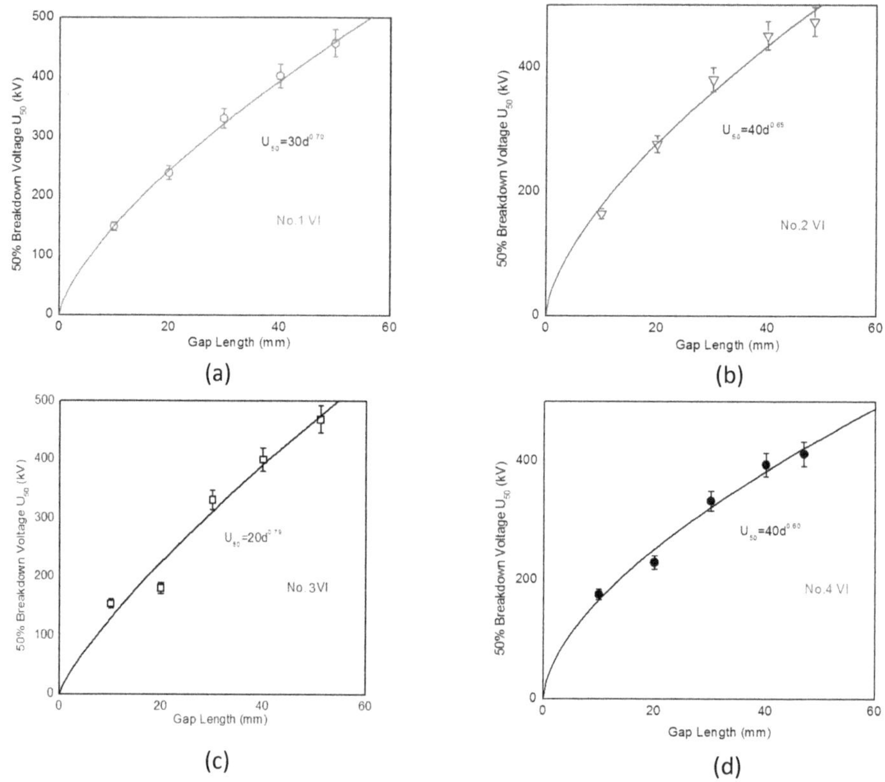

Fig. 3.5 U_{50} as a function of the gap length for the four vacuum interrupters shown in Table 3.1. **a** No. 1 vacuum interrupter is a benchmark; **b** No. 2 vacuum interrupter. The contact edge radius is 2 mm, comparing with 6 mm in the benchmark; **c** No. 3 vacuum interrupter. The surface roughness is 3.2 μm, comparing with 1.6 μm in the benchmark; **d** No. 4 vacuum interrupter. The contact diameter is 75 mm, comparing with 60 mm in the benchmark

vacuum interrupter with contact edge radius of 2 mm was higher than that of No. 1 vacuum interrupter with contact edge radius of 6 mm. The results can be explained by the concept of effective area. Okawa et al. [5] studied the area effect on electric breakdown in vacuum. They found that all breakdown traces with different electrodes were concentrated only in the region with electric field strength higher than 90% of the maximum field strength. They defined this region as an effective area S_{eff}. They found a relationship between the breakdown field strength E_{b} and S_{eff}, as shown in Eq. (3.7).

$$E_b = K_1 S_{eff}^{-n} \tag{3.7}$$

where K_1 and n are constants. So, the breakdown voltage decreased with an increasing of S_{eff}. The experimental results shown in Fig. 3.6a supported the finding of Okawa

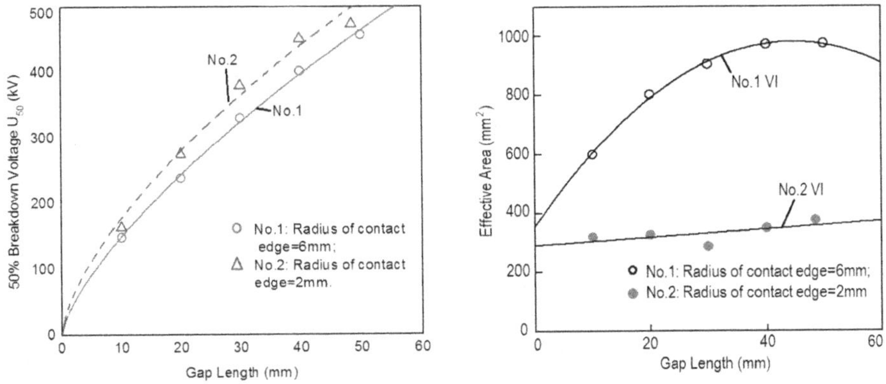

(a) relationship between U_{50} and d (b) effective area

Fig. 3.6 Influence of contact edge radius on breakdown characteristics within the contact gaps of 10–50 mm

et al. [5]. An observation of the electrode surfaces after tests revealed that breakdown traces in both of the two vacuum interrupters with different contact edge radius concentrated in the region with electric field strength higher than 90% of the maximum field. Figure 3.6b shows the effective area of the two vacuum interrupters. It shows S_{eff} of No. 1 vacuum interrupter was significantly higher than S_{eff} of No. 2 vacuum interrupter. That explained why U_{50} of No. 2 vacuum interrupter was higher than that of No. 1 vacuum interrupter.

Figure 3.7 shows an influence of the contact surface roughness on breakdown characteristics within the contact gaps of 10–50 mm. Figure 3.7a shows a relationship between U_{50} and d with two kinds of contact surface roughness. It shows that U_{50} with contact surface roughness of 1.6 μm was close to that with contact surface roughness of 3.2 μm. Figure 3.7b shows the influence of the contact surface roughness

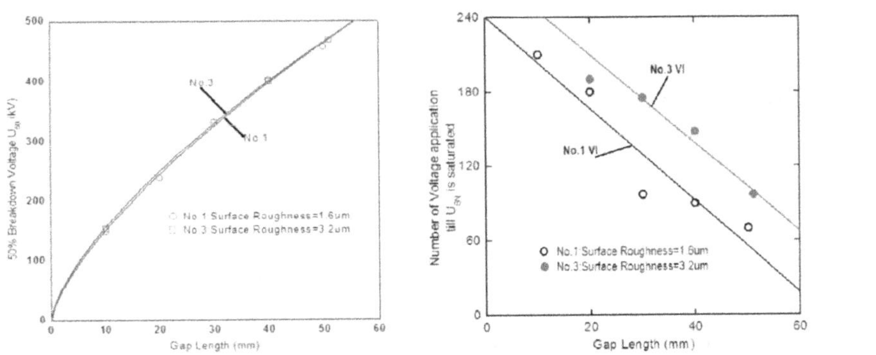

(a) relationship between U_{50} and d (b) influence of the contact surface roughness on conditioning process

Fig. 3.7 Influence of the contact surface roughness on breakdown characteristics within the contact gaps of 10–50 mm

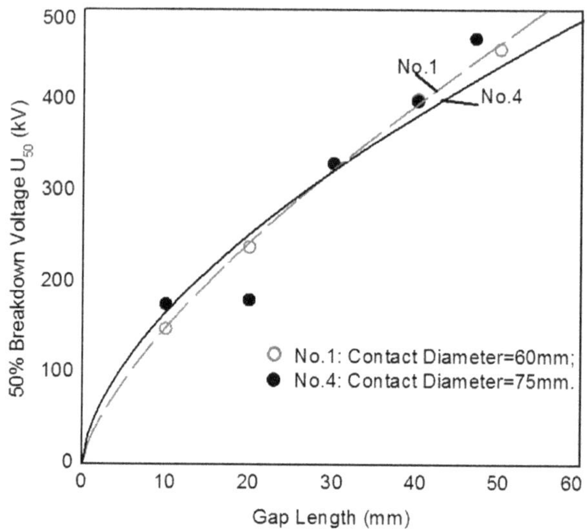

Fig. 3.8 Influence of the contact diameter on breakdown characteristics within the contact gaps of 10–50 mm

on conditioning process. It shows the number of voltage application till breakdown voltage saturated for the two contact surface roughness under various contact gaps. It shows that when the contact surface roughness was higher the conditioning process became slower. The influence of contact surface roughness on conditioning process was significant.

Figure 3.8 shows an influence of the contact diameter on breakdown characteristics within the contact gaps of 10–50 mm. It shows a relationship between U_{50} and d with two kinds of contact diameters of 60 and 75 mm. The U_{50} of the contact diameter 60 mm was close to that of contact diameter 75 mm. The reason may be related to effective area. Although the contact surface areas of the two vacuum interrupters were different, their S_{eff} were still quite close. Thus, the influence of the contact diameter on U_{50} was not significant in a contact gap range of 10–50 mm.

3.1.2 *Breakdowns in High Voltage Vacuum Interrupters with Large Contact Gap*

In Sect. 3.1.1, one can see how the breakdown voltage depended on the contact gap length d and contact design parameters of vacuum interrupters. In the field of vacuum insulation, it is generally accepted that at small gaps breakdown is assumed to be initiated by a field-dependent electron emission mechanism, whereas at large gaps micro-particles processes are thought to take over the dominant role. There seemed no accurate criteria for "small gaps" and "large gaps". For example, Latham [16, 17] proposed small gaps as $d \leq 2$ mm or $d \leq 0.5$ mm and he proposed large gaps as $d \geq 5$ mm or $d \geq 2$ mm. Li et al. [18] found that there was a mechanism transition occurring with an increase of d from 0.5 to 1.0 mm. They reported that the breakdowns were field emission induced breakdown at $d = 0.5$ mm. The breakdowns included both field emission induced breakdown and particles induced breakdown at $d = 0.8$ mm. While the possibility of particles induced breakdown increased at $d = 1.0$ mm. Nevertheless, the contact gap length in transmission voltage level vacuum interrupters (several tens of mm) is definitely large gaps. In the case of large contact gaps, breakdowns become voltage-dependent [14].

The important role of micro-particles in initiating electrical breakdowns in large gaps in vacuum was firstly proposed by Cranberg as a "clump" hypothesis [19]. His hypothesis was that the loosely adhering microscopic particles might be torn from an electrode surface by an applied electric field and, because of their charge, accelerated across the gap to impact on the opposite electrodes as high velocity micro-particles, causing localized fusion and vaporization of electrode material that was sufficient to trigger a breakdown of the gap. This mechanism, under an assumption of a uniform electric field between a pair of electrodes and a DC voltage applied, yielded a relationship in which the breakdown voltage of the gap depended upon the square root of the gap length. The transference of anode material to the cathode after repeated breakdown has been reported in the literature [20]. In commercial vacuum interrupters, micro-particles always presented. Kamikawaji et al. [21] studied the distribution of micro-particles left on contact's surface. To their estimation, there were between 5.7×10^3 and 8.7×10^3 cm^{-2} particles on the contact's surface after contact machining. These particles would be detached from their highly charged parent electrodes in a high electric field in contact gaps. The particles would enter the contact gap as charged particles to be accelerated to high velocities during their transit to the opposite electrode. At impact, the micro-particles could give rise to secondary particles and produce a crater on the contact's surface. This crater could have a characteristic of "crown of thorns" structure on its edges. These sharp points would result in microscopic regions of high field that will in turn produce field emission currents [22]. That would initiate breakdowns of the contact gap.

There have subsequently been many refinements and developments of the original Cranberg model. Slivkov [23] refined the Cranberg hypothesis by proposing two restricting conditions. One was that the micro-particles should be small enough for the impact velocity to reach an impact vaporization limit and the other was that the micro-particles should be large enough for sufficient metal vapor atoms to be released to trigger a localized gas discharge. In Slivkov's hypothesis, the electric field between electrodes was a uniform field, and DC voltage was applied. He found that the breakdown voltage of a vacuum gap depends upon the 0.625 power of the gap length. Farrall [24] extended the Cranberg hypothesis as an impulse voltage and yielded different results. He considered a rise in an impulse voltage occurring during a transit time of a clump flying to the opposing electrode, where the electric field between electrodes was a uniform field. He found that the breakdown voltage of a vacuum gap depended upon the 0.83 power of the gap length in the case of large gap length and large clump radii, with fast rising impulses at a constant rising rate. Based on the previous works, it is meaningful to discuss the breakdown mechanism of impulse voltages breakdown in vacuum interrupters with large contact gaps.

This section is to propose a breakdown mechanism of lightning impulse voltage in transmission voltage vacuum interrupters with large contact gaps by complementing the Cranberg "clump" hypothesis with an impulse voltage and a non-uniform field. Figure 3.9 schematically shows an assumption of an impulse voltage breakdown process in transmission voltage vacuum interrupter under a non-uniform field. The breakdown was due to the loosely adhering microscopic particles as Cranberg's hypothesis described. The microscopic particles might be torn from an electrode surface by the impulse voltage induced high field. Some microscopic particles located at the area with the maximum electric field had the highest possibility to be accelerated, as shows in Fig. 3.9a. The charge carried by microscopic particles was assumed deriving from the field-induced surface charge existing on the original micro-feature immediately before its detachment from the parent electrode. Because of their charge, the microscopic particles were accelerated across the gap by the impulse voltage, as shown in Fig. 3.9b. During the transit process, the voltage rise rate of the impulse voltage was considered. Then the microscopic particles impacted on the opposite electrode as high velocity micro-particles, causing localized fusion and vaporization of electrode material, as shown in Fig. 3.9c. Finally, the vaporization of electrode material could trigger a breakdown of the gap if the impact energy of the micro-particles exceeds critical value, as shown in Fig. 3.9d. This followed the Cranberg's critical energy theory.

Cranberg recognized that in the presence of a strong electric field, the micro-particles would be detached from the electrode by a strong electro-mechanical force. If this occurred, they would enter the gap as charged particles to be accelerated to high velocities during their transit to the opposite electrode [19]. He assumed that the breakdown occurred when the energy per unit area W delivered to an opposite electrode exceeded a constant value that was a characteristic of a given pair of electrodes. W was a product of the applied voltage U_c and the charge density on the micro-particles. The charge density of the micro-particles was proportional to the field E_c on the electrode surface. Thus, the breakdown criterion would be given by Eq. (3.8), where C_1 is a constant value.

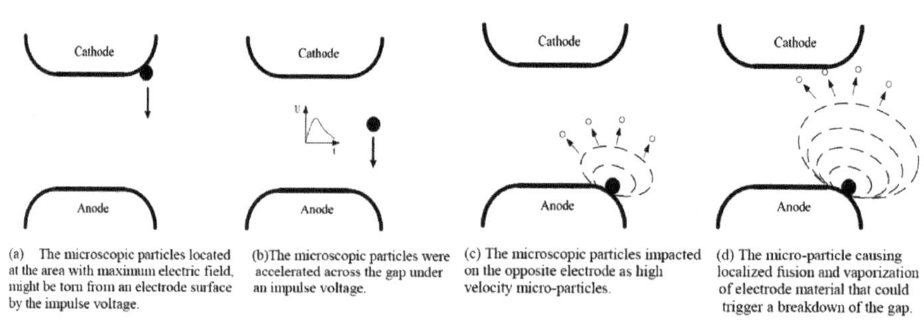

Fig. 3.9 Assumption of a lightning impulse voltage breakdown in transmission voltage vacuum interrupters with a large contact gap

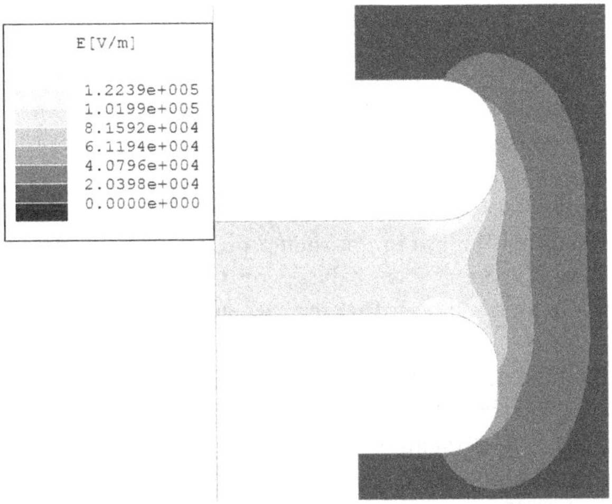

Fig. 3.10 Electric field strength distribution of No. 1 vacuum Interrupter in Sect. 3.1.1. Contact gap-10 mm, contact diameter 60 mm, Contact edge radius 6 mm. Applied voltage-1000 V (stationary contact), 0 V (movable contact). Stationary contact is in the upper position

$$U_c \times E_c \geq C_1 \tag{3.8}$$

Cranberg assumed that the electric field between the electrodes was a uniform field. For the case of a uniform field where $E = U/d$, the breakdown criterion predicted the applied voltage at which the contact gap breakdown was proportional to a square-root of the gap length for a given pair of electrodes, as shown in Eq. (3.9) [19].

$$U_c \geq (C_1 d)^{1/2} \tag{3.9}$$

For the case of a non-uniform field, the relationship between electric field E and the voltage applied between the electrodes U is more complicated than that under a uniform field. The electric field E is not determined by U/d, where U is the voltage impressed across a gap d, but is given by the relationship shown as Eq. (3.10) [25].

$$E = \beta \frac{U}{d} \tag{3.10}$$

where β is an enhancement factor. β includes two components: an enhancement factor resulting from microscopic surface projections β_{m} and a geometric enhancement factor β_{g}. To simplify the breakdown model, the state of the contact surface roughness was not considered. β_{m} was supposed to be a constant. The influence of geometric enhancement factor β_{g} on the breakdown characteristics was considered. In addition, the breakdown was supposed to occur at the point with the maximum electric field. So the critical electric field could then be extended as Eq. (3.11).

$$E_c = \beta_{max} \frac{U}{d} \tag{3.11}$$

where β_{max} was the maximum geometric enhancement factor. Since the state of the contact surface roughness was not considered, β_m was assumed to be a constant. Therefore, β_{max} was only related to β_g. So β_{max} could then be calculated by an electric field simulation.

The electric field distribution in the vacuum interrupters with large contact gaps ranging from 10 to 100 mm, was analyzed by means of a two-dimensional axis-symmetric finite element method. Figure 3.10 shows the electric field distribution between contacts of No. 1 vacuum interrupter (see Sect. 3.1.1) at a contact gap of 10 mm. Then the geometric enhancement factor β_g at various contact gaps can be obtained based on the electric field simulation. Figure 3.11 shows the geometric field enhancement factor β_g as a function of the distance from the center of the contact and of the contact gap. The maximum geometric enhancement factor β_g under various contact gaps can be obtained from Fig. 3.11. With a contact gap of 10 mm, the maximum geometric enhancement factor β_g was found to be 1.22. When the contact gap increased to 20 mm, β_g increased to 1.49. When the contact gap increased to 30 mm, β_g was 1.79. When the contact gap increased to 40 mm, β_g was 2.11. When the contact gap increased to 50 mm, β_g was 2.45. When the contact gap increased to 60, 70, 80, 90 and 100 mm, β_g was 2.68, 2.85, 3.04, 3.14 and 3.20, respectively. Figure 3.12 shows a relationship between βg and contact gap d, with contact gaps ranging from 10 to 100 mm. The relationship is described as Eq. (3.12). The R-square, which was used to confirm the goodness of fit, was 0.99.

Fig. 3.11 The geometric field enhancement factor βg as a function of the distance from the center of the contact and of the contact gap

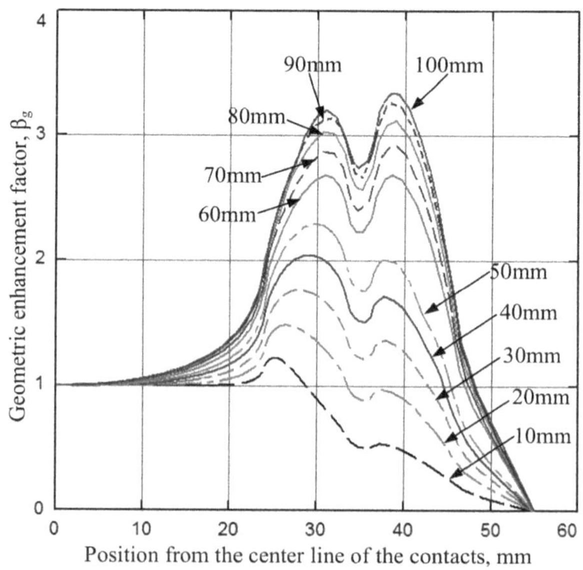

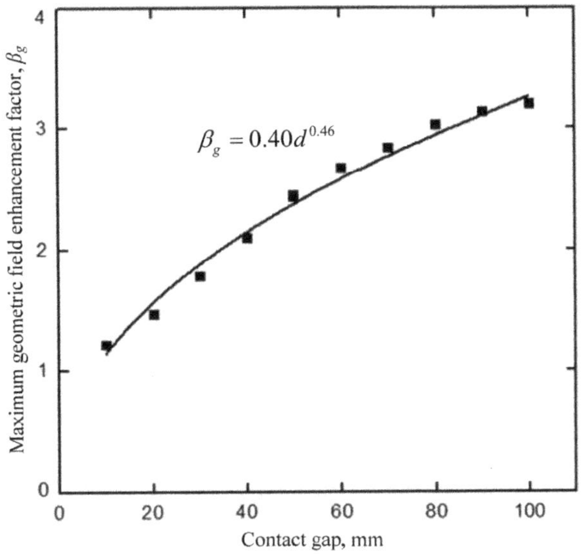

Fig. 3.12 The relationship between the maximum geometric field enhancement factor and the contact gap

$$\beta_g = 0.40d^{0.46} \tag{3.12}$$

The Eqs. (3.11) and (3.12) were solved, and it was found that the critical electric field E_c under a non-uniform field with contact gaps 10–100 mm, can be described by Eq. (3.13).

$$E_c = \beta_g \frac{U_c}{d} = 0.40U_c d^{-0.54} (10\text{mm} \le d \le 100\text{mm}) \tag{3.13}$$

By Eqs. (3.8) and (3.13), one can get Eq. (3.14) for a non-uniform field with contact gaps ranging from 10 to 100 mm, which can be described as.

$$U_c = C_2 d^{0.27} (10\text{mm} \le d \le 100\text{mm}) \tag{3.14}$$

where U_c is the critical voltage and C_2 was a constant. It assumed that, under the non-uniform field, a micro-particle separates from its parent electrode at a critical voltage U_c. In the flight journey, the micro-particle was accelerated by the electric field force, where the acceleration of the micro-particle would be described by the following Eqs. (3.15)–(3.18).

$$m_p \frac{d^2x(t)}{dt^2} = U(t)\frac{Q_p}{d} \tag{3.15}$$

$$x(0) = 0 \tag{3.16}$$

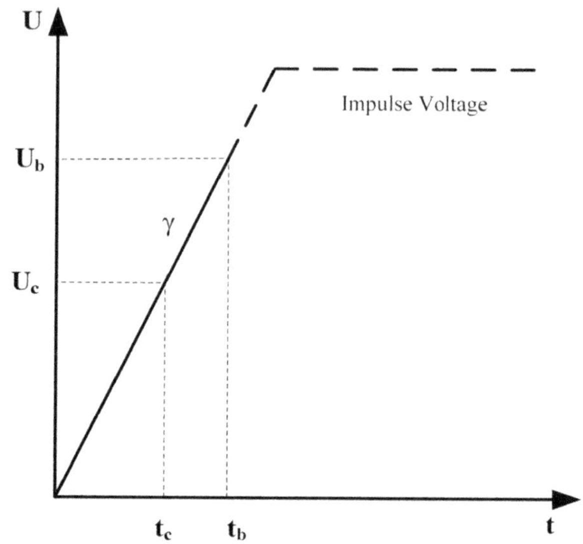

Fig. 3.13 Simplified Impulse voltage waveform. A micro-particle separates from its parent electrode at U_c and a breakdown occurs at U_b. γ is the rate of rise of the impulse voltage

$$x(t_d) = d \tag{3.17}$$

$$\frac{dx}{dt}(x = 0) = 0 \tag{3.18}$$

where m_p is the micro-particle mass. $x(t)$ is the distance between the micro-particle to its parent electrode surface at an instant t. $U(t)$ is the voltage applied on the electrodes at t. Q_p is the micro-particle charge. d is the gap length. t_b is the instant at which the breakdown occurs.

It was assumed that the impulse voltage rose linearly with time. Figure 3.13 shows a simplification of the impulse voltage applied on the electrodes. γ was the rate of rise of the impulse voltage. So, $U(t)$ could be expressed by:

$$U(t) = U_c + \gamma t \tag{3.19}$$

$$U(0) = U_c \tag{3.20}$$

$$U(t_b) = U_b \tag{3.21}$$

where U_c is the critical voltage determined by Eq. (3.14). U_b is the instant breakdown voltage. t_b is the instant at which a breakdown occurs.

The geometry of the micro-particle was assumed to be a spherical particle that is resting on the electrode surface. The particle mass m_p follows Eq. (3.22). The particle

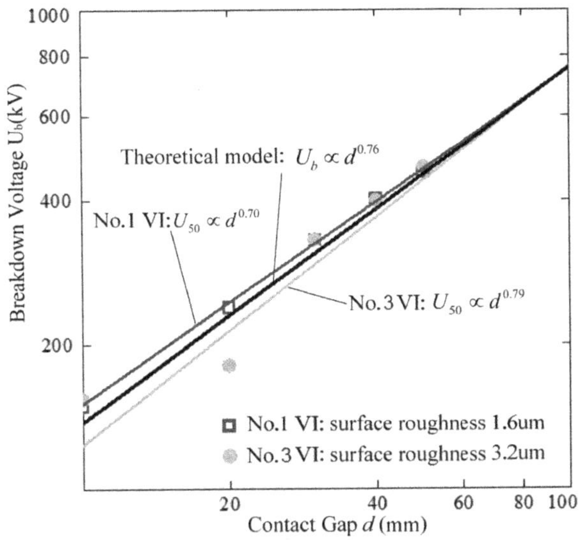

Fig. 3.14 Comparison between the experimental results (Sect. 3.1.1) and theoretical results

charge can be expressed by Eq. (3.23) [26].

$$m_p = \frac{4\pi \rho_p r_p^3}{3} \tag{3.22}$$

$$Q_p = 6.6\pi \varepsilon_0 {r_p}^2 E \tag{3.23}$$

where ρ_p is the density of the micro-particle metal. r_p is the radii of the micro-particles. ε_0 is the vacuum permittivity. E is the electric field on the electrode surface.

In the case of a non-uniform field, the strength of the electric field E is expressed by Eq. (3.11). So the particle charge Q_p would be given by

$$Q_p = 6.6\pi \varepsilon_0 {r_p}^2 E_c \tag{3.24}$$

The Eqs. (3.15)–(3.24) were solved, and one can find that the breakdown voltage U_b follows Eq. (3.25).

$$U_b^3 - 3U_b U_c^2 + 2U_c^3 = 2\Gamma \tag{3.25}$$

Γ follows Eq. (3.26).

$$\Gamma = \frac{3 r_p \rho_p \gamma^2 d^{2.54}}{2\varepsilon_0 U_c} \tag{3.26}$$

where r_{p} is the radii of the micro-particles. ρ_{p} is the density of the micro-particle metal. γ is the rate of rise of the impulse voltage. d is the gap length. ε_0 is the vacuum permittivity. The solution to Eq. (3.25) follows Eq. (3.27).

$$U_b = \left(-\frac{1}{2}Q + \Delta^{1/2}\right)^{1/3} + \left(-\frac{1}{2}Q - \Delta^{1/2}\right)^{1/3} \tag{3.27}$$

where,

$$\frac{1}{2}Q = U_c^3 - \Gamma \tag{3.28}$$

$$\Delta = U_c^6[\Theta(\Theta - 2)] \tag{3.29}$$

$$\Theta = \frac{3r_p\rho_p\gamma^2 d^{2.54}}{2\varepsilon_0 U_c^4} \tag{3.30}$$

For $\varepsilon_0 = 8.85 \times 10^{-12} F/m$, $\gamma \approx 100\,\mathrm{kV}/1.2\,\mu s$, $\rho_p = 8.9 \times 10^3 \mathrm{kg}/\mathrm{m}^3$, $U_c \approx 100\,\mathrm{kV}$, $r_p \approx 10\,\mu m$ one can get $\Theta \gg 2$. So, from the Eqs. (3.27)–(3.30), (3.31) appears.

$$U_b = \left(\frac{3\rho_p\gamma^2 r_p}{\varepsilon_0 C^{1/2}}\right)^{1/3} d^{0.76} (10\mathrm{mm} \le d \le 100\mathrm{mm}) \tag{3.31}$$

Therefore, one can find that, in the case of an impulse voltage breakdown under a non-uniform field in high voltage vacuum interrupters, the breakdown voltage was proportional to the 0.76 power of the gap length. One should note that the theoretical deduction was corresponding to the No. 1 and No. 3 vacuum interrupter in Sect. 3.1.1.

Figure 3.14 shows a comparison between the experimental results indicated in Sect. 3.1.1 and the theoretical result shown in Eq. (3.31). The experimental results of No.1 and No.3 vacuum interrupter was shown, in which the breakdown voltage was proportional to 0.70 and 0.79 power of the gap length, respectively. The experimental results supported the proposed model based on the Cranberg hypothesis, which led to an "expected" slope of 0.76. The difference between the theoretical result and the experimental results may reflect that the proposed model did not consider the contact surface roughness.

3.1.3 Cathode Arc-Melted Layer and Vacuum Insulation

It is important to ensure the reliability of the insulation of transmission voltage level vacuum interrupters. Voltage conditioning and current conditioning have been demonstrated to guarantee vacuum insulation [27]. Okawa et al. [28] found that 5 kA and 10 kA (rms) current conditioning improved the breakdown voltage by 20% and 60%, respectively. They attributed the improvement to a melted layer. Asari et al. [29] measured the breakdown characteristics of CuCr electrodes after electron beam irradiation in vacuum. They found that electron beam irradiation could form a fine Cr layer on the contact surface that could improve the breakdown voltage. Okubo et al. [30] discovered an electrode conditioning mechanism based on the pre-breakdown current in a non-uniform electric field in vacuum with a designed rod-plane electrode pair. They found that the ignition site started on the rod tip and then went upward and finally back to the rod tip. Ding et al. [31] investigated the DC breakdown in vacuum with CuCr electrodes. They found that breakdown conditioning increased the dielectric strength, which they attributed to the production of a small melted layer on the electrode surface. The melted layer had a much finer microstructure and more homogeneous micro-composition. The above-mentioned results implied that melted layer had an important influence on vacuum insulation in vacuum interrupters.

However, there is no quantitative report on the influence of the melted layer on vacuum insulation. This section is to quantitatively determine the influence of the arc-melted layer of the cathode on the vacuum insulation. A special experimental setup was designed to achieve the objective. A pair of rod-plane electrodes was prepared. The rod electrode was arranged as cathode in the dielectric test. Before the dielectric test, an arc-melted layer was generated on the rod electrode by a DC arc. In the DC arc, the rod electrode was arranged as anode. A high-current anode mode was easily to achieve with the rod as an anode, which led to an arc-melted layer on the rod anode. The depth of the arc-melted layer was adjusted by arcing time. After that, the influence of various depth of cathode arc-melted layer on vacuum insulation can be investigated.

Figure 3.15 shows a vacuum interrupter with a pair of rod—plane electrodes. Both the rod and plane electrode was made of CuCr25 contact material (25% of Cr). The envelope of the vacuum interrupter was made of glass and the main shield was removed, so that the vacuum arc behaviors could be observed by a high-speed camera. The diameter of the glass envelope was 110 mm. The rod electrode was a hemisphere of 4 mm diameter, and the diameter of the plane electrode was 48 mm. The surface roughness of the electrodes was 1.6 μm. The vacuum degree of the vacuum interrupter was 2×10^{-4} Pa.

Figure 3.16 shows experimental setup to determine the influence of the arc-melted layer of the cathode on the vacuum insulation. Figure 3.16a shows the experimental procedure. Four vacuum interrupters, labelled A, B, C and D, were used in the experiment. Vacuum interrupter A was used for dielectric test. There was no arcing process with it. It acted as a benchmark for the other three vacuum interrupters. Vacuum interrupters B, C and D were subjected to arcing for 10 ms, 46 ms and 73 ms, respectively. After the arcing test, all of the four vacuum interrupters endured standard 1.2/50 μs lightning impulse voltage breakdown tests. After the lightning impulse voltage breakdown tests, the four vacuum interrupters were subjected to a power frequency voltage to measure their field emission current. Finally, the four vacuum interrupters were opened and scanned by an electron microscope. The cross

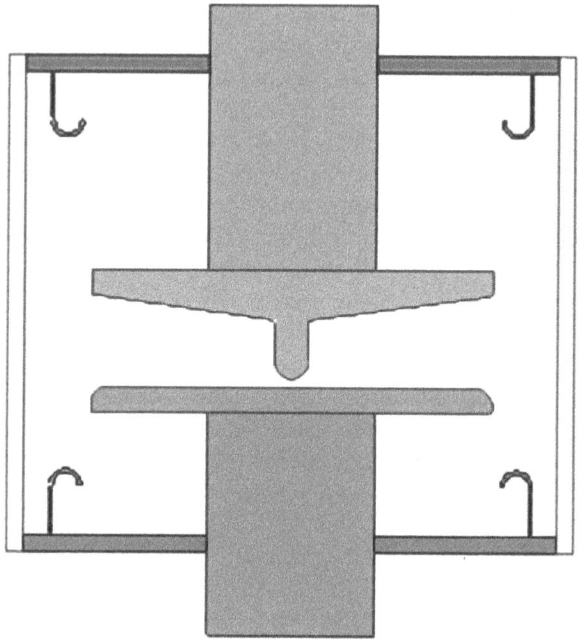

Fig. 3.15 A vacuum interrupter with rod-plane electrodes. In the arc-melted layer formation, the rod electrode was the anode. In the dielectric test, the rod electrode was the cathode

section of the cathode surface was observed, so that the depth of the arc-melted layer was determined. Figure 3.16b shows the arc melted layer forming experiment circuit. The test vacuum interrupter was connected to a DC current source and an air contactor. An external Helmholtz coil provided a 60 mT axial magnetic field to control a drawn vacuum arc. Figure 3.16c shows the arcing time control for the arc melted layer forming. Two relays controlled the arcing time. The air contactor closed 20 ms before test vacuum interrupter opened, and then the air contactor interrupted the drawn DC arc after a pre-set arcing time. A high-speed camera observed the vacuum arcs during the arcing. The camera recording speed was set as 10,000 frames/s with the aperture fixed at 4, and the exposure time was set as 2 μs. The DC arc current was 250 A. In the standard lightning impulse voltage test, the gap of the test vacuum interrupter was set as 1.0 mm. The rod acted as a cathode. The tests were carried out by the up-and-down method [12], and the ΔV step was set as ~4 kV. The occurrence of a breakdown was determined from the voltage waveforms on a digital oscilloscope. Figure 3.16d shows a test circuit for the 50 Hz power frequency voltage generation and field emission current measurement. In this circuit, a 2 kΩ resistor Rs was in series with test vacuum interrupter to measure the circuit current. A 15 V transient voltage suppressor (TVS) was in parallel with R_s to protect the oscilloscope when breakdowns occurred. Below the breakdown voltage, the maximum leakage current through the TVS was 1 μA, which was less than 1‰ of the measured circuit current. Thus the leakage current could be neglected. A voltage divider was in parallel with test vacuum interrupter for the voltage measurement. The gap length was also set as 1.0 mm in the field emission current measurement. The power frequency voltage

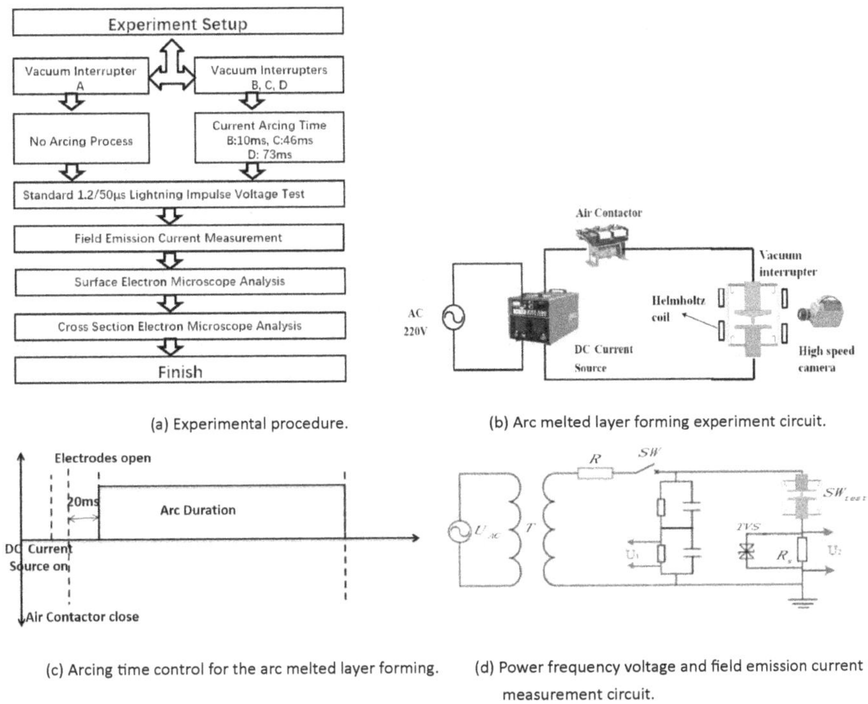

(a) Experimental procedure. (b) Arc melted layer forming experiment circuit.

(c) Arcing time control for the arc melted layer forming. (d) Power frequency voltage and field emission current measurement circuit.

Fig. 3.16 Experimental setup to determine the influence of the arc-melted layer of the cathode on the vacuum insulation

started from 20 kV (peak value). If no breakdown occurred over 1 min, it increased by a step of 0.5 kV until breakdowns occurred. The test was carried out two times for each test vacuum interrupter.

Figure 3.17 shows the arc-melted layer forming experiment. Figure 3.17a shows the vacuum arc voltage of vacuum interrupter B, C and D. Figure 3.17b and c shows photographs of vacuum arcs. In Fig. 3.17a, the arc voltage fluctuated in the first 26 ms, which corresponded to the fluctuating arc shown in Fig. 3.17b. During this period, the vacuum arc rotated randomly, and it was not aiming to the rod anode. Figure 3.17a

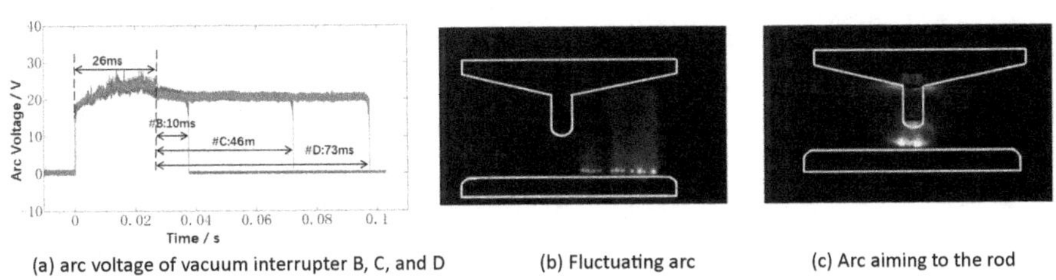

(a) arc voltage of vacuum interrupter B, C, and D (b) Fluctuating arc (c) Arc aiming to the rod

Fig. 3.17 Arc-melted layer forming experiment

shows that, after 26 ms, a stable arc voltage appeared. During this period, the arc aimed to the rod anode, as shown in Fig. 3.17c. Therefore, the arc melting of the rod electrode started at 26 ms. The arcing time of vacuum interrupter B, C, and D was 10 ms, 46 ms and 73 ms, respectively. During a few preliminary tests, it showed that the rod anode deformed after arcing of 110 ms [32]. So the arcing time was controlled to less than 110 ms.

Figure 3.18 shows the conditioning history with up-and-down method for the four test vacuum interrupters. The conditioning process followed Eq. (3.1). The definition of saturation region also followed the 0.95 criterion (see Sect. 3.1.1). The arrows indicate the entering of the saturation region.

Figure 3.19 shows the breakdown probability of the four vacuum interrupters in the saturation region. Figure 3.19a shows the accumulative probability of the

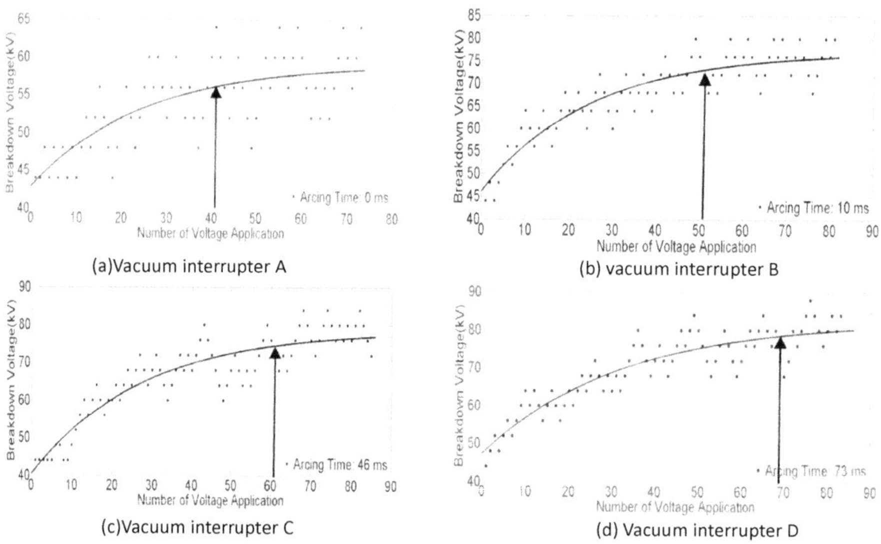

Fig. 3.18 The conditioning history with up-and-down method for the four test vacuum interrupters

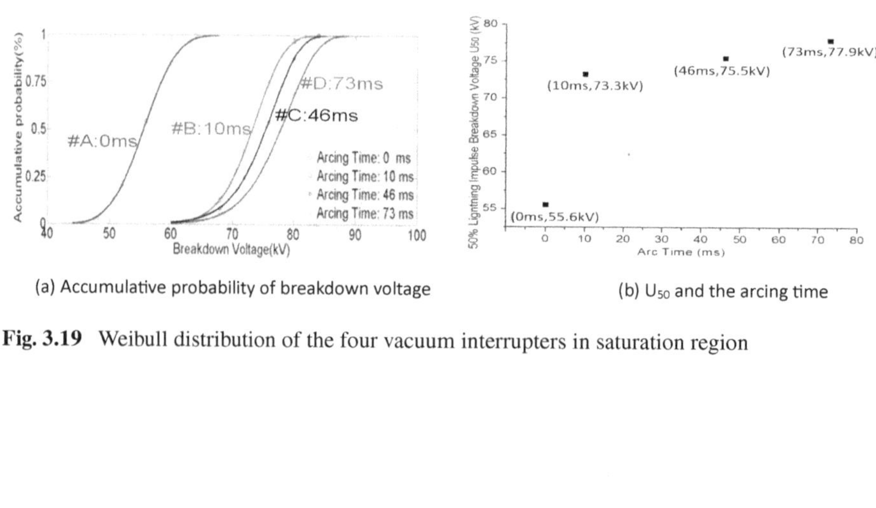

Fig. 3.19 Weibull distribution of the four vacuum interrupters in saturation region

breakdown voltage. It shows that the accumulative breakdown voltage probabilities of the four vacuum interrupters followed a Weibull distribution. Figure 3.19b shows a relationship between the 50% breakdown voltage U_{50} and the arcing time. It shows that with arcing time of 10, 46, and 73 ms, U_{50} was significantly higher than that of the no arcing case. It implied that the arc-melted layer improved the breakdown voltage. The U_{50} of the arcing time of 10, 46, and 73 ms were 73.3, 75.5, and 77.9 kV, respectively. They were quite close to each other. The various arcing time seems to have insignificant influence on the breakdown voltage.

The field emission current and the corresponding voltage followed the Fowler–Nordheim equation, as shown in Eq. (3.32).

$$\ln\frac{I}{V^2} = \ln\left[\frac{1.54 \times 10^{-6} A_e \beta^2 10^{4.52\phi^{-0.5}}}{\phi d^2}\right] - \frac{2.84 \times 10^9 d \phi^{1.5}}{\beta} \cdot \frac{1}{V} \tag{3.32}$$

where I is the field emission current in A; V is the corresponding voltage in V; d is the gap length in m; A_e is the effective emission area in m^2; and β is the electric field enhancement factor; Φ is work function in eV. Here, the work function was taken as 4.6 eV [33]. Table 3.3 shows β of the four vacuum interrupters based on the field emission measurement. Because the field emission current emitted from the rod electrode, β indicated the rod electrode's surface condition. For vacuum interrupter A, C and D, the field emission current was measured before the first occurrence of power frequency voltage breakdown. Table 3.3 shows that β of vacuum interrupters A, C and D were nearly the same, which were 270, 273, and 275, respectively, under the same power frequency voltage of 39.9 kV (peak value). For vacuum interrupter B, until the first occurrence of power frequency breakdown, the field emission current was so low that the calculated β has no physical meaning. Thus, β of vacuum interrupter B was available only before the second breakdown occurred, and it was 340 at a power frequency peak voltage of 42.7 kV.

Figure 3.20 shows the scanning electron microscope photographs of the cross sections of the rod electrodes in the four vacuum interrupters. Figure 3.20a shows no indications of arc-melted layer formation on the rod electrode of vacuum interrupter A. It is expected because there was no arcing with vacuum interrupter A. Figure 3.20b shows that an arc-melted layer formed on the rod electrode surface of

Table 3.3 Electric field enhancement factor of the four vacuum interrupters after lightning impulse voltage test

VI	Peak value of voltage/kV	Electric field enhancement factors	β measured before breakdown no
A	39.9	270	1
B	42.7	340	2
C	39.9	273	1
D	39.9	275	1

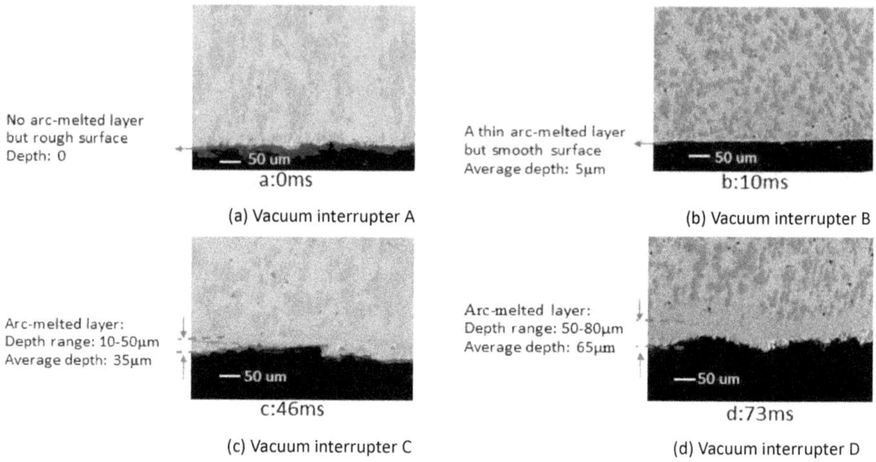

Fig. 3.20 Scanning electron microscope photographs of the cross sections of the rod electrodes in the four vacuum interrupters

vacuum interrupter B, and the depth was approximately 5 μm. Figure 3.20c shows that an arc-melted layer ranging from 10 to 50 μm formed on the rod electrode surface of vacuum interrupter C, and its average depth was approximately 35 μm. Figure 3.20d shows that an arc-melted layer ranging from 50 to 80 μm formed on the rod electrode surface of vacuum interrupter D, and its average depth was approximately 65 μm.

Table 3.4 shows the arcing time, U_{50} and average depth of the arc-melted layer of the four vacuum interrupters. It shows that the arcing time significantly influence the average depth of the arc-melted layers. The arcing time of 0, 10, 46, and 73 ms corresponded average depth of the arc-melted layer of 0, 5, 35, and 65 μm, respectively. It also shows that there was a significant gap of the U_{50}, 17.7 kV, between the 0 and 5 μm arc-melted layer depth. While the gaps of 2.2 and 2.4 kV of U_{50}, among the 5, 35 and 65 μm arc-melted layer depth are relatively small. The results suggested that an arc-melted layer significantly improved vacuum insulation, while a thicker arc-melted layer only slightly improved the vacuum insulation. Table 3.3 shows β of vacuum interrupters A and C, D are very close. While the U_{50} of vacuum interrupters

Table 3.4 The arcing time, 50% breakdown voltage, U_{50}, and average depth of arc-melted layer of the four vacuum interrupters

Vacuum interrupter	Arcing time (ms)	Average depth of arc-melted layer/μm	50% breakdown voltage U_{50}/kV
A	0	0	55.6
B	10	5	73.3
C	46	35	75.5
D	73	65	77.9

A and C, D were quite different. The results implied that it was an appearance of the arc-melted layer, not the the contact surface β value , that determined the vacuum insulation. An appearance of the smooth and fine Cr particles rich arc-melted layer determined the breakdown voltages .

3.2 Vacuum Interrupter Contacts

Contacts design for transmission voltage level vacuum interrupters requires a development of new arc control systems which are effective at the large contact gaps necessary for these high voltages and also have relatively low resistance when closed. A desired contacts design of transmission vacuum interrupters would be that the vacuum arcs are controlled well under large contact gaps and the loop resistance is so low that it meets the requirements of normal rated current and temperature rise in transmission voltage level circuit breakers. A high normal rated current is a challenge for vacuum interrupters, especially for vacuum interrupters at transmission voltage level (see Sect. 4.3). In general, axial magnetic field is preferred in large contact gaps, which forces the high-current anode mode vacuum arc to rapidly become diffuse and better distributed within the contact gap. The diffuse arc shows a superior current interruption performance. In large contact gaps, the vacuum arcs are easier to control by axial magnetic fields than transverse magnetic fields. In most axial magnetic field contacts current flows through a coil structure behind the contact plates. However, such axial magnetic contact should compromise a contradicted requirement that a strong axial magnetic field generation needs a long coil length, while a low loop resistance would like to minimize the coil length. One contact design to follow the compromise is a 2/3 coil-type axial magnetic field contact, which is introduced in Sect. 3.2.1. Another contact design is a horseshoe type axial magnetic field contact, which has no additional coil structure to generate axial magnetic field, thus it helps to reduce loop resistance. Section 3.2.2 addresses the horseshoe type axial magnetic field contact design for transmission voltage level vacuum interrupters. The third idea is to use transverse magnetic field contacts to reduce loop resistance, at the same time the vacuum arcs are subjected to an axial magnetic field. This idea has not realized it to an industrial design. However, as a basic research, it may inspire vacuum interrupter engineers to realize. This research is presented in Sect. 3.2.3.

3.2.1 2/3 Coil-Type Axial Magnetic Field Contact

How strong an axial magnetic field is needed in a successful current interruption? Fenski et al. [34] proposed an effective area criterion, which was a contact surface area where an axial magnetic field was higher than 4 mT/kA. Dullni et al. [35] measured the minimum axial magnetic field necessary to prevent arc constriction and contact surface melting. Their results were that for a 40-mm contact diameter, a current interruption of 12 kA needed an axial magnetic field of 40 mT. Morimiya et al. [36] found a current interruption of 14.8 kA peak value needed an axial magnetic field of ~60 mT for a 45 mm contact diameter and a 30 mm gap space. Stoving et al. [37] proposed that for the best arc stability, the axial magnetic field should be ~70 mT for a current interruption of 12 kA, with a 38 mm contact diameter and a 12 mm contact gap. Schulman et al. [38] established a critical axial magnetic field for a diffuse arc, which was linearly proportional to the peak current. Section 1.2.2 proposes Eq. (1.7) as a criterion of the critical axial magnetic field strength to avoid anode spot formation. However, a diffuse vacuum arc can not gurrantee a successful current interruption. It still need to know the axial magnetic field strength required for successful current interruptions in transmission voltage level vacuum interrupters, which has large contact gaps.

This section is to determine how strong an axial magnetic field is needed in test duty T100a for a 126 kV single-break vacuum circuit breaker as interrupting a rated short-circuit current of 40 kA (rms). To investigate the axial magnetic field needed in such case, a vacuum interrupter with a pair of 2/3 coil-type axial magnetic field contacts, was proposed [39]. Figure 3.21a shows a coil structure of the 2/3 coil-type axial magnetic field contacts. The coil has three branches and the current in each branch goes through 2/3 of the circumference. The contact material was CuCr50 (50 wt% of Cr). There was a double "T" beam made of stainless steel inside the coils and there were three stainless steel nail supports across the coil to increase the structure strength of the coil. A fixed contact was connected with a 390-mm-length copper cylinder rod in which a steel rod was inside. A movable contact was connected with a 500-mm-length steel cylinder rod in which a copper rod was inside. Figure 3.21b shows the full arrangement of the 2/3 coil-type axial magnetic field contacts, including the current carrying rods, the stainless steel support plates and the slotted contact plates. The contact diameter was 100 mm and the full contact gap was 60 mm [40]. The contact plates diameter was same to the coils and the contact plates have three slots with a length of 2/3 of the contact radius. The 2/3 coil-type axial magnetic field contact design provided a strong axial magnetic field and a low loop resistance at the same time.

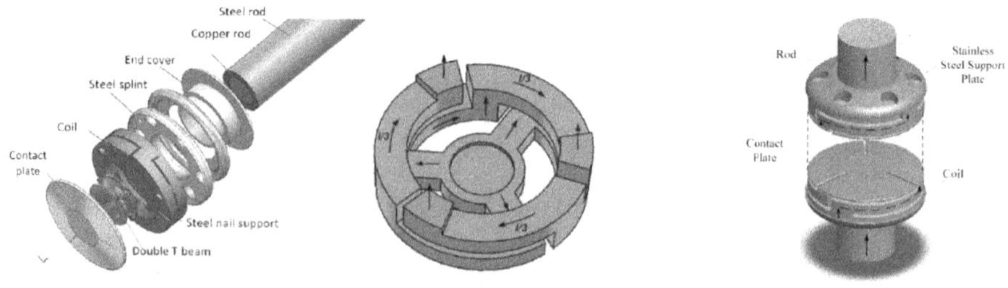

(a) coil structure (b) full arrangement

Fig. 3.21 A 2/3 coil-type axial magnetic field contact

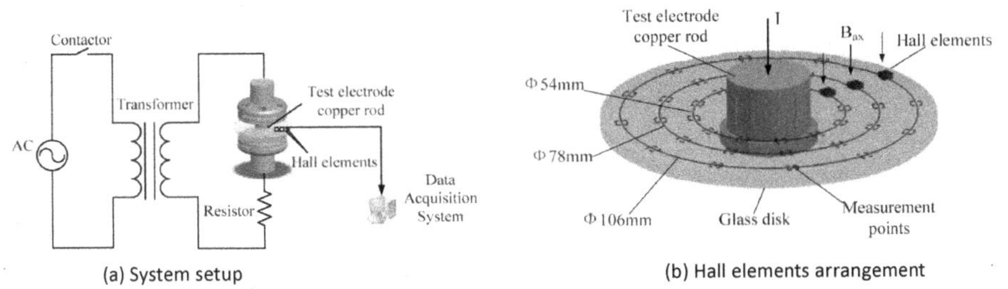

Fig. 3.22 An axial magnetic flux density measurement system setup

To study an axial magnetic field needed for test duty T100a of a 126 kV single-break vacuum circuit breaker, the accuracy of the 3-D finite element simulation of axial magnetic field should be validated by a magnetic measurement experiment. Figure 3.22a shows an axial magnetic flux density measurement system. A current of 1800 A (rms) with a frequency of 50 Hz was supplied to the 2/3 coil-type axial magnetic field contacts from a step-down transformer. The current was measured by a current transformer. The accuracy of the current measuring systems guaranteed the error was less than 10%. The axial magnetic field was measured at a fixed contact gap of 30 mm. The current flowed through the contacts via a pair of 30-mm high and 30-mm diameter copper conductor rods. Figure 3.22b shows an arrangement of Hall elements. Three Hall elements are used to measure axial magnetic flux density distribution. Melexis MLX90215 Hall elements are used to measure the axial magnetic field. The Hall elements are 4.5 × 3.8 × 1.2-mm programmable linear Hall Effect sensors IC fabricated utilizing silicon-CMOS technology. The three Hall elements were set on a glass disk that was installed on the intermediate plane between the two contacts. The Hall elements were set at the diameters of 54, 78, and 106 mm, respectively. The glass disk could rotate around the axis of the disk and the data of the axial magnetic flux density were recorded every 30°. This meant that the distribution of the axial magnetic field between the contacts can be measured from 36 measured points on the intermediate plane.

A 126 kV/2500 A/40 kA single-break vacuum circuit breaker with the 2/3 coil-type axial magnetic field contact was used in the test duty T100a current interruption tests. Figure 3.23a shows a photo of the 126 kV single break vacuum circuit breaker. Its main technical parameters are shown in Table 3.5. The rated voltage is 126 kV. The nominal rated current is 2500 A. The rated frequency is 50 Hz. The rated short-circuit breaking current is 40 kA (rms). To investigate the axial magnetic field needed for the 126 kV single break vacuum circuit breaker in test duty T100a as interrupting a rated short-circuit current of 40 kA (rms), the short circuit current was interrupted 18 times on T100a with various arcing time according to the standard IEC 62,271–101 Table 3.1a [41]. The peak value of the transient recovery voltage (TRV) is 216 kV. The rise rate of TRV is 2 kV/μs. Figure 3.23b shows the 126 kV vacuum circuit breaker's opening displacement curve. The maximum contact gap was 60 mm. The contact gaps at which the arc extinguished under various arcing time, can be calculated

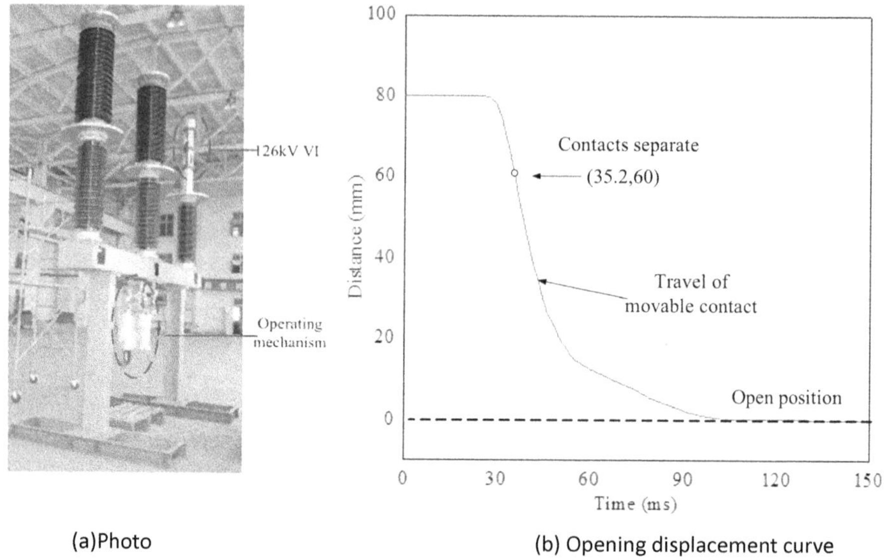

Fig. 3.23 A 126 kV single break vacuum circuit breaker for the test duty T100a current interruption tests

Table 3.5 Main technical parameters of the 126 kV single break vacuum circuit breaker

No.	Technical parameters	Value	Units
1	Rated voltage	126	kV, rms
2	Rated frequency	50	Hz
3	Nominal rated current	2500	A, rms
4	Rated short circuit breaking current	40	kA, rms
5	Rated short circuit making current	100	kA, rms
6	Rated short time withstand current (4 s)	40	kA, rms
7	Peak withstand current	100	kA, rms
8	Power frequency withstand voltage (1 min)	230	kV, rms
9	Impulse withstand voltage (1.2X50us)	550	kV, peak
10	Permissible temperature of terminal	$-40 \leq T \leq 105$	°C
11	Mechanical life	2000	Operations
12	Number of rated nominal current interruption	2000	Operations
13	Contact gap	60	mm
13	Operating mechanism stroke	80	mm
14	Maximum contact erosion	3	mm

from Fig. 3.23b. The T100a tests of the 126-kV single-break VCB were carried out in a synthetic circuit in Xi'an High-voltage Apparatus Research Institute Co., Ltd (XIHARI), Xi'an, China.

An axial magnetic field simulation was carried out with a commercial 3-D finite element analysis software. The simulation included two steps. In the first step, the accuracy of the simulation was validated by the measured results. Figure 3.24a shows the contact model for the validation. There was a copper conductor rod linking the contacts, which hold the glass disk in the axial magnetic field measurement system, as shown in Fig. 3.22b. In the second step, axial magnetic field was analyzed after current interruption tests. After the 18 times T100a tests, axial magnetic field distribution under various arcing contact gaps according to arcing time, were simulated. The simulation results were used to analyze the current interruption results. Figure 3.24b shows the axial magnetic field simulation model for current interruptions. The arc plasma was substituted by a conductor with the same diameter of the contacts. The conductivity of the arc plasma conductor was obtained from the measured arc voltage and the arc current. Table 3.6 shows the material properties in the simulation.

Figure 3.25 shows the axial magnetic field simulation validation results. Figure 3.25a shows the simulation results of the axial magnetic flux density distribution on the intermediate plane of the contact model shown in Fig. 3.24a. The contact gap was 30 mm. The current was 1800 A (rms). The simulation results were used to compare with the measured results. Figure 3.25b shows a comparison of the simulation results and the measured results of the axial magnetic flux density in a radial

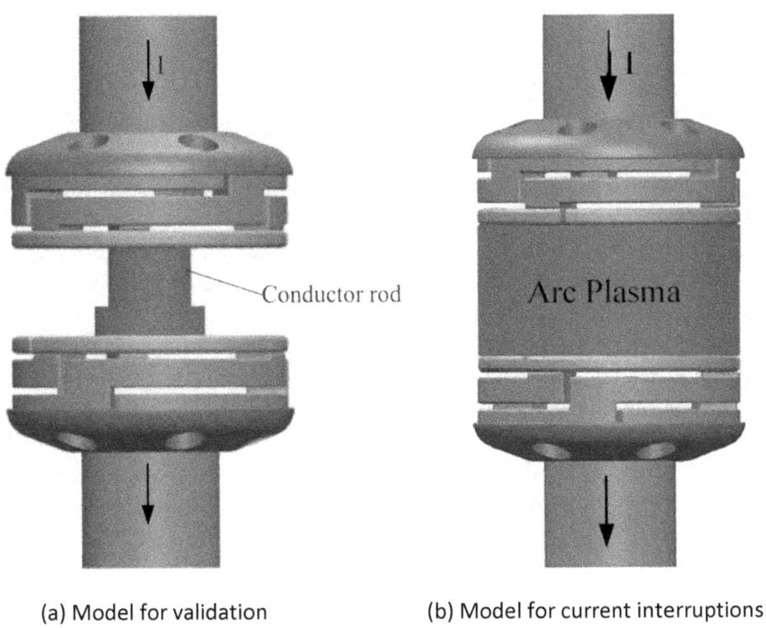

Fig. 3.24 Contact model for axial magnetic field simulation

Table 3.6 Material properties in the axial magnetic field simulation

Parts	Material	Conductivity(Siemens/meter)
Coil	Copper	5.8e+007
Conductor rod	Copper	5.8e+007
Arc plasma	Plasma	1579
Copper arc	Copper	5.8e+007
Contact plate	CuCr50	1.044e+007
Background	Vacuum	0

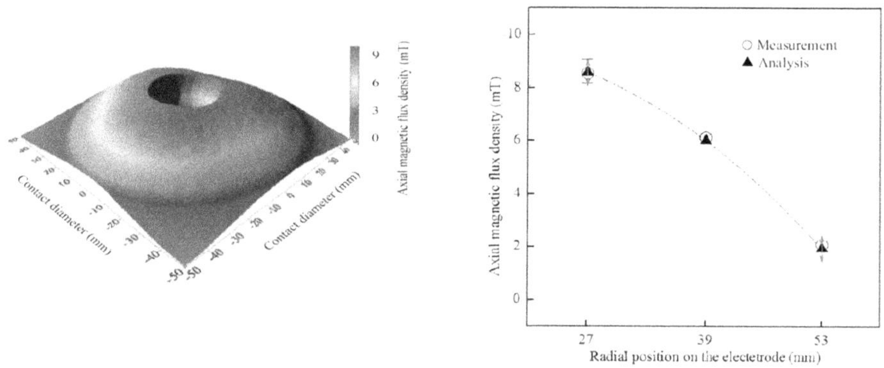

(a) Simulation results (b) Comparison of the simulation results and the measured results.

Fig. 3.25 Axial magnetic field simulation validation results. Distribution of the axial magnetic flux density on the intermediate plane. Current-1800 A (rms) Contact gap-30 mm

direction. It shows the relative errors were less than 10%. The simulation results fitted the measured ones well.

Table 3.7 shows the T100a test results of the 126-kV single-break vacuum circuit breaker. It shows the arcing time, breaking current, contact gap at current zero, and current interruption results. The T100a tests were carried out 18 times with an arcing time ranging from 3.3 to 19.0 ms. The peak of the breaking current ranged from 40.3 to 72.0 kA. The contact gaps at current zero ranged from 12 to 50 mm. The contact gaps at current zero were calculated from the opening displacement curve shown in Fig. 3.23b and arcing time. Figure 3.26 shows three typical current interruption results. Figure 3.26a shows a failure interruption result with an arcing time of 3.3 ms, and breaking current of 40.3 kA peak value. Figure 3.26b shows a successful interruption result with arcing time 10.4 ms, breaking current 67.4 kA peak. Figure 3.26c shows a failure interruption result with arcing time 17.9 ms, breaking current 67.9 kA peak.

The axial magnetic field distribution was simulated with contact gaps ranging from 12 to 50 mm to analyze the current interruption results. Figure 3.27 shows the axial magnetic flux density distribution on the intermediate plane at current peak under the various arcing time. Figure 3.27a shows the axial magnetic flux density

Table 3.7 T100a test results of the 126-kV single-break vacuum circuit breaker

No.	Arc duration (ms)	Brcaking cutent (peak, kA)	Contact gap at current zero (mm)	Switching results
1	3.3	40.3	12	Failure
2	4.4	43.6	16	Failure
3	4.6	41.5	18	Success
4	5.4	44.0	19	Failure
5	5.5	42.1	19	Success
6	6.3	44.7	22	Failure
7	6.4	44.5	23	Success
8	6.6	43.0	23	Success
9	7.4	43.7	26	Success
10	7.6	45.6	27	Success
11	10.4	67.4	34	Success
12	10.7	67.0	38	Success
13	14.6	70.0	44	Success
14	15.8	67.7	47	Failure
15	16.1	71.1	47	Failure
16	17.0	70.1	48	Failure
17	17.9	67.9	49	Failure
18	19.0	72.0	50	Failure

distribution with an arcing time of 3.3 ms and a contact gap of 12 mm. The breaking current was 40.3 kA peak and the maximum axial magnetic field at current peak was 434 mT. Figure 3.27b shows the axial magnetic flux density distribution with an arcing time of 10.4 ms and a contact gap of 34 mm. The breaking current was 67.4 kA peak and the maximum axial magnetic field at current peak was 484 mT. Figure 3.27c shows the axial magnetic flux density distribution with an arcing time of 17.9 ms and a contact gap of 49 mm. The breaking current was 67.9 kA peak and the maximum axial magnetic field at current peak was 384 mT.

There is a phase shift between the source current and the induced axial magnetic field, which is caused by the eddy current induced in conductors, such as the contact plates and the coil structures. The phase shift could have an impact on the current interruption performance, especially its influence at the current zero. It is understood that a residual axial magnetic field after current zero hindered a rapid plasma diffusion outward contact gaps and it reduced the effective mean free path for ionizing collisions under the transient recovery voltage [42]. Table 3.8 shows the phase shift time of the proposed 2/3 coil-type axial magnetic field contact, as an important influencing factor for the current interruption performance. The given phase shift time was at the center of intermediate plane of the contact gap and it was in a range from

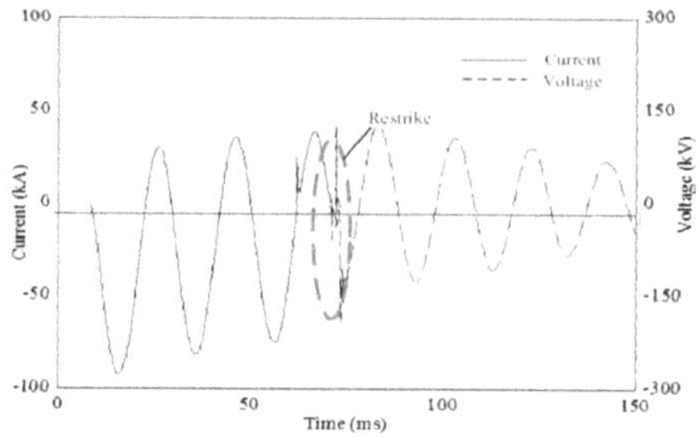

(a) a failure interruption
arcing time 3.3 ms, breaking current 40.3 kA peak

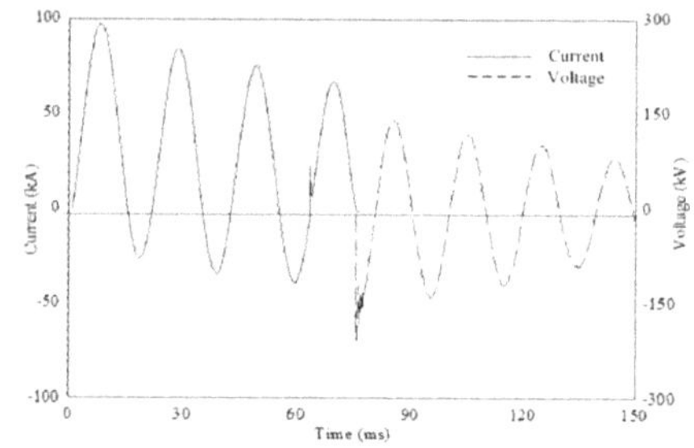

(b) a successful interruption
arcing time 10.4 ms, breaking current 67.4 kA peak

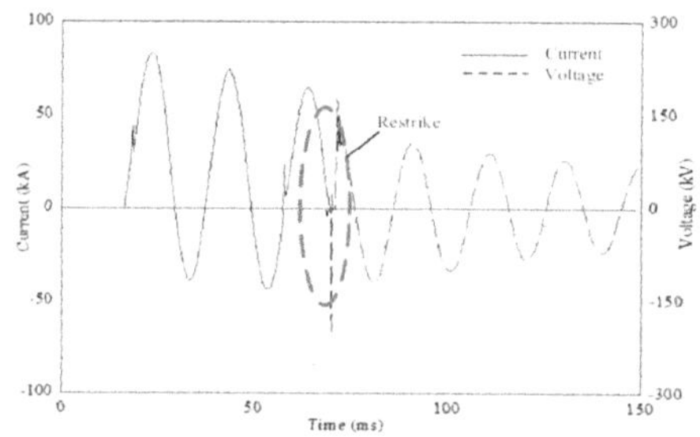

(c) a failure interruption
arcing time 17.9 ms, breaking current 67.9 kA peak

Fig. 3.26 Typical T100a current interruption results of the 126-kV single-break vacuum circuit breaker

1.36 to 0.75 ms according to an increase of the contact gap at current zero from 12 to 50 mm.

Table 3.8 also shows values of the maximum axial magnetic field per kilo-ampere of the breaking current peak for the 18 times of T100a current interruption, with an indication of a failure or a successful interruption. The maximum value of the axial magnetic field was obtained from the maximum value on the intermediate plane of the contact gap at current peak (see Fig. 3.27). The 18 times 40 kA (rms) current interruption tests of test duty T100a for the 126 kV vacuum circuit breaker suggested the minimum arcing contact gap was ranging from 18 to 23 mm and the maximum arcing contact gap was 44 mm (see Table 3.7). The minimum arcing contact gap corresponded to the maximum axial magnetic field per kilo-ampere of the breaking current peak ranging from 9.3 to 8.5 mT/kA. The maximum arcing contact gap corresponded to the maximum axial magnetic field per kilo-ampere of the breaking current peak of 6.1 mT/kA. The results suggested that for a successful current interruption, an axial magnetic field of 9.3–8.5 mT/kA is needed for the minimum arcing contact gap and an axial magnetic field of 6.1 mT/kA is needed for the maximum arcing contact gap.

In a current interruption in vacuum, the minimum arcing time is corresponding to the minimum arcing contact gap. In the period of minimum arcing time, the contact

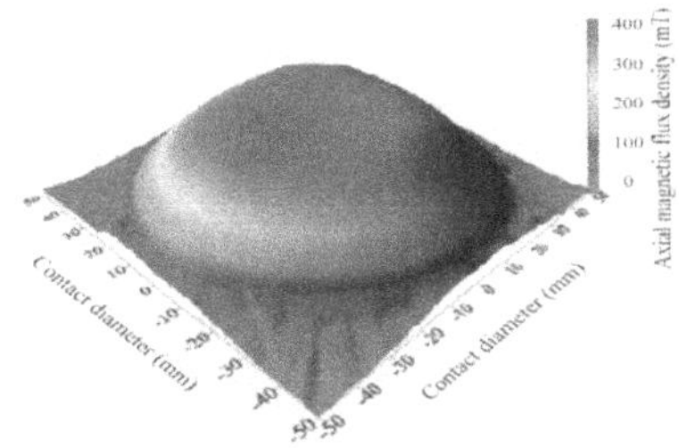

(a) Arcing time: 3.3 ms. Breaking current: 40.3 kA peak
Contact gap at current zero: 12 mm. Maximum AMF: 434 mT.

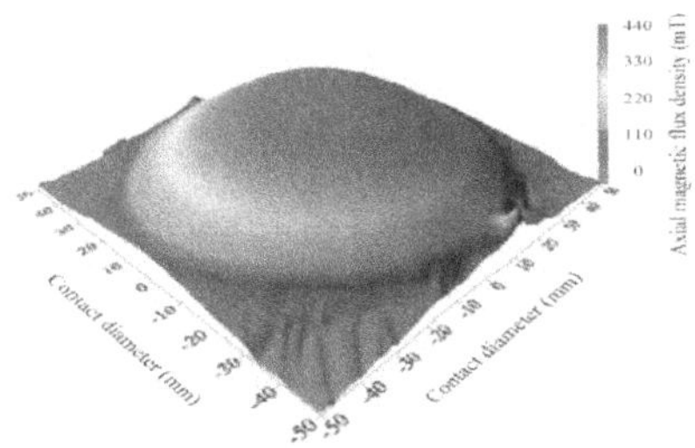

(b) Arcing time: 10.4 ms. Breaking current: 67.4 kA peak.
Contact gap at current zero: 34mm. Maximum AMF: 484 mT.

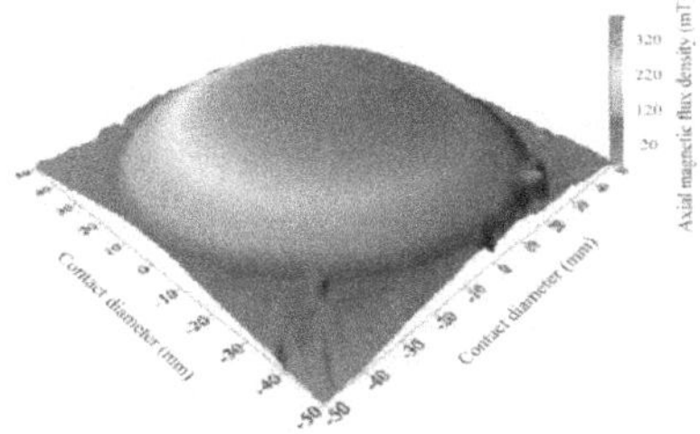

(c) Arcing time: 17.9 ms. Breaking current: 67.9 kA peak.
Contact gap at current zero: 49 mm. Maximum axial AMF: 384 mT.

Fig. 3.27 Axial magnetic field distribution on the intermediate plane at current peak

surfaces suffer a formation and rupture of a molten metal bridge. Then the minimum arcing time evidences a transition from the initial bridge column arc to a diffuse vacuum arc, with a help of the applied axial magnetic field. Thus, at the end of the minimum arcing time at which current zero, the metal vapor density and residual plasma between the closely spaced contacts should be low to withstand a fast-rising transient recovery voltage that reaches a high value (see Sect. 2.4.1). In addition, a solidification process of the local melted region on the contact surface, which is induced by the initial bridge arc, would have an impact on the dielectric recovery process. In addition, the phase shift effect between the axial magnetic field and source current not only causes a time delay for applying the axial magnetic field on the vacuum arc, which has an impact on a transition from the bridge arc to a diffuse arc as the contacts part, but also causes a hinder of the charged particles escaping from the contact gap.

The maximum arcing time of the 126 kV vacuum circuit breaker could be determined by an interaction of vacuum arc and axial magnetic field applied. If the maximum contact gap at which the arc extinguishes is too high, the vacuum arc tends to be unstable burning, and an anode spot has a high possibility to form at a large contact gap. Both an unstable burning arc and an anode spot would easily to lead a reigniting under a high transient recovery voltage value. However, by applying an axial magnetic field keeps a vacuum arc burning more stable and enhances a threshold current of an anode spot formation. Therefore, the maximum arcing time

Table 3.8 Axial magnetic field simulation results for the 126 kV single break vacuum circuit breaker in test duty T100a

No	Arc duration (ms)	Breaking current (peak, kA)	B_{max}/I_p (mT/kA)	Phase shift time (ms)	Switching results
1	3.3	40.3	10.8	1.36	Failure
2	4.4	43.6	9.7	1.27	Failure
3	4.6	41.5	9.3	1.19	Success
4	5.4	44.0	9.1	1.17	Failure
5	5.5	42.1	9.1	1.17	Success
6	6.3	44.7	8.6	1.07	Failure
7	6.4	44.5	8.5	1.07	Success
8	6.6	43.0	8.4	1.07	Success
9	7.4	43.7	8.0	1.02	Success
10	7.6	45.6	7.9	1.00	Success
11	10.4	67.4	7.2	0.90	Success
12	10.7	67.0	6.8	0.86	Success
13	14.6	70.0	6.1	0.80	Success
14	15.8	67.7	5.8	0.78	Failure
15	16.1	71.1	5.9	0.78	Failure
16	17.0	70.1	5.8	0.77	Failure
17	17.9	67.9	5.7	0.76	Failure
18	19.0	72.0	5.6	0.75	Failure

is corresponding to the maximum arcing contact gap and the axial magnetic field strength.

In the test 126 kV vacuum interrupter, the contact material is CuCr50 (50% weight of Chromium). The choice of 50% of Chromium is not an optimum percentage for high current interruption. Contact material of CuCr25 (25% of Cr) or CuCr30 (30% of Cr) showed a better current interrupting performance in medium voltage level. However, CuCr50 contact material could have a better insulation level than that of CuCr25 or CuCr30 contact material. This is also important if one considers a 126 kV single break vacuum interrupter should withstand a much higher transient recovery voltage, power frequency voltage and the basic impulse level (BIL) than its medium voltage counterpart.

3.2.2 *Horseshoe-Type Axial Magnetic Field Contact*

This section focuses on vacuum arc behaviors of horseshoe type axial magnetic field contact used for vacuum interrupters at transmission voltage level. A high normal rated current is a challenge for the transmission voltage level vacuum circuit breakers.

Compared with coil-type axial magnetic field contact, horseshoe-type axial magnetic field contact has much lower loop resistance [43, 44]. Thus, the nominal rated current level of the vacuum interrupters can be promoted. However, horseshoe-type contacts generate a bipolar axial magnetic field. Comparing with the unipolar axial magnetic field generated by the 2/3 coil-type axial magnetic field contacts, the bipolar axial magnetic field weakens faster as the contact gap increases. That means a weak control of the vacuum arcs, which leads to a more risk of current interruption failure at large contact gaps. This section is to determine the vacuum arc behaviors of horseshoe-type axial magnetic field contact at large contact gaps. The results could help horseshoe-type axial magnetic field contact design for vacuum interrupters at transmission voltage level.

Figure 3.28 shows a horseshoe-type axial magnetic field contact. The contact was composed of four components: (1) contact plate; (2) iron plates; (3) terminal plate; and (4) conducting rod. Horseshoe shape iron plates were set behind the contact plate. The directions of the horseshoe shape iron plates slot for stationary contact and movable contact were opposite. Therefore, the magnetic field surround conducting rod was forced out of the iron plates to cross contact space when contact gap was smaller than the slot width. Thus, a bipolar axial magnetic field was generated in contact gap. In order to study the vacuum arc behaviors, a horseshoe-type axial magnetic field contact was designed for vacuum interrupters at transmission voltage level with following parameters. Contact material was CuCr50 (50% of Cr). Contact diameter was 120 mm. Iron plates were with an electrical conductivity of 1×10^5 S/m and a relative permeability of 7000. The slot width of the horse-shoe shape iron plate was 58 mm. The height of the iron plates was 20 mm. Conducting rod was made of oxygen-free high-conductivity copper.

Drawn arc experiments were carried out to observe vacuum arc behaviors. Figure 3.29 shows the experimental setup. A current was generated by a L–C discharging circuit. The capacitance was 55,760 μF and the inductance was 187.5

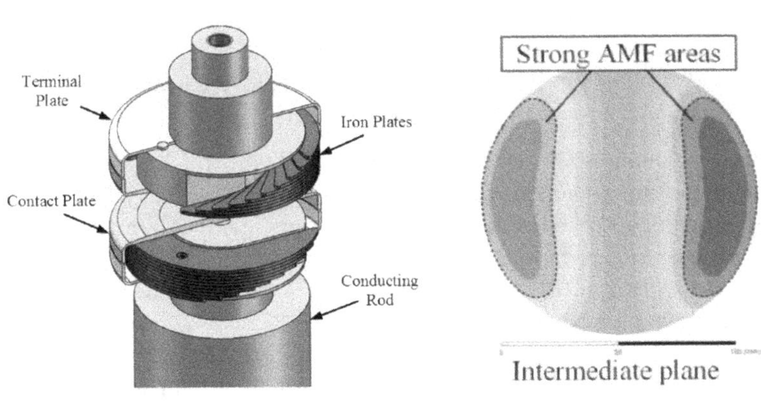

(a) Contact structure (b) bipolar axial magnetic field

Fig. 3.28 A horseshoe-type axial magnetic field contact

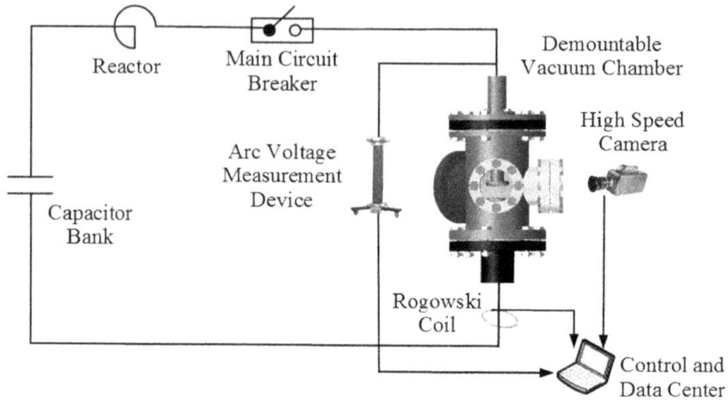

Fig. 3.29 Experimental setup for vacuum arc behaviors observation

μH. The frequency of current was 49.2 Hz. The horseshoe type contacts were arranged in a demountable vacuum chamber. During the experiment, the inner pressure of demountable vacuum chamber was maintained at $\sim 10^{-4}$ Pa through continuous pumping. The moveable contact was driven by a permanent magnetic actuator. A main circuit breaker kept open during charging operation of the capacitor banks. In the experiment, the permanent magnetic actuator opened according to a preset arcing time after main circuit breaker closed. The full contact gap was 60 mm. The average opening velocity within the first 20 mm after contact separation was 2.4 m/s. A high-speed camera was used to record vacuum arc behaviors. Recording velocity of the camera was set at 2×10^4 frames/s. Exposure time was fixed at 4 μs during the entire experiment. Arc current and arc voltage were measured by a Rogowski coil and a high-voltage probe, respectively. Recorded data was sent to a control and data center. The experimental current level ranged from 4 to 40 kA, stepped by 4 kA. Prospective arcing time was set as 9, 12, and 15 ms, respectively.

Figure 3.30 shows that there were eight arc modes observed. The eight arc modes were designated according to arc column appearance [45–47]. The behaviors of the eight arc modes of the horseshoe-type axial magnetic field contact were summarized as following.

(a) Bridge column mode. It is a transient arc column which forms just after the rupture of initial molten metal bridge.
(b) Small-gap constricted column mode. There are distinct luminous boundaries of vacuum arc appearing in this mode when contact gap is small.
(c) Diffuse arc mode. Visible plasma fills contact gap while individual plasma jet is usually not resolved.
(d) Multi-cathode-spot arc mode. This vacuum arc mode is characterized by cathode spots spread over the cathode with independent plasma jets.
(e) Diffuse column mode. A center plasma core extending from anode to cathode with diffuse boundaries observed. The diameter of arc column decreases with

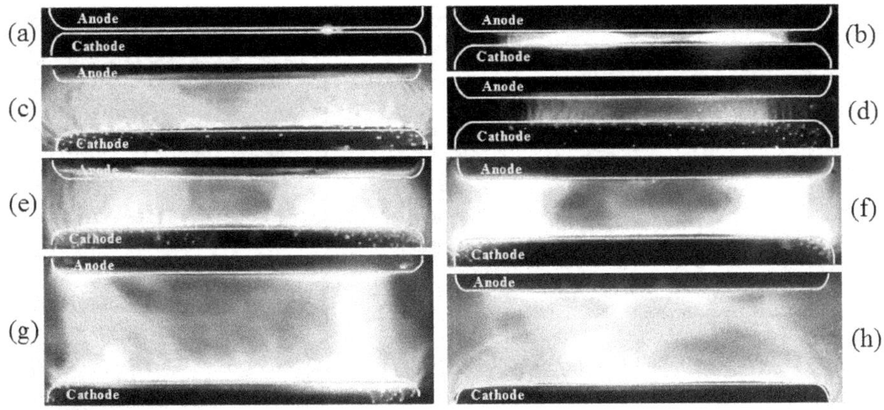

Fig. 3.30 Eight vacuum arc modes of horseshoe-type axial magnetic field contact. **a** Bridge column. **b** Small-gap constricted column. **c** Diffuse arc. **d** Multi-cathode-spot arc. **e** Diffuse column. **f** Large-gap constricted column. **g** Anode jet. **h** Uncontrollable arc

increasing of contact gap. The center of arc column is much brighter than arc periphery. Large quantities of cathode spots move rapidly toward contact edge.

(f) Large-gap constricted column mode. A much brighter plasma core occupies nearly entire column with well-defined boundaries. The diameter of arc column does not change, which is insensitive to instantaneous current value. This mode appears at a relatively large contact gap with high current.

(g) Anode jet mode. Anode jets formed. An anode jet and a cathode jet separate their apexes as contact gap increased.

(h) Uncontrollable arc mode. In this mode, vacuum arc becomes extremely unstable and distorted. Arc column cannot be observed. Vacuum arcs are out of control by the axial magnetic field.

The experiment results revealed a non-synchronization phenomenon of the twin arcs. The non-synchronization has two sides. One side was that the evolution of the twin arcs was not synchronized. There was a time difference with the evolution of the twin arcs. The other side was a non-synchronization behavior of the twin arcs. Figure 3.31 shows a non-synchronized evolution of the twin vacuum arcs. The arc current was 40 kA (rms). Contacts separation moment was start point of arcing evolution. After contacts separation, both the twin arcs appeared in bridge column arc mode. At T1 the twin arcs evolved into small-gap constricted column mode. Then a difference appeared. One arc column evolved into diffuse column mode at T2, while the other arc column evolved into diffuse column mode at T3. The nonsynchronous duration between T2 and T3 was 0.59 ms. After that the twin arcs evolved into diffuse arc mode at T4 and T5, respectively. The nonsynchronous duration between T4 and T5 was 0.61 ms. With arc current further decreased, the twin arcs evolved into multi-cathode-spot mode at T6 and T7, respectively, with a nonsynchronous duration of 0.49 ms. Figure 3.31 also shows the non-synchronization behaviors of the twin arcs as indicated in photos (1)–(5). Photo (1) shows the twin arcs were both in small-gap

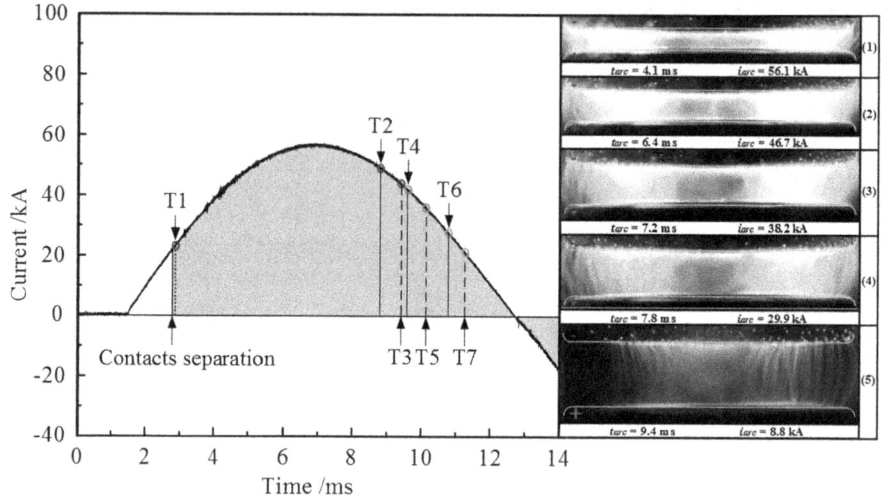

Fig. 3.31 Non-synchronization phenomena of the twin arcs with horseshoe-type axial magnetic field contact

constricted column mode. However, the column diameters were different. The left column and the right column were ~24 and ~27 mm, respectively. Photo (2) shows the left arc column was in diffuse column mode with a reduced column diameter of ~20 mm, while the right column was still in small-gap constricted column mode with a column diameter of ~28 mm. Photo (3) shows the left arc evolved into diffuse arc mode, while the right arc was in diffuse column mode with a column diameter of ~18 mm. Photo (4) shows the twin arcs were in diffuse arc mode and photo (5) shows the twin arcs were in multi-cathode-spot mode. However, the instantaneous current carried by each one of the twin arcs was different. Because the number of cathode spots in each of the twin arcs were different. Non-synchronized arcs would result in uneven arc energy input and contact erosions, which might be adverse to vacuum insulation and current interruption performance of vacuum interrupters with horseshoe-type axial magnetic field contact at transmission voltage level.

Table 3.9 shows the nonsynchronous duration of twin arcs in arc mode transition. The arc current ranged from 24 to 40 kA. The arcing time was 9 ms. It shows that the nonsynchronous phenomena existed in all of the investigated current levels and in each arc mode transition, including the transitions from small-gap constricted column mode to diffuse column mode, from diffuse column mode to diffuse arc mode, and from diffuse arc mode to multi-cathode-spot arc mode. The nonsynchronous duration ranged from 0.41 to 1.55 ms.

Figure 3.32 shows the vacuum arc mode transition diagram of the horseshoe-type axial magnetic field contact at a full contact gap of 60 mm. The arc current ranged from 4 to 40 kA (rms) and there were three arcing time of 9, 12 and 15 ms. It shows the boundaries between the eight arc modes. After contacts separated, small-gap constricted column arc appeared at small contact gaps. When the contact

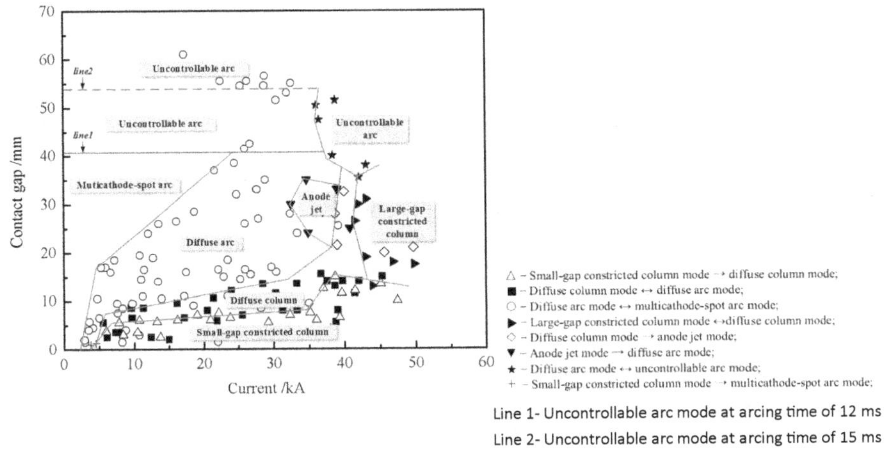

Fig. 3.32 Vacuum arc mode transition diagram of horseshoe-type axial magnetic field contact at a full contact gap of 60 mm

Table 3.9 Nonsynchronous duration of twin arc columns in arc mode transition

Arc current/kA (rms)	Transition from small-gap constricted column mode to diffuse column mode/ms	Transition from diffuse column mode to diffuse arc mode/ms	Transition from diffuse arc mode to multi-cathode-spot arc mode/ms
24	0.71	1.19	1.55
28	0.75	0.95	1.50
32	0.45	0.71	0.45
36	0.98	1.04	0.41
40	0.59	0.61	0.49

gap increased, diffuse column arc appeared. As the contact gap further increased, diffuse arc appeared. When the arc current and contact gap increased, the diffuse arc evolved into anode jet arc mode. When the arc current further increased at a large contact gap, the anode jet arc mode evolved into diffuse column arc and large-gap constricted column arc mode, respectively. If the current decreased, diffuse arc became multi-cathode-spot arc. Uncontrollable arc mode only appeared at long contact gaps, because of the weakened axial magnetic field. At an arcing time of 9 ms, uncontrollable arc mode appeared with contact gap larger than 36 mm. At an arcing time of 12 and 15 ms, uncontrollable arc mode appeared with contact gap of 41 and 54 mm, respectively, as indicated with line 1 and line 2. Therefore, contact gaps should be controlled to avoid vacuum arcs evolving into uncontrollable arc mode. Besides, large-gap constricted column arc mode should also be avoided. Vacuum circuit breaker should control the opening displacement curve carefully when horseshoe-type axial magnetic field contacts are used for vacuum interrupters at transmission voltage level. In this case, it recommended a contact gap at current zero of 36 mm and bellow to avoid uncontrollable arc mode. Section 3.2.1 suggested the maximum arcing contact gap of 44 mm for 2/3 coil-type axial magnetic field contacts used in transmission voltage level vacuum interrupters. However, the loop resistance of the 2/3 coil type axial magnetic field contact was significantly higher than the horseshoe-type axial magnetic field contact. That is why horseshoe-type contact is considered to be used in vacuum interrupters at transmission voltage level. Therefore, vacuum arc mode transition diagram would be helpful in the design of opening displacement curve for vacuum interrupters at transmission voltage levels (see Sect. 4.1).

Figure 3.33 shows a correlation among vacuum arc behaviors, arc erosion on contact surface, and bipolar axial magnetic field distribution on the contact surface. The axial magnetic field was calculated with a contact gap of 20 mm at arc current peak of 40 kA (rms). The vacuum arc was in large-gap constricted column mode. The width of each arc column D was ~25 mm. D corresponded to the area with axial magnetic field strength higher than 75% of the maximum. D also corresponded to severe contact erosion region. Therefore, D seems to be a characteristic parameter for the horseshoe-type axial magnetic field contacts.

3.2.3 Transverse Magnetic Field Contacts Subjected to Axial Magnetic Field

Arc control technology in vacuum is accomplished by the application of a transverse magnetic field or an axial magnetic field on a vacuum arc, and both arc control technologies aim to distribute arc heating homogeneously over the surfaces of the contacts, in order to avoid overheating at single locations [48]. Transverse magnetic field contacts offer an advantage of low loop resistance for nominal current conduction. Axial magnetic field contacts offer an effective control for vacuum arcs until large contact gaps (see Sect. 3.2.1). A new control technology for vacuum arcs appeared recently, which is based on a combination of the transverse magnetic field and an axial magnetic field. Lamara and Gentsch [49] invented a double contact system, which was a coaxial contact consisting of a spiral transverse magnetic field contact and a cup-type axial magnetic field. The double contacts system offered an advantage of low loop resistance for nominal current conduction, as a standard transverse magnetic field contacts, and a vacuum arc control as that of axial magnetic field contacts was expected. Wang et al. [50] developed a double coaxial contact system, which consisted of a pair of "卍" type transverse magnetic field contacts and a pair of cup-type AMF concentric contacts. They indicated an influence of the combined magnetic fields on the vacuum arcs at different stages in the discharging process. The double contact system offers a potential approach for an application in vacuum interrupters at transmission voltage level. However, in the previous studies,

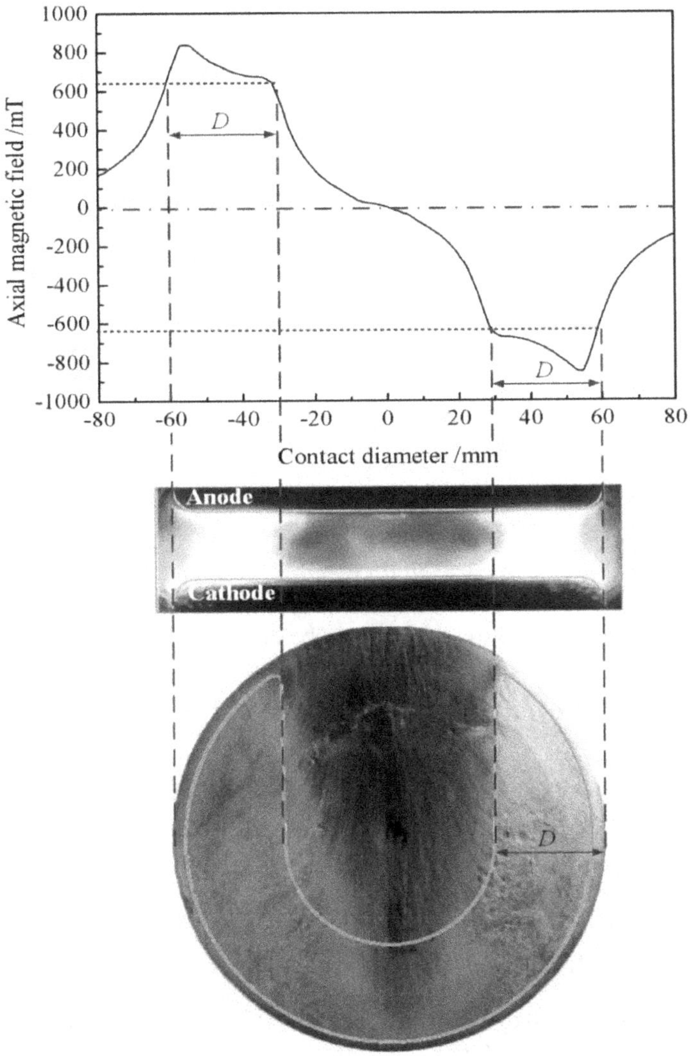

Fig. 3.33 A correlation among vacuum arc behaviors, arc erosion on contact surface, and bipolar axial magnetic field distribution.

a combined magnetic field was generated by the double contact system, in which the strength of each kind of contact's magnetic field could not be artificially adjusted. This section is to reveal an impact of the axial magnetic field strength on the vacuum arc behaviors driven by a transverse magnetic field. The experiment was designed as following. A drawing vacuum arc was initiated between a pair of transverse magnetic field contacts. The vacuum arc was also under a control of an external and adjustable axial magnetic field. The results might be helpful for designing of double contact

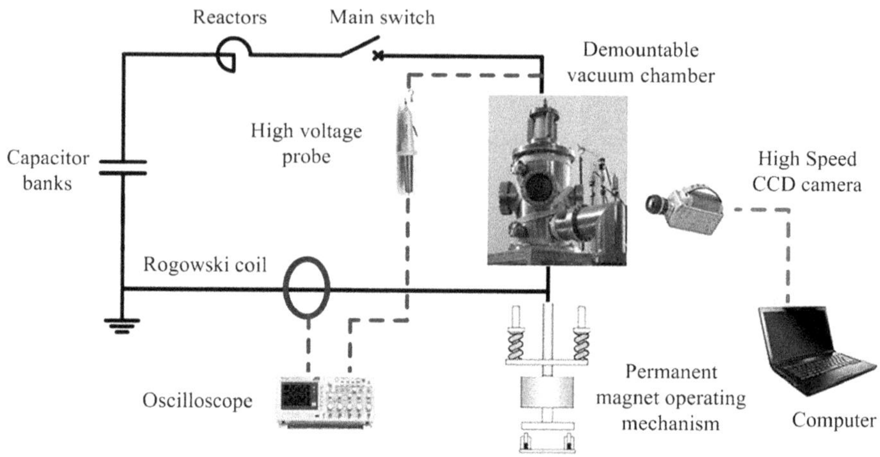

Fig. 3.34 Experimental setup

system with combined magnetic fields in vacuum interrupters at transmission voltage level.

Figure 3.34 shows the experimental setup. The experiment was performed by using an inductor–capacitor (L-C) discharging circuit. The current frequency was 45 Hz, and the arc current reached 30 kA (rms). The arcing time was controlled to be ~11 ms. An oscilloscope was used to record both the arc current and the arc voltage waveform. The arc current was measured by a Rogowski coil. The arc voltage was measured by a high-voltage probe. The contacts opening was using a permanent magnet operating mechanism. The contacts were installed inside a demountable vacuum chamber, which was subsequently evacuated and maintained at a pressure of 10^{-3}–10^{-4} Pa. A high-speed charge-coupled device (CCD) video camera was used to record the vacuum arc behaviors through an observation window. The CCD video camera was set at a recording speed of 10 000 frames/s.

Figure 3.35 shows an arrangement of transverse magnetic field contacts and external Helmholtz coil, both of which were installed inside the demountable vacuum chamber. The Helmholtz coil was connected to a DC current source to generate an axial magnetic field. The Helmholtz coil has 80 turns and an average radius of 160 mm. The DC current source was switched on about 150 ms before contact separated. The applied axial magnetic flux density was uniform and it can be adjusted in a range from 0 to 110 mT. Table 3.10 shows the experimental arrangements. The experimental arc currents were set at 8, 12, 16, 18, and 20 kA (rms). The applied axial magnetic flux density was set at 0, 20, 37, 56, 74, 93, and 110 mT. The diameters of both the anode and the cathode are 60 mm. The average contact opening velocity was set at 1.5 m/s. The full contact gap was set as 20 mm. The contact material of the transverse magnetic field contacts was CuCr25 (25% Cr).

It is meaningful to compare the vacuum arc behaviors only under the transverse magnetic field as a benchmark with that under the combined magnetic fields. The

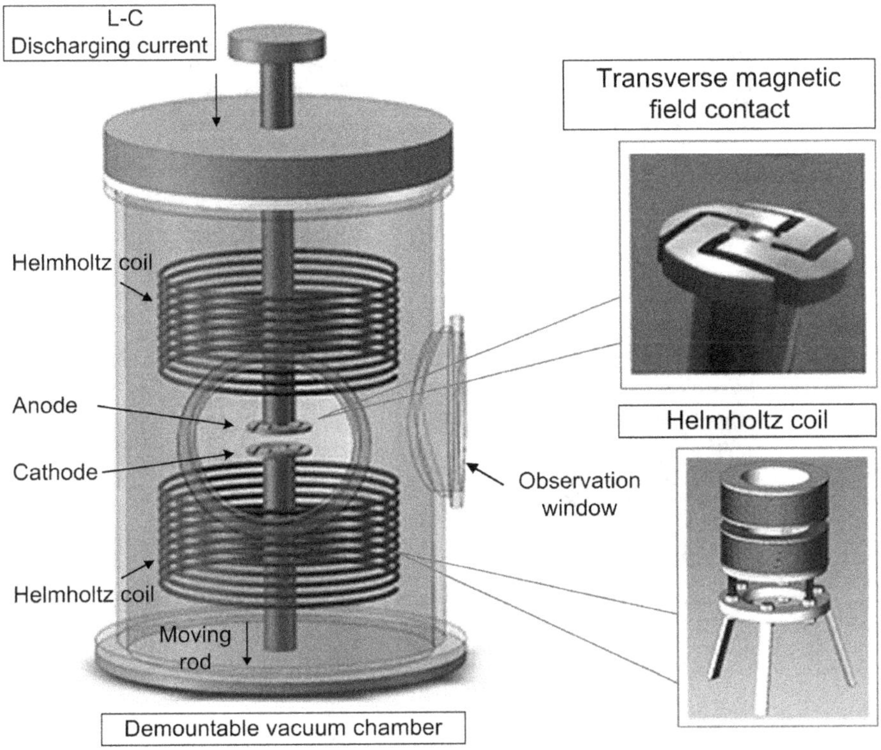

Fig. 3.35 An arrangement of transverse magnetic field contacts and external Helmholtz coil

Table 3.10 Experimental arrangements

Experimental series	Parameter value
Arc current (rms) (kA)	8, 12, 16, 18, 20
AMF fux density (mT)	0, 20,37, 56, 74, 93, 110
Diameter (mm)	60
Opening speed (m/s)	1.5
Contact material	CuCr25

comparison was conducted with two arc currents of 8 kA and 18 kA (rms), which was shown in Figs. 3.36 and 3.37, respectively. Figure 3.36 shows the vacuum arc behaviors with an arc current of 8 kA. Figure 3.36 a and b shows a comparison of arc evolution with the external axial magnetic flux densities of 0 mT and 37 mT, respectively. Figure 3.36c and d shows a comparison of arc voltage with the external axial magnetic flux densities of 0 mT and 37 mT, respectively. Figure 3.36 shows the vacuum arc behaviors consisted of three stages: (1) Arc ignition stage; (2) Constricted arc stage. It usually appears at the arc current peak period with a constricted arc column; and (3) Diffuse arc stage. It is indicated by low and uniform

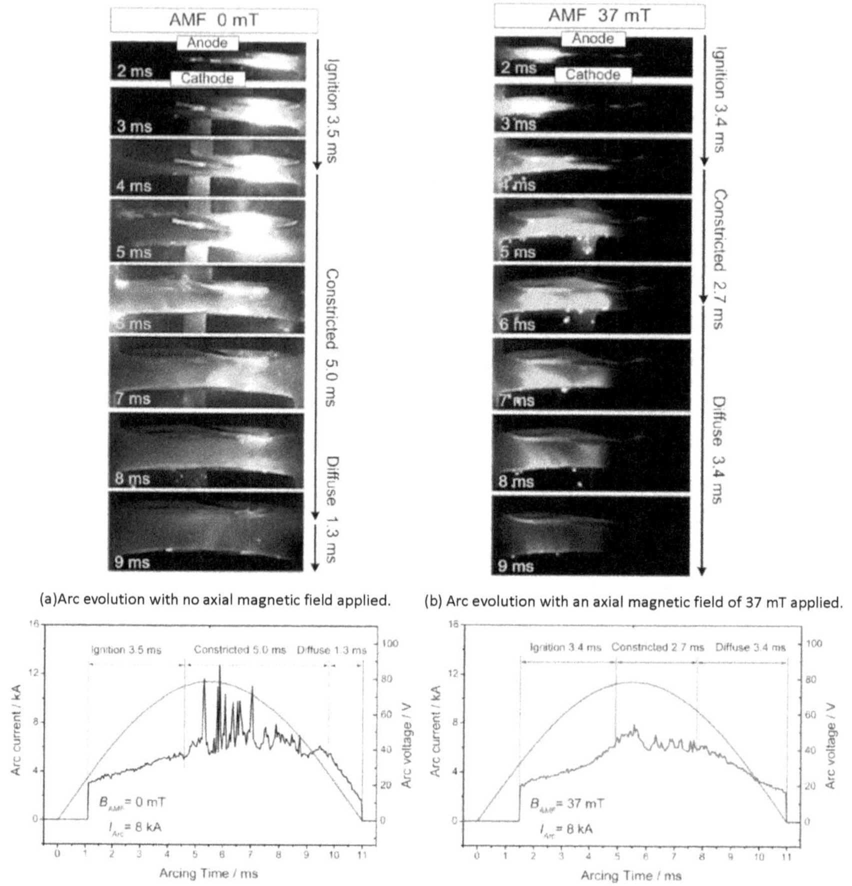

(a) Arc evolution with no axial magnetic field applied. (b) Arc evolution with an axial magnetic field of 37 mT applied.

(c) Arc voltage waveform with no axial magnetic field applied. (d) Arc voltage waveform with axial magnetic field of 37 mT applied.

Fig. 3.36 Vacuum arc evolution and arc voltage with an arc current of 8 kA

brightness at the current decreasing period. Figure 3.36a shows the arc evolution with an external axial magnetic flux density of 0 mT. It shows that the arc ignition stage lasted for 3.5 ms. The arc began to appear on the right side of the contact with an intense arc. The constricted arc stage lasted for 5.0 ms. A constricted arc column was forced to rotate by the transverse magnetic field. During this stage, multiple liquid metal drops ran from the root region of the arc column towards the region surrounding the contacts. The diffuse arc stage lasted for 1.3 ms until arc extinguished. In the diffuse arc stage, a diffuse arc with low brightness appeared. Figure 3.36b shows the arc evolution with an external axial magnetic flux density of 37 mT. The arc ignition stage lasted for 3.4 ms. The constricted arc stage lasted for 2.7 ms. The constricted arc column remained near the arc ignition area without a rotation, and fewer liquid metal drops appeared. The diffuse arc stage lasted for 3.4 ms, with the arc brightness continuously weakening before extinguishing. Figure 3.36c and

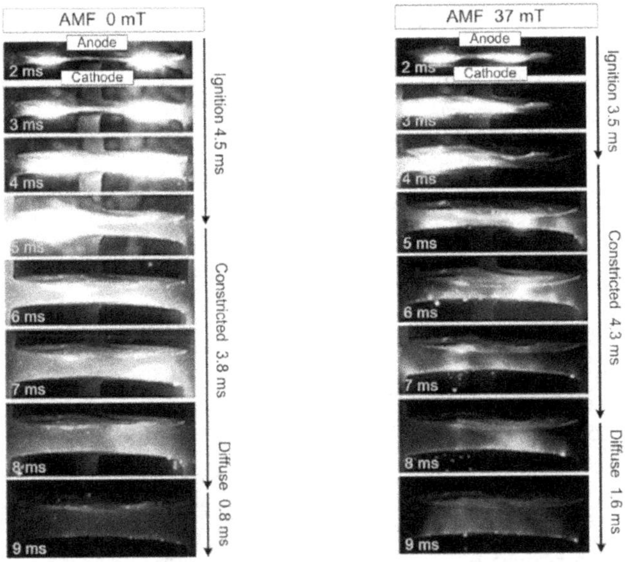

(a) Arc evolution with no axial magnetic field applied. (b)Arc evolution with an axial magnetic field of 37 mT applied.

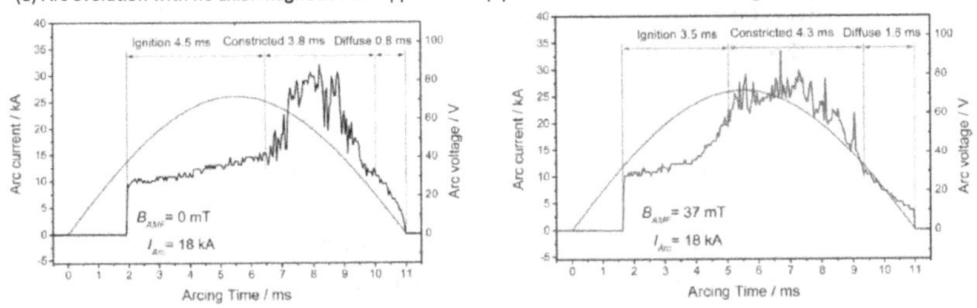

(c) Arc voltage waveform with no axial magnetic field applied. (d) Arc voltage waveform with axial magnetic field of 37 mT applied.

Fig. 3.37 Vacuum arc evolution and arc voltage with an arc current of 18 kA

d also indicated the three stages. Figure 3.36c shows the arc current and arc voltage with an external axial magnetic flux density of 0mT. It shows that the arc voltage increased continuously in the arc ignition stage. In the constricted arc stage, the arc voltage appeared with a high noise component. In the diffuse arc stage, the noise component disappeared and the arc voltage decreased smoothly. Figure 3.36d shows the arc current and arc voltage with an external axial magnetic flux density of 37 mT. The arc voltage also increased smoothly in the arc ignition stage. In the constricted stage, the axial magnetic field effectively reduced the amplitude of the noise component. In the diffuse arc stage, the noise component also disappeared and the arc voltage decreased smoothly. Comparing with the vacuum arc behaviors only under the transverse magnetic field with that under the combined magnetic fields, one can find that the application of the axial magnetic field apparently weakened the arc rotation that is otherwise forced by the transverse magnetic field contacts. In addition, the vacuum arc under an axial magnetic flux density of 37mT was more uniform and has a significantly longer diffuse arc stage before current zero.

Figure 3.37 compares the vacuum arc behaviors only under the transverse magnetic field with that under the combined magnetic fields at a higher arc current of 18 kA (rms). The external AMF flux density applied was 0 mT and 37 mT, respectively. Figure 3.37a and b shows arc evolution. Figure 3.37c and d shows arc voltage. Figure 3.37 also indicated the three stages. Figure 3.37a shows the vacuum arc evolution with an external axial magnetic flux density of 0mT. It shows that the arc ignition stage lasted for 4.5 ms. The initial intense arc began to appear at the center of the contacts. The constricted arc stage lasted for 3.8 ms. A constricted arc column was forced to rotate by the transverse magnetic field, with many liquid metal drops appearing between the contacts. The diffuse arc stage lasted for 0.8 ms until arc extinguished. Figure 3.37b shows the vacuum arc evolution with an external axial magnetic flux density of 37 mT. The arc ignition stage lasted for 3.5 ms. The constricted arc state lasted for 4.3 ms. The arc rotated between the contacts with a high velocity. The diffuse arc stage lasted for 1.6 ms until arc extinguished. Figure 3.37c and d also indicated the three stages. Figure 3.37c shows the arc current and arc voltage with an external axial magnetic flux density of 0 mT. It shows the arc voltage increased continuously in the arc ignition stage. In the constricted stage, the arc voltage increased sharply with a high noise component. In the diffuse arc stage, the noise component disappeared and the arc voltage decreased smoothly. Figure 3.37d shows the arc current and arc voltage waveform with an external axial magnetic flux density of 37mT. In the arc ignition stage, the arc voltage increased smoothly. In the subsequent constricted arc stage, the arc voltage showed a high noise component. In the diffuse arc stage, the noise component disappeared and the arc voltage decreased smoothly. Comparing with Fig. 3.37a and b, it is found that an application of axial magnetic field of 37 mT weakened the light intensity in the constricted stage and it decreased the number of liquid metal drops, which ran towards the surrounding regions of the contacts. In addition, the axial magnetic field extended the diffuse arc stage from 0.8 ms to 1.6 ms. Comparing with Figs. 3.36b and 3.37b, one can find an influence of the combined magnetic field on vacuum arcs under a high current and a low current. The vacuum arc rotated with a high current of 18 kA while it did not rotate under a low current of 8 kA. The arc rotation seemed to be determined by a competition of the transverse magnetic field and the axial magnetic field, while the transverse magnetic field drives the rotation of the arc and the axial magnetic field keeps the arc remained locally and diffuse it.

Now it becomes interesting to understand an impact of the applied axial magnetic field strength on the vacuum arc behaviors between the transverse magnetic field contacts. Seven levels of axial magnetic flux density and five levels of arc current were chosen. The axial magnetic flux density was 0, 20, 37, 56, 74, 93, and 110 mT, respectively. The arc current was 8, 12, 16, 18, and 20 kA (rms). Four typical results were given in the following, which were vacuum arc behaviors with transverse magnetic field contact at an arc current of 16 kA (rms) under external applied axial magnetic flux densities of 0, 37, 74, and 110mT, respectively. Figure 3.38a, b, c, and d shows the arc evolutions of the four cases. Figure 3.38e, f, g, and h shows the arc

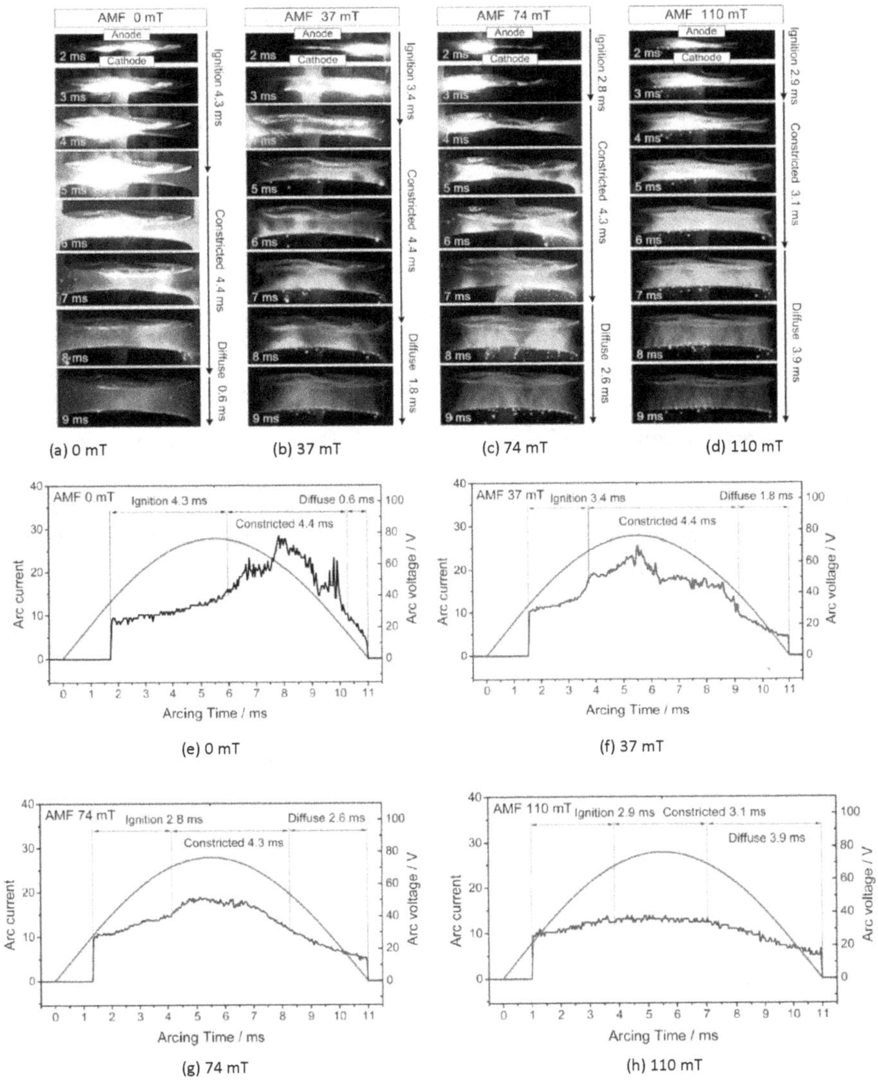

Fig. 3.38 Vacuum arc evolution and arc voltage with various external applied axial magnetic field flux densities arc current: 16 kA (rms)

voltage waveforms of the four cases. Figure 3.38a shows the arc evolution at an axial magnetic flux density of 0 mT. It shows that the arc ignition stage lasted for 4.3 ms. The constricted arc stage lasted for 4.4 ms. The constricted arc rotated rapidly, with multiple liquid metal drops running towards the surrounding regions. The diffuse arc stage lasted for 0.6 ms until arc extinguished. Figure 3.38b shows the arc evolution at an axial magnetic flux density of 37 mT. It shows that the arc ignition stage lasted for 3.4 ms. The constricted arc stage lasted for 4.4 ms. The constricted arc rotated and the number of liquid metal drops was much reduced comparing to Fig. 3.38a. The diffuse arc stage lasted for 1.8 ms. Figure 3.38c shows the arc evolution at an axial magnetic flux density of 74 mT. It shows that the arc ignition stage lasted for 2.8 ms. The constricted arc stage lasted for 4.3 ms. The constricted arc rotated rapidly. The diffuse arc stage lasted for 2.6 ms. Figure 3.38d shows the arc evolution at an axial magnetic flux densities of 110 mT. It shows that the arc ignition stage lasted for 2.9 ms. The constricted arc stage lasted for 3.1 ms. The constricted arc remained in the center region without rotations. The diffuse arc stage lasted for 3.9 ms. Figure 3.38e, f, g, and h shows that, with an increasing of axial magnetic flux density, both the arc voltage amplitude and its noise component decreased significantly. The arc voltage waveform becomes increasingly smooth, especially during the constricted stage. In addition, it is notable that the vacuum arc became more diffuse and stable before current zero with an increase of the axial magnetic flux density. The diffuse arc stage duration increased from 0.6 to 3.9 ms when the axial magnetic flux density increased from 0 to 110 mT.

Figure 3.39 shows an influence of the axial magnetic flux density on the diffuse arc duration before current zero with the seven levels of axial magnetic flux density and five levels of arc current. The axial magnetic flux density ranged from 0 to 110 mT. The arc current ranged from 8 to 20 kA. Figure 3.39 shows the diffuse arc duration increased linearly with an increase of axial magnetic flux density. Additionally, the lower the arc current, the higher the slope of the linear relationship. It implied that the axial magnetic field had a more significant influence on the lower current vacuum arcs.

Next it investigated an impact of the arc current on the vacuum arc behaviors between the transverse magnetic field contacts under an external applied axial magnetic field. Seven levels of axial magnetic flux density and five levels of arc

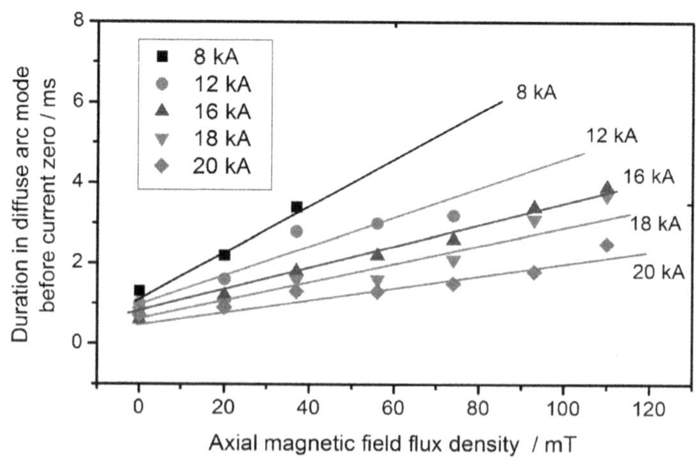

Fig. 3.39 A relationship between diffuse arc duration before current zero and the external applied axial magnetic flux density

current were chosen. The axial magnetic flux density was 0, 20, 37, 56, 74, 93, and 110 mT, respectively. The arc current was 8, 12, 16, 18, and 20 kA (rms). Four typical results were given in the following, which were vacuum arc behaviors with transverse magnetic field contact under an external applied axial magnetic flux density of 74 mT at arc currents of 12, 16, 18, and 20 kA (rms), respectively. Figure 3.40a, b, c and d shows the arc evolutions of the four cases. Figure 3.40e, f, g, and h shows the arc voltage waveforms of the four cases. Figure 3.40a shows the arc evolution at an

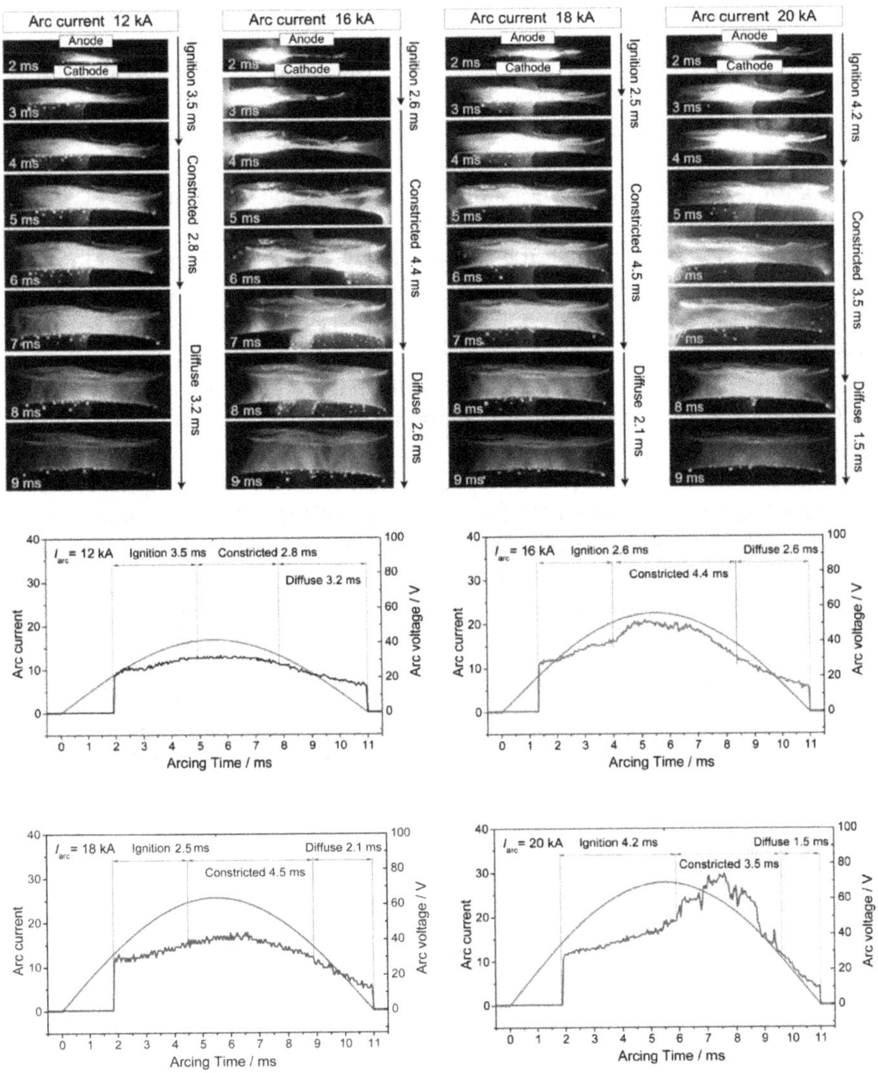

Fig. 3.40 Vacuum arc evolution and arc voltage with various arc currents external applied axial magnetic field flux densities: 74 mT

arc current of 12 kA (rms). It shows that the arc ignition stage lasted for 3.5 ms. The constricted arc stage lasted for 2.8 ms. The constricted arc remained in the center region of the contacts without rotation. The diffuse arc stage lasted for 3.2 ms. The diffuse arc distributed uniformly between the contacts. Figure 3.40b shows the arc evolution at an arc current of 16 kA (rms). The arc ignition stage lasted for 2.6 ms. The constricted arc stage lasted for 4.4 ms. The constricted arc rotated rapidly. Multiple liquid metal drops can be seen between the contacts. The diffuse arc stage lasted for 2.6 ms. Figure 3.40c shows the arc evolution at an arc current of 18 kA (rms). The arc ignition stage lasted for 2.5 ms. The constricted arc stage lasted for 4.5 ms. The constricted vacuum arc remained in the center region of the contacts without rotation. The diffuse arc stage lasted for 2.1 ms. The diffuse arc distributed uniformly between the contacts. Figure 3.40d shows the arc evolution at an arc current of 20 kA (rms). The arc ignition stage lasted for 4.2 ms. The constricted arc stage lasted for 3.5 ms. The constricted vacuum arc rotated rapidly, with multiple liquid metal drops running towards the surrounding region. The diffuse arc stage lasted for 1.5 ms. It is notable that, at the arc current of 18 kA (rms) and under an external applied axial magnetic flux density of 74 mT, the vacuum arc constricted and remained locally at the center region between the transverse magnetic field contacts, without rotations. This is not desired because a local contact surface will be seriously eroded by the constricted arc column, which has a negative impact for successful current interruptions. Figure 3.40e, f, g, and h shows that, with an increasing of arc current from 12 to 20 kA (rms), both the arc voltage amplitude and its noise component increased significantly. In addition, the diffuse arc stage duration decreased from 3.2 to 1.5 ms when the arc current increased from 12 to 20 kA.

Figure 3.41 shows an influence of the arc current on the diffuse arc duration before current zero with the seven levels of axial magnetic flux density and five levels of arc current. The axial magnetic flux density ranged from 0 to 110 mT. The arc current ranged from 8 to 20 kA. It shows that the diffuse arc duration decreased linearly with

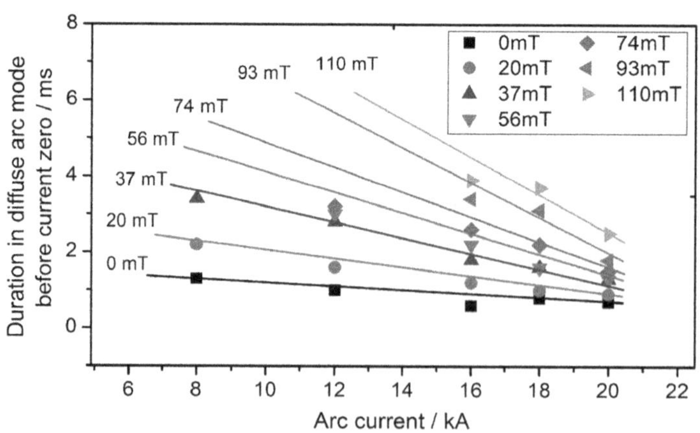

Fig. 3.41 A relationship between diffuse arc duration before current zero and the arc currents

an increasing of arc current. Additionally, the higher the axial magnetic flux density, the more quickly decrease of the diffuse arc duration with increasing of arc current. It suggested the same meaning with that of Fig. 3.39, in which the axial magnetic field had a more significant influence on the lower current vacuum arcs.

This section discussed a combination of transverse magnetic field and axial magnetic field to control vacuum arcs. The transverse magnetic field contacts offer an advantage of low loop resistance for nominal current conduction. The axial magnetic field contacts offer an effective control of vacuum arc, particularly in large contact gaps. A combined magnetic fields contact is potentially to give the advantages of both low loop resistances for nominal current conduction and effective vacuum arc control at large contact gaps. Lamara and Gentsch [49] and Wang et al. [50] also demonstrate experimentally that the vacuum arcs evolved into a diffuse arc under the combined magnetic fields. However, an industrial contact design should be careful for vacuum interrupters at transmission voltage levels. It is observed that, in the case of high externally applied axial magnetic flux densities, the vacuum arcs driven by the transverse magnetic field failed to rotate and remained in the center region of the contacts. In this case, the Lorentz force generated by the transverse magnetic field cannot overcome the control of the axial magnetic field. Therefore, the axial magnetic field should not be too strong to impede the rotation of the arc. On the other side, the axial magnetic field should be higher than a critical level, especially in large contact gaps, to guarantee successful current interruptions (see Sect. 3.2.1). In addition, the axial magnetic field proved its influence on the diffuse arc duration before current zero. A diffuse arc duration of 1 or 2 ms before current zero is expected for successful current interruptions [51]. The higher the diffuse arc duration is, the more metal vapors and residual plasma dissipate from the contact gap. Thus, the density of metal vapor and residual plasma would be low, which is desired for the dielectric recovery (see Sect. 2.4.1). Moreover, there is more time for contact surface to cool down, which also benefits for the successful current interruptions (see Sect. 1.4.3 and 1.4.4).

3.3 X-Radiation for 126 kV Vacuum Interrupters

It is well known that X-rays emit when a pair of opened electrodes are subjected to high voltages in vacuum. X-radiation results from electrons emitted from the cathode by means of field emission, obtaining energy under the effect of the voltage drop across the electrodes, and releasing the energy when impacting with the anode. The intensity and energy of the X-radiation depend on the applied voltage and the level of electron emission. Vacuum interrupters at transmission voltage level would be subjected to significantly higher voltages comparing to their medium voltage counterpart. Therefore, it would emit more X-rays. Vacuum interrupters should comply with the limits for X-radiation according to standards, such as IEC [52] and ANSI [53] standards. According to the IEC standard [52], measurement of X-radiation levels should be carried out on new vacuum interrupters, and the X-radiation emitted from a vacuum interrupter does not exceed the following: (a) 5 μSv/h at rated voltage with 1 m distance; (b) 150 μSv/h at rated power–frequency withstand voltage with 1 m distance. Renz and Gentsch [54] demonstrated that the X-radiation dose emitted from 7.2 to 36 kV vacuum interrupters under a power–frequency voltage of 36–70 kV at 0.1 m distance from the touchable surface was not higher than 1 μSv/h, which is specified by Council Directive 96/29/Euratom [55]. Li et al. demonstrated that X-radiation emitted from 12 to 40.5 kV vacuum interrupters was higher than 1 mSv/a during a certain voltage conditioning procedure with applying a power frequency voltage of 60–75 kV at the 50% rated contact gap [56]. The value of 1 mSv/a is specified in a Chinese national standard for the public [57]. The value of 20 mSv/a is specified for professionals [57]. This section is to measure the X-radiation level for a kind of 126 kV vacuum interrupter and the influence of three factors including applied power frequency voltage, contact gap and voltage conditioning. The results will be helpful to verify whether the X-radiation dose emitted from the 126 kV vacuum interrupter exceeds the standards' limit or not.

Figure 3.42 shows an experimental setup for the X-radiation measurement. A 126 kV vacuum interrupter with an opened gap was mounted at a test fixture. A power frequency voltage was applied to the movable terminal of the vacuum interrupter while the stationary terminal was earthed. If the applied voltage did not exceed the external flashover value of the test vacuum interrupter, the test vacuum interrupter was placed in the air. If the applied voltage exceeded the external flashover value, the test vacuum interrupter was assembled in an insulating medium, such as SF_6 gas, to ensure the insulation performance of the external surface of the test vacuum interrupter. A X-radiation meter shall meet the following minimum specifications [52]: (1) Accuracy: Capable of measuring 150 μSv per hour with an accuracy of ±25% and a response time less than 15 s. (2) Energy response: From 12 keV to 0.5 MeV. (3) Sensitive area: Less than 100 cm^2. In the experiment, a radiometer with type FJ-347A was chosen for surveying the X-radiation emission level. Its technical parameters were as following: a maximum dose of 1000 μSv per hour

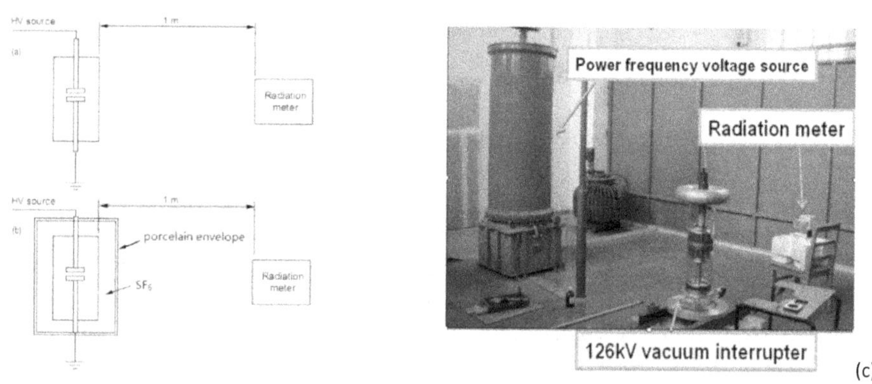

Fig. 3.42 Experimental setup for measuring X-radiation emission level for 126 kV vacuum interrupters. **a** vacuum interrupter in air **b** vacuum interrupter in SF_6 gas **c** photo for the measurement

with an accuracy of 10% and a response time less than 8 s; an energy response of 10–10 MeV; sensitive area is less than 50 cm^2. Thus, it met the specifications in the IEC standard [52]. The sensor element of a radiation meter should be placed in the middle plane of the opened contacts and it pointed at the contacts with 1 m distance from the touchable surface of the test vacuum interrupter. In the measurement, the X-radiation level was recorded for three minutes. The atmospheric environmental background radiation in the laboratory was measured by a JB4000 (A) intelligent X-radiation meter.

There were three groups of experiments. The first one was to measure the X-radiation level. The second one was to determine the influence of applied power frequency voltage and contact gap on X-radiation level. The third one was to determine the influence of voltage conditioning on X-radiation level. In the first group, five 126 kV vacuum interrupters were measured. Two levels of power frequency voltage were chosen, which was 126 kV and 230 kV, respectively. The contact gap was 60 mm. In the second group, six 126 kV vacuum interrupters were measured. The six vacuum interrupters were measured under three levels of power frequency voltage including 126, 150, and 170 kV. The contact gaps were 20 mm and 30 mm, respectively. Subsequently, the six vacuum interrupters were measured under three levels of contact gap including 10, 20, and 30 mm. The applied power frequency voltage was 126 kV and 150 kV, respectively. In the third group, three 126 kV vacuum interrupters were measured. Table 3.11 shows the voltage conditioning procedures of the three vacuum interrupters. There were seven steps and each step lasted for 3 min. The X-radiation dose was measured in the 5th and 7th step. When the applied power frequency voltage was lower than 180 kV, the test vacuum interrupters were put in air. When the applied voltage was higher than 180 kV, a test vacuum interrupter was

Table 3.11 Voltage conditioning procedures

Step	Power frequency voltage (kV)	Duration of voltage application (min)	Contact gap (mm)	Note
1	126	3	10	Conditioning in air
2	73	3	5	Conditioning in air
3	82	3	5	Conditioning in air
4	126	3	10	Conditioning in air
5	230	3	60	X-radiation measurement* Vacuum interrupter in SF_6
6	240	3	60	Conditioning in SF_6
7	230	3	60	X-radiation measurement* Vacuum interrupter in SF_6

*The X-radiation dose was measured in three minutes

Table 3.12 Measured X-ray radiation dose for five 126 kV vacuum interrupters at 1 m distance in 3 min

No.	Dose in 3 min (μSv)	Dose rate (μSv/h)	Threshold value (μSv/h)
(a) Applied power frequency voltage 126 kV			
1	0.15	2.9	5
2	0.15	2.9	5
3	0.15	2.9	5
4	0.15	2.9	5
5	0.15	2.9	5
(b) Applied power frequency voltage 230 kV			
1	6.27	125.5	150
2	7.40	147.9	150
3	3.70	73.9	150
4	2.07	41.3	150
5	2.35	46.9	150

put into an insulator envelope sealed with SF_6 gas in order to ensure the external insulation of the test vacuum interrupter.

In the first group of experiment, five new 126 kV vacuum interrupters were measured for the X-radiation level. The atmospheric environmental background radiation dose was measured as 0.1 μSv/h. Table 3.12 shows the measured X-radiation dose for the five 126 kV vacuum interrupters at 1 m distance in three minutes. Table 3.12a and b shows the results under an applied power frequency voltage of 126 kV and 230 kV, respectively. The contact gap was 60 mm. Table 3.12a shows that the X-radiation dose was 2.9 μSv/h at an applied power frequency voltage of 126 kV ($\sqrt{3}$ times of rated voltage of each phase) for the five vacuum interrupters. It met the requirement of 5 μSv/h at rated voltage specified by IEC standard [52]. Table 3.12b shows that the maximum X-radiation dose for the five 126 kV vacuum interrupters was 147.9 μSv/h at an applied power frequency voltage of 230 kV. It also met the requirement of 150 μSv/h at rated power frequency withstand voltage specified by IEC standard [52].

In the second group of experiment, six new 126 kV vacuum interrupters were measured the X-radiation dose to determine the influence of applied power frequency voltage and contact gap. Table 3.13 shows the measured X-radiation dose for the six 126 kV vacuum interrupters under an applied power frequency voltage of 126, 150 and 170 kV, respectively. Table 3.13a and b shows the measured results with contact gaps of 20 and 30 mm, respectively. The X-radiation dose was measured in three minutes. It shows that the X-radiation dose increased significantly when the applied voltage increased from 126 to 150 and then 170 kV.

Figure 3.43 shows the average values of measured X-radiation dose of the six vacuum interrupters under the three levels of power frequency voltages with contact

Table 3.13 Measured X-radiation dose* for six 126 kV vacuum interrupters under three levels of power frequency voltage

Voltage (kV)	No. 1 (μSv)	No. 2 (μSv)	No. 3 (μSv)	No. 4 (μSv)	No. 5 (μSv)	No. 6 (μSv)
(a) Contact gap of 20 mm						
126	0.8	32	1.6	0.1	4.2	28
150	12	160	9.2	5.3	16.8	250
170	42	275	36	14.3	42	610
(b) Contact gap of 30 mm						
126	2.7	10.5	0.2	0.2	0.6	0.9
150	4.2	29.0	1.0	0.6	2.4	4.7
170	9.0	145.0	96.0	0.7	2.4	14.0

*The X-radiation dose was measured in 3 min

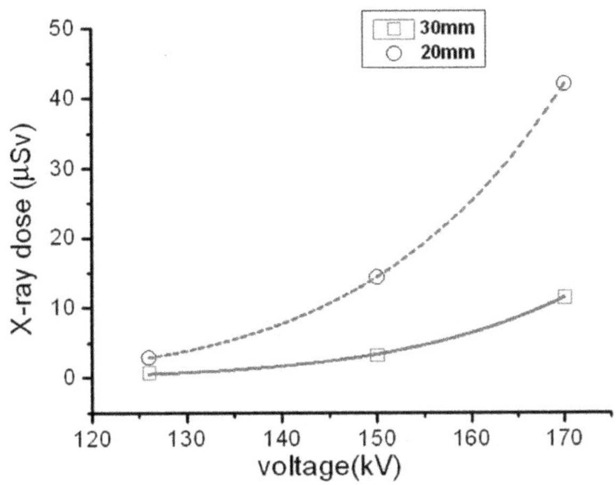

Fig. 3.43 Average values of measured X-radiation dose under three levels of power frequency voltages

gaps of 20 and 30 mm. The average values increased from 0.75 to 3.3 and then 11.5 μSv at the contact gap of 30 mm when the power frequency voltage increased from 126 to 150 and then 170 kV, respectively. The average values follow the Eq. (3.33). The average values increased from 2.9 to 14.4 and then 29.0 μSv at the contact gap of 20 mm when the power frequency voltage increased from 126 to 150 and then 170 kV, respectively. The average values follow the Eq. (3.34).

$$H = 3.0 \times 10^{-6} \cdot U^4 \cdot e^{-916.9/U} \quad (3.33)$$

$$H = 3.6 \times 10^{-6} \cdot U^4 \cdot e^{-724.6/U} \quad (3.34)$$

where H (μSv) is the X-radiation dose emitted from a vacuum interrupter. U (kV) is the power frequency voltage. The X-radiation dose at the contact gap of 20 mm was significantly higher than that of 30 mm. The reason might be that a higher electric field led to an increase of energy of the field emission electrons. The electrons with higher energy impacted with the anode and emitted more dose of X-radiation.

Then the six 126 kV vacuum interrupters were used to measure the X-radiation dose with three levels of contact gaps including 10, 20, and 30 mm. Figure 3.44a and b shows the measured results under applied power frequency voltage of 126 kV and 150 kV, respectively. The X-radiation dose was measured in three minutes. It shows that both the X-rays radiation dose and its scattering significantly decreased with an increase of contact gap from 10 to 20 and then 30 mm.

Figure 3.45 shows the average values of the measured X-radiation dose of the six vacuum interrupters with the three levels of contact gaps under power frequency voltages of 126 and 150 kV. The average values decreased from 120.0 to 3.0 and then 0.42 μSv under the power frequency voltage of 126 kV when the contact gap

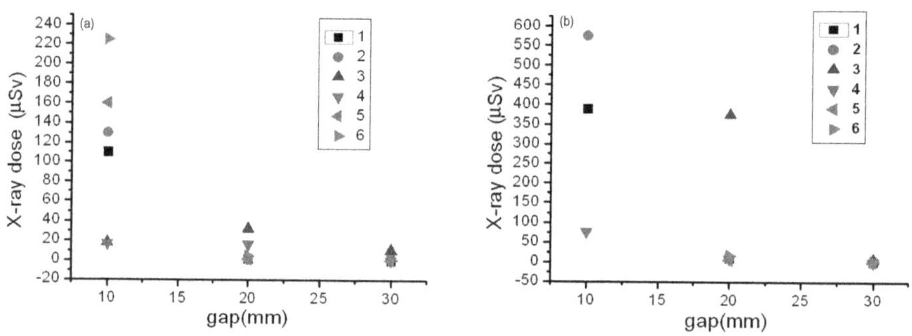

Fig. 3.44 Measured X-radiation dose for six 126 kV vacuum interrupters under three levels of contact gaps **a** Power frequency voltage of 126 kV **b** Power frequency voltage of 150 kV. The X-radiation dose was measured in 3 min

Fig. 3.45 Average values of measured X-radiation dose under three levels of contact gaps

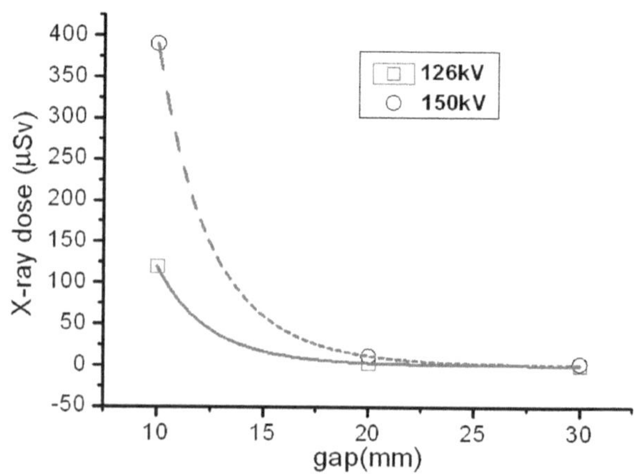

increased from 10 to 20 and then 30 mm, respectively. These average values followed the Eq. (3.35). The average values decreased from 390.0 to 11.5 and then 2.25 μSv under the power frequency voltage of 150 kV when the contact gap increased from 10 to 20 and then 30 mm, respectively. These average values follow the Eq. (3.36).

$$H = \frac{118982.4}{D^2} \cdot e^{-0.2D} \tag{3.35}$$

$$H = \frac{325711.8}{D^2} \cdot e^{-0.2D} \tag{3.36}$$

where H (unit: μSv) is the X-radiation dose emitted from a vacuum interrupter and D (unit: mm) is the contact gap.

For commercial vacuum interrupters, conditioning procedures were performed to guarantee the insulation levels. Therefore, it is desirable to know the influence of the conditioning on the emission of X-ray radiation. In the third group of experiment, three new 126 kV vacuum interrupters were measured the X-ray radiation dose with a voltage conditioning procedure shown in Table 3.11. Table 3.14 shows the measured results. It shows that the voltage conditioning significantly decreased the X-radiation dose. There was a large scattering of the data among the three vacuum interrupters before conditioning. However, the scattering became much lower after conditioning. Therefore, power frequency voltage conditioning has a significant effect to reduce the X-radiation.

The X-radiation emission level measured for the 126 kV vacuum interrupters supported the equations given by Renz and Gentsch [54]. They pointed that the relationship between H and U followed the Eq. (3.37) and the relationship between H and the D followed the Eq. (3.38).

$$H = a \cdot U^4 \cdot e^{b/U} \tag{3.37}$$

$$H = \frac{a}{D^2} \cdot e^{bD} \tag{3.38}$$

where H is X-radiation dose; U is applied power frequency voltage; D is contact gap; a and b are constants. Therefore, the X-radiation level proved the equations also fit for the 126 kV vacuum interrupters.

Table 3.14 Measured X-ray radiation dose with a voltage conditioning procedure shown in Table 3.11

No.	Before conditioning[a] (μSv)	After conditioning[b] (μSv)
1	15	3.7
2	2.3	2.1
3	5.6	2.4

Note [a]Step 5 in Table 3.11; [b]Step 7 in Table 3.11; The X-radiation dose was measured in 3 min

The influence of three factors including applied power frequency voltage, contact gap and voltage conditioning is summarized as follows. The X-radiation dose decreased as the applied power frequency voltage decreased or the contact gap increased. These phenomena may be related to the macroscopic electric field strength in the contact gap. It suggested that a low electric field strength corresponded to a low X-radiation dose and a low scattering of the distribution. The voltage conditioning also significantly decreased the X-radiation and the scattering of the distribution. The conditioning procedure reduced the micro-protrusions on the contact surfaces, thus it reduced microscopic electric field strength.

The measured X-radiation dose was less than 5 μSv/h at a power frequency voltage of 126 kV and it was less than 150 μSv/h at a power frequency voltage of 230 kV. So that the X-radiation dose met the IEC standard [52] for 126 kV vacuum interrupters. However, a precaution to protect personals is necessary, especially when the vacuum interrupters are conditioned at high voltages.

References

1. Shioiri T, Kamikawaji T, Yokokura K, Kaneko E, Ohshima I, Yanabu S, Dielectric Breakdown Probabilities for Uniform Field Gap in Vacuum, 19th Int'l. Sympos. Discharge Electr. Insul. Vacuum, xi'an, China, pp.17–20, 2000.
2. Sato S and Koyama K, Effect of mechanical polish of electrode on several breakdown characteristics of vacuum gap, 20th Int'l. Sympos. Discharge Electr. Insul. Vacuum, Tours, France, pp. 427-430, 2002.
3. Miyazaki F, Inagawa Y, Kato K, Sakaki M, Ichikawa H and Okubo H, Electrode conditioning characteristics in vacuum under impulse voltage application in non-uniform electric field. IEEE Trans. Dielectr. Electr. Insul., Vol. 12, pp.17-23, 2005.
4. Budde M and Kurrat M. Dielectric investigations on micro discharge currents and conditioning behaviors of vacuum gaps. 22th Int'l. Sympos. On Discharges Electr. Insul. Vacuum, Matsue, Japan, pp. 75–78, 2006.
5. Okawa N, Shioiri T, Okubo H and Yanabu S, Area Effect on Electric Breakdown of Copper and Stainless Steel Electrodes in Vacuum, IEEE Trans. Electr. Insul., Vol. 23, pp. 77-81,1988.
6. Toya H, Ueno N, Okada T, and Murai Y, Statistical property of breakdown between metal electrodes in vacuum, IEEE Trans. Power App. Syst., Vol. 100, pp. 1932-1939,1981.
7. Fukuoka Y, Yasuoka T, Kato K, Saitoh H, Sakaki M and Okubo H, Breakdown Conditioning Characteristics of Long Gap Electrode in Vacuum, 22th Int'l. Symp. Discharge Electr. Insul. Vacuum, Matsue, Japan pp. 59-62, 2006.
8. Fukuoka Y, Yasuoka T, Kato K, and Okubo H, Breakdown Conditioning Characteristics of Long Gap Electrodes in a Vacuum, IEEE Trans. Dielectr. Electr. Insul., Vol.14, pp. 577-582, 2007.
9. Kārner H C and Bender H G, Breakdown of Large Vacuum Gaps under Lightning Impulse Stress, 12th Int'l. Sympos. Discharge Electr. Insul. Vacuum, Shoresh, Israel, pp. 54-58, 1986.
10. Spolaore P, Bisoffi G, Cervellera F, Pengo R, Scarpa P, Binelle G and Zanon A, The large gap case for high voltage insulation in vacuum, 17th Int. Symp. On Discharges and Electrical Insulation in Vacuum, Berkeley, California, USA, pp. 523-526,1996.
11. Giere S, Kārner C. and Knobloch H, Dielectric Strength of Double and Single Break Vacuum Interrupters, IEEE Trans. Dielectr. Electr. Insul., Vol.8, pp. 43–47, 2001.
12. Schümann U, Giere S and Kurrat M, Breakdown voltage of electrode arrangement in vacuum circuit breakers, IEEE Trans. Dielectr. Electr. Insul., Vol.10, pp. 557-562, 2003.

13. Shioiri T, Kamikawaji T, Kaneko E, Homma M, Takahashi H and Ohshima I, Influence of electrode area on the conditioning effect in vacuum, IEEE Trans. Dielectr. Electr. Insul.,Vol. 2, pp. 317-320, 1995.
14. Latham R V, High voltage Vacuum Insulation: The Physical Basis, Chapter2: The Operational Characteristics of a High Voltage Vacuum Gap, London, UK, Academic Press, 1981, pp.13-50.
15. Zhang Y, Liu Z, Cheng S, Yang L, Geng Y and Wang J., Influence of Contact Contour on Breakdown Behavior in Vacuum under Uniform Field, IEEE Trans. Dielectr. Electr. Insul., Vol.16, pp. 1717-1723, 2009.
16. Latham R V, High voltage Vacuum Insulation: The Physical Basis, London, UK, Academic Press, 1981, pp.7.
17. Latham R V, High voltage Vacuum Insulation: Basic Concepts and Technological Practice, London, UK, Academic Press, 1995, pp.29.
18. Li S, Geng Y, Liu Z, Wang J, A breakdown mechanism transition with increasing vacuum gaps, IEEE Transactions on Dielectrics and Electrical Insulation Vol. 24, No. 6; December 2017, pp.3340-3346
19. Cranberg L, "The initiation of electrical breakdown in vacuum", J. Appl. Phys., Vol. 23, pp. 518-522, 1952.
20. Kojima H, Yasukawa H, Nishimura R, Okubo H, Sato H, Noda Y, "Improvement of Cu-Cr electrode surface under impulse voltage conditioning in vacuum", High Voltage Engineering and Application, Louisiana, USA, pp. 100-103, 2010.
21. Kamikawaji T, Shioiri T, Funahashi T, Okawa T, Kaneko M and Ohshima I, Generation of micro-particles from copper-chromium contacts and their influence on insulating performance in vacuum, 23rd Int'l. Conf. Electr. Contact phenomena, pp. 895–902, 1994.
22. Slade P G, The Vacuum Interrupter Theory, Design and Application. Boca Raton, London, New York, CRC Press, Taylor & Francis Group, pp. 49–58, 2008.
23. Slikov I N, "Mechanism for electrical discharge in vacuum", Sov. Phys-Tech Phys., Vol. 2, pp.1928-1934, 1957.
24. Farrall G A, "Cranberg hypothesis of vacuum breakdown as applied to impulse voltages", J. Appl. Phys., Vol. 33, No.1, pp. 96-99, 1962.
25. Alpert D, Lee D, Lyman E and Tomaschke H, Initiation of electrical breakdown in ultra high vacuum , J. Vacuum Sci. Techn., Vol. 1, pp.35-50, 1964.
26. Latham R V, High voltage Vacuum Insulation: The Physical Basis, Chapter 4: Breakdown Initiating Mechanisms II – Microparticle Based, London, UK, Academic Press, 1981, pp.83-115.
27. Slade P G, The Vacuum Interrupter Theory, Design and Application. Boca Raton, London, New York, CRC Press, Taylor & Francis Group, pp. 90–91, 2008.
28. Okawa M, Yanbu S and Kaneko E, The Investigation of Copper-Chromium Contacts in Vacuum Interrupters Subjected to an Axial Magnetic Field, IEEE Transactions on Plasma Science, Vol. I5, No. 5, pp. 533-537, 1987.
29. Asari N, Sasaki Y and Shioiri T, Breakdown Characteristics of CuCr Electrode after Electron beam irradiation in Vacuum, 1st Int'l. Conf. Electric Power Equipment-Switching Technology (ICEPE-ST), xi'an, China, pp.102–105, 2011.
30. Okubo H, Yasuoka T, Kato T and Kato K, Electrode conditioning mechanism based on pre-breakdown current under non-uniform electric field in vacuum, 23rd Int. Sympos. Discharges and Electr. Insul. Vacuum, Bucharest, Romania, pp.13-16, 2008.
31. Ding B J, Yang Z M and Wang X T, Influence of Microstructure on Dielectric Strength of CuCr Contact Materials in a Vacuum, IEEE Trans CPMT, Vol.19, No. 1, pp. 76−81, 1996.
32. Li S, Geng Y, Liu Z and Wang J, The Influence of Melted layer on Vacuum Insulation, 27th Int'l. Sympos. Discharges and Electr. Insul. Vacuum, pp.17–20, Suzhou, China, 2016.
33. Haynes W M, (Ed.), CRC Handbook of Chemistry and Physics, 97th Edition, CRC Press, Taylor & Francis Group, LLC, 2017.
34. Fenski B, Heimbach M, Lindmayer M, and Shang W, Characteristics of a vacuum switching contact based on bipolar axial magnetic field, IEEE Trans. Plasma Sci., vol. 27, no. 4, pp. 949–953, Aug. 1999.

35. Dullni E, Plessl A, and Reininghaus U, Research for vacuum-circuit breakers, ABB Rev., vol. 3, pp. 11–18, Mar. 1989.
36. Morimiya O, Sohma S, Sugawara T, and Mizutani H, High current vacuum arcs stabilized by axial magnetic fields, in Proc. IEEE PES Winter Meeting, Sep. 1973, pp. 1723–1732.
37. Stoving P and Bestel F, Finite element analyses of AMF vacuum contacts, in Proc. IEEE 18th Int. Symp. Discharges Electr. Insul. Vac., 1998, pp. 522–529.
38. Schulman M B and Bindas J A, Evaluation of AC axial magnetic fields needed to prevent anode spots in vacuum arcs between opening contacts, IEEE Trans. Compon., Packag., Manuf. Technol. A, vol. 17, no. 1, pp. 53–57, Mar. 1994.
39. Liu, Z, Wang, J, Wang J, Geng Y, Yao J, Zheng, Y, a 252kV single break vacuum interrupter, Chinese patent, 200710018001.9, August 19, 2009.
40. Zhang Y, Liu Z, Geng Y, and Yao J, A new axial magnetic field contact for 126 kV single break vacuum interrupters, in Proc. 23th Int. Symp. Discharges Electr. Insul. Vac., 2008, pp. 280–283.
41. High-Voltage Switch-Gear and Control-Gear—Part 101: Synthetic Testing, First Edition, IEC Standard 62271–101, 2006.
42. Fenski B, Lindmayer M, Vacuum interrupters with Axial Field contacts. 3-d finite element simulation and switching experiments", IEEE Trans. On Dielectrics and Electrical Insulation, Vol.4, No.4, pp.407–412, 1997.
43. Schellekens H, Shang W, and Lenstra K, Vacuum interrupter design based on arc magnetic field interaction for horseshoe electrode, IEEE Trans. Plasma Sci., vol. 21, no. 5, pp. 469–473, Oct. 1993.
44. Shang W, Schellekens H, and Hilderink J, Experimental investigations into the arc properties of vacuum interrupters with horseshoe electrode, four pole electrode, and their applications, IEEE Trans. Plasma Sci., vol. 21, no. 5, pp. 474–477, Oct. 1993.
45. Heberlein J V R and Gorman J G, The high current metal vapor arc column between separating electrodes, IEEE Trans. Plasma Sci., vol. 8, no. 4, pp. 283–288, Apr. 1980.
46. Schulman M B and Slade P G, Sequential modes of drawn vacuum arcs between butt contacts for currents in the 1 kA to 16 kA range, IEEE Trans. Compon., Packag., Manuf. Technol., A, vol. 18, no. 2, pp. 417–422, Jun. 1995.
47. Schulman M B and Schellekens H, Visualization and characterization of high-current diffuse vacuum arc on axial magnetic field contacts, IEEE Trans. Plasma Sci., vol. 28, no. 32, pp. 443–451, Apr. 2000.
48. Schade E, Physics of high-current interruption of vacuum circuit breakers, IEEE Trans. Plasma Sci., vol. 33, no. 5, pp. 1564–1575, Oct. 2005.
49. Lamara T and Gentsch D, IEEE Trans. Plasma Sci. 41, 2043 (2013).
50. Wang L, Deng J, Wang H, Jia S, Qin K, and Shi Z, Phys. Plasmas 22, 103512 (2015).
51. Slade P G, The Vacuum Interrupter Theory, Design and Application. Boca Raton, London, New York, CRC Press, Taylor & Francis Group, pp. 282–295, 2008.
52. IEC 62271–1. 2007, High-Voltage Switchgear and Controlgear-part1: Common Specifications. IEC Publication.
53. ANSI C37.85. 1989, Alternating-current high-voltage power vacuum interrupters-safety requirements for X-radiation limits. American National Standard.
54. Renz R, Gentsch D., Permissible X-ray radiation emitted by vacuum interrupters/devices at rated operating conditions. XXIVth International Symposium on Discharges and Electrical Insulation in Vacuum, Braunschweig, Germany, 2010, pp.133–137
55. Council Directive 96/29/Euratom, Basic safety standards for the protection of the health of workers and the general public against the dangers arising from ionising radiation, 1996.
56. Li W, Guo Y, Wang C, et al. Influence of X-rays in Vacuum Interrupter on Human Health, High Voltage Engineering, 2007, Vol.33, No.12, pp.101-104 (in Chinese)
57. GB18871–2002, Basic Standards for Protection against Ionizing Radiation and for the Safety of Radiation Sources. Standards Press of China, Beijing, 2002, (in Chinese).

Chapter 4
Vacuum Circuit Breakers at Transmission Voltage Level

Previously, there was a general idea that all technologies in developing transmission voltage level vacuum circuit breakers were based on medium voltage vacuum circuit breakers. No essentially new technical features were necessary, while it was only an extrapolation in geometry. However, later stories proved that the idea was not true. As much severe electrical and mechanical stresses applied on transmission voltage level vacuum circuit breakers than their medium voltage counterpart, many failure modes emerged because less margin left. For example, a successful current interruption needs not only a good design of vacuum interrupter, but also an appropriate opening displacement curve. Many high current interruption failures promoted that opening curves should follow the vacuum arc behaviors, with which the interruption failures can be avoided. In medium voltage level the opening displacement curve seemed not so important because it was covered by a good margin in the vacuum interrupters. Another failure was the welding in making operations. Such failure suggested that an appropriate closing curve was needed, which minimized the arc duration in making current, which also minimized the welding force. By such way the operating mechanism should output the appropriate displacement curves to follow the vacuum arc behaviors, so that good performances can be achieved. At the same time, the requirements of mechanical life called for a guarantee of the mechanical reliability. The temperature-rise at a high nominal rated current was also a headache for the transmission voltage level vacuum circuit breakers. This urged a numerical thermal analysis of the vacuum circuit breakers.

This chapter addresses three aspects in transmission voltage level vacuum circuit breakers. The first one is to determine the opening and closing velocities for transmission voltage vacuum circuit breakers. The second one is to develop operating mechanism and to increase its mechanical reliability. The third one is to enhance the nominal rated current. Finally, based on the three aspects, a 126 kV vacuum circuit breaker was successfully developed and its type test results were demonstrated.

Z. Liu et al., *Switching Arc Phenomena in Transmission Voltage Level Vacuum Circuit Breakers*, https://doi.org/10.1007/978-981-16-1398-2_4

4.1 Determination of Opening and Closing Velocities

It is vacuum arc behaviors that determines the opening and closing displacement curves. Different kind of vacuum interrupters have different vacuum arc behaviors. Each kind of vacuum interrupter shall be mateched with its own opening and closing displacement curves. This section discusses how to determine the opening and closing characteristics by vacuum arc behaviors.

4.1.1 Anode Mode Diagram: A Determination of Opening Displacement Curve

An appropriate opening velocity characteristic proved to be a contributor for improving the high-current interruption performance of vacuum interrupters [1–3]. This implies that an optimum displacement curve exists, which improves the current interruption capability of vacuum interrupters. From the point of view of avoiding high-current anode mode, an appropriate opening displacement curve may be determined based on a vacuum arc anode discharge mode diagram. This section is to propose an opening displacement curve for vacuum interrupters based on vacuum arc anode discharge mode diagram.

Figure 4.1 shows a schematic of the test vacuum interrupters. In order to observe vacuum arcs, the envelope of the test vacuum interrupter was made of glass and no shield was used. Therefore, the anode phenomena can be observed by high-speed photography. Figure 4.2 shows that two kinds of CuCr25 (25% by weight of Cr) contact materials were prepared in the test vacuum interrupters. One was nanocrystalline CuCr25 which has a size of Cr particles less than 120 nm, as shown in Fig. 4.2a. The nanocrystalline CuCr25 contact material was examined by a transmission electron microscopy (TEM). The other one was microcrystalline CuCr25 in which the Cr particle size was less than 100 μm, as shown in Fig. 4.2b. The microcrystalline CuCr25 contact material was examined by a scanning electron microscopy (SEM). The details of processing method of contact materials and the manufacturing of the vacuum interrupters were indicated in reference [4]. For each contact material, it prepared three vacuum interrupters with 12 mm contact diameter and three vacuum interrupters with 25 mm contact diameter, respectively.

Figure 4.3 shows the experimental setup. A L-C discharging circuit was used to provide the arc current. By pre-charging the capacitor banks to an appropriate voltage, a certain power frequency (50 Hz) current flowed through the test vacuum interrupter by passing through reactors. The amplitude of the discharging current was controlled by the charging voltage of the capacitor banks. The test vacuum interrupters were operated by a permanent-magnet operating mechanism. The electrodes of the test vacuum interrupters were controlled to be separated by the operating mechanism. A drawn vacuum arc initiated after contacts separated. The arcing time was electronically adjusted in a range from 1 to 10 ms. The arc voltage was measured by means of

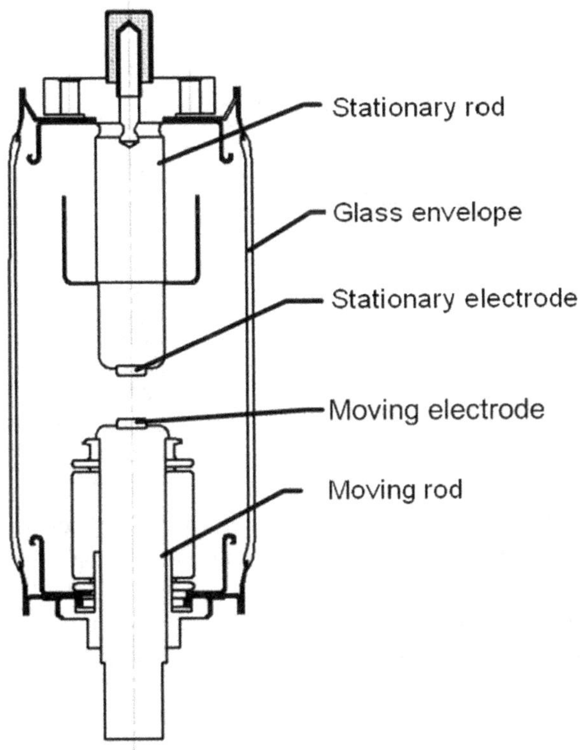

Fig. 4.1 Schematic of the test vacuum interrupters

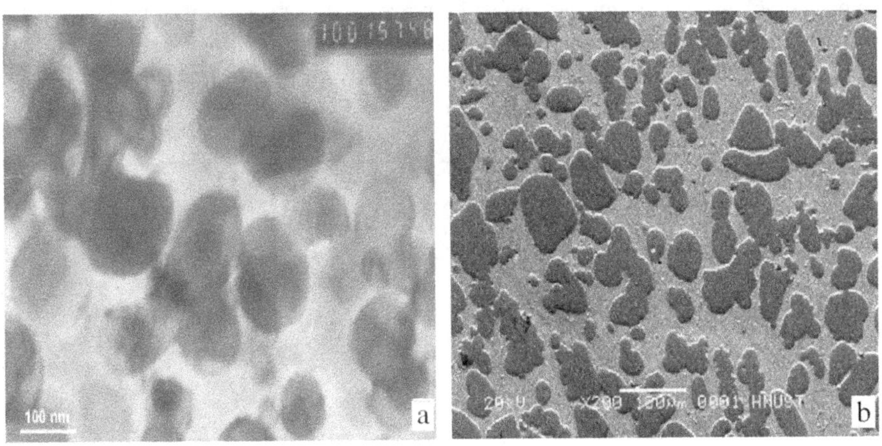

Fig. 4.2 Microstructures of **a** nanocrystalline (scale: 100 nm per unit) and **b** microcrystalline CuCr25 (scale: 100 m per unit) alloys (25% by weight of Cr)

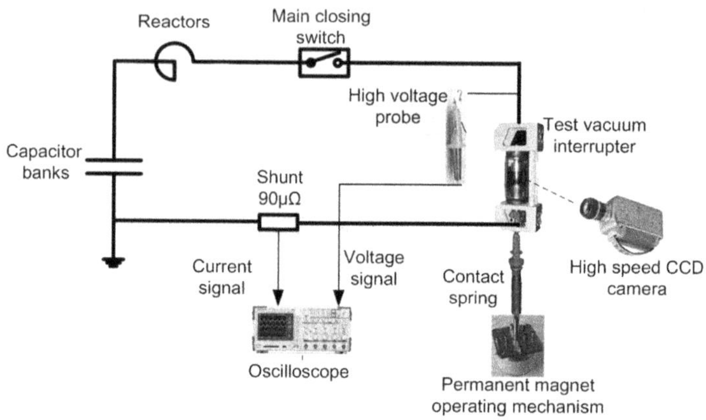

Fig. 4.3 Experimental setup

a Tektronix high-voltage probe 6015 A (1000:1) and the arc current was measured by a shunt of a 90 μΩ resistor. The vacuum arc phenomena were observed with a high-speed charge-coupled device (CCD) video camera Phantom V10 at a speed of 10 frames/ms. Exposure time of each individual frame was 2 μs. The high-speed CCD camera kept focusing on the upper electrode (stationary electrode), which acted as an anode in arcing period by controlling the opening instant in the experiment, as shown in Figs. 4.4 and 4.5 shows arc current and opening displacement curves. The stroke between the contacts was determined by comparing it with the contact diameter (12 or 25 mm) in each frame of a camera movie. The diameter of anode spots can also be determined by this method. A frame where the first bright spot appeared was identified as the beginning of arcing and it correlated to the instant of a voltage jump of 20 V on the arc voltage waveform, as shown in Fig. 4.4. Two opening velocities (1.1 m/s and 2.0 m/s) were chosen, which was realized by adjusting a contact spring

Fig. 4.4 The contacts open at the third half-wave, therefore the stationary electrode is an anode

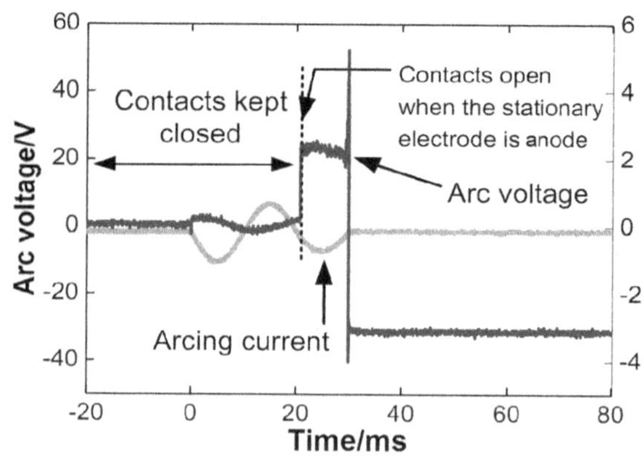

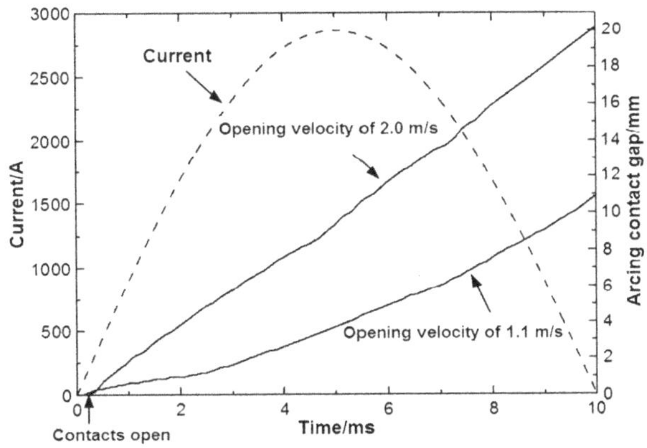

Fig. 4.5 Arc current and opening displacement curves with opening velocity of 2.0 and 1.1 m/s

stroke of the operating mechanism. The opening velocity was defined as an average velocity during 10 ms after contacts separated.

The anode discharge mode diagram was discussed in details in Sect. 1.1. The anode discharge mode is associated with arc current and contact gaps. Figure 4.6 shows four typical anode discharge modes observed in the experiments. Figure 4.6a–d shows the diffuse arc, footpoint, anode spot, and intense arc, respectively, which includes arc photos and arc voltage.

Figure 4.7 shows the anode discharge diagram of the test vacuum interrupters. Figure 4.7a is with microcrystalline CuCr25 contact material and 12 mm contact diameter. Figure 4.7b is with microcrystalline CuCr25 contact material and 25 mm contact diameter. Figure 4.7c is with nanocrystalline CuCr25 contact material and

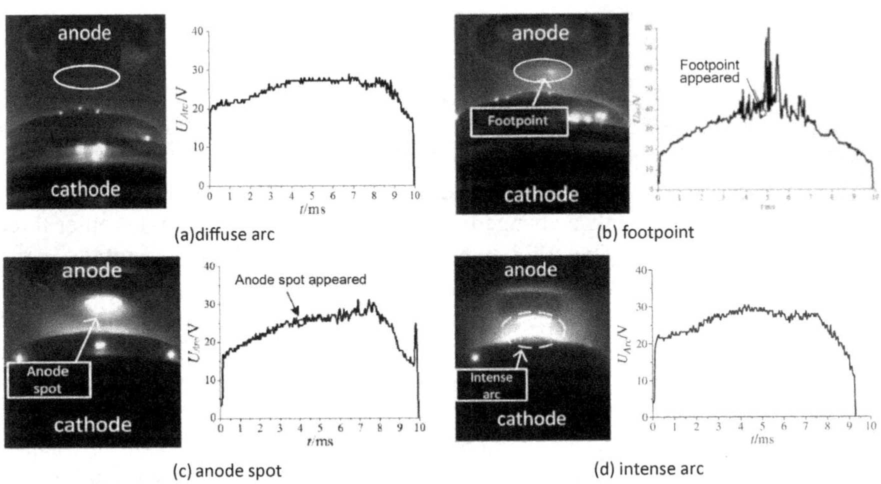

Fig. 4.6 Four anode discharge modes observed in the experiments

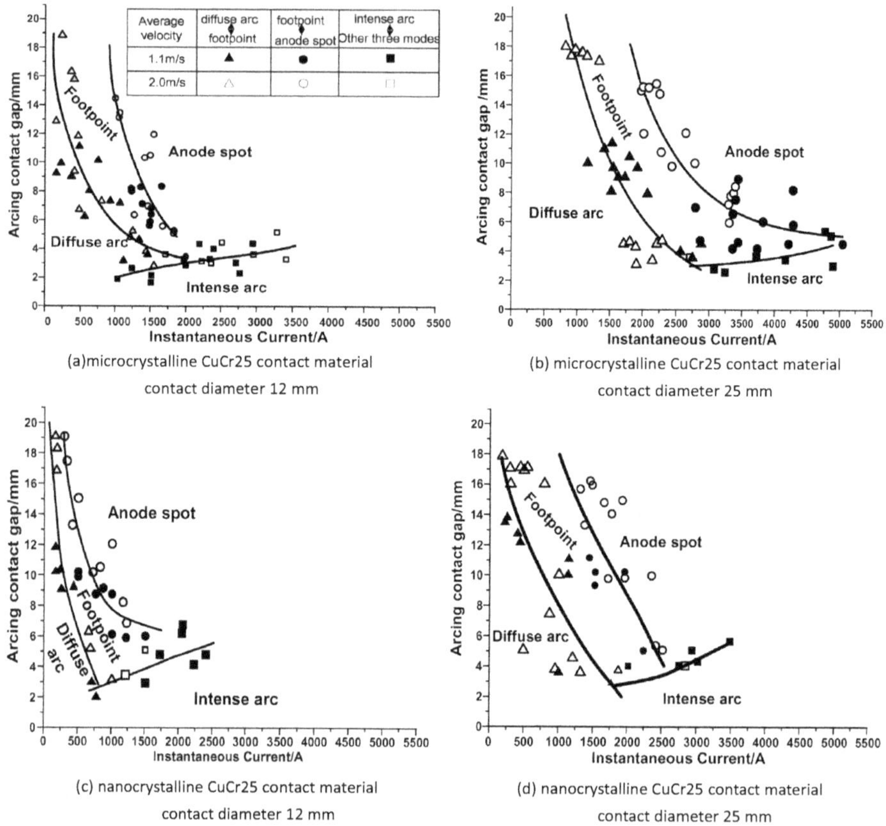

(a)microcrystalline CuCr25 contact material contact diameter 12 mm

(b) microcrystalline CuCr25 contact material contact diameter 25 mm

(c) nanocrystalline CuCr25 contact material contact diameter 12 mm

(d) nanocrystalline CuCr25 contact material contact diameter 25 mm

Fig. 4.7 Anode discharge diagram

12 mm contact diameter. Figure 4.7d is with nanocrystalline CuCr25 contact material and 25 mm contact diameter. Symbol "triangle" represents a change between a "diffuse arc" and a "footpoint" anode discharge modes. Symbol "circle" represents a change between a "footpoint" and an "anode spot" anode discharge modes. Symbol "square" represents a change between an "intense arc" and the other three anode discharge modes. The solid symbols "▲,•,■" and the hollow symbols "△,□,○" represent anode discharge mode transition with the average opening velocity of 1.1 and 2.0 m/s, respectively. With a large number of transition points between anode discharge modes, one can establish boundaries between anode discharge modes. The anode discharge mode diagrams with the two contact materials and two contact diameters suggested that the opening velocity of 1.1 m/s and 2.0 m/s has no significant influence on the boundaries among anode discharge modes. Each kind of vacuum interrupter has its unique anode discharge mode diagram, which is related to the contact material, contact diameter and applied magnetic fields etc. The difference between various anode discharge mode diagrams is the boundaries among the anode

discharge modes. For example, the high-current anode mode of the nanocrystalline CuCr25 contact material took place at a lower current than that of the microcrystalline one. In addition, the boundaries shifted to right with the increase of contact diameters from 12 to 25 mm.

Once a vacuum interrupter is interrupting a certain current, there is an arcing curve that is a function of arc current and arcing contact gap in the anode discharge diagram. For example, Fig. 4.8 shows three arcing curves in the anode discharge diagram for the vacuum interrupter with 12 mm diameter butt-type contacts and microcrystalline CuCr25 contact material. Two of the three arcing curves are with opening velocity of 1.1 and 2.0 m/s, respectively. The third one is a proposed arcing curve. There were eight key transition points associated with the three arcing curves, which are described in Table 4.1. The three arcing curves were with the same arc current of 1650 A (peak value) and with the same arcing time of 9.5 ms. When the contacts separated with an average opening velocity of 2.0 m/s, the arcing curve was represented by a dashed line with the key transition points 1–4. It shows that a diffuse arc mode formed initially. Thereafter, at a contact gap of 4.0 mm and the instantaneous arc current of 984 A (point 1), a footpoint formed. Thereafter, the vacuum arc evolved into an anode spot mode when the instantaneous arc current reached 1490 A and

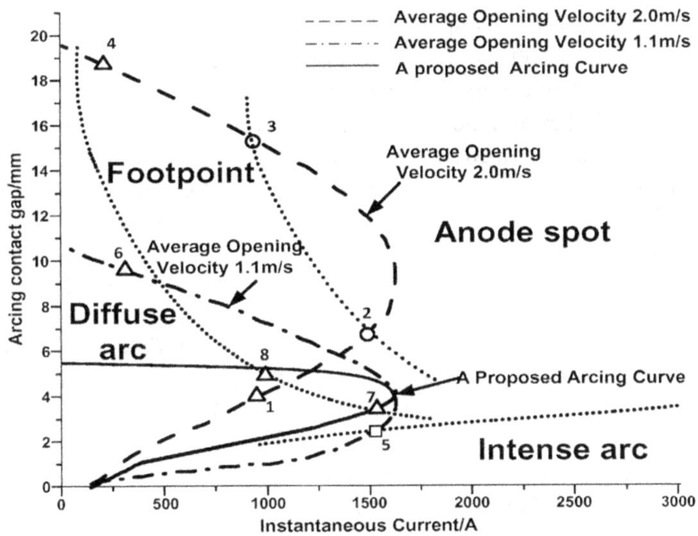

Fig. 4.8 Three arcing curves in the anode discharge mode diagram with contact material of microcrystalline CuCr25 and contact diameter of 12 mm. The arcing curve shows a relationship between the contact gap and the arc current in an anode discharge diagram. Dashed line (------) is the opening curve with an opening velocity of 2.0 m/s; Dash dot line (- • - • - • -) is the opening curve with an opening velocity of 1.1 m/s; Solid line (—) is the proposed opening curve. The descriptions of points 1–8 are shown in Table 4.1. Points 1–4 show the key transition points in the arcing curve with an opening velocity of 2.0 m/s; Points 5–6 show the key transition points in the arcing curve with an opening velocity of 1.1 m/s; Points 7–8 show the key transition points in the proposed arcing curve

Table 4.1 Descriptions of points 1–8 in Fig. 4.8

No.	(Current, Gap)	Arc mode	Opening velocity
1	(984A, 4.0 mm)	Diffuse arc -> Footpoint	2.0 m/s
2	(1490A, 6.7 mm)	Footpoint -> Anode spot	
3	(1006A, 15.0 mm)	Anode spot -> Footpoint	
4	(214A, 1 8.7 mm)	Footpoint -> Diffuse arc	
5	(1528A, 2.4 mm)	Intense arc -> Footpoint	1.1 m/s
6	(315A, 9.6 mm)	Footpoint -> Diffuse arc	
7	(1574A, 3.3 mm)	Diffuse arc -> Footpoint	Proposed Arcing curve
8	(1006A, 4.9 mm)	Footpoint -> Diffuse arc	

the contact gap reached 6.7 mm (point 2). When the arc current falled below an instantaneous value of 1006 A, the anode spot converted into a footpoint mode with a contact gap of 15.0 mm (point 3). The footpoint disappeared when the arc current became less than 214 A and the contact stroke reached 18.7 mm (point 4). Finally, the vacuum arc became a diffuse arc. When the contacts separated with an average opening velocity of 1.1 m/s, the arcing curve represented by a dash dot line with the key transition points 5–6. It shows that an intense arc mode formed initially. At the contact gap of 2.4 mm, the intense arc transfered into a footpoint mode when the instantaneous arc current reached 1528 A (point 5). Then the footpoint mode disappeared when the arc current decreased to 315 A and the contact gap reached 9.6 mm (point 6). Finally, a diffuse vacuum arc formed. One can find that, with the arcing curve of 2.0 m/s, the intense arc mode was avoided and with the arcing curve of 1.1 m/s, the anode spot mode was avoided. In such way, one can find an optimum arcing curve, which avoids entering both the anode spot mode region and the intense arc mode. The solid line shown in Fig. 4.8 is an example of such arcing curves. The arcing curve has two key transition points 7–8. Following this curve, a diffuse arc mode forms at the beginning. Then, the arc enters a footpoint mode when the instantaneous arc current reaches 1574 A and the contact gap reaches 3.3 mm (point 7). As the arc current further increases, the arc remains in a footpoint mode. When the arc current decreases to 1006 A and the contact gap reaches 4.9 mm (point 8), the footpoint disappears. In the remaining duration, the vacuum arc keeps a diffuse arc mode. Therefore, the proposed arcing curve is better than that of the other two curves, because the anode spot arc mode and the intense arc mode can be avoided. Both the anode spot mode and the intense arc mode has a negative impact on successful current interruption. The proposed arcing curve is expected to improve the high-current interruption performance of vacuum interrupters.

The arcing curves can be converted into opening displacement curves. Figure 4.9 shows three opening displacement curves according to the three arcing curves shown in Fig. 4.8. The points 1′–8′ shown in Fig. 4.9 corresponds to the key transition points 1–8 shown in Fig. 4.8. Table 4.2 gives the descriptions of points 1′–8′. Figure 4.9 and Table 4.2 also indicates the duration of each arc mode. When the opening velocity was 2.0 m/s, the duration of anode spot mode lasted from 3.6 to 7.8 ms and the overall duration in which the anode was active was from 1.9 to 9.5 ms. When the average opening velocity was 1.1 m/s, the duration of intense arc mode was from 0 to 3.8 ms, and the anode kept active until a footpoint disappeared at 9.3 ms. Following the proposed opening displacement curve, neither anode spot mode nor intense arc mode would be presented on the anode. The anode is only active in a period of footpoint mode. The duration of the footpoint mode is from 4.1 to 7.8 ms. Table 4.3 summarizes the high-current anode phenomena durations of the three cases. The proposed opening displacement curve shows a minimum duration of active anode among the three cases, which is desirable because the anode suffers the minimum time of high heat energy input. It would have a positive impact on a high-current interruption in the vacuum. The proposed opening displacement curve shows a high opening velocity (2.2 m/s) at the beginning of contacts separation (0.9 ms), which keeps the vacuum arc away from intense arc mode. Then the opening velocity slows

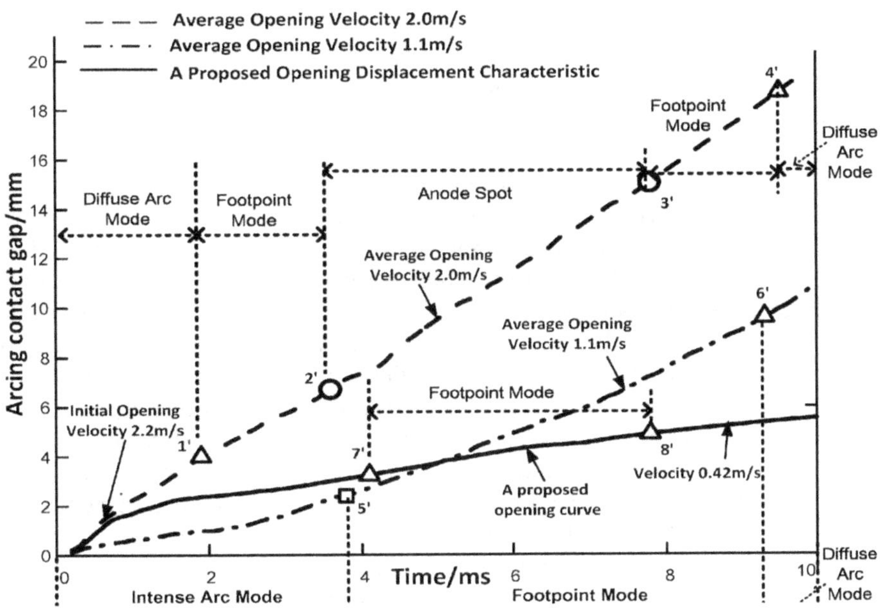

Fig. 4.9 Three opening displacement curves corresponding to the three arcing curves shown in Fig. 4.8

Table 4.2 Descriptions of points 1′–8′ in Fig. 4.9

No.	(Time, Gap)	Arc mode	Opening velocity
1′	(1.9 ms, 4.0 mm)	Diffuse arc -> Footpoint	2.0 m/s
2′	(3.6 ms, 6.7 mm)	Footpoint -> Anode spot	
3′	(7.8 ms, 15.0 mm)	Anode spot -> Footpoint	
4′	(9.5 ms, 18.7 mm)	Footpoint -> Diffuse arc	
5′	(3.8 ms, 2.4 mm)	Intense arc -> Footpoint	1.1 m/s
6′	(9.3 ms, 9.6 mm)	Footpoint -> Diffuse arc	
7′	(4.1 ms, 3.3 mm)	Diffuse arc -> Footpoint	Proposed Opening curve
8′	(7.8 ms, 4.9 mm)	Footpoint -> Diffuse arc	

Table 4.3 Summarization of the high-current anode phenomena durations among that of the three opening displacement curves

Opening velocity	Duration of intense arc mode	Duration of anode spot mode	Duration of active anode (including footpoint mode, anode spot mode and intense arc mode)
2.0 m/s	None	4.2 ms	7.6 ms
1.1 m/s	3.8 ms	None	9.3 ms
Proposed opening curve	None	None	3.7 ms

down to 0.42 m/s during the arcing time from 0.9 to 10.0 ms in order to avoid anode spot mode.

The proposed opening displacement curve shown in Fig. 4.9 was in a condition with an arcing time of 9.5 ms. However, the arcing time of a current interruption is random in field as contacts separate. Figure 4.10a shows a case that has an arcing time of 5 ms or 15 ms in the anode mode diagram. Figure 4.10b shows the corresponding opening displacement curve and the duration of each kind of anode mode appearing. The contact diameter was 12 mm and contact material was microcrystalline CuCr25. The arcing curve with a dashed line is with the arcing time of 5 ms. The contacts separated at the current peak of 1650 A. In the arcing curve, an intense arc mode formed as contacts separated. Then, the arc escaped from the intense arc mode and it entered the diffuse arc mode at an instantaneous current of 1380 A and a contact gap of 2.3 mm. The duration of intense arc was 1.8 ms. If the arc reignited at the first current zero, a long arcing time occurs. The arcing curve of the long arcing time of 15 ms is shown as the dashed line followed by the solid line. In this arcing curve, the contacts also separated at the current peak. In the first 5 ms, the arcing curve was as same as the previous one. As the current continued after the first current zero, the cathode in the arcing period of the first 5 ms became the new anode in the next current half-wave. Assuming that the anode-mode diagram of the new anode kept same with the previous one, a footpoint appeared on the new anode at an instantaneous current

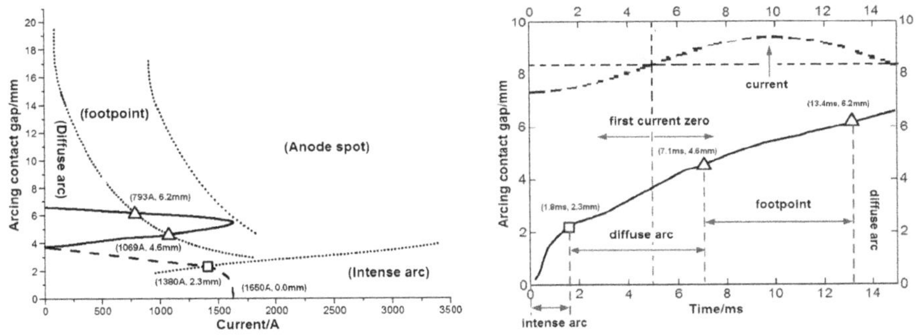

(a) arcing curve (b) opening displacement curve

Fig. 4.10 Arcing curve and opening displacement curve with a short arcing time of 5 ms and a long arcing time of 15 ms. The anode discharge mode diagram is with contact material of microcrystalline CuCr25 and contact diameter of 12 mm

of 1069 A and a contact gap of 4.6 mm. After that, a diffuse arc mode appeared when the current decreased to 793 A and the contact gap increased to 6.2 mm. In this arcing curve, the intense arc lasted 1.8 ms in the first 5 ms, and a footpoint mode lasted 6.3 ms in the second current half wave. However, some interaction between anode and cathode phenomena should be a factor that influences the anode discharge mode diagram of the new anode in long arcing time cases. Such influence will be considered in Sect. 4.1.2. In addition, it also considers an impact of axial magnetic field.

4.1.2 Determination of Opening Velocity Characteristic for a 126 kV Vacuum Interrupter with 2/3 Coil-Type Axial Magnetic Field Contact

Based on the method proposed in Sect. 4.1.1, this section is to experimentally determine an appropriate opening velocity characteristic for a 126 kV single-break vacuum interrupter with 2/3 coil-type axial magnetic field contact (see Sect. 3.2.1). The experiment was performed in a L-C discharging circuit. The capacitance was 55760 μF and the inductance was 187.5 μH, which determines a current frequency of 49.2 Hz. The discharging current was ranging from 4 to 40 kA. The current waveform was measured by a Rogowski coil and the arc voltage was measured by a RC divider. A demountable vacuum chamber was used to observe the vacuum arc behaviors. The demountable vacuum chamber was kept pumping and maintained to a pressure level of 1×10^{-4} Pa. A high-speed charge coupled device (CCD) camera was adopted to record the vacuum arc. The exposure rate of the CCD was 10 frame/ms and the aperture was fixed at 4. A pair of 2/3 coil-type axial magnetic field contacts were assembled in the demountable vacuum chamber. The contact material was CuCr50

(50% weight of Cr). The diameter of the contact was 100 mm. In the experiment, the full contact gap was set to 60 mm. Table 3.7 shows the minimum arcing time and maximum arcing time of the 2/3 coil-type axial magnetic field contacts, respectively. It also shows that the minimum arcing contact gap was ranging from 18 to 23 mm and the maximum arcing contact gap was 44 mm. The average opening velocity in the minimum arcing contact gap is defined as V_1. The average opening velocity in the maximum arcing contact gap is defined as V_2.

Figure 4.11 shows the arc modes in the case of minimum arcing gap. Figure 4.11a, b show two cases that V_1 was 1.8 and 2.4 m/s, respectively. In the case of Fig. 4.11a, the arc current was 28.4 kA (rms) and the arcing time was 9.3 ms. In the case of Fig. 4.11b, the arc current was 28.1 kA (rms) and the arcing time was 10.2 ms. Figure 4.11a shows the contacts separated at an instantaneous current of 23.9 kA. When the vacuum arc extinguished, the contact gap was 15.36 mm. There were two anode discharge modes observed, which were intense arc mode and diffuse arc mode, respectively. The intense arc mode started from contact separation. The arc photos at 1.3 and 3.2 ms indicated the intense arc. The inter-electrode gap was filled with luminous arc plasma. At 3.2 ms, the arc current reached its peak of 40.2 kA. After that the arc began to diffuse with the decreasing of arc current. The inter-electrode plasma became less luminous and more diffuse until arc extinguished. The arc voltage was relatively low (<40 V) and quiet without noise component. Figure 4.11b shows the contacts separated at an instantaneous current of 15.78 kA. The contact gap was 25.7 mm at the instant of arc extinction. The vacuum arc evolution was also related to the two anode discharge modes: intense arc mode and diffuse arc mode. The intense arc mode initiated as contacts separated. It began to diffuse at 4.3 ms, which was at the peak of the arc current. The arc plasma entirely diffused at 7.3 ms. The arc voltage was also less than 40 V and kept quiet.

Figure 4.12 shows the arc modes in the case of maximum arcing gap. Figure 4.12a, b show two cases that V_2 was 3.0 and 2.7 m/s, respectively. Figure 4.12a shows that the arc did not extinguish at the first current zero and the contact gap reached 57 mm

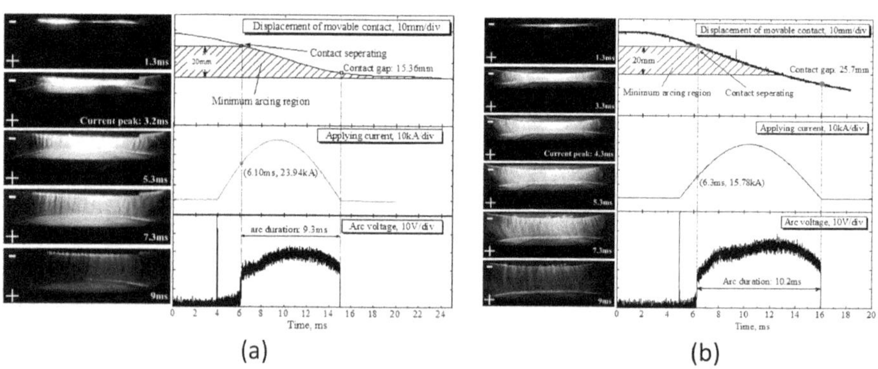

Fig. 4.11 Arc modes in the minimum arcing gap. **a** Arc current—28.4 kA. Arcing time −9.3 ms. V_1 −1.8 m/s. **b** Arc current −28.1 kA. Arcing time −10.2 ms. V_1 −2.4 m/s

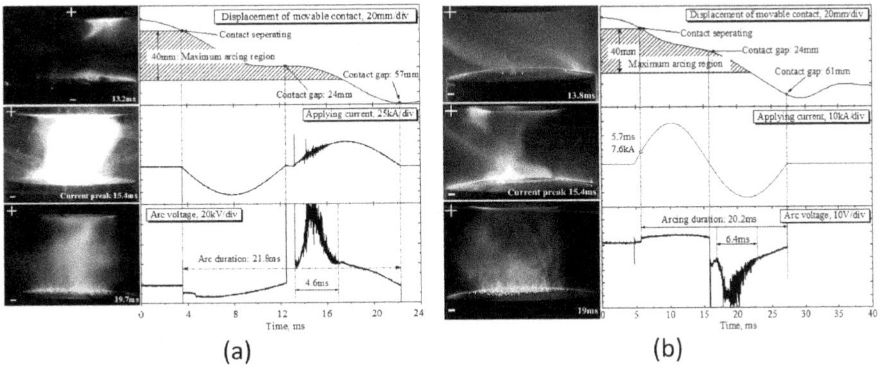

Fig. 4.12 Arc modes in the maximum arcing gap. **a**Arc current −39.5 kA (rms) for the 1st half-wave and 35.6 kA (rms) for the 2nd half-wave, Arcing time −21.8 ms, V_2 −3.0 m/s. **b**Arc current −20.2 kA (rms) for the 1st half-wave and 16.5 kA (rms) for the 2nd half-wave, Arcing time −20.2 ms, V_2 −2.7 m/s

at the second current zero. The arc current was 39.5 kA (rms) for the 1st half-wave and it was 35.6 kA (rms) for the 2nd half-wave. The arcing time was 21.8 ms. There were three anode discharge modes observed, which were footpoint mode, anode spot mode and diffuse arc mode, respectively. After the first current zero, a footpoint appeared on the anode at 13.2 ms. An anode spot formed at 14.0 ms, in which an anode plasma jet and a cathode plasma jet met. The two plasma jets mixed together at 15.4 ms, which corresponded to the current peak of the second half-wave. It lasted until 19.7 ms, which corresponded to a contact gap of 55.7 mm. After that the arc became diffuse. A remarkable noise component can be observed in the arc voltage, which lasted about 4.6 ms. The maximum value was over 100 V. Figure 4.12b also shows that the arc reignited at the first current zero and the contact gap reached 61 mm at the second current zero. The arc current was 20.2 kA (rms) for the 1st half-wave and it was 16.5 kA (rms) for the 2nd half-wave. The arcing time was 20.2 ms. The vacuum arc evolution was related to two anode discharge mode: footpoint mode and diffuse arc mode. The contacts separated at 5.7 ms with an instantaneous current of 7.6 kA. Just after the initiation of the second current half-wave, a diffuse arc appeared at 13.8 ms. However, the arc plasma distorted significantly. At 14.5 ms, a footpoint appeared on the edge of the anode, which lasted for 6.4 ms. One photo of the footpoint at 15.4 ms was given, which corresponded to current peak of the second half-wave. After that, the arc became diffuse at 19 ms and it lasted until the second current zero. A notable noise component can be observed in the arc voltage, with the maximum value higher than 80 V. The duration of the high noise component lasted 6.4 ms.

Figure 4.13 shows the anode discharge mode diagrams of the 2/3 coil-type axial magnetic field contacts. Symbol "Δ" represents a transfer between diffuse arc mode and footpoint mode. Symbol "o"represents a transfer between footpoint mode and anode spot mode. Symbol "□" represents a transfer between intense arc mode and

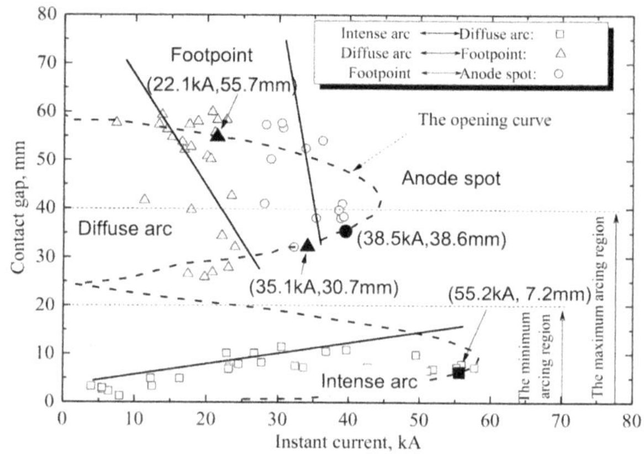

Fig. 4.13 Anode discharge mode diagrams of 2/3 coil-type axial magnetic field contacts

the other three modes. There is a solid line as a boundary among each mode in Fig. 4.13. The dashed line shows an arcing curve for the case shown in Fig. 4.12a, in which the current was 39.5 kA (rms) in the 1st half wave and it was 35.6 kA (rms) in the 2nd half-wave. V_1 was 2.7 m/s and V_2 was 3.0 m/s. The contacts separated at an instantaneous current of 25.1 kA as an intense arc. Then, it transferred to the diffuse arc mode 3.5 ms later, where the instantaneous current was 55.2 kA and the contact gap was 7.2 mm. The diffuse arc maintained until the first current zero, then it reignited. In the second half-wave, it is still a diffuse arc until the instantaneous current reached 35.1 kA and the contact gap reached 30.7 mm, where it transferred to footpoint mode. The footpoint transferred to anode spot when the instantaneous current increased to 38.5 kA and the contact gap increased to 38.6 mm. Finally, it became a diffuse arc when the instantaneous current decreased to 22.1 kA and the contact gap reached 55.7 mm.

Figure 4.14 shows a preferred opening velocity characteristic based on the anode discharge mode diagram, as indicated as the line 1. The preferred opening velocity

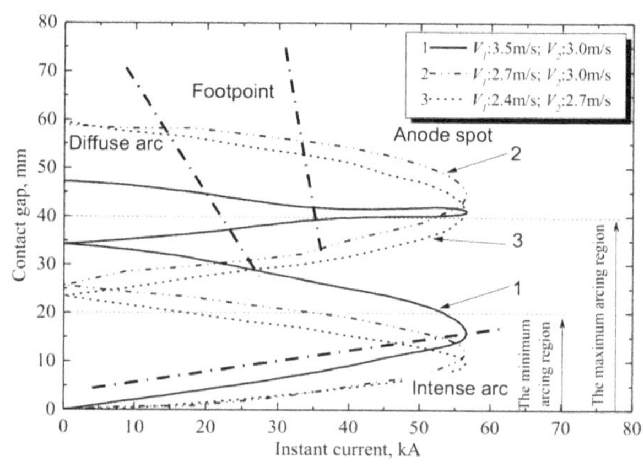

Fig. 4.14 Determination of opening velocity characteristic based on anode discharge mode diagram for 126 kV single-break vacuum interrupter with 2/3 coil-type axial magnetic field contact

characteristic is with a V_1 of 3.5 m/s in the minimum arcing gap and it is with a V_2 of 3.0 m/s in the maximum arcing gap, respectively. A high V_1 of 3.5 m/s would escape the intense arc mode quickly comparing with the lower V_1 of 2.7 m/s (line 2) or 2.4 m/s (line 3). Consider the high V_1 of 3.5 m/s, the V_2 of 3.0 m/s means a low opening velocity in the contact gap ranged between the minimum arcing gap and the maximum arcing gap. This is desirable because the preferred opening velocity characteristic has a shorter duration in the footpoint mode and anode spot mode than that of line 2 ($V_2 = 3.0$ m/s) and line 3($V_2 = 2.7$ m/s). Therefore, the opening velocity characteristic of $V_1 = 3.5$ m/s and $V_2 = 3.0$ m/s is chosen for the 126 kV single-break vacuum interrupter with the 2/3 coil-type axial magnetic field contact.

4.1.3 Determination of Opening Velocity Characteristic for a 126 kV Vacuum Interrupters with Horseshoe-Type Bipolar Axial Magnetic Field Contact

Compared with coil-type axial magnetic field contact, the horseshoe-type bipolar axial magnetic field contact has no artificial coils, which helps to reduce loop resistance of contact system. In addition, the irons behind contact would provide a strong axial magnetic field. The advantages suggest an application of the horseshoe-type axial magnetic field contact for vacuum interrupters at transmission voltage level, which require both high normal rated current level and strong short-circuit current interruption ability. Section 3.2.2 determined the vacuum arc mode transition diagram for a horseshoe-type axial magnetic field contact. It recommended a contact gap at current zero bellow a critical value to avoid uncontrollable arc mode. This section is to experimentally determine an opening velocity characteristic for a 126 kV single-break vacuum interrupter with the horseshoe-type axial magnetic field contact.

Figure 4.15 shows a schematic of the experimental circuit. A L-C discharging circuit was used to generate a sinusoidal current. The current frequency was 45 Hz. A pair of horseshoe-type contacts designed for 126 kV vacuum interrupters were installed in a demountable vacuum chamber. The demountable vacuum chamber was continuously pumped at a pressure of 10^{-4} Pa. The structure of the horseshoe-type axial magnetic field contact was as same as that shown in Sect. 3.2.2. Contact material was CuCr50 (50% of Cr). The contact diameter was 120 mm. Before an experiment began, a main circuit breaker was kept open and the horseshoe-type contacts closed. The experiment began by closing the main circuit breaker. Then, the horseshoe-type contacts separated, which was driven by a spring operating mechanism. Therefore, a drawn vacuum arc was initiated. The opening velocity could be adjusted through the spring. The arc currents amplitude was controlled by adjusting the charging voltages of the capacitor banks. Arc current level ranged from 12 to 40 kA (rms), stepped by 4 kA. Prospective arcing time was chosen as 2, 5, and 9 ms in the first half-wave and the total arcing time was 13, 16, and 20 ms with a variation of ± 1 ms. The vacuum

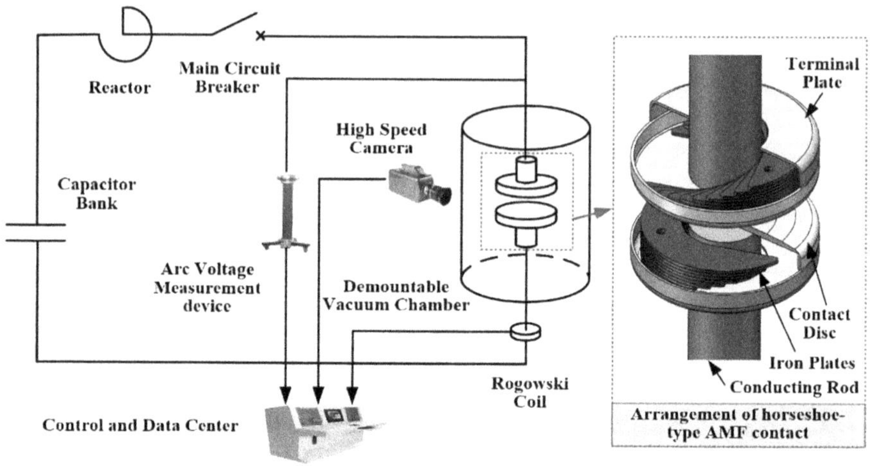

Fig. 4.15 Schematic of experimental circuit for horseshoe-type axial magnetic field contact

arc behaviors were recorded by a high-speed camera. Arc voltage and arc current were measured by a high voltage divider and a Rogowski coil, respectively. The full contact gap was set as 60 mm. A linear displacement sensor with a resolution of 0.01 mm was used to measure the displacement curve. Two contact gap intervals were defined, which were interval I and II, respectively. Internal I was from 0 to 20 mm. Interval II was from 0 to 40 mm. Average opening velocities in interval I (v_1) ranged from 1.9 to 2.6 m/s. Average opening velocities in interval II (v_2) ranged from 2.6 to 2.8 m/s. The variation of average opening velocity was ±0.05 m/s.

Figure 4.16 shows the axial magnetic field characteristics of the horseshoe-type contact. The finite-element method was employed in the analysis. The arc current

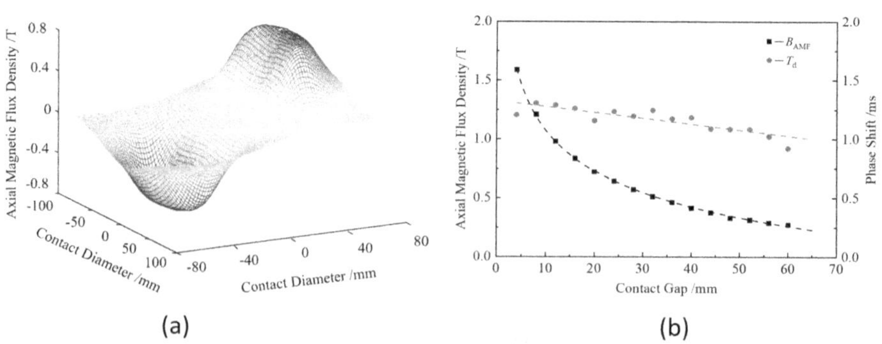

Fig. 4.16 Axial magnetic field characteristics of the horseshoe-type contact. **a** Axial magnetic field distribution on the intermediate plane of contact gap. Contact gap was 24 mm. Arc current was 40 kA (rms). **b** Dependence of maximum value of axial magnetic field on the intermediate plane and phase shift time on the contact gap

was 40 kA. Figure 4.16a shows a bipolar axial magnetic field distribution on the intermediate plane of contact gap. When the contact gap was 24 mm, the maximum value of axial magnetic flux density (B_{AMF}) was 645 mT on the intermediate plane of contact gap. Figure 4.16b shows a dependence of B_{AMF} and phase shift time T_d on contact gap. T_d is a typical value at the point with the maximum B_{AMF} on the intermediate plane of contact gap. It shows that B_{AMF} exponentially decreased from 1.589 to 0.276 T with increasing of contact gap from 4 to 60 mm. At the same time, T_d decreased linearly from 1.19 to 0.92 ms.

Figure 4.17 shows one typical experimental result including vacuum arc photos, opening displacement curve, arc current, and arc voltage. The arc current was 40 kA (rms). The arcing time was 10 ms in the first current half-wave and the total arcing time was 21 ms. v_1 and v_2 were 2.3 and 2.7 m/s, respectively. Definitions of the vacuum arc modes referred to that in Sect. 3.2.2. When contacts separated, a small-gap constricted column mode (termed as T1) first appeared after liquid metal bridge explosion. T1 started at contact gap of 0.65 mm with an instantaneous arc current of 26.28 kA. The vacuum arcs still remained in the small-gap constricted column mode when arc current reached its peak value of 56.5 kA. When the arc current decreased to an instantaneous value of 42.81 kA, the vacuum arcs evolved into diffuse column mode (termed as T2). The contact gap at the arc mode transition

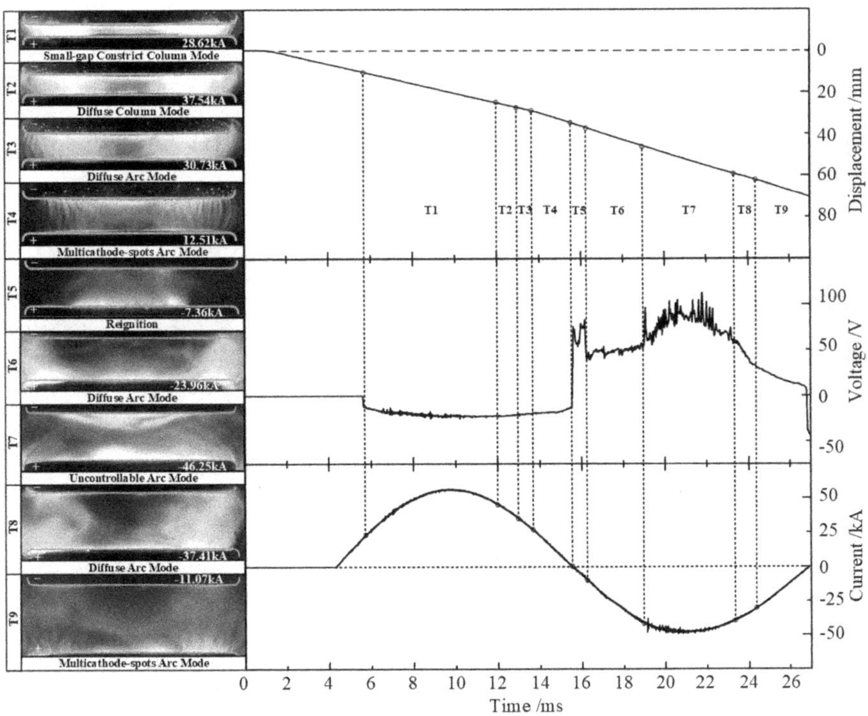

Fig. 4.17 A typical experimental result with arc current of 40 kA (rms)

point was 14.0 mm. In the diffuse column mode, two arc columns were observed. Each column had a central plasma core between anode and cathode. With decreasing of current, diffuse arc mode (termed as T3) and multi-cathode-spots arc mode (termed as T4) appeared in succession until current zero. The arc voltage during T1–T4 was low (~23 V) with little noise. The vacuum arcs reignited at current zero. An unstable vacuum arc (termed as T5) appeared. The arc voltage in this period jumped to ~75 V. Then, the vacuum arcs evolved into diffuse arc mode (termed as T6) at an instantaneous arc current of 15.11 kA. As vacuum arc became stable in this period, arc voltage decreased to ~50 V with low noises. With further increasing of arc current and contact gap, an uncontrollable arc mode (termed as T7) appeared with severely distorted arc plasma jets. The arc voltage increased to 85 V with significant noise components. After arc current passed the peak of the second half-wave and decreased to an instantaneous value of 36.71 kA, the vacuum arc evolved into diffuse arc mode (termed as T8) again. The arc voltage decreased from 57 to 28 V with little noise in this period. With current further decreased, multi-cathode-spots arc mode (termed as T9) appeared. The arc voltage was smooth and it decreased gradually until current zero.

Figure 4.18 shows the vacuum arc mode transition points from diffuse column mode to diffuse arc mode. The arc current ranged from 12 to 40 kA (rms). Figure 4.18a–c show three cases with arcing time of 2, 5 and 9 ms in the first

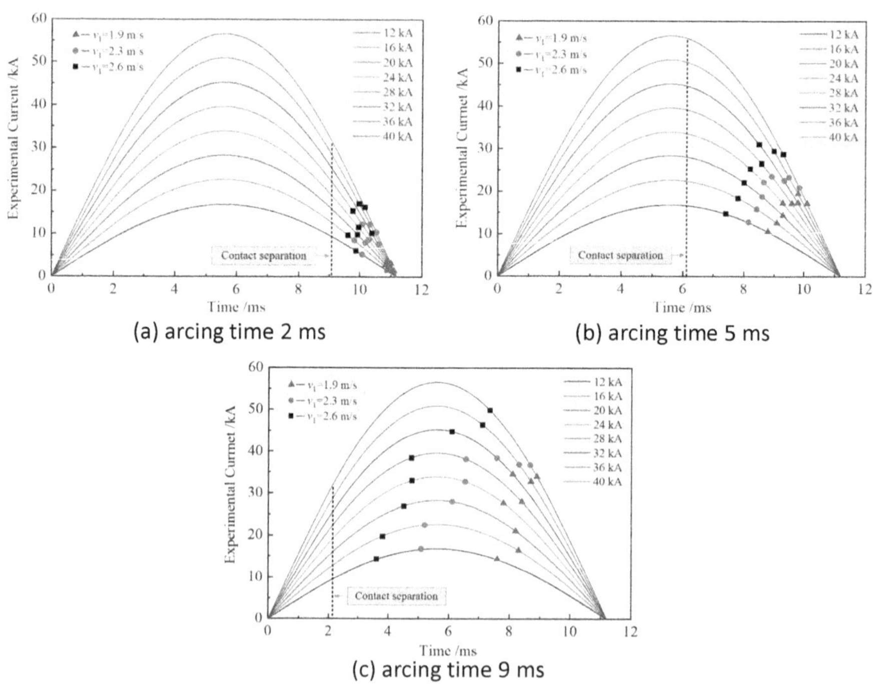

Fig. 4.18 Vacuum arc mode transition points from diffuse column mode to diffuse arc mode

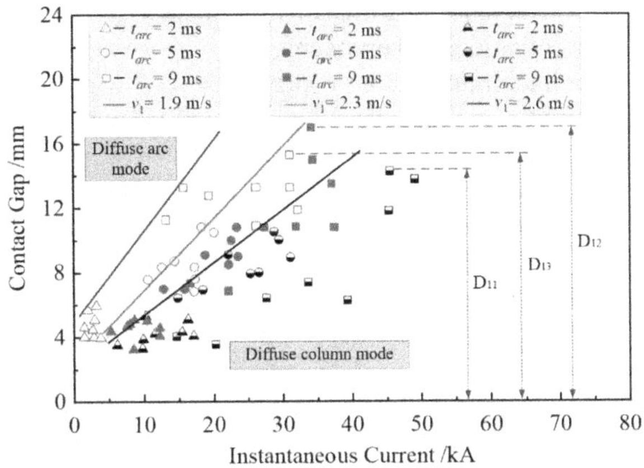

Fig. 4.19 Determination of minimum arcing contact gap

half-wave, respectively. As increasing of v_1 from 1.9 to 2.3 and 2.6 m/s, the arc mode transition points approached to the contact separation instant (dash line). This indicated that a high v_1 led to a quick evolution into diffuse arc mode.

The minimum arcing contact gap (D_1) was chosen as the highest instantaneous contact gap, at which vacuum arcs evolved from diffuse column mode to diffuse arc mode. Figure 4.19 shows a dependence of the contact gaps as vacuum arc evolved from diffuse column mode to diffuse arc mode on instantaneous current. The arc current ranged from 12 to 40 kA (rms). In general, with increasing of arc current, the minimum arcing contact gaps increased. When v_1 was 2.6, 2.3, and 1.9 m/s, the minimum arcing contact gap D_{11}, D_{12}, and D_{13} were determined as 14.3, 16.8, and 15.3 mm, respectively.

In addition, the duration from contacts separation to the minimum arcing contact gap $T_{\min}$ was obtained according to the experimental data shown in Fig. 4.18. A dependence of $T_{\min}$ on arc current I, v_1, and arcing time in the first half-wave t_{arc} was fit as Eq. (4.1) by using least-square method.

$$T_{\min} = 0.038I - 1.877v_1 + 0.616t_{arc} + 3.389 \tag{4.1}$$

where $T_{\min}$ is in ms. Velocity is in m/s. Displacement is in mm. Current is in kA. The fitting errors of Eq. (4.1) were within 27%, 6%, and 8% when t_{arc} was 2, 5, and 9 ms, respectively. According to Eq. (4.1), $T_{\min}$ decreases by increasing v_1, decreasing I or t_{arc}. A short $T_{\min}$ is desirable, which would lead to a slight contact surface erosion.

In a certain condition, uncontrollable arc mode appeared in the second current half-wave. When arc current was higher than 32 kA (rms) and the arcing time in the first current half-wave was 9 ms, extremely unstable and distorted vacuum arcs were observed in uncontrollable arc mode during the second current half-wave. However, when the arcing time in the first current half-wave was 2 and 5 ms, uncontrollable arc mode did not appear in the second current half-wave, in which vacuum arcs were well controlled by the bipolar axial magnetic field. Therefore, the experimental results with the arcing time of 9 ms in the first current half-wave were used to analyze uncontrollable arc phenomena.

The maximum arcing contact gap (D_2) was chosen as the lowest instantaneous contact gap, at which vacuum arcs evolved from diffuse arc mode to uncontrollable arc mode. Figure 4.20 shows a dependence of instantaneous contact gap at which arc mode evolved from diffuse arc mode to uncontrollable arc mode on arc current levels. There were three current levels, which was 32, 36 and 40 kA (rms), respectively. The arcing time in the first current half-wave was 9 ms. When v_2 was 2.8, 2.7, and 2.6 m/s, the maximum arcing contact gaps D_{21}, D_{22}, and D_{23} were determined as 36.1, 36.7, and 38.5 mm, respectively. Besides, higher arc current led to lower D_2. Therefore, D_{21}, D_{22}, and D_{23} were determined at arc current 40 kA. When instantaneous contact gap did not reach D_2, vacuum arcs can be well controlled. While instantaneous contact gap was higher than D_2, uncontrollable arc mode appeared at arc current of 40 kA. Table 4.4 shows the D_2 at the three levels of v_2 at 2.6, 2.7 and 2.8 m/s and three arcing time in the first current half-wave of 2, 5, and 9 ms.

Figure 4.21 shows three typical arc voltage waveforms with uncontrollable arc mode in second current half-wave. The arc currents were 32, 36, and 40 kA (rms), respectively. It indicated the duration of uncontrollable arc mode Δt, in which vacuum arcs were extremely unstable and there was a significant arc voltage noise component. The arc voltage amplitude and noise component decreased with decreasing of arc current. Δt was 1.91, 3.35, and 4.35 ms when arc current was 32, 36, and 40 kA rms, respectively. Table 4.5 shows Δt values at the three levels of arc current (32, 36, and

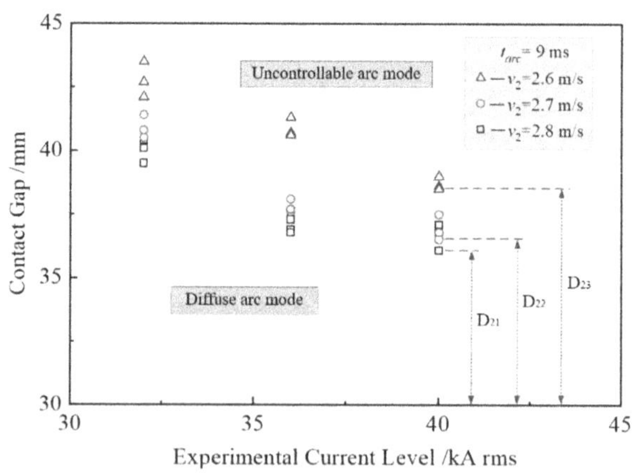

Fig. 4.20 Dependence of instantaneous contact gap at which arc mode evolved from diffuse arc mode to uncontrollable arc mode on arc current levels. The arc currents were 32, 36 and 40 kA (rms). The arcing time in the first current half-wave was 9 ms. v_2 was 2.8, 2.7, and 2.6 m/s

Table 4.4 D_2 at the three levels of v_2 at 2.6, 2.7 and 2.8 m/s and three arcing time in the first current half-wave of 2, 5, and 9 ms

v_2/m/s	Maximum arcing contact gap/mm		
	t_{arc} = 2 ms	t_{arc} = 5 ms	t_{arc} = 9 ms
2.6	34.1	41.9	38.5
2.7	35.2	43.5	36.7
2.8	37.1	46.1	36.1

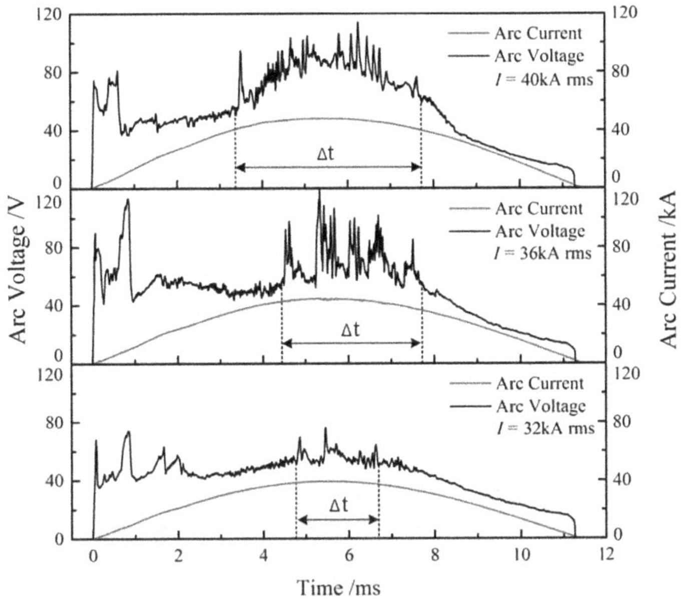

Fig. 4.21 Three typical arc voltage waveforms with uncontrollable arc mode in second current half-wave. Arc currents were 32, 36, and 40 kA (rms)

Table 4.5 Uncontrollable arc mode duration

Current level	Δt/ms	Δt/ms	Δt/ms
/kA	(v_2 = 2.6 m/s)	(v_2 = 2.7 m/s)	(v_2 = 2.8 m/s)
32	1.78	1.91	2.55
36	2.86	3.35	3.71
40	3.98	4.35	4.43

40 kA rms) and three levels of v_2 (2.6, 2.7, and 2.8 m/s). It shows that Δt increased with increasing of arc current or increasing of v_2. Note that not only a higher arc current but also a higher v_2 led higher duration of the uncontrollable arc mode. It suggested that a lower v_2 is desirable.

Figure 4.22 shows an opening velocity characteristic proposed for vacuum interrupters with the horseshoe-type bipolar axial magnetic field contact. It shows that the opening displacement curve was divided into two regions, which was arcing region and no arcing region, respectively. In the arcing region, vacuum arcs should be effectively controlled and it should avoid an appearance of uncontrollable arc mode. It was designed that vacuum arcs extinguish in the arcing region. Thus, the minimum arcing time T_{min} and the maximum arcing time T_{max} in current interruptions should be within the arcing region. In the no arcing region, vacuum arcs were not expected. The no arcing region was considered only for vacuum insulation and mechanical damping. The boundary between the arcing region and the no arcing

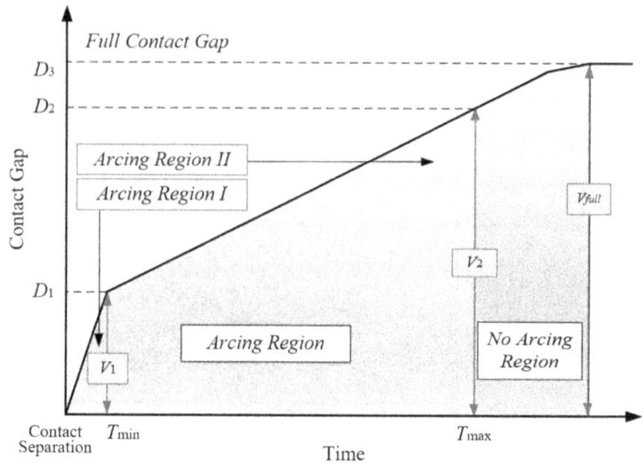

Fig. 4.22 Opening velocity characteristic proposed for vacuum interrupters with horseshoe-type bipolar axial magnetic field contact

region was the maximum arcing contact gap D_2. In addition, the arcing region was divided into arcing region I and arcing region II. Arcing region I was characterized by the minimum arcing contact gap D_1 and T_{min}. Arcing region II was characterized by the maximum arcing contact gap D_2 and T_{max}. Therefore, the opening velocity characteristic had three velocities v_1, v_2, and v_{full}, which were the average velocity from contacts' separation to D_1, D_2, and full contact gap, respectively. A favorable opening velocity characteristic for the 126 kV vacuum interrupter with the horseshoe-type bipolar axial magnetic field contact would be that, there is a high v_1, a low v_2, and a lower v_{full}. A high v_1 led to a short duration in small-gap constricted column mode and diffuse column mode, which has a benefit of less contact erosion and less evaporation of metal vapors and liquid droplets. A low v_2, would help to control the vacuum arcs well by the axial magnetic field and the uncontrollable arc mode would be avoided. A lower v_{full} is good for mechanical damping. The proposed opening velocity characteristic would be beneficial for successful current interruptions.

The determination of v_1 is as following. Multiplying Eq. (4.1) by v_1, the minimum arcing contact gap D_1 can be obtained as Eq. (4.2).

$$D_1 = v_1 T_{min} = -1.877v_1^2 + (0.038I + 0.616t_{arc} + 3.389)v_1 \tag{4.2}$$

Equation (4.2) could be used to estimate D_1. According to Eq. (4.2), the curve of D_1 has a peak at a certain arc current and an arcing time. There is a critical v_1 corresponding to the peak value of D_1. The critical v_1 is determined as Eq. (4.3), based on the arc current I and arcing time in the first current half-wave t_{arc}.

$$v_{1\_crit} = \frac{0.038I + 0.616t_{arc} + 3.389}{3.754} \tag{4.3}$$

In designing the opening velocity characteristic of the vacuum interrupters with the horseshoe-type bipolar axial magnetic field contact, v_1 is suggested to be faster than v_{1_crit} in order to reduce the arc duration of small-gap constricted column mode and diffuse column mode. A high v_1 is preferred.

Next v_2 is determined as following. v_2 was designed to avoid uncontrollable arc mode. The formation of the uncontrollable arc mode may be due to the transverse magnetic field component in the contact gap. At a large contact gap, the magnetic field flux would not go through the contact gap path and it tended to go through a self-closed path between the two branches of the horseshoe-shaped iron plates. In such case, a strong transverse magnetic field component contributed to distorted arcs, which may lead to uncontrollable arc mode. Instantaneous current and contact gap determined a unique magnetic field distribution in the arcing gap, which had a significant influence on the formation of uncontrollable arc mode. Figure 4.23 shows the arc mode transition points at which vacuum arc evolved into uncontrollable arc mode. The contact gaps of the arc mode transition points $D_{\text{uncontrol}}$ was fitted as Eq. (4.4).

$$D_{uncontrol} = -0.486i + 61.01 \tag{4.4}$$

where, the instantaneous arc current i follows Eq. (4.5)

$$i = \sqrt{2}I \sin[2\pi f \cdot (\frac{D}{v_2} - t_{arc}) \cdot 10^{-3}] \tag{4.5}$$

where t_{arc} is arcing time in the first current half-wave. D is the contact gap. f is current frequency. I is the arc current in rms. The uncontrollable arc mode can be avoided if Eq. (4.6) is satisfied. Equations (4.4)–(4.6) could be used to determine v_2.

$$D < D_{uncontrol} \tag{4.6}$$

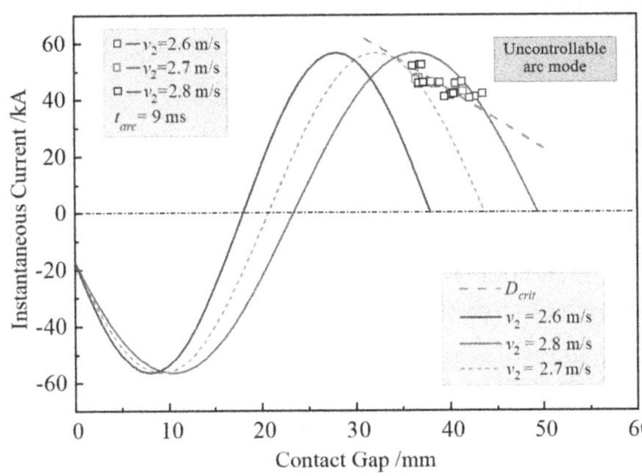

Fig. 4.23 Arc mode transition points at which vacuum arc evolved into uncontrollable arc mode. Arcing time in the first current half-wave is 9 ms

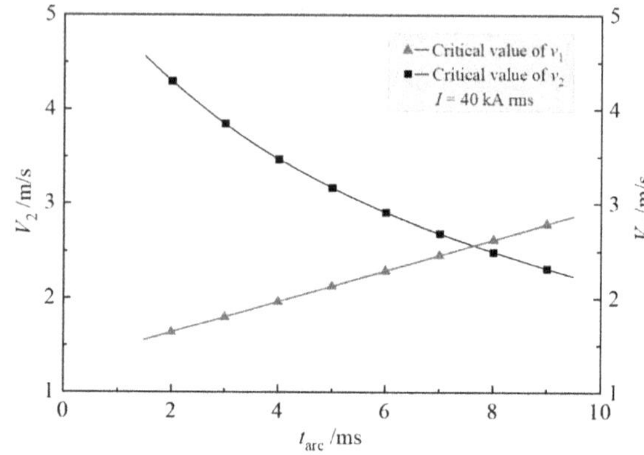

Fig. 4.24 Design of opening velocity characteristic (v_1 and v_2) for the 126 kV vacuum interrupter with the horseshoe-type bipolar axial magnetic field contact Test duty T100s of 40 kA

Based on the above methods, one can determine v_1 and v_2 for 126 kV vacuum interrupter with the horseshoe-type axial magnetic field contact. A test duty T100s of 40 kA (rms) was considered. The critical v_1 was determined by Eq. (4.3). The critical v_2 was determined by Eqs. (4.4)–(4.6), and it was fitted as Eq. (4.7).

$$v_2 = 3.96e^{-0.17t_{arc}} + 1.52 \tag{4.7}$$

Figure 4.24 shows a dependence of the critical v_1 and critical v_2 on the arcing time in the first half-wave t_{arc}. It can be used to design the opening velocity characteristic. For example, if t_{arc} is 7 ms, v_1 should be higher than 2.5 m/s, and v_2 should be lower than 2.9 m/s. Besides, v_1 should be faster than v_2. In high-current interruptions, such design of opening velocity characteristic based on arc behaviors could not only weaken the constrict arc (diffuse column mode) effect within minimum arcing time but also avoid uncontrollable arc phenomena within maximum arcing time. Thus, contact surface erosion and metal vapor evaporation would be weakened, which help to withstand transient recovery voltage.

4.1.4 Determination of Closing Velocity for Minimizing Percussion Welding Force

The phenomenon of contact welding is an important consideration when designing vacuum circuit breakers [5]. Percussion welding might occur during circuit breaker's making operation when a high-current passes through the contacts. An electrical arc appears across a pair of contacts when two contacts are closing enough or bouncing apart [6]. This arc will cause contact surfaces to melt locally. If the local contact surface melting is serious enough, the contacts may be welded together when they are finally forced together [7]. If a percussion welding is too strong, it may cause a failure to open the contacts of vacuum interrupters when the operating mechanism cannot break the weld [8]. Energy to form the percussion welding mainly comes from the prestrike arc and the contact bouncing arcs. It was reported that the percussion welding was decreased by reducing the arc duration [9–12]. An appropriate closing velocity design is important to reduce the total arc duration in making operations of vacuum circuit breakers. The total arc duration in making operations includes a prestriking arc duration and a bouncing arc duration.

In the prestrike stage, a vacuum circuit breaker with a high closing velocity expects a short prestrike duration before contacts mate. Slade et al. [13] found that as the contact gap at prestriking increased, a longer prestrike arc duration caused the contacts to weld. McBride and Sharkh [14] also showed that a higher applied voltage across a pair of contacts caused a longer arc duration for the same closing velocity, which indicated that the prestrike arc duration was influenced by the prestrike gap and the voltage of power system. Garzon [15] pointed out that increasing the closing velocity decreased not only the prestrike arc duration but also the magnitude of the current at the instant of contact engagement.

A high closing velocity also leads to a long contact bouncing duration in vacuum circuit breakers. With a long bouncing duration, the probability of contact welding also increases as a constricted arc forms between the opening contacts and the arc melts the contact surfaces locally [8]. In transmission voltage level vacuum circuit breakers, contact bouncing in current making operation becomes more significant than that in medium voltage. Therefore, a long contact bouncing duration could damage contacts by severe arc erosion [16]. Both Greenwood [17] and Barkan [18] gave excellent discussions on how to reduce the bouncing duration.

However, there was no attempt so far to balance the prestrike arc and bouncing arc from the perspective of closing velocity of transmission voltage level vacuum circuit breakers to improve their percussion welding performance. This section is to find a closing velocity design method for a 126 kV vacuum circuit breaker from the perspective of percussion welding characteristics. The design method aims at minimizing the total arc duration in making operations.

The main difficulty in the study of making operation of vacuum circuit breakers is the significant dependence on accompanying arcing phenomena. Kharin et al. [19–22], in a series of papers, explored the complexity of making vacuum interrupter contacts in an electrical circuit. As a simplification, this section considers only the mechanical closing process without arcs. Therefore, the following assumptions was made for a convenience for the analysis. Note that in current making operations prestrike arcs will occur along with bouncing arcs.

(1) Once a breakdown occurs at a prestrike gap (d), a prestrike arc is initiated and the circuit current will flow immediately and continuously until contacts mate. Then the prestrike duration (T_P) will be a function of a closing velocity (v_P) in the prestrike gap (d).
(2) The bouncing arc duration (T_B) is considered as the sum of all of the bouncing-open durations during contact bouncing.

Figure 4.25 shows an example to determine the bouncing arc duration T_B. In this example, T_B was 2 ms, which was the sum of twice bouncing-open durations.

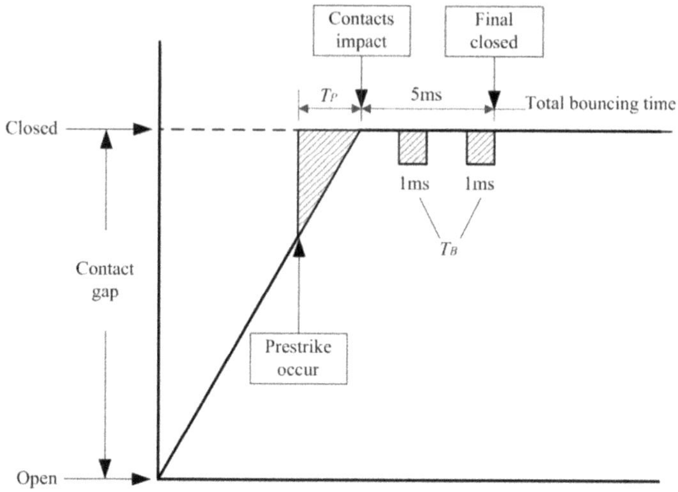

Fig. 4.25 Example of bouncing arc duration T_B

Note that T_B is different to the total bouncing time, which starts from the contacts impacting to finally closing. The total bouncing time was 5 ms in this example. The bouncing arc duration (T_B) is chosen in the following analysis because the arc energy to weld the contacts during making mainly comes from the prestrike arc and bouncing arcs. While the energy input during contact closing is much lower than that during arcing duration. Therefore, contact bouncing arc duration T_B is considered for determination of the closing velocity. Equation (4.8) shows a dependence of the welding force on the arc energy during a making operation [8].

$$F_W = K W_c^{2/3} \tag{4.8}$$

where F_W is the weld force. K is a constant related to contact material. W_C is the arc energy used to melt the contact, which follows Eq. (4.9).

$$W_c = \int U_C i(t) dt \tag{4.9}$$

where U_C is the voltage across the contacts. $i(t)$ is the instantaneous current.

(3) All the energy of the prestrike arcs and bouncing arcs is concentrating to melt the contacts, without heat radiation and heat losses. Thus, the total arc duration for welding (T_W) is considered as the sum of the prestrike arc duration T_P and the bouncing arc duration T_B.

An average closing velocity in the prestrike gap (d) before the contacts mate each other was defined as a *closing velocity* (v_P), thus the prestrike arc duration (T_P) and the bouncing arc duration (T_B) can be correlated together. T_P can be determined by

$T_P = d/v_P$ during the prestriking duration. T_B can be determined by measuring a relationship between T_B and v_P in the making operations. The relationship can be expressed as $T_B = T(v_P)$. Thus, the total arc duration for welding T_W follows a function of v_P as Eq. (4.10).

$$T_W = T_P + T_B = \frac{d}{v_p} + T\left(v_p\right) \tag{4.10}$$

Next, it discusses how to measure the prestrike gap, the closing velocity and the contact bouncing duration, respectively.

Figure 4.26 shows an experiment setup to measure the prestrike gap for vacuum interrupters. The experiments were performed on two samples of 126 kV vacuum interrupters with the 2/3 coil-type axial magnetic field contact structure (see Sect. 3.2.1). Before testing, the two vacuum interrupters were conditioned by a dc current of 120 A with an arcing time of 2 s. Thirty conditioning operations were performed: fifteen operations at positive polarity and fifteen at negative polarity. Then, the vacuum interrupter was oriented vertically to the ground. The movable contact was in the upper position and the stationary contact was in the lower position. A 50 Hz power frequency voltage was applied between the two contacts at a fixed contact gap. The voltage across the contact gap in vacuum increased until a breakdown occurred. The voltage at the instant when breakdown occurred was

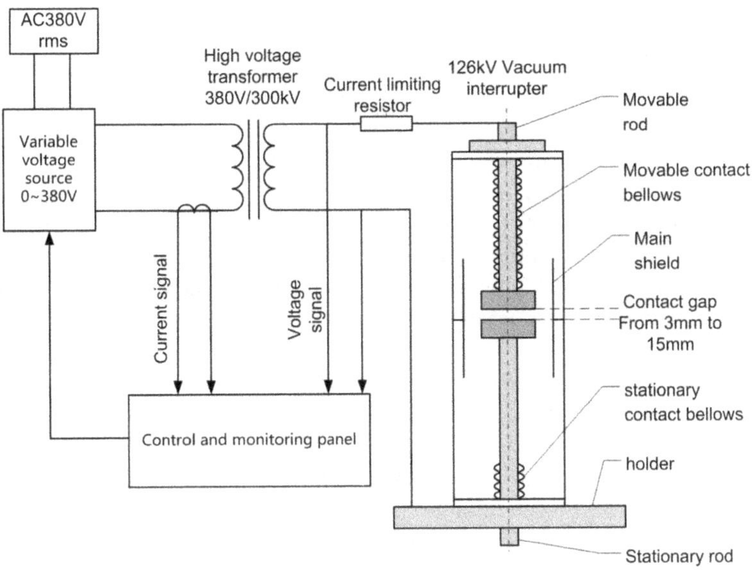

Fig. 4.26 High-voltage power frequency voltage source for measuring of prestrike gap of vacuum interrupters

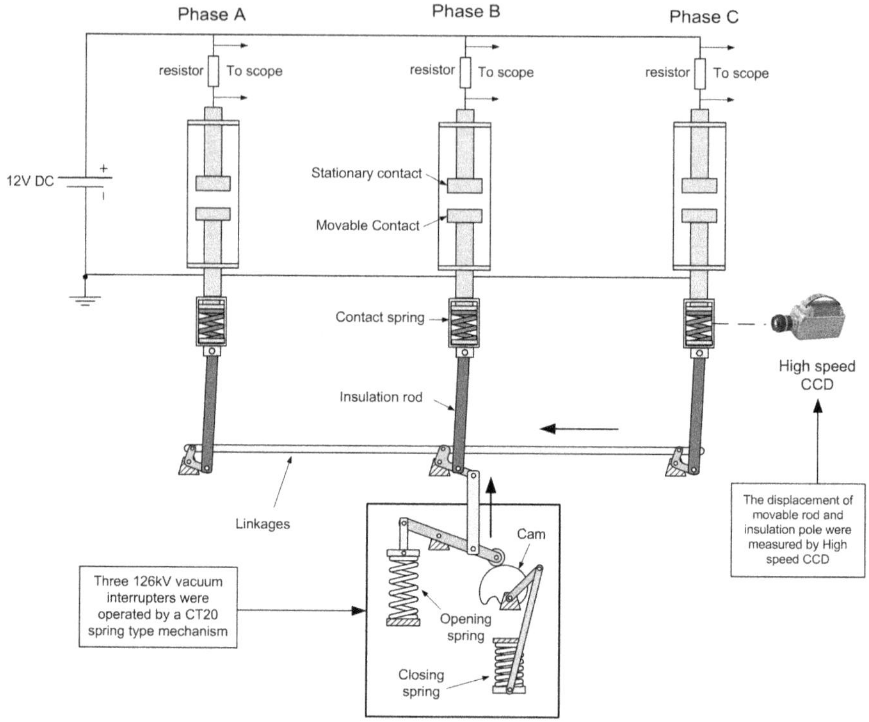

Fig. 4.27 Experimental setup to measure the closing displacement characteristic

recorded as a breakdown voltage. Each vacuum interrupter was tested more than three times at a contact gap, which ranged in 3–15 mm.

Figure 4.27 shows an experimental setup to measure the closing displacement curve. The experiments were performed on a 126 kV/2500 A/40 kA single-break vacuum circuit breaker. Three vacuum interrupters were equipped in the 126 kV vacuum circuit breaker, which was operated by a spring-type operating mechanism. Figure 4.27 also shows a schematic of the operating mechanism, which included a closing spring, an opening spring, a cam, and linkages. In a closing operation, the closing spring drives the vacuum interrupter's moving contacts to the close position and latches it through the cam. In the closing operation the opening spring is fully charged. For the 126 kV vacuum circuit breaker, one can adjust the closing velocity from 0.7 to 2.0 m/s by adjusting the preload of the closing spring in a range from 27, 000 to 38, 000 N. A high-speed charge coupled device (CCD) camera (Phantom V10) recorded the displacement of the movable contact and the operating mechanism. The CCD camera focused on a contact spring. The recording velocity was set at 4000 frames/s and its exposure time was 247 μs. The upper position of the contact spring was connected to a movable conducting rod, and therefore it represented the displacement of the movable contact of the 126 kV vacuum interrupter. The

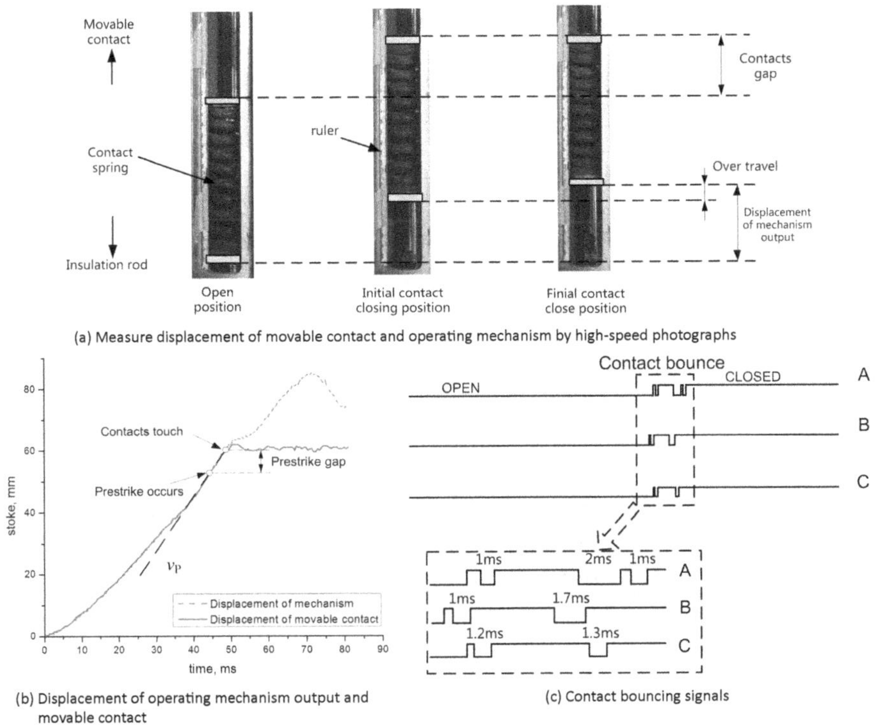

Fig. 4.28 Measurement of the displacement of movable conact and operating mechanism, and the contact bouncing duration

bottom position of contact spring was connected to an insulation rod, and therefore it represented the output displacement of the operating mechanism.

Figure 4.28 shows how to measure the displacement of movable conact and operating mechanism, and the contact bouncing duration. Figure 4.28a shows a method to measure displacement of movable contact and operating mechanism by high-speed photographs. It shows three typical positions of the vacuum circuit breaker, including the open position, initial contact close position, and final contact close position. Therefore, the displacements of the operating mechanism output and the movable contact were measured by using a series of photographs. Figure 4.28b shows a typical displacement curve of the movable contact and the operating mechanism output. The closing velocity v_P was determined from the displacement curve of the movable contact. Figure 4.28c shows typical contact bouncing signals. The contact bouncing duration was measured when the vacuum circuit breaker closed a 12 V dc signal, as shown in Fig. 4.27. The open/close status of each vacuum interrupter was recorded by measuring a sharing voltage of a resistor in series with a vacuum interrupter. In Fig. 4.28c, the low-voltage level indicated that the contacts were in open position. The high-voltage level indicated that the contacts were in close position.

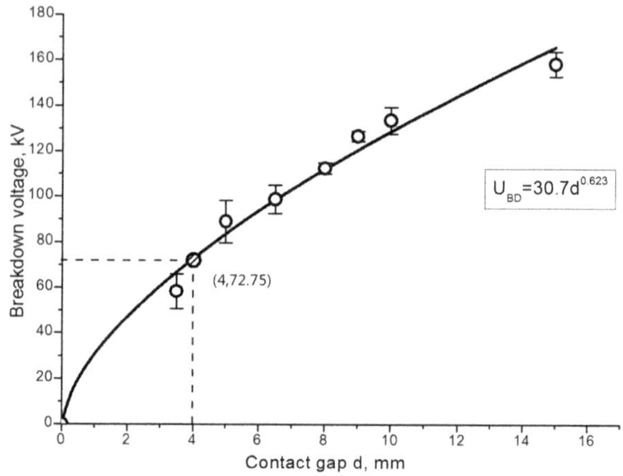

Fig. 4.29 Power frequency breakdown voltage as a function of contact gap for the 126 kV vacuum interrupter

In this case, it shows the total contact bouncing arc duration of phase A was 4 ms, phase B was 2.7 ms, and phase C was 2.5 ms.

Figure 4.29 shows the measured relationship between the power frequency breakdown voltage U_{BD} and the contact gap d. It shows that U_{BD} followed Eq. (4.11) in the range of contact gap from 3 to 15 mm.

$$U_{BD} = 30.7 \times d^{0.623} \tag{4.11}$$

where U_{BD} is the breakdown voltage (kV), and d is the contact gap (mm). Considering a 126 kV grounded system, the phase voltage across each vacuum interrupter's contacts follows Eq. (4.12).

$$U_{BD} = 126/\sqrt{3} = 72.75(kV) \tag{4.12}$$

Assume a vacuum breakdown occurs when the contacts are approaching close enough, the breakdown occurs at the contact gap d, which can be calculated by Eqs. (4.13) and (4.14).

$$U_{BD} = 72.75 = 30.7 \times d^{0.623} \tag{4.13}$$

$$d \approx 4(\text{mm}) \tag{4.14}$$

Thus, the prestrike gap was calculated as 4 mm for the 126 kV vacuum interrupter in a ground system. The prestrike arc duration was determined as Eq. (4.15).

$$T_P = 4/v_p \tag{4.15}$$

Figure 4.30 shows the measured contact bouncing arc duration of the three phases

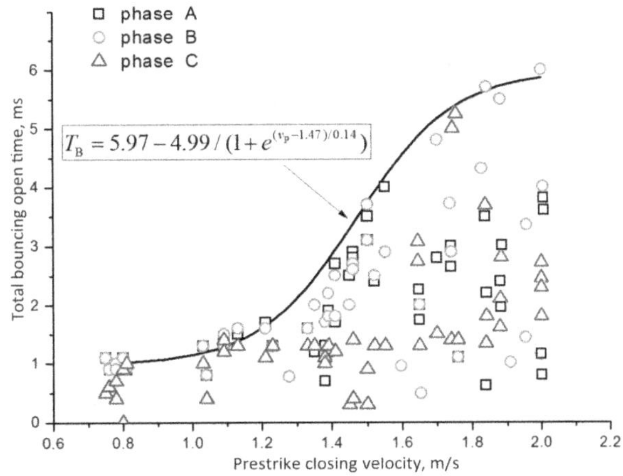

Fig. 4.30 Contact bouncing arc duration of the three phases of the 126 kV vacuum circuit breaker as a function of closing velocity

of the 126 kV vacuum circuit breaker based on 34 closing operations. Although there was a significant variation in the experimental data, the maximum values of the duration increased as the closing velocity increased from 0.7 to 2.0 m/s. The maximum bouncing arc duration followed a function of the closing velocity as Eq. (4.16).

$$T_B = 5.97 - 4.99/\left(1 + e^{(v_P - 1.47)/0.14}\right) \tag{4.16}$$

where T_B is the contact bouncing arc duration, and v_P is the closing velocity.

As assumed, the arc energy to form the percussion welding comes from the prestrike arc and the bouncing-open arc duration. To minimize the percussion welding force, one should minimize the total arc duration. Substituting Eqs. (4.15) and (4.16) into (4.10), then the total arcing duration T_A is a function of the closing velocity v_P in the prestrike gap, as shown in Eq. (4.17).

$$T_A = 4/v_p + 5.97 - 4.99/\left(1 + e^{(v_P - 1.47)/0.14}\right) \tag{4.17}$$

where T_P is the prestrike duration (ms). T_B is the contact bouncing arc duration (ms). Figure 4.31 shows the relationship between T_A and v_P. It shows that, as v_P increased, T_A decreased at first, until it reached the minimum value of approximately 4.91 ms at a v_P of 1.15 m/s. Thereafter, T_A increased as v_P increased further. Therefore, the minimum percussion welding duration for the 126 kV vacuum circuit breaker can be achieved at the closing velocity of 1.15 m/s.

The prestrike arcing phenomena are complex. Kharin et al. [19] showed that the formation of the prestrike arc required a development of a self-sustaining cathode spot. Slade [5, 13] pointed that the contact surface was critical for the initiation of a vacuum breakdown of the closing contact gap. Moreover, the phase angle of system voltage when the prestrike occurred was random during a making operation.

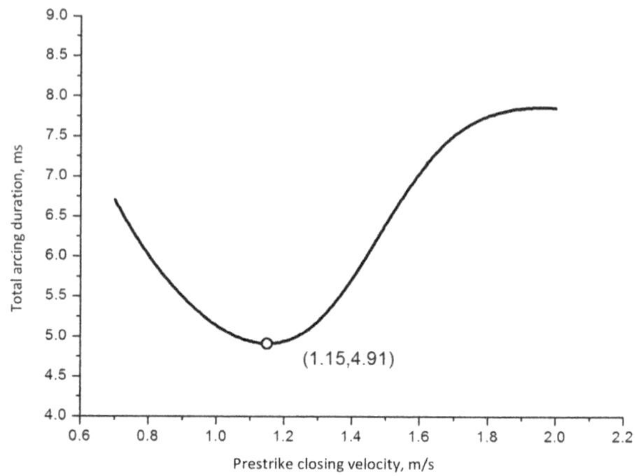

Fig. 4.31 Total arcing duration as a function of closing velocity of the 126 kV vacuum circuit breaker

Therefore, the prestrike gap and the prestrike arc duration are usually random from one operation to another. Figure 4.32 shows a relationship between total percussion arc duration and the closing velocity with a variation of prestrike gaps in the range from 3 to 7 mm. For each prestrike gap, there was an optimized closing velocity corresponding to a minimized duration for percussion arc welding. The optimized closing velocity increased from 1.12 to 1.21 m/s as the prestrike gap increased from 3 to 7 mm.

The above-mentioned results were limited to the situation that contacts close and open mechanically. However, in high-power circuits, both prestrike arcs and contact bouncing arcs may carry the full short circuit current, which may reach tens of thousands of amperes. The high current modifies the mentioned analysis and results. In the prestrike period, remembering that the prestrike arc occurs prior to contacts mate, there is a force to attract the coils in both contacts and there is a blow-off force to pull out the contacts. Assuming a condition of maximum prestrike arcing time, the

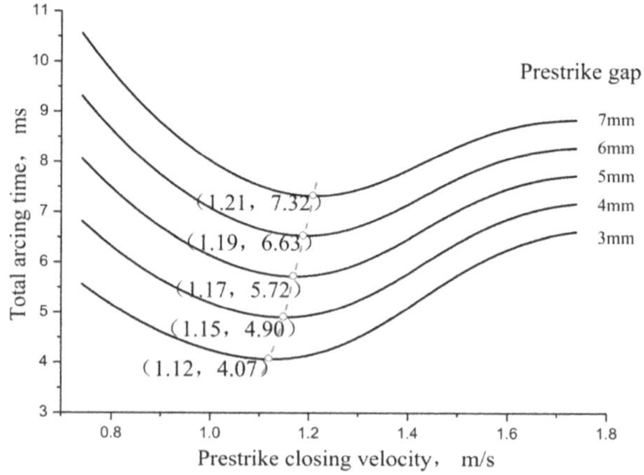

Fig. 4.32 Total percussion arc duration as a function of the closing velocity with a variation of prestrike gaps

prestrike occurs at the peak value of voltage, and a symmetrical short-circuit current flows through the contacts. The attractive force coming from a pair of co-axial coils was given by Christian [23], as shown in Eq. (4.18).

$$F_A = \mu_0(nI)^2\left(\frac{R+r}{2D}\right) \tag{4.18}$$

where μ_0 is the magnetic permeability in vacuum. I is the instantaneous current. n is the equivalent coil turns of contact coil. R is the outside radius of the contact coil. r is the inside radius of the contact coil. D is the distance between the center of movable contact coil and the center of stationary contact coil. In this case, the current was 40 kA (rms), so its peak value was 56.56 kA, at which the maximum attractive force was calculated as 2234 N. The blow-off force is produced by the constriction of the current. If the blow-off force exceeds the holding-in forces, the contacts could blow apart to form bounces. When a high-current flow through a pair of the closed contacts, the blow-off force can be calculated by Eq. (4.19). [24]

$$F_B = \frac{\mu_0}{4\pi}I^2 In\frac{b}{a} = \frac{\mu_0}{4\pi}I^2 In\sqrt{\frac{\xi H\pi b^2}{F}} \tag{4.19}$$

where μ_0 is the magnetic permeability in vacuum. I is the instantaneous current. b is the contact radius. a is the average radius of the constriction. ξ is a constant (it is ≈ 1 for plastic deformation). H is the hardness of contact materials. F is the total force on the contacts, which is represented by Eq. (4.20). [25]

$$F = F_S + F_A - F_B \tag{4.20}$$

where F_s is the spring force holding the contacts together. F_A is the attractive magnetic force resulting from the current flowing in the coils. In the case that prestrike occurred 5 ms before contact touch and with the following parameters: $\xi = 1$, $H =$ 100 HB, $b = 50$ mm, $F_S = 5500$ N, $F_A = 2234$ N, the blow-off force F_B was 1508 N. Therefore, the net force was 726 N, which tended to pull the contacts together.

Some other parameters also influence the making operations of vacuum circuit breakers at high currents. The metal vapor pressure in the contact gap during arcing at closure may cause contacts to quake and the phenomena of contact floating, which increased arc duration and contact erosion [20]. However, the prestrike arc can also anneal and soften the contacts, which may decrease the contact bouncing duration and blow-off force [18]. It is complex that once the contacts close under the condition of high current. Their subsequent behaviors will differ from the no current prediction. Therefore, further work should be done to obtain an optimal closing velocity under high-current conditions. Nevertheless, investigation of mechanical closing operation is still important to give a starting point for approaching the minimum percussion welding in making operations. It constituted a desirable conservative

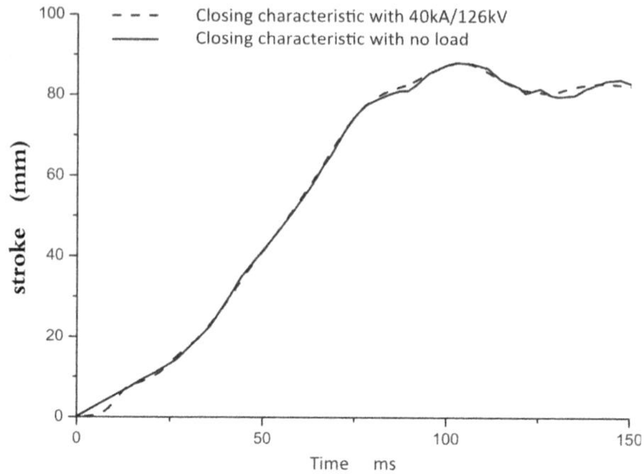

Fig. 4.33 Closing displacement curves of operating mechanism output in the conditions of no load and making short circuit current

basis for designing an operation mechanism for vacuum circuit breakers at transmission voltage level. A full voltage synthetic making test has been done in a high-power lab for a 126 kV vacuum circuit breaker with the 2/3 coil-type axial magnetic field contact. Figure 4.33 shows two closing displacement curves of the 126 kV vacuum circuit breaker, in which one was mechanical closing operation with no-load, and the other was making short-circuit current. One can find that the two curves almost coincided with each other. Moreover, the vacuum circuit breaker opened successfully after the making test.

Spring-type operating mechanism is most popularly adopted for vacuum circuit breakers. Spring-type operating mechanisms interposes a cam between a spring for closing operation and a cam follower [26], because the cam can easily control the dynamic behavior of spring type operating mechanisms. Based on the spring driven, cam follower systems design method proposed by Barkan et al. [26, 27], one can take into account the closing velocity of 1.15 m/s on the design for a spring-type operating mechanism of the 126 kV vacuum circuit breaker. The cam transmission rate at each point of the closing displacement curve can be calculated based on the relationships among the closing velocity of movable contact, the load characteristic, and the energy of the closing spring. Then, a cam outline can be designed according to the cam transmission rate. The method has the merits of requiring only simple hand calculations and the designed cam can directly meet any given closing displacement characteristic of high voltage vacuum circuit breakers. In addition, a cam was designed by this method for the 126 kV vacuum circuit breaker (see Sect. 4.2.1). Dynamic simulations and measurement results confirmed the adequacy of the cam design method. Therefore, a spring-type operating mechanism can be designed to follow the determined closing velocity, which minimized the percussion welding strength.

4.2 Operating Mechanism and Mechanical Reliability

In vacuum circuit breakers, operating mechanism is designed to operate contacts in the vacuum interrupters to open or close. A good vacuum circuit breaker operating mechanism exhibits two merits. The first one is to realize the opening/closing velocities following the vacuum arc behaviors, such as that shown in Sect. 4.1. The second one is a high mechanical reliability. The circuit breaker's reliability is essential to its fulfillment in power network. This section first addresses realizing the opening/closing displacement curve by spring type operating mechanism. Then it addresses the contacts impact phenomena, which are important for the mechanical reliability of circuit breakers. Thereafter, components design improvement is discussed to enhance the mechanical endurance of a 126 kV vacuum circuit breaker. Next it proposes a concept of reset time difference to improve the reliability of closing operation of spring type operating mechanism. Finally, it introduces a design of permanent magnetic actuator for 126 kV vacuum circuit breakers.

4.2.1 Opening/Closing Displacement Curve and Spring Type Operating Mechanism Design

Spring type operating mechanisms are popular in vacuum circuit breakers because of high reliable performance and low cost. In order to achieve appropriate opening/closing velocities, such as that mentioned in Sect. 4.1, several key components of spring type operating mechanisms should be considered, which include contact spring, opening spring, damper and cam et al. This section discusses how the opening/closing velocities of movable contact are influenced by contact spring, opening spring, damper and cam.

Figure 4.34 shows a mechanical dynamic analysis simulation model of a 126 kV vacuum circuit breaker with a spring type operating mechanism. Its main components include stationary contacts, movable contacts, contact springs, insulating poles, linkages, opening spring, closing spring, cam, damper and trip unit. Opening spring and contact spring energy was stored in closing operation by the closing spring. The damper was installed inside of the opening spring component. The damper provided a force to adjust the moving velocity. The simulation was done by a multi-bodies dynamics simulation software ADAMS (Automatic Dynamics Analysis of Mechanical System) where mechanical design parameters can be varied easily as will to analyze its influence on the opening/closing displacement characteristics.

Section 4.1.2 determined an opening velocity of $V_1 = 3.5$ m/s and $V_2 = 3.0$ m/s for a 126 kV single-break vacuum interrupter with 2/3 coil-type axial magnetic field contact. It can be achieved by adjusting the opening spring, the contact spring and the damper, as shown in Fig. 4.35a. The opening velocity can be calculated by Eq. (4.21).

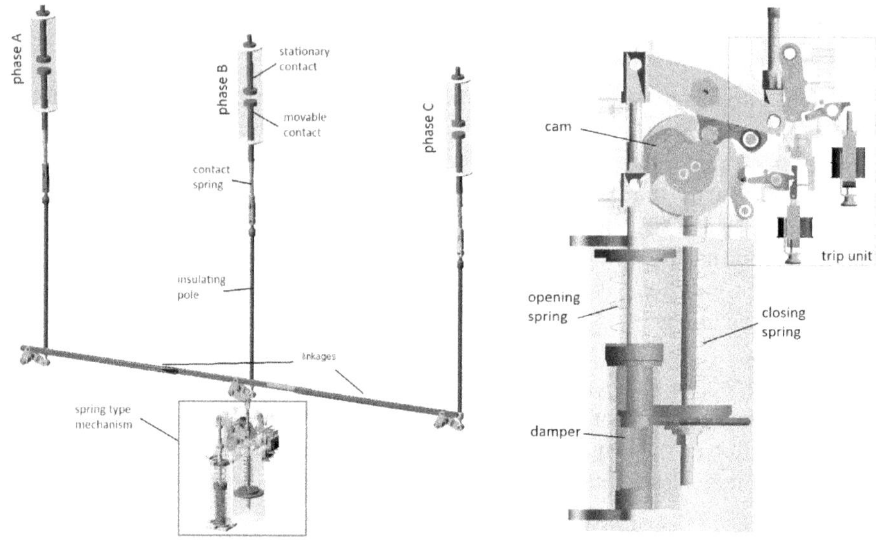

(a) Vacuum circuit breaker model (b) Spring type operating mechanism model

Fig. 4.34 Simulation model of a 126 kV vacuum circuit breaker with a spring type operating mechanism

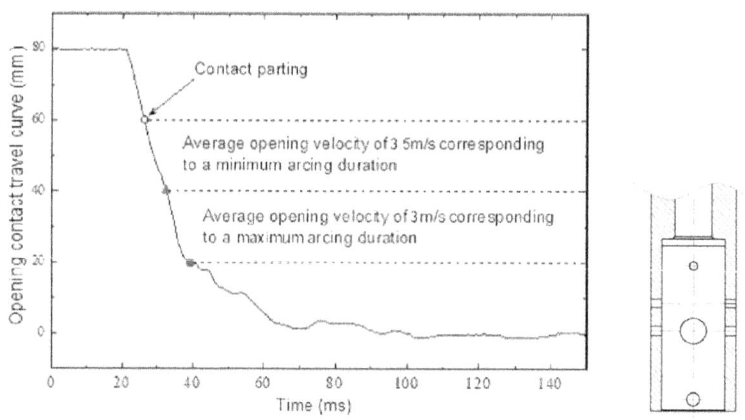

(a) Opening displacement curve (b) Damper

Fig. 4.35 An opening displacement curve of a 126 kV vacuum circuit breaker achieved by adjusting the opening spring, the contact spring and damper

$$V_o = \sqrt{\frac{2 \cdot \eta \cdot (W_o + W_{ct} - W_d)}{m}} \tag{4.21}$$

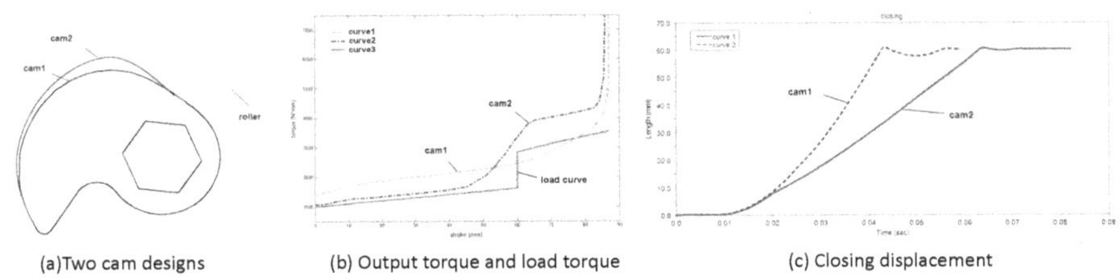

Fig. 4.36 Cam design and the closing velocity characteristics

where W_o is the energy stored in opening spring; W_{ct} is the energy stored in contact spring; W_d is the energy absorbed by the damper; m is the equivalent dynamic mass of the vacuum circuit breaker; V_o is the opening velocity; η is the friction coefficient. The energy stored in the opening spring and contact spring can be calculated by Eq. (4.22).

$$W = F_1 \times h - k \times \frac{h^2}{2} \tag{4.22}$$

where F_1 is the preload of the spring; k is the stiffness coefficient of the spring; h is the stroke of spring. Damper has a significant influence on the opening velocities characteristics. Figure 4.35b shows a kind of oil damper with a number of holes. The opening displacement curve shown in Fig. 4.35a can be achieved by designing the number, the size and the position of holes.

Section 4.1.4 suggested a closing velocity of 1.15 m/s, which can be achieved by designing a cam in the spring-type operating mechanism. Figure 4.36 shows that cam outline has a significant influence on closing velocity characteristics. Figure 4.36a shows two designs of the cam outline. They influence the output torque characteristic significantly, as shown in Fig. 4.36b. The output torque of the operating mechanism should match that of load torque characteristic of the circuit breaker. It shows that the torque curve of cam2 matched the load torque curve better than that of cam1. Figure 4.36c shows that the cam2 guaranteed not only a closing velocity of 1.15 m/s, but also a lower bouncing duration and a lower bouncing amplitude. High contact impact force could damage contacts, conducting rod, and vacuum interrupter insulator envelope. Contact impact phenomena will be discussed in next Sect. 4.2.2.

4.2.2 Contacts Impact Phenomena

A closing operation of a vacuum circuit breaker is invariably accompanied by impact phenomena. The contact impact force is significantly higher than the contact force. A high contact impact force has a negative impact for the mechanical reliability of vacuum circuit breakers. Some parts of vacuum circuit breakers may be destroyed under an excessive impact force. Therefore, the impact, as a basic phenomenon, should be well understood in designing vacuum circuit breakers. Many researchers investigated the impact behavior of electrical contacts [7, 9, 28]. Barkan [18] developed a mathematic model of the dynamics in the impact. He also measured the dynamic force by using a strain gage during no-load closing operations. Mcbride [29] investigated the electrical contacts impact phenomena passing current of vacuum interrupters. All the mentioned researches focused in a field of low and medium voltage circuit breakers. In developing transmission voltage level vacuum circuit breakers, the operating energy increased comparing to their medium voltage counterpart. However, there is lack of contact impact force data in transmission voltage level vacuum circuit breakers. This section is to measure the contact impact force for a 126 kV vacuum circuit breaker. The results would be a reference for the mechanical design of the 126 kV vacuum circuit breaker.

Figure 4.37 shows a measurement system for contacts impact of a 126 kV vacuum interrupter. It is difficult to measure an impact force of contacts directly in a vacuum circuit breaker because the contacts are sealed in a vacuum interrupter with a glass or a ceramic envelope. However, one can use a pair of contacts for the measurement. The 2/3 coil-type axial magnetic field contact was used in the measurement. The contact design and its parameters referred to Fig. 3.21. The mass of movable part was about 9.6 kg. A PVDF (Polyvinylidene Fluoride) piezoelectric sensor was put on the fixed contact to record the impact force during each impact. The fixed conducting rod was fixed vertically on the ground. The movable contact moved towards the fixed contact via a free falling. The drop height of the contact gap was 120 mm. A displacement

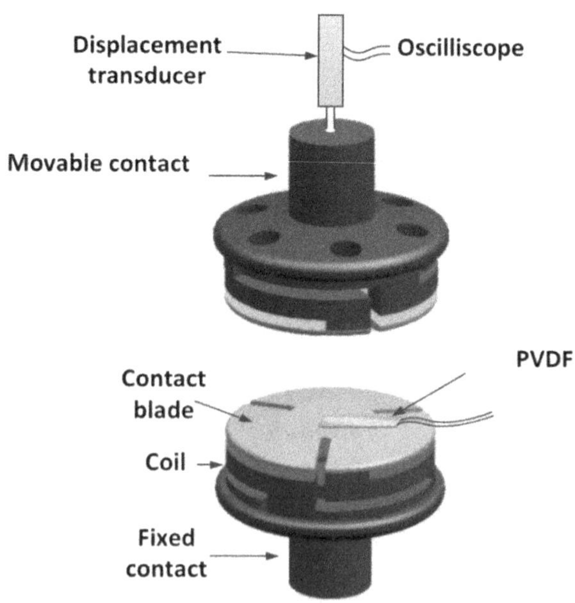

Fig. 4.37 Contacts impact measurement system

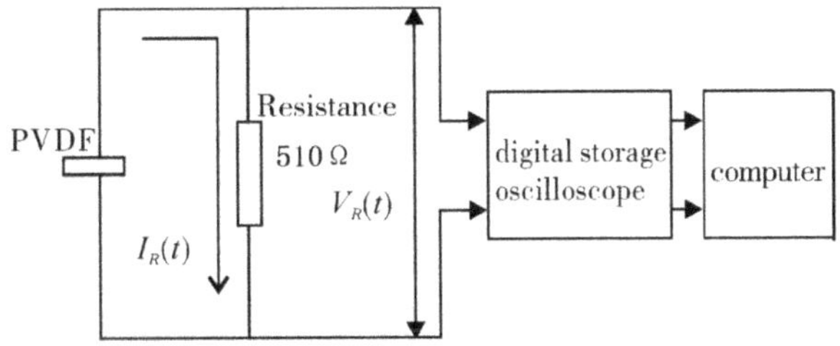

Fig. 4.38 Measurement circuit of PVDF sensor

transducer was used to record a velocity characteristic of the movable contact, which was connected on the top of the movable contact.

The PVDF piezoelectric sensor measured the impact force between two contacts. Figure 4.38 shows a measurement circuit of PVDF sensor. The PVDF sensor produces a certain electrical charge as applied a pressure. The charge passed through a 510 Ω non-inductive resistor during the impact and an oscilloscope was used to monitor the voltage of the resistor. The charge of the PVDF piezoelectric sensor during an impact follows Eq. (4.23).

$$Q(t) = \int_0^t \frac{V(t)}{R} dt \tag{4.23}$$

where *R* is the resistor, which is 510 Ω. *V*(t) is the voltage across the resistor. The PVDF sensor has an intrinsic dynamic sensitivity coefficient *K*, which reflects a function between the pressure and the charge of per unit area. Thus, the impact force measured by PVDF sensor follows Eq. (4.24).

$$F = \sigma \cdot A + K \cdot Q \cdot A \tag{4.24}$$

where σ is the stress of the PVDF piezoelectric sensor. *A* is the contact area. *K* is the coefficient. *Q* is the charge during the impact.

The coefficient of sensor *K* needs to be calibrated before using the PVDF piezoelectric sensor. Figure 4.39 shows how to calibrate the coefficient. Figure 4.39a shows that the calibration was done in a standard split Hopkinson compressive bar. One can compare the measurement results between strain gages and the PVDF piezoelectric sensor. In an impact test, the stress was measured by the strain gauges on the input and output bars. At the same time, the total charge of the PVDF sensor can be measured as Eq. (4.23). By changing the input velocity of bullet, the stress of strain gauges linearly increased with the charges of the PVDF sensor, as shown in Fig. 4.39b. Therefore, the slope of the linear relationship was the dynamic sensitivity coefficient *K*, which was determined as 5.0721 (100 MPa/μC/cm^2).

Figure 4.40 shows the measured contact impact force. Figure 4.40a shows the measured voltage across the resistor of PVDF piezoelectric sensor and the displacement of the movable contact. The average velocity during 1 ms before contacts

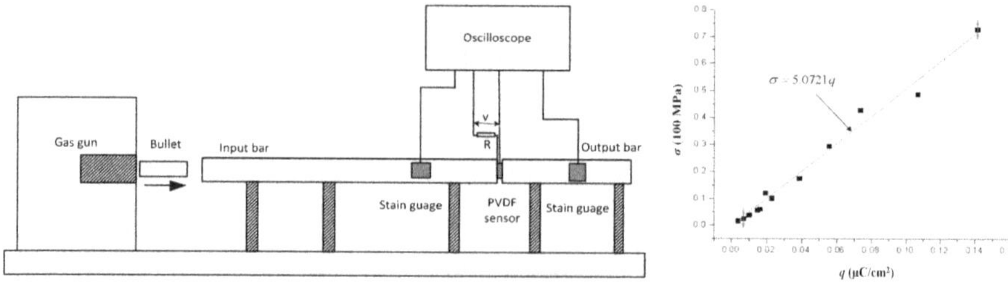

(a) a standard split Hopkinson compressive bar (b)

Fig. 4.39 Calibration of the coefficient of PVDF sensor

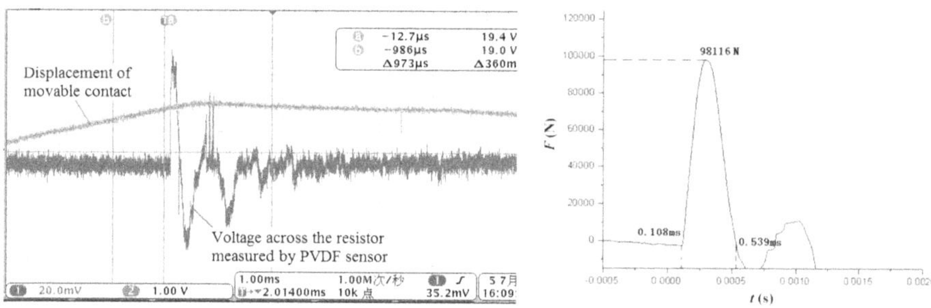

(a) voltage across the resistor measured by PVDF piezoelectric sensor and the displacement of the movable contact

(b) Calculated impact force

Fig. 4.40 Measured contact impact force

impact was 1.5 m/s. Figure 4.40b shows the calculated impact force by Eqs. (4.23) and (4.24). It shows that the peak value of the impact force was 98116 N at the impact velocity of 1.5 m/s, and the duration of first impact was 0.431 ms.

In order to analyze the impact phenomena in details, a finite element analysis simulation model was set up by using ANSYS/LS-DYNA software. The static and dynamic response of contact structure deformation were analyzed. A contact force was also applied to the movable contact from a contact spring during an over-travel. The contact force was a function of the over-travel displacement of movable contact, as shown in Eq. (4.25).

$$F_{\text{contact}} = F_0 + 50d \tag{4.25}$$

where F_{contact} is the contact force. F_0 is the initial force of contact spring, which was set as 5150 N in the simulation. d is the over-travel displacement of movable contact. The impact velocity was 1.5 m/s. The contact material was CuCr50 (50% wt. Cr). The CuCr50 contact material has the following material parameters: a density of 8000 kg/m^3, a modulus of elasticity of 10^{11} Pa, a Poisson's rate of 0.32, a yield stress of 2.5×10^8 Pa, and a shear modulus of 10^8 Pa. The material of conducting rod and coil was copper. The material of the double T beam, steel nail, steel splint, end cover and steel rod was stainless steel. The fixed rod is fixed on ground.

Figure 4.41 shows the simulation result of the contacts impact force. It shows that there were a number of contact impacts during a closing operation. The first contact impact force had a peak value of 108247 N and the duration was 0.345 ms. The peak values of contact impact forces decreased with the numbers of contact impact because of energy dissipation. Compared with the measured data and the simulated data of the peak value of the first impact, the error was 10.3%. The simulation was acceptable.

Figure 4.42 shows the simulated displacement of movable contact. The total bouncing duration was 12.5 ms with the closing velocity of 1.5 m/s. The following parameters were measured in Fig. 4.42, including contacts bouncing displacement Z_i, the contacts open duration T_i, and the separation velocity u_i. The coefficient of restitution e is a dissipation ratio of contact separation velocities [30]. This assumes

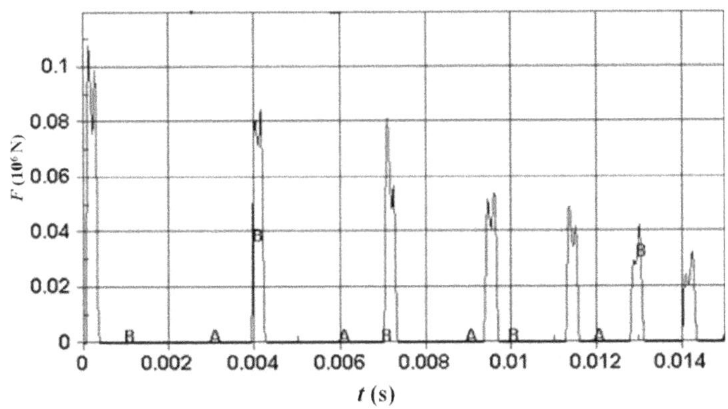

Fig. 4.41 Simulation result of contact impact force

Fig. 4.42 Simulated displacement of movable contact bouncing

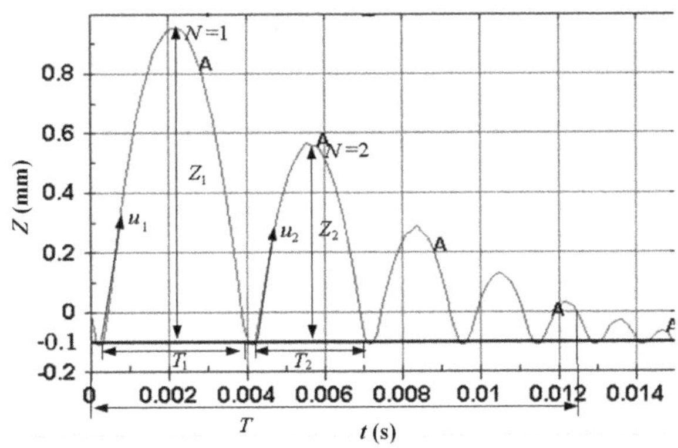

Table 4.6 Contact bouncing characteristic parameters

i	1	2	3	4	5
Z_i (mm)	1.0604	0.6692	0.3933	0.2380	0.1326
T_i (ms)	3.75	2.85	2.25	1.80	1.35
u_i (m/s)	1.1310	0.9392	0.6992	0.5288	0.3928
e_i	0.7540	0.8304	0.7445	0.7563	0.7428

that e is independent of impact velocity. The above mentioned bouncing characteristic parameters were summarized in Table 4.6. It shows that e was about 0.75. The restitution coefficient was useful for predicting the contact impact and bouncing phenomena based on the first impulse.

The contact impact phenomena can be analyzed by adjusting the design parameters in the simulation, such as the impact velocity and the contact spring force. Equation (4.26) shows a relationship between the peak value of the first impulse of the contacts impact force F_1 and the impact velocity V. V was in a range from 1.0 to 2.0 m/s. F_0 is the contact force, which was in a range from 2000 to 10000 N.

$$F_1 = 55621.0 F_0^{0.02} V + F_0 \tag{4.26}$$

Equation (4.27) shows the contacts bouncing displacement of the first impulse Z_1 as a function of V and F_0. Equation (4.28) shows the contacts open duration of the first impulse T_1 as a function between V and F_0.

$$Z_1 = \frac{2440.0}{F_0} V^2 \tag{4.27}$$

$$T_1 = \frac{13433.0 \times V}{F_0} \tag{4.28}$$

By using the equations, one can describe the contacts impact phenomena including the impact force, the bouncing displacement and the open duration.

The stress distribution of contact components was also investigated. Figure 4.43 shows the stress distribution of the contact plate, the coil, the steel nail support and the double T beam support during the contacts impact. Table 4.7 shows the peak

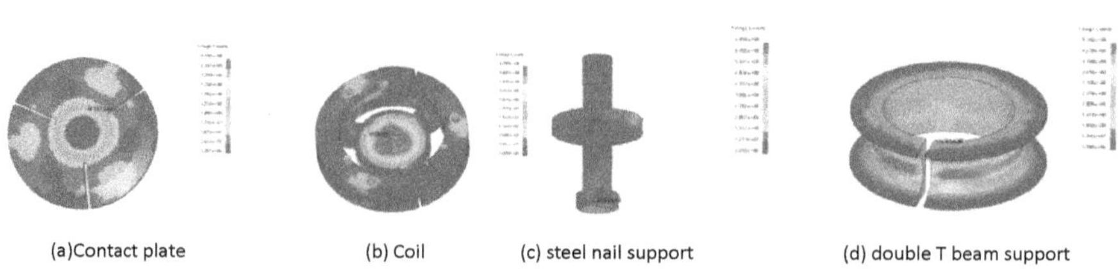

Fig. 4.43 Stress distribution of four contact components

Table 4.7 Peak stress of four components

	Contact plate	Coil	Steel nail support	Double T beam support
Materials	CuCr50 (50% of Cr)	Copper	Stainless steel	Stainless steel
Stress peak value (MPa)	250	178	689	514

stress of the four components. Such data can be used as a reference in the contact designs.

4.2.3 Mechanical Endurance of a 126 kV Single-Break Vacuum Circuit Breaker

Circuit breakers act as control and protection devices, play important roles in power systems. A major failure of high-voltage switchgear and controlgear may cause a cessation of one or more of its fundamental functions [31], and thus it increases the risk to the overall reliability of the power system [32]. The major failure modes mostly include deterioration of electrical degradation, fatigue of a mechanical component, and plastic deformation or corrosion of a sealed component [33, 34]. The International Council on Large Electric Systems (CIGRE) conducted three international surveys on high-voltage circuit-breaker failures and defects in service, and the enquiries covered failures and defects of circuit-breakers with a rated voltage of 63 kV and above. It revealed that the mechanical performance needs most attention from the point of view of reliability [32]. CIGRE also summarized the state-of-the art regarding the impact of the application of vacuum switchgear at voltages above 52 kV [35], which indicated that mechanical failure of the operating mechanism was a main failure type for vacuum circuit breakers and gas circuit breakers in 72 kV power systems.

Mechanical endurance testing is a mandatory test duty for circuit breakers according to the IEC 62271-100 standard with regard to the mechanical performance [36]. IEC distinguishes two classes of circuit breakers, which is class M1 with normal mechanical endurance (2000 opening-closing operations) and class M2 with extended mechanical endurance (10000 opening-closing operations), respectively. During a mechanical endurance test of a 126 kV vacuum circuit breaker, three mechanical parts failed, which included a pin in a contact spring component, a joint bearing and an inter-locking pawl. This section is to demonstrate the failure modes of the three mechanical parts and show how they can be improved to extend the mechanical endurance.

Figure 4.44 shows a prototype a 126 kV vacuum circuit breaker together with its contact spring component and its spring type operating mechanism. Because vacuum interrupters use butt-type contacts, a contact spring component is required to keep the contacts mated even though suffering an electrical dynamic force. The contact

Fig. 4.44 A 126 kV vacuum circuit breaker with spring type operating mechanism. A contact spring component was shown in details

spring component included contact springs, spring cushions, separated cushions, a pin, a conducting rod and a guide sleeve. The initial force and final force of the contact spring were 4,010 N and 4,938 N, respectively. The over-travel was 20 mm. The closing energy of the spring mechanism was 2,900 J, which corresponded to a compressed closing spring length of 100 mm. The initial forces of the opening spring and the closing spring were 11,000 N and 21,000 N, respectively. The final forces of the opening spring and the closing spring were 19,500 N and 38,500 N, respectively. The closing velocity was 1.15 m/s (see Sect. 4.1.4). The opening velocities were 3.5 $\pm$ 0.2 m/s during minimum arcing contact gap and 3.0 $\pm$ 0.2 m/s during maximum arcing contact gap (see Sect. 4.1.2).

Figure 4.45. shows a displacement curve of the guide sleeve in the contact spring component when the circuit breaker performed a closing and opening operation cycle. In the closing operation, the contact spring component drove the movable contact to move towards the stationary contact. When the movable contact went through the whole contact gap of 60 mm, it mated with the stationary contact. Then, the contact spring compressed from the initial force to the final force. The contact over-travel of 20 mm was the contact spring compression length during the closing operation, and it was also the relative displacement of the pin in the contact spring component. The

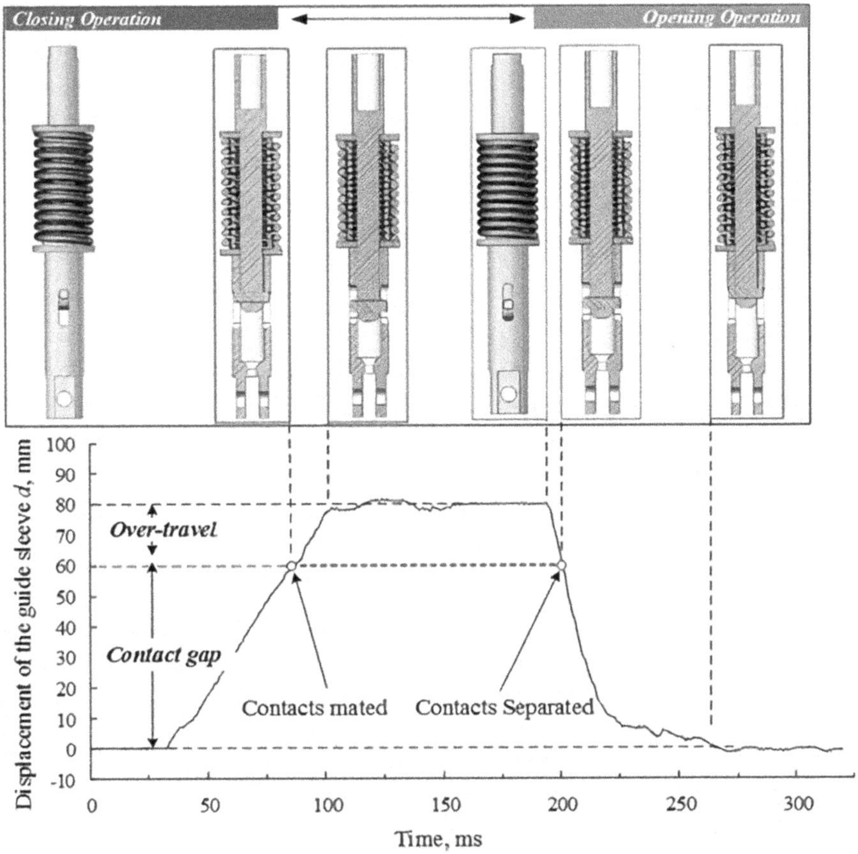

Fig. 4.45 Displacement of the guide sleeve in the contact spring component during a closing and opening operation cycle

pin was to maintain the initial force of the contact spring component. In the opening operation, the movement of the contact spring component followed the opposite process.

Figure 4.46 shows three failure parts of the 126 kV vacuum circuit breaker that failed during the mechanical endurance operations. Figure 4.46a shows the pin in the contact spring component. The pin material was 20CrNiMo stainless steel. After 178 mechanical endurance operations, the pin suffered a brittle fracture. A crack was initiated and propagated across the aperture on the pin. Figure 4.46b shows the joint bearing. The joint bearing was to transfer force and movement from an insulation pole to a movable contact. The material of the joint bearing was 1Cr18Ni9Ti stainless steel. After 1500 operations, the joint bearing deformed significantly and then fractured. Figure 4.46c shows the inter-locking pawl in the spring operating mechanism. The inter-locking pawl was to maintain the vacuum circuit breaker in the closing position

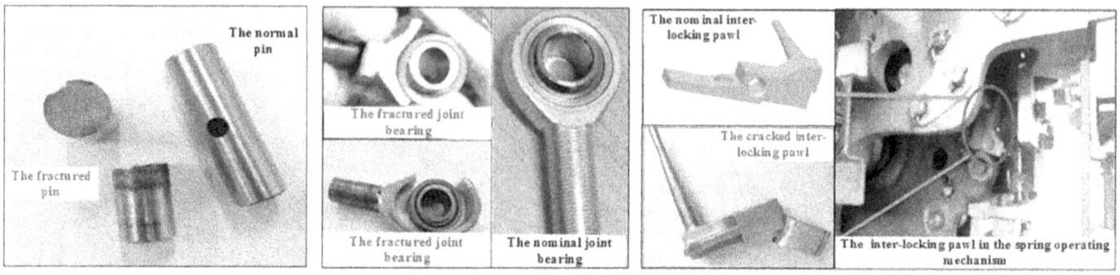

(a) the pin in the contact spring component (b) the joint bearing (c) the inter-locking pawl

Fig. 4.46 Three failure parts compared with their normal states

reliably. The material of the inter-locking pawl was Q235 steel. After 2100 operations, the joint bearing suffered a brittle fracture.

Figure 4.47 shows simulated forces imposed upon the pin during a closing and opening operation cycle. The simulation was done by a multibody dynamic simulation software MSC-ADAMS. In the closing operation, the pin was subjected to an initial force of 4010 N before the contacts mated. The force decreased to zero

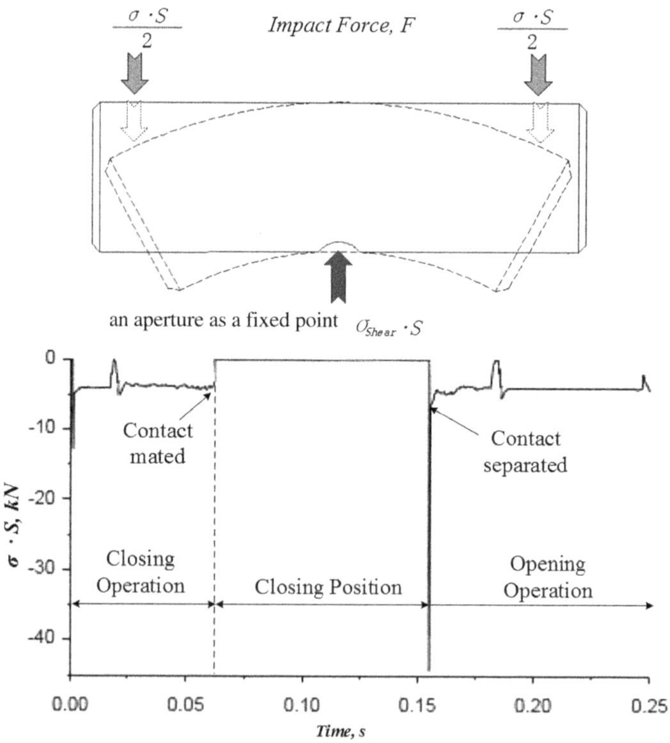

Fig. 4.47 The impact force imposed upon the pin during a closing and opening operation cycle

immediately after the contacts mated. Then, in the opening operation, the pin was subjected to an impact force immediately after a release of the over-travel of 20 mm. The amplitude of the force was 44 kN. There was an aperture on the middle position of the pin, which was used as a fix point of the pin. The impact force imposed upon both terminals of the pin. Therefore, the impact force caused a shear force on the pin.

Figure 4.48 shows a measured impact force imposed upon the pin during an opening operation. The impact force was measured via a PVDF piezoelectric sensor (see Sect. 4.2.2). The maximum impact force was 30 kN. The impact force waveform showed a crest, because the PVDF piezoelectric sensor cracked when the pin struck the guide sleeve.

A finite element analysis simulation model was built up to estimate the mechanical endurance of the pin. The analysis was performed by using the Ncode-DesignLife software. The analysis based on both the strain-stress relationship and the S-N characteristics of the materials. Figure 4.49 shows the fatigue life analysis results. The simulation suggested that, after 81 operations, the pin would fail. The weak point located at the opposite side of the aperture position. The analysis result agreed with the failure one shown in Fig. 4.46a.

Figure 4.50 shows the simulations for the failure of the joint bearing. Figure 4.50a shows simulated force imposed upon the joint bearing during a closing and opening operation cycle. During the closing operation, the joint bearing subjected a force of F_{closing}, which drove the movable contact approaching the stationary contact. During the opening operation, the joint bearing subjected to a force of F_{opening}, which forced the movable contact to open. The amplitude of F_{opening} was 34 kN. Figure 4.50b shows the fatigue life analysis for the joint bearing. The simulation suggested that, after 1141 operations, the joint bearing would fail. The weak point position was in accordance with the failure one shown in Fig. 4.46b.

Figure 4.51 shows the simulations for the failure of the inter-locking pawl.

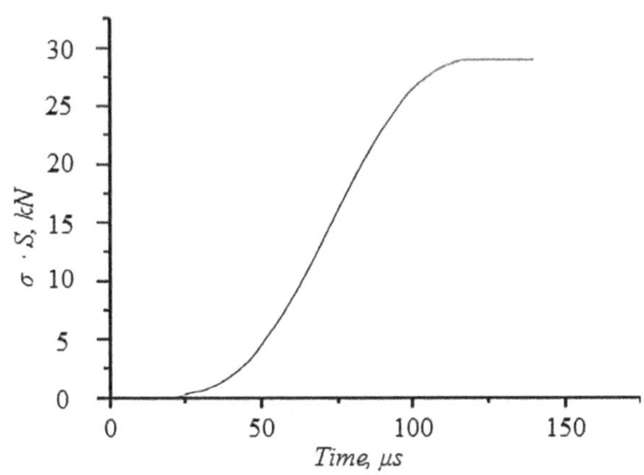

Fig. 4.48 Measured impact force imposed upon the pin in the contact spring component during an opening operation

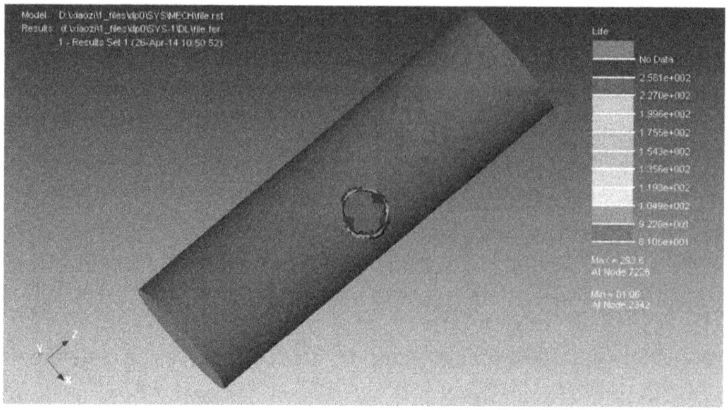

Fig. 4.49 Fatigue life analysis the pin

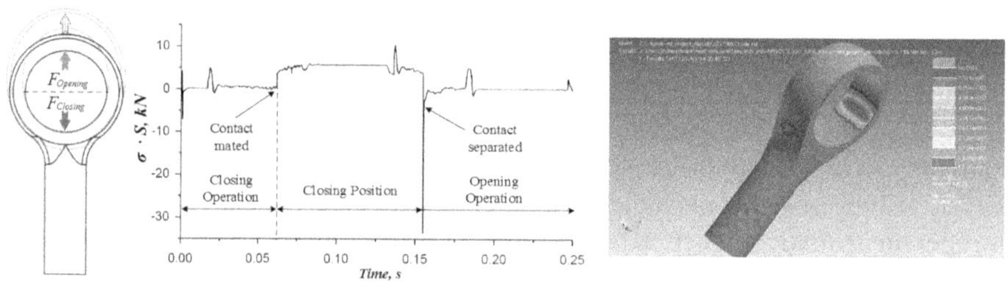

(a) Forces simulation (b) Fatigue life analysis

Fig. 4.50 Failure simulation of the joint bearing

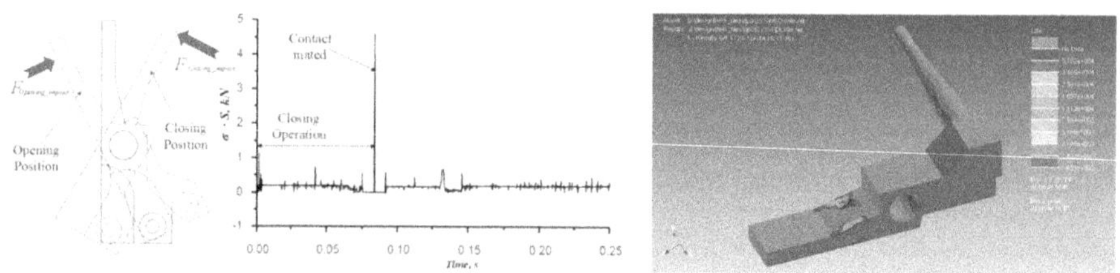

(a) Forces simulation (b) Fatigue life analysis

Fig. 4.51 Failure simulation of the inter-locking pawl

Figure 4.51a shows simulated force imposed upon the inter-locking pawl during a closing and opening operation cycle. During the closing operation, immediately as the vacuum circuit breaker reached a closing position, the inter-locking pawl turned to lock a closing pawl in the operating mechanism. At this moment, the inter-locking pawl was subjected to an impact force of $F_{\text{closing\_impact}}$. When the vacuum circuit breaker was opened, the inter-locking pawl was unlocked to release the closing pawl. Then, the interlocking pawl was subjected to a force of $F_{\text{opening\_impact}}$. The amplitude of $F_{closing\_impact}$ was 4.7 kN. Figure 4.51b shows the fatigue life analysis for the inter-locking pawl. The simulation suggested that, after 1654 operations, the inter-locking pawl would fail. The weak point position was in accordance with the failure one shown in Fig. 4.46c.

Figure 4.52 shows the improved designs for the three failed parts. The improvement was based on the design of experiments (DOE) method by considering the geometry and material [37]. Table 4.8 shows the improvement of the mechanical endurance. First, one can find that the simulated mechanical endurance was generally close to that of the experimental ones for the three parts. Second, the simulation results showed that the improved designs shown in Fig. 4.52 enhanced the mechanical endurance significantly. Third, mechanical endurance test with the three improved

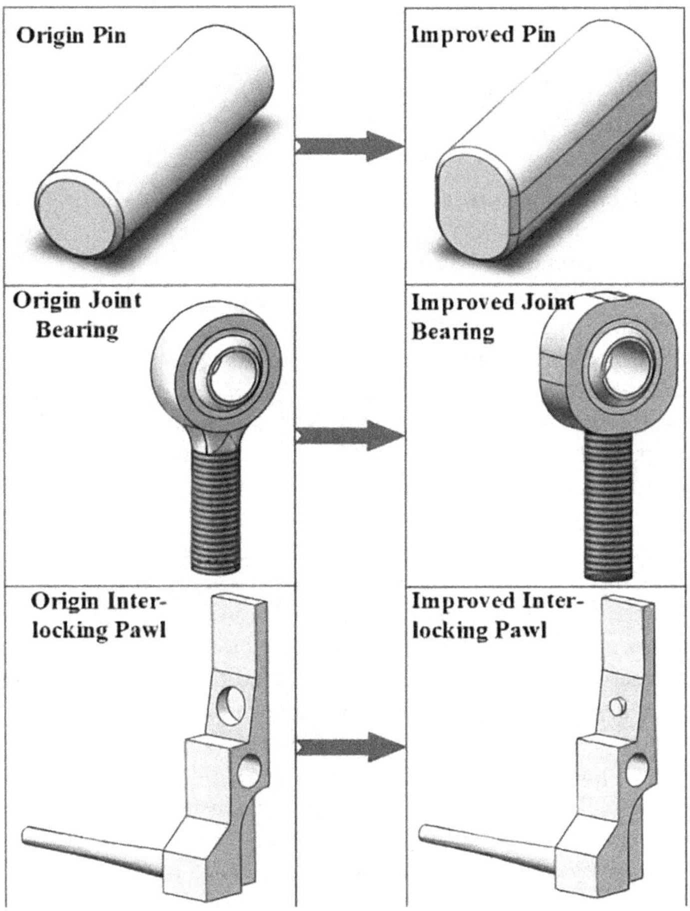

Fig. 4.52 Original and improved design of the three failed parts

Table 4.8 Improvement of the mechanical endurance

Part	Experiment, original design	Simulation, original design	Simulation, improved design	Experiment, improved design
Pin in the contact spring component	178	81	4000	>2000
Joint bearing	1500	1141	12200	>2000
Inter-locking pawl	2100	1654	7290000	>2000

parts showed that, the vacuum circuit breaker passed M1 class (2000 opening-closing operations). The proposed methodology enhanced mechanical reliability of the 126 kV vacuum circuit breaker.

4.2.4 *Reliability in Closing Operation of Spring Type Operating Mechanism*

It was found that closing operation failures occasionally occurred when a spring type operating mechanism was used to operate a high voltage circuit breaker with a high load. The circuit breaker exhibited a failure mode of "Does not close on command". This phenomenon was found to be attributed to a trip-open latch unit of the operating mechanism. Once the trip-open latch unit was unable to hold the high voltage circuit breaker in its close position, it led to the failure of closing operation. This section proposes a concept of *reset time difference*. By setting an appropriate reset time difference, the failure mode disappeared and the reliability of closing operation improved.

A 126 kV high voltage circuit breaker was used for testing. The circuit breaker was operated by a spring type operating mechanism. Figure 4.53 shows its spring-type operating mechanism and trip-open latch unit. Figure 4.53a shows that the spring type operating mechanism consisted of a closing spring, a closing cam, an opening spring, a damper, a charging motor, various toggles and linkages, and trip latch units. The closing spring was charged by an electric motor. A closing signal released a latch of the charged closing spring, which allowed the closing spring to drive the circuit breaker into the close position. At the same time, the opening spring was fully charged and a trip-open latch unit was engaged to maintain the circuit breaker in the close position. Figure 4.53b shows the detail of the trip-open latch unit. There was a big latch and a small latch. When the operating mechanism reached the close position, the big latch began to rotate clockwise by a spring of big latch. Then the small latch rotated clockwise by a spring of small latch and reached its closing position. In this way the closing position was maintained by the trip-open latch unit.

In order to observe the movements of the trip-open latch unit, one may open an observation window on an envelope of the spring type operating mechanism without distorting its operations. A high-speed CCD camera (Phantom V10) was used to

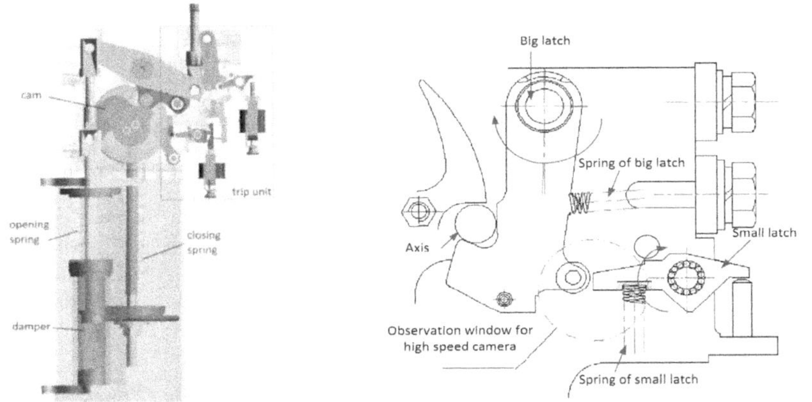

(a) Simulation model of spring type operating mechanism (b) A trip-open latch unit

Fig. 4.53 A spring-type operating mechanism and its trip-open latch unit

record the movement of the big and small latches during closing operations. The exposure time of the camera was set as 197 μs, and its recording speed was set as 4000 frames/s. Figure 4.54 shows the trip-open unit at the open position and at the close position. One can match these photographs with the drawings of the trip-open latch unit in an AutoCAD software. Therefore, the actual movement of the big latch and the small latch can be measured by using a series of photographs.

Table 4.9 shows four designs of the trip-open latch unit. They were type I to type IV. Type I was an original design used in a commercial spring operating mechanism; Type II increased the preload of big latch spring by twice compared to Type I; Type III decreased the height of the small latch compared to Type II. Type IV designed

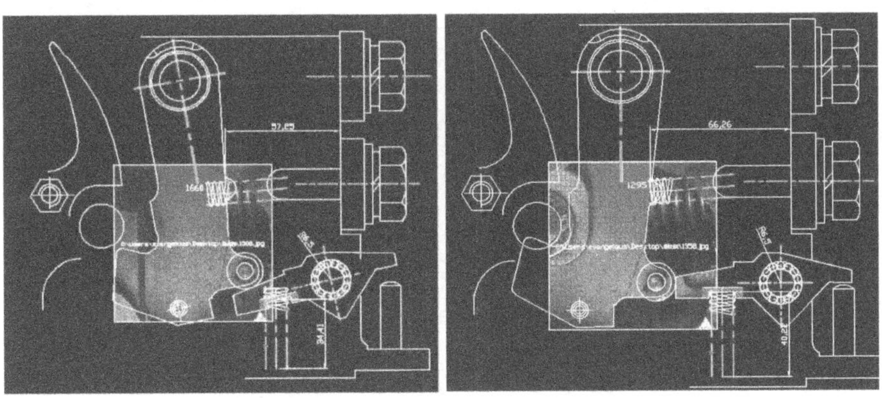

a) At open position b) At close position

Fig. 4.54 The trip-open unit at the open position and at the close position

Table 4.9 Four designs of the trip-open latch unit

Type	Preload of the spring of big latch (N)	Height of small latch platform (mm)	Diameter of the groove in small latch (mm)	The shape of small latch
I	350	3.00	0	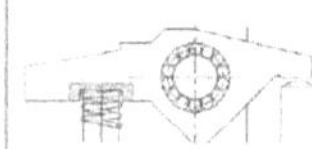
II	700	3.00	0	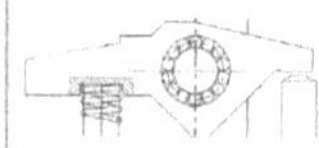
III	700	2.00	0	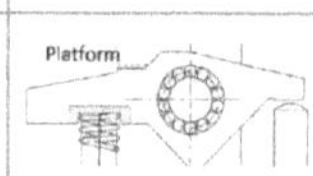
IV	700	2.00	10.89	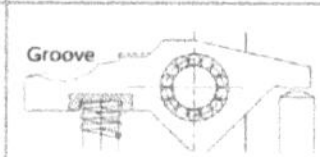

a groove on the platform of the small latch compared to Type III. The movement of each type of trip-open latch unit was measured by the above-mentioned method three times.

Figure 4.55 shows a measured travel of the big and small latches in a closing operation. It was found that both latches switched from initial position to final latch position. The small latch reached its final latch position at t_A, while the big latch reached its final latch position at t_B. There is a time difference ($t_B - t_A$) between the two latches. This is a time difference for resetting the two latches, which was defined as a *reset time difference* T_R. If T_R becomes negative, the small latch cannot reach its final latch position. Thus, it is impossible to hold the load of the circuit breaker, which led to the failures of "Does not close on command". If T_R is positive, the small latch reaches its final latch position first, then it waits for the big latch. Thus, the circuit breaker can be latched at the close position. Therefore, the appropriate sequence of the small latch and the big latch is important for the reliability of the

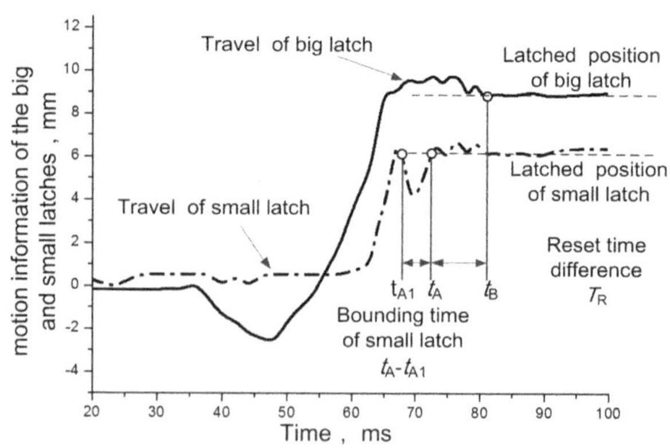

Fig. 4.55 Measured travel of the big and small latches in a closing operation and definition of reset time difference

closing operation. Figure 4.55 also shows that there was a bounce of the small latch before it reached the final latch position. It occurred during an interval from t_{A1} to t_A. The bouncing phenomenon of the small latch may also influence the reliability of closing operation.

Figure 4.56 shows the characteristics of the four types of trip-open latch unit. Figure 4.56a shows the reset time difference T_R among the four designs of the trip-open latch types. T_R of Type I is the lowest one (10.47 ms) and T_R of Type II, III and IV increased to ~12 ms. The reason of T_R increasing may be due to the increase of the preload of the big latch spring. Thus, the big latch moved faster. A larger preload of the big latch spring also prevented the big latch returning back. In addition, Type III showed a significant variation of T_R than the other three types. Figure 4.56b shows the bouncing time of the small latch (t_A-t_{A1}) among the four designs of the trip-open latch types. It shows Type III and Type IV had a lower bouncing time than that of Type I and Type II. Considering the results in Fig. 4.56a, b, Type IV design won among the four designs. It had a large reset time difference and a low bouncing time of the small latch. Therefore, Type IV design of the trip-open latch unit was adopted to a 126 kV vacuum circuit breaker and the failure mode disappeared. The reliability of closing operations was improved.

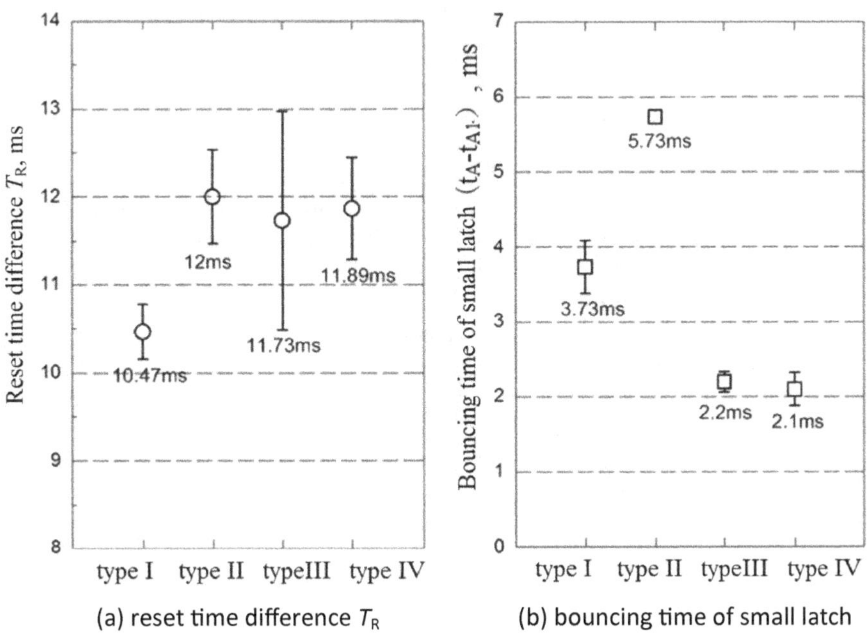

Fig. 4.56 The characteristics of the four types of trip-open latch unit

4.2.5 A Permanent Magnetic Actuator for 126 kV Vacuum Circuit Breakers

A permanent magnetic actuator is a type of easy to control operating mechanism. It offers an electromagnetic force to drive a movable contact. Then a permanent magnetic force holds it in a close or open position. Permanent magnetic actuators have been widely used in the medium-voltage vacuum circuit breakers due to their simple structure, high reliability, good controllability, long life cycle, and so on [38]. There are mainly two types of permanent magnetic actuator used, which includes a bi-stable type permanent magnetic actuator [39] and a mono-stable type permanent magnetic actuator [40]. Typically, in the medium-voltage vacuum circuit breakers only short travels (<30 mm) are required for the permanent magnetic actuators.

Extensive research works have been done to improve the performance of permanent magnetic actuators. Lequesne [41] suggested a long stroke solenoid with two springs to satisfy the velocity requirements. Ro et al. [42] proposed a permanent magnetic actuator with separated magnets to improve the efficiency of coil currents, which provide magnetic force to drive the movable contact. Cai et al. [43] designed additional auxiliary coils to a traditional bi-stable permanent magnetic actuator. Kang et al. [44] adopted three links to connect a permanent magnetic actuator to a vacuum interrupter. Wang et al. [45] used both a closing coil and an opening coil, which were discharged simultaneously to promote an opening velocity. Moreover, many simulations were done to optimize the permanent magnetic actuators [46–50]. However, it is still difficult to satisfy the requirements of transmission voltage level vacuum circuit breakers, such as large travels and high velocities. For example, a 126 kV vacuum circuit breaker had a travel of 80 mm, which was much longer than that in the medium-voltage vacuum circuit breakers. In addition, it required a higher opening velocity than that of a medium voltage vacuum circuit breaker (see Sects. 4.1.2 and 4.1.3). This section is to propose a multi-magnetic-circuit permanent magnetic actuator for a 126 kV vacuum circuit breaker.

Figure 4.57 shows the proposed permanent magnetic actuator. As a comparison, a conventional bi-stable permanent magnetic actuator is shown in Fig. 4.57a. In the conventional bi-stable permanent magnetic actuator, permanent magnets are imbedded in a yoke. A closing coil and an opening coil are arranged above and below the permanent magnets, respectively. The closing and opening coils and the permanent magnets exert magnetic forces on an armature, which drives the movable contact through a pole. Moreover, the magnetic flux from the magnets and the closing/opening coils share the same magnetic circuit in the yoke and the armature. Figure 4.57b shows the structure of proposed permanent magnetic actuator. It also adopted a bi-stable form, as the conventional one did. But it consisted of two independent magnetic circuits, which carried out two functions of holding and driving, respectively. The two functions were realized by two independent components: a driving component and a holding component, which were shown in the two dotted boxes in Fig. 4.57b. The holding and the driving components had their own magnetic circuits rather than sharing the same magnetic circuit. Moreover, a non-magnetic

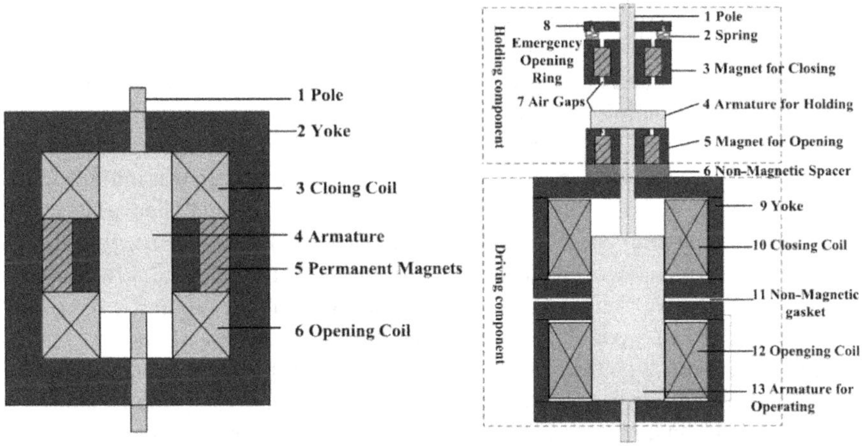

(a) Conventional permanent magnetic actuator (b) Proposed permanent magnetic actuator

Fig. 4.57 A proposed permanent magnetic actuator and a conventional permanent magnetic actuator

spacer attempted to prevent electromagnetic interference between the holding and the driving components. Therefore, their magnetic fluxes separated. The holding component consisted a magnet for closing and a magnet for opening. There was an armature for holding. It transferred a holding force from the magnet for closing to the movable contact through a pole. There designed air gaps to increase the initial opening velocity. An emergency opening ring was designed at the top of the holding component. It carried out manually opening operations through springs in emergency events. The driving component consisted of an armature for operating, a yoke, a closing coil and an opening coil. It was different from the conventional bi-stable structure, because its magnetic circuits for closing and opening operations separated by a non-magnetic gasket, which divided the yoke into two parts. Thus, the magnetic fluxes of the closing and opening coils have their independent magnetic circuits. In addition, the armature of operating was connected to the pole. It transferred the magnetic force from the closing or opening coils to the movable contact. The closing and opening operations energy came from pre-charged capacitors, which were manipulated by a controller through power semiconductors insulated gate bipolar transistor (IGBT).

The proposed permanent magnetic actuator can increase opening or closing velocities comparing to that of conventional permanent magnetic actuator, which was desirable for transmission voltage level vacuum circuit breakers (see Sect. 4.1). First, one can design the holding force at open or close position separately because the magnet for opening and the magnet for closing were designed separately. In this way, a large holding force can be guaranteed because air gaps were designed in magnets for opening or magnets for closing. Second, a large holding force needs a large electromagnetic force to overcome, thus a large coil current was needed to guarantee the electromagnetic force. So that the opening or closing velocities can be increased. Third, the magnetic loop of the opening coil and closing coil separated. Therefore, the opening velocity and closing velocity can be controlled as required separately.

The operation principle of the proposed permanent magnetic actuator was similar to that of conventional permanent magnetic actuator. In an opening operation of the conventional bi-stable permanent magnetic actuator, its armature was initially kept in the close position by the holding force of the permanent magnets. Then, a DC current flowed into the opening coil. Then the holding force that kept the armature in close position gradually decreased with the increasing of the current. As the current reached a certain value, the armature began to move. Once the armature left the close position, an air gap formed between the armature and the yoke. The new formed air gap had a significant influence on the force coming from the permanent magnets. Therefore, the armature accelerated until it reached the open position. Finally, the armature kept in the open position by the permanent magnets. Then the current in the opening coil was cut off. For the proposed bi-stable permanent magnetic actuator, in an opening operation the armature also initially kept in the close position by the holding force of the permanent magnet for closing. Then, a DC current flowed into the opening coil. In such way, the current generated an electromagnetic force with a direction opposite to the holding force. The electromagnetic force increased with the increasing of the current. Meanwhile, the holding force decreased once the armature leaving the closing position. The air gaps designed in the holding component enhanced the decreasing of the holding force. When the electromagnetic force exceeded the holding force, the armature for operating accelerated to the open position together with the pole, the armature for holding, and the movable contact. When the armature for operating arrived at the open position, it kept in the opening position by the force of the magnet for opening. For emergency situations, there will be no discharged current in the opening coil and it can operate manually. The emergency opening ring can be punched down to decrease the holding force in the closing position and it helped emergency devices to realize a manual operation. In both permanent magnetic actuators, the closing process was reverse to the opening process.

A mechanical motion simulation of a permanent magnetic actuator is a multiple physical fields simulation. It coupled electrical circuit, magnetic fields, and mechanical motions. Generally, a commercial software package focused on a certain kind of physical field. Faruque et al. [51] showed a multi-physical-fields simulation method that applied multiple software packages, such as software packages of ANSOFT and ADAMS. Their simulation was realized by programming in ADAMS, which solved a set of differential equations, in which the magnetic field was obtained from ANSOFT.

Here, the mechanical motion simulation followed three steps. First, the static magnetic characteristics of the permanent magnetic actuator, such as magnetic force and flux linkage, were calculated by ANSOFT package based on the Maxwell's equations. The magneto-static Maxwell equations transformed into a Poisson equation in terms of the magnetic vector potential A, which follows Eq. (4.29).

$$\nabla \times \frac{1}{\mu} \nabla \times \boldsymbol{A} = -\boldsymbol{J} \tag{4.29}$$

where μ is the permeability of the material and $\boldsymbol{J}$ is the current density vector. For a current-free region, it can be reduced to a Laplace Eq. (4.30).

$$\nabla \times \frac{1}{\mu} \nabla \times \boldsymbol{A} = 0 \tag{4.30}$$

In addition, in the magnet region it follows Eq. (4.31).

$$\nabla \times \frac{1}{\mu} (\nabla \times \boldsymbol{A} - \boldsymbol{B}_r) = 0 \tag{4.31}$$

where B_r is the residual flux density of the permanent magnets. Considering the ferromagnetic material's non-linearity, the permeability related to the magnetic flux density $\boldsymbol{B}$ in the calculated regions can be expressed as Eq. (4.32)

$$\mu = \mu(\boldsymbol{B}) = \mu(\nabla \times \boldsymbol{A}) \tag{4.32}$$

Second, a simplified 126 kV vacuum circuit breaker model was established in ADAMS package. Figure 4.58 shows the ADAMS model of a 126 kV vacuum circuit breaker for dynamic simulations. The model consisted of a vacuum interrupter, a permanent magnetic actuator, and a contact spring connecting the two components. In closing operations, the contact spring did not work until the movable contact mated the stationary contact. Once the contacts mated, the contact spring acted as a resisting force to the actuator. In opening operations, the contact spring acted as an assisting force to the actuator.

Third, a set of partial differential equations were solved to obtain a time-depending force. The force coming from the permanent magnets was independent with that coming from the coil current. The permanent magnets force was a function of distance between the magnets and the armature for holding. This force was calculated by ANSOFT at various positions of the armature for holding and the results were stored in a data table. The magnetic force generated by the coils was related to both the coil current and the armature position. The two factors should be solved together. The movable contact, the armature for holding, the armature for operating, and the pole were connected rigidly, so they were considered to be a lumped moving mass in the calculation. The partial differential equations are shown as Eq. (4.33), which included the voltage balance equation in circuit, D'Alembert equations in motion and Maxwell equations in magnetic field [52, 53].

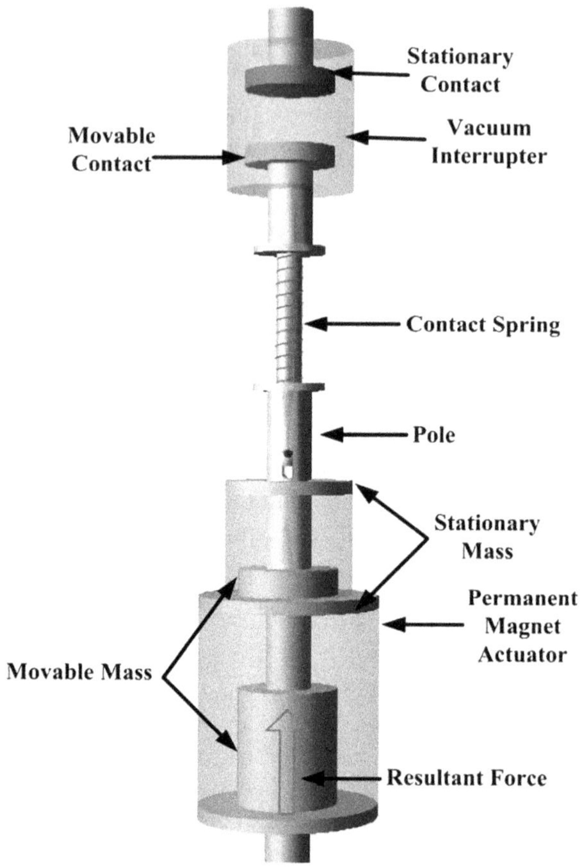

Fig. 4.58 An ADAMS model for a 126 kV vacuum circuit breaker for dynamics simulation

$$\begin{cases} \frac{d\psi}{dt} = U_C - iR \\ \frac{dv}{dt} = \frac{F(x, \frac{dx}{dt})}{m} \\ \frac{dx}{dt} = v \\ \frac{dU_C}{dt} = -\frac{i}{C} \\ \psi|_{t=0} = \psi_0 \\ v|_{t=0} = v_0, x|_{t=0} = x_0 = 0, U_C|_{t=0} = U_{C0} \end{cases} \tag{4.33}$$

where ψ is the flux linkage through the coils in the permanent magnetic actuator; U_C is the voltage across the discharging capacitors; R is the coil resistance; i is the coil current; v is the velocity of the lumped moving mass; F is the total force of the permanent magnets and the coils currents; m is the mass of movable mass; x is the displacement of the movable mass; and C is the capacitance of capacitors.

To solve Eq. (4.33), a user-defined function SFOSUB was developed in the ADAMS software package. Figure 4.59 shows a flow chart of the function SFOSUB.

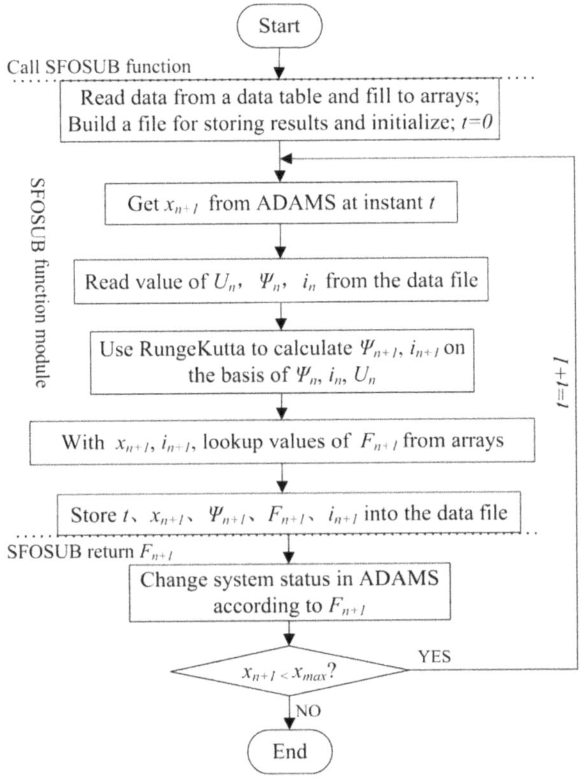

Fig. 4.59 Flow chart of the user-defined function SFOSUB

The dynamic force F drove the movable mass m during the simulations. The calculation in SFOSUB was an iterative process. At time $t = 0$, all parameters were initialized. Then, the function SFOSUB read data from a data table created in a static magnetic simulation. The data table included data of x, i, F, and flux linkages ψ. Both F and ψ were functions of x and i as $\Psi = \Psi(x, i)$ and $F = F(x, i)$. The user-defined function read the values of U_{n}, ψ_{n}, i_{n} at time step n from the data table, then it used a Runge–Kutta method to calculate the values of $\psi_{\mathrm{n+1}}$ and $i_{\mathrm{n+1}}$ at the next time step n +1 based on Eq. (4.33). Next, the user-defined function estimated $F_{\mathrm{n+1}}$ by an interpolation algorithm and $F_{\mathrm{n+1}}$ was returned to ADAMS for subsequent calculation. In addition, the calculated results at time step n + 1 were stored in another data file. The iterative process terminated until the contacts mated.

Table 4.10 shows the simulation parameters for the 126 kV vacuum circuit breaker. With the parameters, both static and dynamic characteristics were obtained, as shown in Fig. 4.60. Figure 4.60a shows the holding force at the open and close positions. In the open position, the holding force was ~2000 N. In the close position it was 7300 N. When the armature approached to the close position, the holding force increased rapidly. It shows that once the armature left the open position or close position, the holding force decreased to zero quickly. In such way, a high initial velocity of the

Table 4.10 Simulation parameters for the 126 kV vacuum circuit breaker

Item	Value	Unit
Capacitor	0.13	F
Initial voltage of capacitor	220	V
Closing coil resistor	1.52	Ω
Opening coil resistor	1.52	Ω
Number of turns of closing coil	500	turn
Number of turns of opening coil	500	turn
Contact stroke	64	mm
Over travel of armature	20	mm
Mass of movable parts	37	kg
Pre-load force	3900	N
Spring stiffness coefficient	80	N/mm

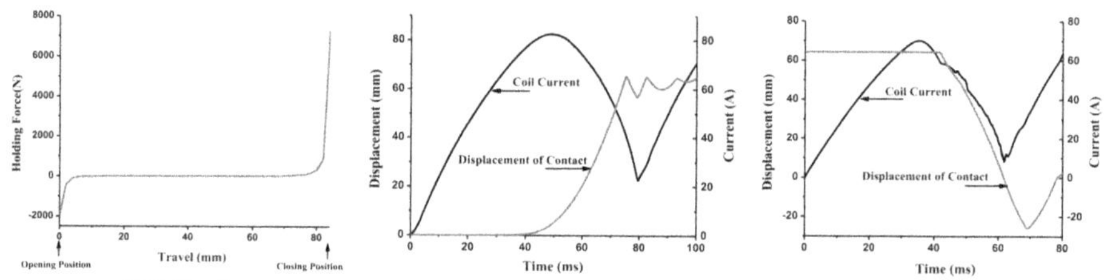

(a) Holding force (b) Closing displacement curve and closing coil current (c) Opening displacement curve and opening coil current

Fig. 4.60 Simulated static and dynamic characteristics of a 126 kV vacuum circuit breaker with a proposed permanent magnetic actuator

armature can be achieved. In addition, the holding force can be adjusted by changing the air gaps in the magnet for closing or in the magnet for opening. Figure 4.60b shows the movable contact closing displacement curve and closing coil current. The closing velocity can reach 2.0 m/s. The closing velocity was enough for a 126 kV vacuum circuit breaker, which was determined as 1.15 m/s (see Sect. 4.1.4). The closing coil current came from pre-charged capacitors. When the coil current approached to the peak value, the contact started to move accompanying a decrease of the current. The maximum value of the coil current was ~82 A. Figure 4.60c shows the movable contact opening displacement and opening coil current. The opening velocity of 3.0 m/s met the requirement in the maximum arcing time (see Sect. 4.1.2). The maximum current in the opening coil was ~70 A.

Figure 4.61 shows the experimental results of the 126 kV vacuum circuit breaker prototype with the proposed permanent magnetic actuator. Figure 4.61a shows a photo of the prototype. A damper was assembled at the bottom of the actuator to damp the impacting in opening operations. Figure 4.61b shows a driving circuit for the permanent magnetic actuator. Figure 4.61c shows a measured displacement curve of the movable contact in closing operations. A simulation result was put together for

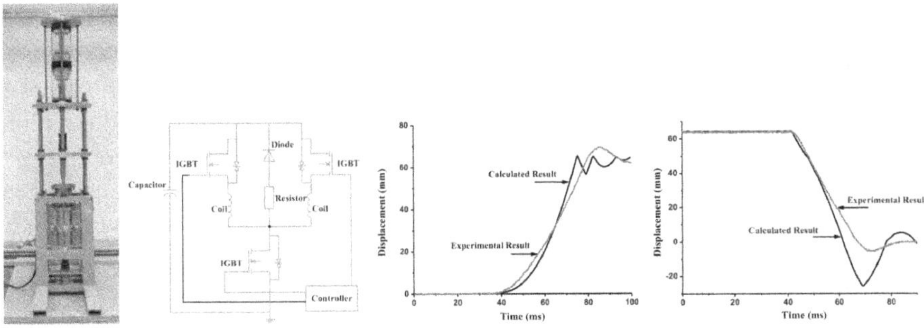

Fig. 4.61 Experimental results of a 126 kV vacuum circuit breaker with the proposed permanent magnetic actuator

comparison. The experimental result validated the simulation. Figure 4.61d shows a measured displacement curve of the movable contact in opening operations. Again, a simulation result was put together for comparison. The simulation result agreed with the experimental result well at the first 2/3 contact gap. However, the experimental velocity was slower than the simulated one in the last 1/3 contact gap. This was because the damper took effect. Therefore, the experimental results validated the simulation results. From the point of view of the velocity characteristics, the proposed permanent magnetic actuator proved to be a potential candidate operating mechanism for the 126 kV vacuum circuit breaker.

4.3 Nominal Rated Current and Temperature Rise

In transmission voltage level vacuum circuit breakers, it is a challenge to enhance the nominal rated current level. In oil circuit breakers and gas circuit breakers, this is not so challengeable. This is because they have a pair of main contact that is independent with the arc contact. The main contact is responsible for carrying the nominal rated current. In addition, their heat transfer includes not only heat conduction but also heat convection in oil or in gas. In such way, a high nominal rated current is achievable. However, in vacuum circuit breakers, there is only a pair of contacts, which not only carries the nominal rated current, but also interrupts current. At the same time, the only effective heat transfer in vacuum interrupter is heat conduction. There is no heat convection in vacuum and the heat radiation is quite low. Thus, it is quite difficult to enhance the nominal rated current for vacuum circuit breakers. For transmission voltage level vacuum circuit breakers, the high insulation requirement leads to a long heat conduction path. So it becomes more difficult for thermal design of transmission voltage level vacuum circuit breakers to meet the standards' temperature rise requirements. A high temperature rise of conductors and insulators leads to reduction of their mechanical strength. Besides, an overheating of insulators leads to insulating degradation. Therefore, temperature rise of circuit breakers should meet the requirements of international standards [36] or standards in each country.

It is a good way to understand temperature rise distribution of a circuit breaker in carrying nominal rated current by an electromagnetic- thermal field coupling analysis. Paulke et al. [54] thermally analyzed a switchgear by using a commercial software and the heat conduction/convection in the current path was obtained. Heat transfer from all parts/components exposed to the surrounding gas was taken as thermal coefficients for radiation and conduction/convection, and they were applied to the exposed surfaces. Wang et al. [55] analyzed an electromagnetic-thermal-flow field to understand thermal performance of a 12 kV medium-voltage switchgear. The temperature distribution of the switchgear was obtained under both natural and forced convection conditions. This section is to analyze the thermal performance of a 126 kV vacuum circuit breaker in carrying nominal rated current. It was done by a three-dimensional electromagnetic-thermal analysis [56, 57].

Figure 4.62 shows a porcelain type 126 kV vacuum circuit breaker. Figure 4.62a shows a pole. It consisted of an interrupter part and a supporting part. Each part has a porcelain bushing. The interrupter part conducted the nominal rated current. Figure 4.62b shows the interrupter part. The current path included the upper connection terminal, the upper flange, the upper finger box, the upper fingers, the vacuum interrupter, the lower fingers, the lower finger box, the cylinder conductor, the lower flange and the lower connection terminal. Figure 4.62c shows the vacuum interrupter. The current path included a stationary conducting rod, a stationary contact, a moving contact, and a moving conducting rod. The 2/3 coil type axial magnetic field contacts were used (see Sect. 3.2.1). There was a sliding contact between the upper fingers and the static conducting rod. There was another sliding contact between the lower fingers and the movable conducting rod. The diameter of the static conducting rod was 40 mm and the diameter of the movable rod was 45 mm. The length of the static rod was 406 mm and the length of the movable rod was 383 mm.

The following assumptions were made for the electromagnetic-thermal coupling analysis of the 126 kV vacuum circuit breaker. The heat generation included the ohmic heat and eddy current loss. The heat transfer was mainly the heat conduction

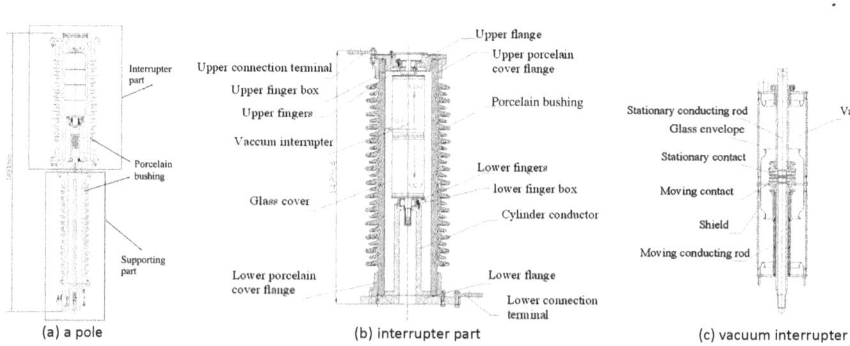

Fig. 4.62 A porcelain type 126 kV vacuum circuit breaker

inside the vacuum circuit breaker. When the heat transferred to the external surfaces of the breaker, it dissipated through both natural convection and thermal radiation. Heat generation and dissipation from the porcelain bushing and the operating mechanism was negligible. Therefore, the 126 kV vacuum circuit breaker shown in Fig. 4.62 can be simplified based on the assumptions. Figure 4.63 shows a simplified thermal model of the 126 kV vacuum circuit breaker. An electromagnetic-thermal coupling analysis was done based on the model [56, 57].

The analysis included two parts, in which one was electromagnetic part and the other was thermal part. The two parts were coupled each other. The electromagnetic part calculated the ohmic heat and eddy current heat. Then they were used as an input data for thermal part to calculate the temperature. As the temperature increased, the conductor's resistance and heat transfer coefficients changes. Therefore, there are iterations inside.

In the electromagnetic model, the ohmic heat loss Q_o was defined by Eq. (4.34).

$$Q_o = I^2 R \tag{4.34}$$

where I is the nominal rated current. R is the electrical resistances of conductors. R included electrical contact resistances R_c and conductors bulk resistance in current

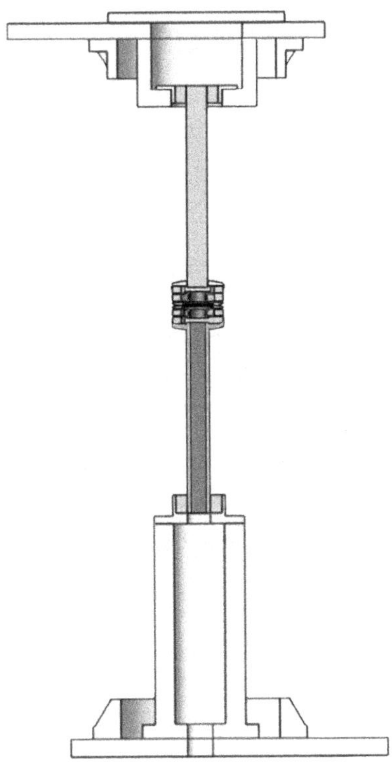

Fig. 4.63 A simplified model of the 126 kV vacuum circuit breaker using for electromagnetic-thermal coupling analysis

flowing path R_b, as shown in Eq. (4.35).

$$R = R_c + R_b \tag{4.35}$$

R_b varies with temperatures, effects of which are considered as Eq. (4.35).

$$R_b = \rho_0(1 + \alpha T)\frac{L}{A} \tag{4.36}$$

where ρ_0 is the conductor resistivity at 0 °C. α is the temperature coefficient of electrical resistance. T is the temperature of conductors. L/A is the ratio of length to section area of conductors.

R_c follows Eq. (4.37) [58].

$$R_c = \sqrt{\frac{\rho^2 \eta \pi H}{4F}} \tag{4.37}$$

where ρ is the resistivity of the conductor. η is an empirical coefficient of order unity for clean interfaces. H is hardness of material. F is load applied to the electrical interfaces.

The eddy current heat is calculated by using Maxwell's equations as shown in Eqs. (4.38) to (4.40).

$$\nabla \times \boldsymbol{H} = \boldsymbol{J} \tag{4.38}$$

$$\nabla \times \boldsymbol{E} = -\frac{\partial \boldsymbol{B}}{\partial t} \tag{4.39}$$

$$\nabla \cdot \boldsymbol{B} = 0 \tag{4.40}$$

Magnetic vector potential **A** follows Eq. (4.41).

$$\boldsymbol{B} = \nabla \times \boldsymbol{A} = \mu \boldsymbol{H} \tag{4.41}$$

From Eqs. (4.38) to (4.41), one can get a governing equation as Eq. (4.42).

$$\nabla \times \frac{1}{\mu}(\nabla \times \boldsymbol{A}) = \boldsymbol{J} = \boldsymbol{J}_s + \boldsymbol{J}_e \tag{4.42}$$

where $\boldsymbol{J}_s$ is the source current and $\boldsymbol{J}_e$ is the eddy current. The eddy current in the conductor follows Eq. (4.43).

$$\boldsymbol{J}_e = \sigma_e \boldsymbol{E} = -\sigma_e \frac{\partial \boldsymbol{A}}{\partial t} \tag{4.43}$$

From Eqs. (4.42) and (4.43), the Eq. (4.44) was derived.

$$\nabla \times \frac{1}{\mu}(\nabla \times \boldsymbol{A}) = \boldsymbol{J}_s - \sigma_e \frac{\partial \boldsymbol{A}}{\partial t} \quad (4.44)$$

A can be solved by Eq. (4.44). Then $\boldsymbol{J}_e$ can be solved by Eq. (4.43). The eddy current loss in the conductor per unit length follows Eq. (4.45).

$$q_e = \int \frac{J_e^2}{\sigma_e} dS \quad (4.45)$$

In the thermal model, thermal conduction in the conductors is calculated by the heat conduction differential equation, as shown in Eq. (4.46).

$$\frac{\partial^2 T}{\partial x^2} + \frac{\partial^2 T}{\partial y^2} + \frac{\partial^2 T}{\partial z^2} + \frac{\Phi}{\lambda} = 0 \quad (4.46)$$

where T is temperature. x, y, z is the three-dimensional coordinates. Φ is the heat generation in the unit volume, which is caused by ohmic heat and eddy current heat of conductors. λ is the thermal conductivity of conductors. The boundary conditions were set by heat transfer coefficients. The heat transfer coefficients incorporated both heat radiation and convection. The heat transfer coefficient was affected by the surface temperatures and hence they calculated via iterations. One can find more details in determination of the heat transfer coefficients in references [56, 57]. The electromagnetic-thermal coupling analysis was done by the finite element method.

Figure 4.64 shows a simulation result of temperature distribution of the 126 kV vacuum circuit breaker. The simulation was based on the above mentioned electromagnetic-thermal coupling analysis. The nominal rated current was 2500 A. The environment temperature was 28 °C. The simulation result indicated that the highest temperature 528 K located in the contacts region.

In order to validate the simulation result, an experiment was done to measure the temperature. Figure 4.65 shows an experimental setup for temperature measurement. An interrupter part of the 126 kV vacuum circuit breaker shown in Fig. 4.62b was used as a test object. The applied contact force was 5500 N, which was supplied by a spring. A high current generator, which can supply current of 0–3000 A, provided an electrical current of 2500 A to the test object. The electrical current was measured by a current transformer. The temperatures were measured by using copper-constantan thermocouples (T type), which have an accuracy of $\pm$ 0.75%t of reading ($\pm$0.75%t °C, t is the measured temperature). There were twelve positions of temperature measuring points, as shown in Fig. 4.65. For each measuring point, a 0.5-mm diameter hole was drilled in the conductor. In order to minimize heat loss due to the radiation of the thermocouples, the thermocouples were firstly wrapped by the tinfoil. Then the thermocouples were inserted into the hole and fixed with a high-conductivity epoxy. The temperatures were recorded by a high-precision data

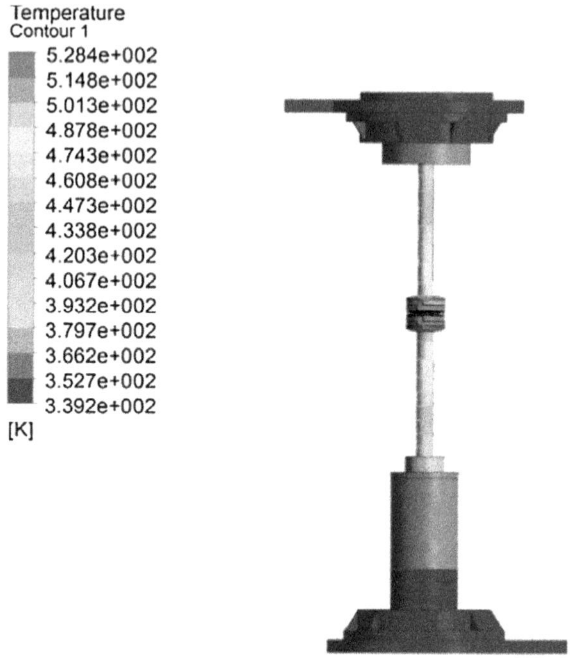

Fig. 4.64 A schematic for electrical contact resistance

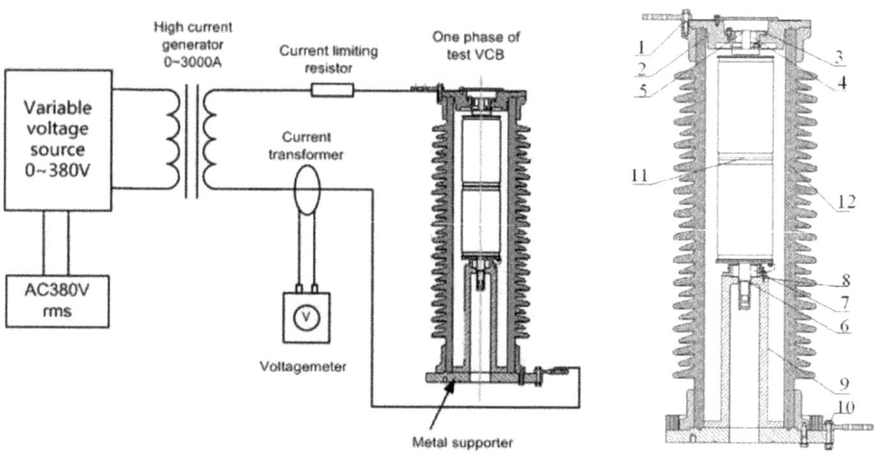

Fig. 4.65 Experimental setup for temperature measurement

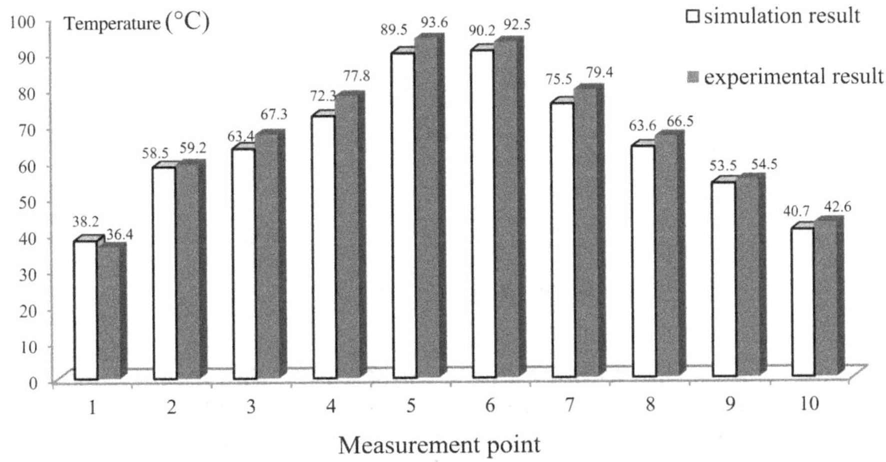

Fig. 4.66 Comparison between simulated and measured temperature. The ambient temperature was 28 °C

acquisition system. Typically, the test object reached a thermal steady state less than 9 h when the change of the temperatures was less than 1 °C per hour. In order to avoid the electromagnetic interference to the reading of the thermocouple and the data acquisition system, the temperatures were recorded at the moment once the electrical current was cut off. The measurements were repeated at least three times to minimize the measurement error. The ambient temperature was 28 °C.

Figure 4.66 shows a comparison between simulated and measured temperatures. It shows that the measured maximum temperature of the 126 kV vacuum circuit breaker with the nominal rated current of 2500 A was 93.6 °C. Considering the ambient temperature, the maximum temperature rise was 65.6 K, which met the standards' requirement of 75 K at the maximum temperature point [36, 63]. Overall, the simulation results showed a good agreement with the measured temperature rise. Figure 4.66 shows a comparison of measurement point 1–10. That was because point 11 and 12 had no simulation results, which located at the external surface of the vacuum interrupter and at the inner wall of the porcelain bushing, respectively. The temperature rises above ambient temperature of point 11 and 12 were 22.7 °C and 10.1 °C, respectively. At measurement point 1–10, the maximum difference of the temperature was 5.5 °C, which was at the measurement point 4. Its relative error was 5.2%. This comparison proved that the electromagnetic-thermal coupling analysis can accurately predict the temperature distribution of the 126 kV vacuum circuit breaker.

Based on the electromagnetic-thermal coupling analysis, it was found that the vacuum interrupter played an important role thermally. It can be reflected by two thermal parameters, which are power loss and thermal resistance, respectively. The total power loss of a pole of the 126 kV vacuum circuit breaker was calculated as 473.5 W, in which the ohmic loss was 436 W and the eddy current loss was 37.5 W.

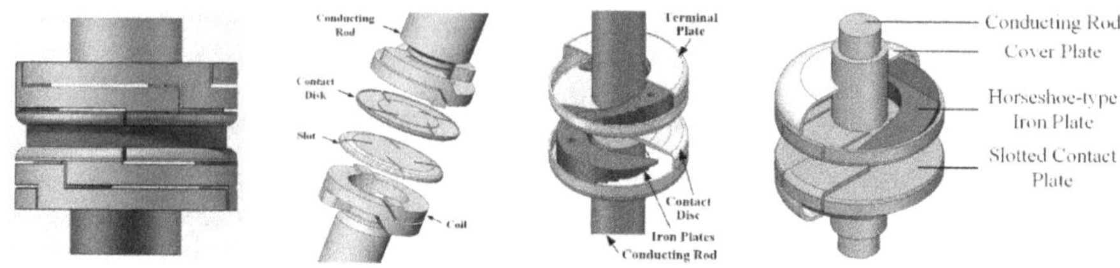

(a) 2/3 coil-type (b) 3/4 coil-type (c) horseshoe-type (d) slotted horseshoe-type

Fig. 4.67 Four kinds of axial magnetic field contact designs

The power loss of the vacuum interrupter occupied 63.3% of that of the 126 kV vacuum circuit breaker. The total thermal resistance of a pole of the 126 kV vacuum circuit breaker was calculated as 2.92 K/W. The thermal resistance of the vacuum interrupter occupied 79%. In the vacuum interrupter, the thermal resistance of the contacts occupied 40%. Therefore, the thermal design of the contacts is important.

Four kinds of axial magnetic field contact designs were compared for their thermal performance as they were applied in the 126 kV vacuum circuit breaker with a nominal rated current of 3150 A. [59] Fig. 4.67 show the four contact designs. It included a 2/3 coil-type axial magnetic field contact (see Sect. 3.2.1), a 3/4 coil-type axial magnetic field contact [60, 61], a horseshoe-type AMF contact (see Sect. 3.2.2), and a slotted horseshoe-type AMF contact [62], respectively. The electromagnetic-thermal coupling analysis was adopted to compare their thermal performance. Their simulation results showed that the maximum temperature rise point was point 6 in Fig. 4.65. The temperature rise of the 126 kV vacuum circuit breaker with the 2/3 coil-type, the 3/4 coil-type, the horseshoe-type and the slotted horseshoe-type axial magnetic field contact at point 6 were 103.71, 85.59, 72.46, and 66.04 K, respectively. According to standards [36, 63], the allowable temperature rise at the point 6 was 75 K. Therefore, the simulation results suggested that the two kinds of horseshoe-type contacts were thermally better than the two kinds of coil-type contacts. The two kinds horseshoe-type contacts have lower resistance than the two kinds of coil-type contact designs. In addition, the slotted horseshoe-type axial magnetic field contact has the best thermal performance among the four types of contact designs. If a 126 kV vacuum circuit breaker would reach a nominal rated current of 3150 A, the slotted horseshoe-type axial magnetic field contact provided a good choice. The slotted horseshoe-type axial magnetic field contact designed two pairs parallel slots on each contact plate. The AC current induces eddy currents in the contact plates, while the DC current does not have such effect. So, the contact's DC resistance would not be same as its AC resistance. One should note this because typically the measured vacuum interrupter's resistance was the DC resistance, which may not necessarily reflect the thermal performance. The slots on the contact plate regulated the current distribution, so that most current flowed through the middle part of the contact plate, which reduced the AC resistance from 23.98 $\mu\Omega$ (horseshoe-type) to 16.25 $\mu\Omega$ (slotted horseshoe-type). That was why the slotted horseshoe-type axial magnetic field contact had a better thermal performance at high nominal rated currents.

4.4 Type Test of a 126 kV Single-Break Vacuum Circuit Breaker

Extensive efforts have been devoted to the development of transmission voltage level vacuum circuit breakers. Slade et al. [64] successfully retrofitted an oil circuit breaker to a 121 kV–145 V/40-kA double-break vacuum circuit breaker by combining two 76 kV/40-kA vacuum interrupters in series. Schellekens et al. [65] proposed a 72.5 kV/31.5 kA/2000 A vacuum circuit breaker with a compact structure by arranging multiple floating shields in the vacuum interrupter. The current interrupting performance of the prototype was verified. Japan AE Power [66] developed a 145-kV single-break vacuum circuit breaker by optimizing the electrical field in the vacuum interrupter and by using a new composite insulator. Japan AE Power [67] also developed a 72/84-kV SF_6 gas-free vacuum circuit breaker, a 120-kV single-break tank-type vacuum circuit breaker, and a 168-kV double-break tank-type vacuum circuit breaker, based on the axial magnetic field contact technique and copper-chromium contact materials. Siemens company [68] developed two kinds of clean air insulated live tank single break vacuum circuit breakers, which was 72,5 kV and 145 kV, respectively. China developed a single-break 126 kV/40 kA/2500 A vacuum circuit breaker. This section is to introduce the type test results of the 126 kV vacuum circuit breaker.

Figure 4.68 shows the 126-kV single-break vacuum circuit breaker and its technical parameters. The external insulation of 126 kV vacuum interrupter was provided by SF_6 gas at 0.1 MPa, which was a gauge pressure. The main technical parameters of the 126 kV vacuum circuit breaker included rated voltage of 126 kV, rated current of 2500 A, rated short-circuit breaking current of 40 kA, rated short-circuit

Voltage	kV	126
Normal current	kA	2.5
Number of poles		3
Frequency	Hz	50
Operating sequence	O-0.3s-CO-3min-CO	
Short-time withstand current	kA	40
Peak withstand current	kA	100
Duration of short-circuit	s	4
Short-circuit making current	kA	100
Short-circuit breaking current	kA	40
First-pole-to-clear factor	1.5	
Capacitive voltage factor	1.4	
Line-charging breaking current	A	31.5
Cable-charging breaking current	A	140
Presure for insulation SF6 at 20℃	Mpa	0.1
Supply voltage of closing and opening device	V	220
Class	E2, C1, M1	

Fig. 4.68 A 126-kV single-break vacuum circuit breaker and its technical parameters

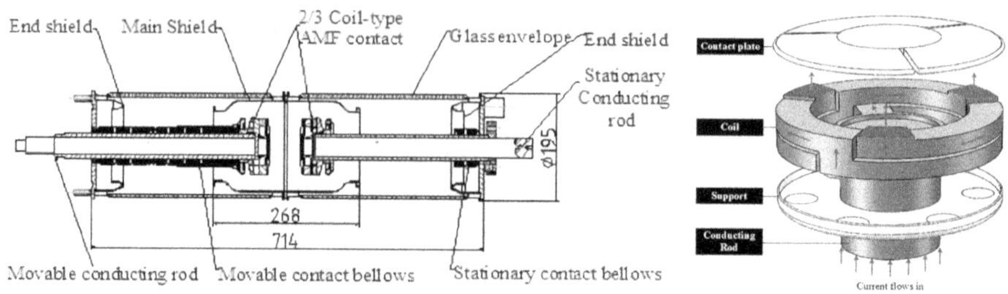

(a) vacuum interrupter (b) 2/3 coil-type contact structure

Fig. 4.69 126 kV vacuum interrupter and its 2/3 coil-type axial magnetic field contact

making current of 100 kA, rated short-time withstand current of 40 kA, 4 s and peak withstand current of 100 kA. The 126 kV vacuum circuit breaker passed type tests performed according to an IEC 62271-100 standard [36] and a Chinese standard GB 1984–2014 [69].

Figure 4.69 shows the 126 kV vacuum interrupter and its 2/3 coil-type axial magnetic field contact. The vacuum interrupter length between the two terminal flanges was 714 mm, and the diameter of the glass envelope was 195 mm. The length of the main shield was 268 mm. The full contact gap was 60 mm. A pair of 2/3 coil-type axial magnetic field contacts, with a diameter of 100 mm, was chosen (see Sect. 3.2.1).

In the 126 kV vacuum circuit breaker, a spring-type operating mechanism was used (see Sect. 4.2.1). A contact spring provided a force of 4938 N when the contacts are fully closed. Figure 4.70 shows the opening displacement curve and the closing displacement curve of the 126 kV vacuum circuit breaker. In the opening contact travel curve, the average velocity corresponding to the minimum arc duration was 3.5 m/s and the average velocity corresponding to the maximum arc duration was 3.0 m/s (see Sect. 4.1.2). The minimum arcing contact gap was in a range between 18 and 23 mm and the maximum arcing contact gap was 44 mm (see Sect. 3.2.1). While

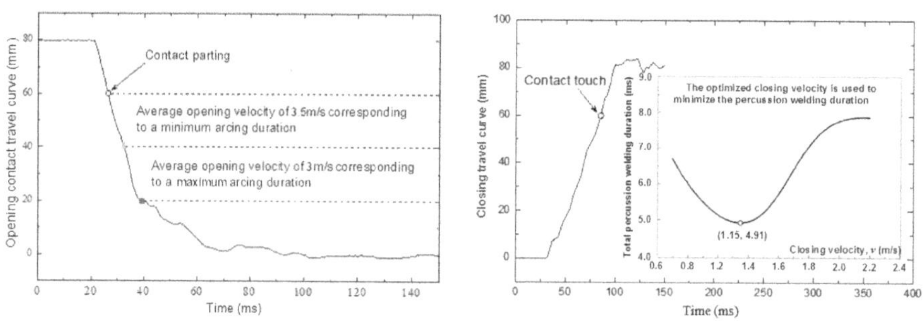

(a) Opening displacement curve (b) Closing displacement curve

Fig. 4.70 Opening displacement curve and the closing displacement curve of the 126 kV vacuum circuit breaker

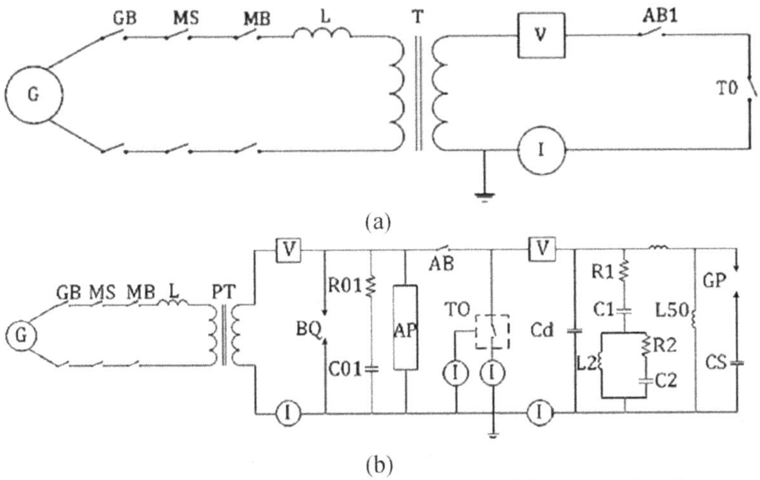

G: generator TO: test VCB MB: master breaker
GB: generator breaker T: short-circuit transformer L: reactor
MS: make switch PT: power transformer CS: voltage source
AB: auxiliary breaker I,V: current and voltage measurements
PQ: protective gap in current source circuit.

Fig. 4.71 Type-test circuits for the 126 kV vacuum circuit breaker making and breaking test. **a** Direct test circuit and **b** single-phase current injection synthetic test circuit

the remaining gap slowing the contact motion. The closing velocity was 1.15 m/s, which was to minimize the percussion welding (see Sect. 4.1.4). The 1.15 m/s was an average velocity during 4-mm contact gap before contact mating.

Figure 4.71 shows two test circuits for making and breaking tests. Figure 4.71a shows a direct test circuit and Fig. 4.71b shows a single-phase current injection synthetic test circuit. The two test circuits were recommended by the IEC standards, which imposed appropriate stresses on the test object (TO) during the type test. The circuits were used to verify the short-circuit current making and breaking capabilities of the 126 kV vacuum circuit breaker. A direct test method, as shown in Fig. 4.71a, was preferred because it was regarded as being identical to the actual power system conditions. The test current and voltage were supplied by a short-circuit generator. However, a synthetic test method, as shown in Fig. 4.71b, acted as an equivalent method when the required power exceeded the available power of the test laboratory for direct test. In the synthetic test circuit, the test current and voltage were imposed on the 126 kV vacuum circuit breaker individually. Before the test procedure, both circuits were adjusted to an appropriate value according to the parameters required for each test duty.

The 126 kV single-break vacuum circuit breaker passed the type test, which included the dielectric test, short-time withstand current and peak withstand current tests, terminal fault test, double earth fault tests, short-line fault test, out-of-phase test, capacitive current switching test of class C1, electrical endurance test of class E2, temperature rise test, and mechanical endurance test. Each test item is introduced in details as following.

a. **Dielectric Test**

The vacuum circuit breaker passed the dielectric tests without any disruptive discharge according to the standards. The test duties of the dielectric tests included a short-duration power frequency withstand voltage test, a lightning impulse test, and a series of dielectric condition assessment tests after each test duty cycle of the short-circuit current breaking test. The power frequency voltage withstand test was performed on the three phases of the 126 kV vacuum circuit breaker with a 50 Hz voltage of 230 kV and it maintained for 1 min. The test procedure consisted of three cases: phase-to-phase, phase-to-earth, and across open contacts. During this test procedure, no disruptive discharge appeared either on the external self-restoring insulation or on the inner vacuum insulation. The standard lightning impulse (1.2/50 μs) test applied to the 126 kV vacuum circuit breaker, was performed with 15 consecutive standard lightning impulses on each of the three poles. The applied voltage was 550 kV at peak with positive and negative polarities. There appeared no disruptive discharge on either the self-restoring insulation or the non-self-restoring insulation during each of the series of 15 impulses.

The dielectric condition assessment tests were performed after the making test, the breaking test, the electrical endurance test, and the mechanical endurance test, because the insulation properties across the open contacts of the test 126 kV vacuum circuit breaker cannot be verified by visual inspection with sufficient reliability. Consequently, the power frequency withstand voltage test under dry conditions across the open contacts was appropriate. The applied voltage was 80% of the rated short-duration power frequency withstand voltage, that was, 184 kV. In addition, the lightning impulse withstand voltage test was performed at 60% of the rated value, that was, 330 kV at peak, which was used to verify the insulation conditions after the electrical endurance testing. The 126 kV vacuum circuit breaker passed all of the required dielectric condition assessment tests after each type test.

b. **Short-Time Withstand Current and Peak Withstand Current Tests**

Table 4.11 shows the test results. It verified that the 126 kV vacuum circuit breaker successfully passed the short-time withstand current (40 kA, 4 s) test and the peak withstand current (100 kA peak) test. Figure 4.72 shows a diagram of this test. The test current was 40.5 kA on average and the duration was equivalent to 4.2 s. Immediately after the test, a no-load opening operation was performed on the test vacuum circuit breaker. The test vacuum circuit breaker opened successfully at the first attempt, which proved that the contacts had not welded together. Comparison of the resistance of the main circuit, as shown in Table 4.11, before and after testing, showed that the resistance decreased in each pole. According to the IEC standards, it specified that the resistance must not increase by more than 20%, which proved

Table 4.11 Test results of short-time withstand current and peak withstand current tests

Phases			A	B	C
Peak value of current		kA	−104.6	73.0	95.3
Peak current duration		ms	317.0	322.0	322.0
Symmetrical current		kA	39.3	42.9	39.2
Symmetrical current duration		s	4.105	4.105	4.100
Average current value		kA	40.5		
Thermal equivalent		s	4.2		
Resistance of main circuit	Before test	μΩ	50.8	54.4	50.2
	After test	μΩ	46.4	47.2	49.3

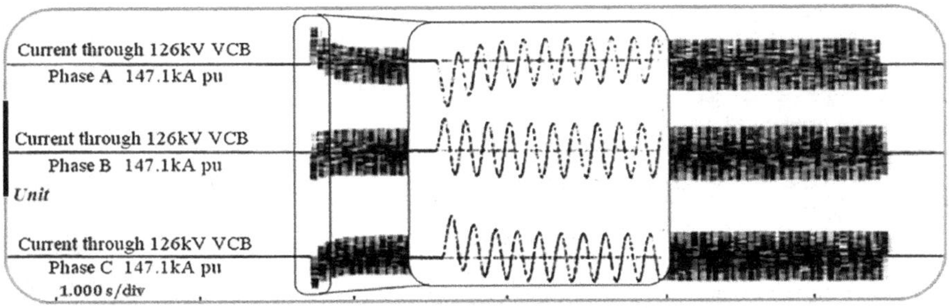

Fig. 4.72 Three-phase short-time withstand current test and peak withstand current test on the 126 kV vacuum circuit breaker

that the 126 kV vacuum circuit breaker was capable of sustaining both the short-time withstand current and the peak withstand current.

c. **Terminal Fault Test**

A terminal fault is a fault that is located directly at the terminals of the circuit breaker, which gives the maximum short-circuit current. According to IEC62271-100, this test is performed using four symmetrical short-circuit test duties, related to the symmetrical terminal fault conditions (T10, T30, T60, T100s), and one asymmetrical test duty (T100a). Table 4.12 shows the results of the terminal fault tests for the 126 kV vacuum circuit breaker, which were carried out by using the synthetic test circuit. All test duties were conducted on phase C (i.e., an outer pole of the 126 kV vacuum circuit breaker), with a rated operating sequence, that was, O-0.3 s-CO-180 s-CO. Successful current interruption results of the test duty T30, T60, T100s, and T100a are shown in Table 4.12. In Table 4.12, "I_p" is the value of prospective interrupting current; "I_t" is the real interrupting current value; "t_2" or "t_3" are the time for the TRV rising from the initiation to the peak value; "u_1/t_1" or "u_C/t_3" is the TRV rate of rise. Because the breaking current is 10% of the rated short circuit current of 40 kA, the T10 test duty successfully passed a direct test with an "O-0.3 s-CO-180 s-CO" switching operation. The prospective value of the current and the transient recovery

Table 4.12 Test results of the terminal fault tests

Items		Unit	Test duty			
			T30	T60	T100s	T100a
I_P		kA	12	24	40	40.74
U_C		kV	236	231	216	216
t_2 or t_3		μs	47	234	232	232
u_1/t_1 or u_C/t_3		kV/μs	5.0	3.0	2.0	2.0
Minimum arcing	I_t	kA	12.3	23.7	40.0	40.2
	t_{arc}	ms	3.4	4.5	8.5	4.6
Maximum arcing	I_t	kA	12.0	22.9	40.1	40.4
	t_{arc}	ms	14.2	13.5	17.8	15.9
Medium arcing	I_t	kA	12.4	24.4	40.1	40.7
	t_{arc}	ms	8.5	10	13.2	12.4
Passed or not			Passed	Passed	Passed	Passed

"I_P" is the prospective interrupting current value
"I_t" is the real interrupting current value
"t_2" or "t_3" is the time for the TRV rising from the initiation to the peak value
"u_1/t_1" or "u_C/t_3" is the TRV rate-of-rise

voltage (TRV) for the T10 test duty were 4 kA and 236 kV peak, respectively. Therefore, the results proved that it successfully passed the terminal fault test.

d. **Double Earth Fault Tests**

The double earth fault (DEF) is related to an ineffectively earthed neutral system, when two faults are located on different phases, or one of the faults is located on one side of the circuit breaker and the other fault is on the other side of the circuit breaker (on a different phase). Although a double earth fault is extremely unlikely to occur, it is a severe condition for the circuit breaker, because one pole of the three phases must interrupt a short-circuit current of 87% of the rated short-circuit current, under a phase-to-phase voltage. The test duty cycle of the double earth fault was performed on the phase C of the 126 kV vacuum circuit breaker. During each test procedure, the TRV rate of rise was 2 kV, as same as for test duty T100s. The double earth fault test of the 126 kV vacuum circuit breaker contained two valid "O" operations and one valid "CO" operation, where the peak TRV value was not less than 249 kV. Figure 4.73 shows a record of the double earth fault test duty, in which an "O-CO" operation was performed. On the first breaking operation, a fault current of 35.2 kA was interrupted after an arc duration of 5.5 ms; however, no particular TRV stress was applied across the test vacuum circuit breaker in order to apply the full TRV in the second operation. For the second breaking operation, the breaking current was 35.1 kA after an arc duration of 13.5 ms, in which the peak TRV value was 250 kV.

e. **Short-Line Fault Test**

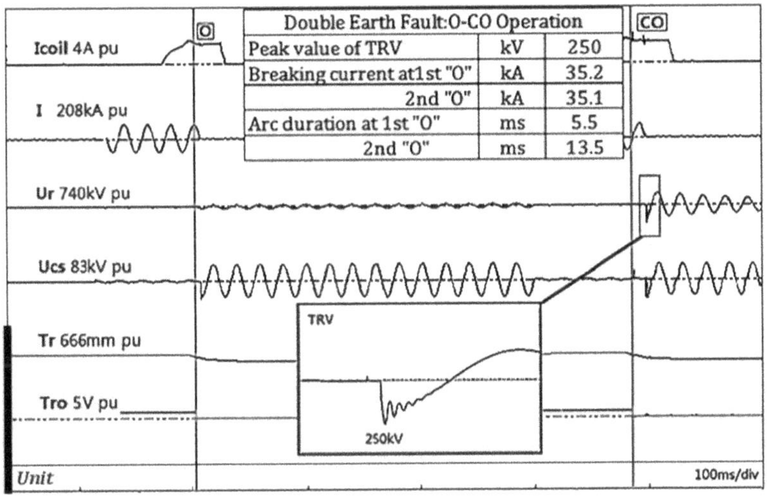

Fig. 4.73 Double earth fault test: "O-CO" operation, where the 1st breaking current was 35.2 kA and the arc duration was 5.5 ms; the 2nd breaking current was 35.1 kA, and the arc duration was 13.5 ms. "Icoil" is driven current of the closing and opening coil; "I" is the test current; "Ur" is the test voltage applied to the 126 kV vacuum circuit breaker; "Ucs" is the voltage of the current source. "Tr" is the travel curve of the test vacuum circuit breaker; "Tro" is the digital signal for illustrating the closing or opening position of the test vacuum circuit breaker

Short-line fault (SLF) is a single-phase-to-earth fault of an overhead line. The short-line fault occurs at some distance, typically hundreds of meters up to a few kilometers, from the circuit breaker terminal. The short-line fault test consists of L90 and L75 test duties. The tests were carried out on phase C of the 126 kV vacuum circuit breaker. Table 4.13 shows the test results of the L75 and L90 test duties. It showed that in the L75 test duty, a minimum arcing duration of 3.2 ms was attained in an "O" operation with an applied current of 30.2 kA and the applied TRV peaked at 139 kV. A maximum arcing duration of 12.8 ms was obtained in the second "O" operation in a reclose operation. In the second "O" operation, the breaking current was 30.2 kA and the applied TRV peaked at 138 kV. For L90, a prospective current of 36 kA was applied. The minimum arcing duration of 5.2 ms was obtained in an "O" operation. After extinction of the vacuum arc, the TRV peaked at 166 kV and there was no restrike occurring. The maximum arcing duration of 14.3 ms obtained in the second breaking operation of the "O-CO" operation. The TRV peaked at 170 kV.

f. **Out-of-Phase Test**

An out-of-phase fault caused by asynchronous operation of different parts of the power system. An out-of-phase fault would stress an extremely high voltage to a circuit-breaker connecting the different parts. However, the breaker would be subjected to a relatively low current. The 126 kV vacuum circuit breaker successfully cleared the out-of-phase faults, following a series of "CO" and "O" operations.

Table 4.13 Test results of the L75 and L90 test duties for short-line fault

L75	Operating Sequence	I_{SC}	DC component (%)	U_{peak}	t_{arc}
		kA		kV	ms
	O	30.1	0.31	139	5.6
	O	30.2	0.23	138	5
	O	30.2	0.72	139	3.2
	O-CO-180 s CO	30.7	9.31		2.1
		30.2	2.33	138	12.8
		30.1	1.15	139	8.1
	O	30.8	0.13	139	9.2
L90	O	36.1	3	166	5.2
	O-CO	37	0		2.4
		36.9	15	170	14.3
	CO	36.3	2	171	8.8

The breaking current was 10% of the rated short-circuit breaking current, and the prospective recovery voltage was 1.44 times the rated voltage. The OP2 test duty was carried out on the phase C. Figure 4.74 shows the out-of-phase test results. It included the test duties OP2(a) and OP2(b). Figure 4.74a shows the test duty OP2(b). It shows that a peak TRV of 328 kV, which is much higher than the basic short-circuit test duties, was applied. When the TRV disappeared, a remained power frequency recovery voltage of 175 kV applied on the test vacuum circuit breaker. The breaking current capability of test duty OP2(b) was successfully performed in a direct test circuit, with "O" operations. Figure 4.74b shows the test duty OP2(a). It demonstrated a symmetrical current-making capability. The making current was 6.5 kA

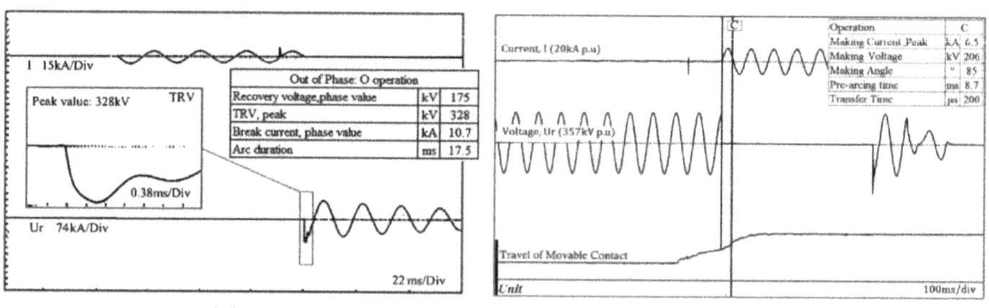

Fig. 4.74 Results of out-of-phase test duty of the 126 kV vacuum circuit breaker. **a** Test duty OP2 **b**, where the applied voltage was 175 kV, the breaking current was 10.7 kA, and the maximum arc duration of 17.5 ms was attained in this "O" operation; "I" was the test current; "Ur" was the test voltage applied to the test object. **b** Test duty OP2(a): a current making test, where the making voltage was 206 kV and the making electrical angle was 85°

and the making electrical angle was 85°. The prestrike time, from the breakdown of the applied voltage to the initiation of the circuit current, was 200 μs.

g. Capacitive Current Switching Test of Class C1

The capacitive current switching operations of the 126 kV vacuum circuit breaker consider the switching of unloaded transmission lines and the unloaded cables et al. A test, specified for class C1, was applied on the phase C of the 126 kV vacuum circuit breaker. The test combined with the line-charging (LC1 and LC2) and cable charging (CC1 and CC2) current switching tests. The assigned line-charging current rating was 50 A, and the cable-charging current rating was 140 A, with a capacitive voltage factor k_C of 1.4. There were 24 "O" operations for the CC1/LC1 test duty and there were 24 "CO" operating cycles for the CC2/LC2 test duty. There was no maintenance being required during the entire testing process. This proved that the 126 kV vacuum circuit breaker passed the tests successfully. To meet the CC1/LC1 test duty, a total of 24 "O" operations were performed in a synthetic test circuit. A test current of 15 A was chosen to fit within 10% and 40% range of both rated currents of 50 A and 140 A. The test current for the CC1/LC1 tests must be between 14 A and 20 A. The standard specified number of nine operations at the minimum arcing time was achieved. Figure 4.75 shows capacitive current breaking test results. A valid CC1/LC1 test procedure, which was performed as an "O" operation, is shown in Fig. 4.75a. The applied TRV peak and the applied current were 322 kV and 16.1 A, respectively. The arc duration in Fig. 4.75a was 0.9 ms. The combined CC2/LC2 test duty was performed in a test circuit equivalent to that of CC1/LC1 (except for the values of the circuit impedances). The test current of 141 A was used so that it was not lower than the rated currents of 31.5 A and 140 A. This test duty comprised a total of 24 "CO" operating cycles. Figure 4.75b shows one of the 24 valid CC2/LC2 "CO" operations. The breaking current was 152 A, and the TRV was 315 kV peak. The arc duration was 1 ms. There was no reignition, restrike, or nonsustained disruptive discharge (NSDD) occurring.

h. Electrical Endurance Test of Class E2

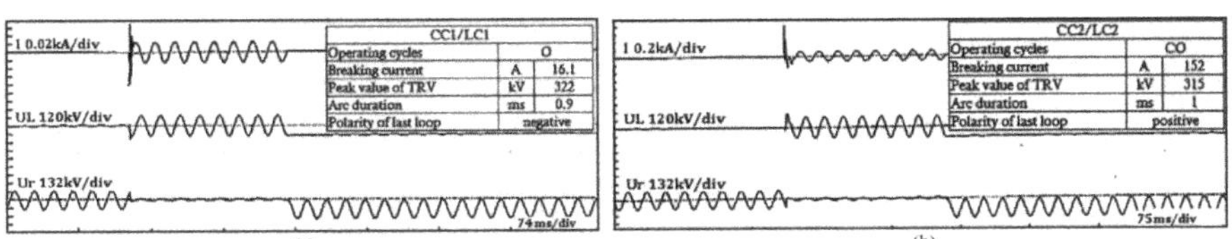

Fig. 4.75 Capacitive current breaking test results. **a** Test duty CC1/LC1: the test current was 16.1 A, the stressed TRV was 322 kV peak, and the arc duration was 0.9 ms; **b** Test duty CC2/LC2: the test current was 152 A, the stressed TRV was 315 kV peak, and the arc duration was 1 ms. "I" is the test current, "UL" is the voltage on the capacitive load side, and "Ur" is the test voltage applied to the test object

The electrical endurance test is a withstand test of repeated arc stresses against circuit breaker contacts and other components. Class E2 is an extended electrical endurance. The verification of the extended electrical endurance capability of the 126 kV vacuum circuit breaker was carried out by additional tests on the phase C with T100s test duty. According to the standard GB 1984–2014 [69], it allowed to perform the electrical endurance test on a single pole of a three-phase circuit breaker. During the tests, the vacuum circuit breaker was stressed by 24 accumulated breaking operations. The specified total number of operations (20) was exceeded by 4 to meet the specified number of operation sequences and to meet that required by a Chinese standard JB/T 3855–2008 [70]. The JB/T 3855–2008 specified that the arcing times should distribute uniformly in the arcing window of the terminal fault test duty T100s. In addition, the average arcing time should not be less than the middle arcing time. After successfully completing the 24 current breaking tests, a standard lightning impulse test, at 60% (i.e., 330 kV) of the rated value of 550 kV, was carried out on the phase C, and there was no disruptive discharge during the insulation assessment test. This proved that the 126 kV vacuum circuit breaker passed the Class E2.

i. **Temperature Rise Test**

The 126 kV vacuum circuit breaker passed the temperature rise test while carrying its nominal rated current of 2500 A. The test was performed on phase C with a new vacuum interrupter. The external insulation of the vacuum interrupter was provided by SF_6 gas with a pressure of 0.1 MPa. Figure 4.76 shows sixteen temperature measurement points of the 126 kV vacuum circuit breaker. The test performed over

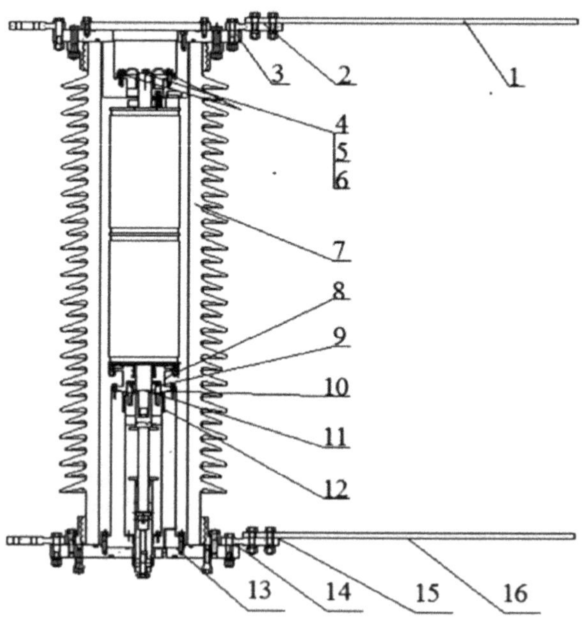

Fig. 4.76 The temperature measurement points of the 126 kV vacuum circuit breaker

Table 4.14 Temperature rise test results of the 126 kV vacuum circuit breaker

No.	Measurement position	T	L	Coatings
		°C	°C	
1	Bus bar	50.8	/	/
2	Connecting terminal to bus	47.6	65	Silvering
3	Solid connection	47.3	75	Silvering
4	Connection of VI stationary terminal	67.4	75	Silvering
5	Connection of VI stationary terminal	63.1	75	Silvering
6	Connection of VI stationary terminal	62.5	75	Silvering
7	Shell of porcelain pole	15.5	30	/
8	Connection of VI movable terminal	63.2/61.6	65	Silvering
9	Flexible connection	60.6/60.4	75	Silvering
10	Solid connection	51.3/51.8	75	Silvering
11	Movable conduct rod to conductive clip	68/67.9	75	Silvering
12	Flexible connection	54.5/54.8	75	Silvering
13	Conducting seat to lower flange	36.5/35.5	75	Silvering
14	Solid connection	35.2	75	Silvering
15	Connecting terminal to bus	38.2	65	Silvering
16	Bus bar	42.1	/	/

"L"expresses temperature rise limitation according to IEC 62271-100
"T"expresses temperature rise

a period of 9 h. When the temperature increases at each measurement point did not exceed 1 K in 1 h, it indicated that the temperature rise reached a stable state. The temperature was measured by using K-type thermocouples. Table 4.14 shows the measured results at each point. The maximum temperature rise 68 K appeared at point 11, which was the terminal of the movable conducting rod. It was lower than the limit of 75 K specified for silver coated sliding contacts. The temperature rises at the remaining measurement points were much lower than the limits allowed by the IEC standards. The loop resistance and the ambient air temperature were measured before and after thermal testing. The ambient temperatures were 17.5 °C and 17.0 °C before and after the test, respectively. Before the test, the loop resistance of the phase C was 44.9 $\mu\Omega$, and it became to 46.2 $\mu\Omega$ after the temperature rise test, which met the standard's requirement.

The temperature rise test of the auxiliary and control equipment was performed since the resistance of the energized circuit increased during switching operations. The resistances of the opening coil and the closing coil in the spring operating mechanism were measured after 10 operations. The interval between each operation was 10 s. Table 4.15 shows the temperature rise test results for the auxiliary and control equipment. It shows that after 10 operations the temperature rises of the

Table 4.15 Temperature rise and coil resistance test results of the auxiliary and control equipment of the 126 kV vacuum circuit breaker

TO	R1	T1	R2	T2	Tr	Tr.L
	mΩ	°C	mΩ	°C	K	K
Closing coil	112.22	21	112.8	21	1.3	50
Opening coil	54.96	21	55.55	21	2.7	50

Tr.: Temperature rise at ambient air temperature not exceeding 40 °C
Tr.L: Limitation of temperature rise at ambient air temperature of 40 °C
R1: Resistance before test
Tl: Ambient temperature before test
R2: Resistance after test
T2: Ambient temperature after test

closing coil and the opening coil increased 1.3 K and 2.7 K, respectively. Neither the temperature rise of the closing coil nor that of the opening coil exceeded the limitation of 50 K. The resistances of the closing coil and the opening coil increased by 0.58 mΩ and 0.59 mΩ after 10 operations, respectively.

j. **Mechanical Endurance Test**

The mechanical endurance-type test performed on the 126 kV vacuum circuit breaker was specified as class M1, which represents normal mechanical performance of 2000 opening and closing operations for the circuit breaker. Table 4.16 shows the test results. It consisted of four parts, based on a rated operation sequence. In the first 500 operations, the opening and closing operations performed under the rated voltage of the operating mechanism (i.e., 220 V dc). For the second 500 operations, the opening operations performed with an applied dc voltage of 143 V and the closing operations performed with a dc voltage of 176 V, respectively. For the third 500 operations, the driven voltages of the opening and closing operations were 264 V dc and 242 V dc, respectively. The fourth series of operations consisted of 250 reclosing operations driven by a rated operating voltage of 220 V dc. The 126 kV vacuum circuit breaker demonstrated class M1 mechanical endurance test performance. After the specified 2000 operations, the test vacuum circuit breaker was inspected and it was capable of operating normally. A power frequency voltage withstand test, where the applied

Table 4.16 Mechanical endurance tests results of the 126 kV vacuum circuit breaker

Operating sequence	Control voltage (DC, V)		Operations
	C	O	
C-t1-O-t1	176	143	500
	220	220	500
	242	264	500
O-t2-CO-t3-C-t4	220	220	250

tl: 40 s; t2:0.3 s; t3: 40 s; t4: 40 s

voltage was 230 kV and it maintained for 1 min, performed on all three phases without any disruptive discharge. In addition, the opening time and the closing time before and after the test did not exceed the predetermined values, i.e., 40 ± 15 ms and 80 ± 15 ms, respectively. Before the mechanical endurance test, the opening time, and the closing time at the rated operating voltage were 27.4–27.8 ms and 82.1–84.9 ms, respectively. After the test, the opening time and the closing time at the rated operating voltage were 27.2–27.8 ms and 82.3–85.2 ms, respectively.

References

1. Yuan S, Wang Y, and Wang J, Optimal moving curve of electrode to interrupt a short current for vacuum circuit breaker, in Proc. 2nd International Conference on Electrical Contacts, Arcs, Apparatus and their Applications (IC-ECAAA), Xi'an, China, 1993, pp. 248–251.
2. Kusserow J and Renz R, Method for Opening the Contact Gap of a Vacuum Interrupter, U.S. Patent 7 334 319, Feb. 26, 2008.
3. Kong G, Liu Z, Wang D, and Rong M, High-current vacuum arc: The relationship between anode phenomena and the average opening velocity of vacuum interrupters, IEEE Trans. Plasma Sci., vol. 39, no. 6, pt. 1, pp. 1370–1378, Jun. 2011.
4. Yu L, Wang J, Geng Y, Kong G, and Liu Z, High current vacuum arc phenomena of nano-crystalline CuCr25 contact materals, IEEE Trans. Plasma Sci., vol. 39, no. 6, pp. 1418–1426, Jun. 2011.
5. Slade P G, Taylor E D, and Haskins R E, Effect of short circuit current duration on the welding of closed contacts in vacuum, in Proc. 51st IEEE Holm Conf. Electr. Contacts, Sep. 2005, pp. 69–74.
6. Kisliuk P, Arcing at electrical contacts on closure. Part V. The cathode mechanism of extremely short arcs, J. Appl. Phys., vol. 25, no. 7, pp. 897–900, 1954.
7. Koren P P, Slade P G, and Lin C Y, Welding characteristics of Ag-W contacts under high-current conditions, IEEE Trans. Compon. Hybrids Manuf. Technol., vol. 3, no. 1, pp. 50–55, Mar. 1980.
8. Slade P G, The Vacuum Interrupter: Theory, Design, and Application. Boca Raton, Fl, USA: CRC Press, 2008, pp. 304–318.
9. Shaw N C, Investigations of contact bounce and contact welding, in Proc. 2nd Int. Conf. Contact Phenom., vol. 15, no. 12, pp. 286–302, 1964.
10. Slade P G, Electrical Contacts: Principles and Applications. Boca Raton, FL, USA: CRC Press, 1999, pp. 514–519.
11. Neuhaus A R, Rieder W F, and Hammerschmidt M, Influence of arc duration and current on contact welding in low power switches, in Proc. 43rd IEEE Holm Conf. Electr. Contacts, Dec. 2002, pp. 17–20.
12. Neuhaus A R, Rieder W F, and Hammerschmidt M, Influence of electrical and mechanical parameters on contact welding in low power switches, IEEE Trans. Compon. Packag. Technol., vol. 27, no. 1, pp. 4–11, Mar. 2004.
13. Slade P G, Smith R K, and Taylor E D, The effect of contact closure in vacuum with fault current on prestrike arcing time, contact welding and the field enhancement factor, in Proc. 53rd IEEE Holm Conf. Electr. Contacts, Sep. 2007, pp. 32–36.
14. J. W. McBride and S. M. Sharkh, "Electrical contact phenomena during impact," IEEE Trans. Compon. Hybrids Manuf. Technol., vol. 15, no. 2, pp. 184–192, Apr. 1992.
15. Garzon R D, High Voltage Circuit Breaker Design and Applications. New York, USA: Marcel Dekker, 2002, pp. 202–217.
16. Liu Z, Wang J, Xiu S X, Wang Z, Yuan S, Jin L, Zhou H, and Yang R, Development of high-voltage vacuum circuit breakers in China, IEEE Trans. Plasma Sci., vol. 35, no. 4, pp. 856–865, Aug. 2007.

17. Greenwood A, Vacuum Switchgear. London, U.K.: Inst. Eng. Technol., 1994, pp. 146–149.
18. Barkan P, A Study of contact bounce phenomenon, IEEE Trans. Power App. Syst., vol. 86, no. 7, pp. 231–240, Jul. 1967.
19. Kharin S N, Nouri H, and Amft D, Dynamics of arc phenomena at closure of electrical contacts in vacuum circuit breakers, IEEE Trans. Plasma Sci., vol. 33, no. 5, pp. 1576–1581, Oct. 2005.
20. Kharin S N, Nouri H, and Amft D, Dynamics of electrical contact floating in vacuum, in Proc. 48th IEEE Holm Conf. Electr. Contacts, Dec. 2002, pp. 197–205.
21. Kharin S N and Ghori Q K, Influence of the pre-arcing bridging on the duration of vacuum arc, in Proc. 19th Int. Symp. Discharges Electr. Insul. Vacuum, vol. 1. 2000, pp. 278–285.
22. Kharin S N, Nouri H, and Davies T, Effect of metallic vapour pressure on the vibration of electrical contacts in vacuum, Modern Pract. Stress Vibrat. Anal., vol. 440, pp. 505–512, Nov. 2003.
23. Christian F, Electrodynamic repulsion between electric contacts with arbitrary current density distribution, IEEE Trans. Compon., Hybrids, Manuf. Technol., vol. 9, no. 2, pp. 188–189, Feb. 1986.
24. Slade P G, The Vacuum Interrupter: Theory, Design, and Application. Boca Raton, FL, USA: CRC Press, 2008, pp. 127–129.
25. Barkan P, A new formulation of the electromagnetic repulsion phenomenon in electrical contacts at very high current, in Proc. 11th Int. Conf. Electr. Contact Phenomenon, 1982, pp. 185–188.
26. Barkan P and McGarrit R V, A spring-actuated cam-follower system–design theory and experimental results, J. Eng.Ind., vol. 87, no. 3, pp. 279–276, 1965.
27. Dresner T L and Barkan P, New methods for the dynamic analysis of flexible single-input and multiinput cam-follower systems, J.0 Mech. Design, vol. 117, no. 1, pp. 150–155, 1995.
28. Kisliuk P, Arcing at Electrical Contacts on Closure Part. 5. the Cathode Mechanism of Extremely Short Arcs, Journal of Applied Physics, vol. 25, pp. 897–900, 1954.
29. McBride J W, An Experimental Investigation of Contact Bounce in Medium Duty Contacts. IEEE Trans. on Components, hybrids, and Manufacturing Technology, 1991, 14: 319–326.
30. McBride J W. Electrical Contact Bounce in Medium-duty Contacts[J]. IEEE Trans. on Components, hybrids, and Manufacturing Technology, 1989, 12(1): 82–90.
31. IEC 60694, "Common Specifications for High-voltage Switchgear and Controlgear Standards," 2002, pp. 31.
32. Smeets Rene, Sluis L V D, Kapetanovic M, Peelo D, and Janssen A, Switching in Electrical Transmission and Distribution Systems, West Sussex, UK, John Wiley & Sons, Ltd, pp. 347–364, 2015.
33. Sato Y, Takahashi K, Otsuka T, Minagawa T, Kohyama H, Improvement of lifetime estimation of aged HV switchgear based on the consideration of applied technologies, design and conditions in service, SCA3-251, CIGRE, Auckland, New Zealand, Sep. 16-17, 2013.
34. Nakada T, Takahashi K, Iwabuchi K, Ishizeki S, Obayashi K, Survey and Research on Deterioration of 300 and 550 kV GCB, SCA3-252, CIGRE, Auckland, New Zealand, Sep. 16-17, 2013.
35. CIGRE Working Group A3.27, The Impact of the Application of Vacuum Switchgear at Transmission Voltages. CIGRE Technical Brochure 589, 2014.
36. IEC 62271-100, "High-voltage Switchgear and Controlgear: High-voltage Alternating-current Circuit Breaker," International Standard, 2006, pp. 151-153.
37. Wang H, Zhang B, Li X, Yao X, Zhang H, Liu Z, Geng Y, Wang J, Cheng J, Yang H, An improvement of a contact spring pin of a 126 kV vacuum circuit breaker by orthogonal experimental design, 4th International conference on electric power equipment-switching technology, xi'an, China, 2017, pp. 947–950.
38. Slade P G, The Vacuum Interrupter: Theory, Design, and Application, Cleveland, OH, USA: CRC Press, 2008, pp. 1–2.
39. Dullni E, A vacuum circuit-breaker with permanent magnetic actuator for frequent operations, in Proc. 18th Int. Symp. Discharges Electr. Insul. Vacuum, 1998, pp. 688–691.

40. Cao Y D, Liu X M, Wang E Z, and Liu Z P, Design and research on novel permanent magnetic actuator in low voltage and high current vacuum circuit breaker, in Proc. 20th Int. Symp. Discharges Electr. Insul. Vacuum, 2002, pp. 471–474.
41. Lequesne B, Fast-acting, long-stroke solenoids with two springs, IEEE Trans. Ind. Appl., vol. 26, no. 5, pp. 848–856, Sep./Oct. 1990.
42. Ro J S, Hong S K, and Jung H K, Characteristic analysis and design of a novel permanent magnetic actuator for a vacuum circuit breaker, IET Electr. Power Appl., vol. 7, no. 2, pp. 87–96, 2013.
43. Cai Z Y, Ma S H, and Wang J M, An approach of improve permanent magnetic actuator of vacuum circuit breaker, in Proc. 23rd Int. Symp. Discharges Elect. Insul. Vacuum, 2008, pp. 165–168.
44. Kang J H, Kwak S Y, Kim R E, and Jung H K, The optimal design and dynamic characteristics analysis of electric actuator (EMFA) for 170 kV/50kA VCB based on three-link structure, in Proc. 23rd Int. Symp. Discharges Elect. Insul. Vacuum, 2008, pp. 177–180.
45. Wang Z X, Yan P, Geng Y S, and Yu L, Simulation of an improved operating method for vacuum circuit breakers with permanent magnetic actuators, Int. J. Appl. Electromagnet. Mech., vol. 33,Nos. 3–4, pp. 1373–1381, 2010.
46. Woo K I and Kwon B I, Characteristic analysis and modification of PM-type magnetic circuit breaker, IEEE Trans. Magn., vol. 40, no. 2, pp. 691–694, Mar. 2004.
47. Fang S, Heyun L, Ho S L, Wang X, Ping J, and Hengchuan L, Characteristics analysis and simulation of permanent magnet actuator with a new control method for air circuit breaker, IEEE Trans. Magn., vol. 45, no. 10, pp. 4566–4569, Oct. 2009.
48. Batdorff M A and Lumkes J H, High-fidelity magnetic equivalent circuit model for an axisymmetric electromagnetic actuator, IEEE Trans. Magn., vol. 45, no. 8, pp. 3064–3072, Aug. 2009.
49. Seok C J, Izui K, and Nishiwaki S, Multi-material optimization of magnetic devices using an Allen-Cahn equation, IEEE Trans. Magn., vol. 48, no. 11, pp. 3579–3582, Nov. 2012.
50. Sang-in P and Seungjae M, Design of magnetic actuator with nonlinear ferromagnetic materials using level-set based topology optimization, IEEE Trans. Magn., vol. 46, no. 2, pp. 618–621, Feb. 2010.
51. Faruque M O, Dinavahi V, Steurer M, Monti A, Strunz K, Martinez J A, et al., Interfacing issues in multi-domain simulation tools, IEEE Trans. Power Delivery, vol. 27, no. 1, pp. 439–448, Jan. 2012.
52. Srairi K and Feliachi M., Numerical coupling models for analyzing dynamic behaviors of electromagnetic actuators, IEEE Trans. Magn., vol. 34, no. 5, pp. 3608–3611, Sep. 1998.
53. Melgoza E and Rodger D., Comparison of table models of electromagnetic actuators, IEEE Trans. Magn., vol. 38, no. 2, pp. 953–956, Mar. 2002.
54. Paulke J, Weichert H, Steinhaeuser P, Thermal simulation of switchgear, IEEE Trans. Compon. Pack. Technol. 25 (3) (2002) 434–439.
55. Wang L J, Zheng W S, Wang L H, Electromagnetic-thermal-flow field coupling simulation of 12-kV medium-voltage switchgear, IEEE Trans. Compon., Pack. Manuf. Technol. 6 (8) (2016) 1208–1219.
56. Yu X, Liu Z, Chen Y, Feng Q, Heat transfer in a high-voltage vacuum circuit breaker, Proc. Inst. Mech. Eng., Part A: J. Power Energy 225 (8) (2011) 1099–1110, https://doi.org/10.1177/0957650911421060.
57. Lv Q, Yu X, Tan Y, Liu Z, Wang X, Structure improvement of a 126 kV vacuum circuit breaker using electromagnetic-thermal field coupling simulation, Applied Thermal Engineering 160 (2019) 114076.
58. Slade P G, Electrical Contacts: Principles and Applications. Boca Raton, FL, USA: CRC Press, 1999, pp. 12.
59. Liu S, Geng Y, Wang Z, Liu Z, Li H, Wang J, Li Y, Temperature-Rise Performance of 126 kV Single-Break Vacuum Circuit Breakers with 4 Types of AMF Contacts, 28th International Symposium on Discharges and Electrical Insulation in Vacuum (ISDEIV), Greifswald, Germany, 2018, pp. 547–550.

60. Wang Z, Li H, Shi D, Wu J, Xue C, Wang J, Liu Z, Design Optimization of a 3/4 Coil-Type Axial Magnetic Field Contact for 126 kV Vacuum Interrupter, IEEE International Conference on Electric Power Equipment-Switching Technology (ICEPE-ST), Xi'an, China 2017:170–174.
61. Li H, Wang Z, Han G, Li Y, Bi Y, Geng Y, Liu Y, Minimum Arcing Interruption Performance of a 126 kV Single-break Vacuum Circuit Breaker with 3/4 Coiltype Axial Magnetic Field Contacts, IEEE International Conference on Electric Power Equipment-Switching Technology (ICEPE-ST), Xi'an, China 2017:175–178.
62. Liu Z, Li H, Wang Z, Geng Y, Wang J, A kind of axial magnetic field contact vacuum interrupter with high nominal rated current. Chinese Patent. 201710419825.0, 2017.
63. China National Standards. Technical requirements for high voltage circuit breakers and control equipments. China National Standards, GB/T 11022-1999, 1999.
64. Slade P G et al., The development of a vacuum interrupter retrofit for the upgradeing and life extension of 121 kV–145 kV oil circuit breakers, IEEE Trans. Power Delivery, vol. 6, no. 3, pp. 1124–1131, Jul. 1991.
65. Schellekens H and Gaudart G, Compact high-voltage vacuum circuit breaker, a feasibility study, IEEE Trans. Dielectr. Electr. Insul., vol. 14, no. 3, pp. 613–619, Jun. 2007.
66. Saitoh H, Ichikawa H, Nishijima A, Matsui Y, Sakaki M, Honma A, Okubo H, Research and development on 145 kV/40 kA one break vacuum circuit breaker, in Proc. IEEE/PES Transmission and Distribution Conference and Exhibition, Vol.2, 2002, pp. 1465–1468.
67. Matsui Y, Nagatake K, Takeshita M, Katsumata K, Sano A, Ichikawa H, Saitohu H, Sakaki M, Development and Technology of High Voltage VCBs; Breaf History and State of Art, International Symposium on Discharges and Electrical Insulation in Vacuum, Matsue, Japan, 2006, pp. 253–256.
68. Heinz T, Helbig D, Teichmann J, Wethekam S, Prucker U, Konermann M, Engel M, Wolf C, 145 kV Vacuum Circuit Breaker and Clean Air Instrument Transformer—Performance, Installation-and Operational Experience, VDE High Voltage Technology 2018; ETG-Symposium, Berlin, Germany, 2018, pp. 1–6.
69. High-Voltage Alternating-Current Circuit-Breakers, GB standard 1984-2014, 2014.
70. High-Voltage Alternating-Current Vacuum Circuit-Breaker, JB/T standard 3855, 2008.

Chapter 5
Capacitive Current Switching in Vacuum

Capacitive current switching is energization or de-energization of no-load cables/overhead lines, capacitor banks, or filters in ac circuits at distribution or transmission voltages. The switching process may involve high voltage and high current transients. Capacitor banks are widely used for the compensation of reactive power in the power system; The capacitor banks are switched frequently, perhaps once or twice a day [1–3]. It shows that 60% of all capacitor banks are switched up to 300 times a year and a further 30% are switched up to 700 times a year [1]. Therefore, capacitive current switching is an essential operation for a reliable power grid and it is a crucial requirement for circuit breaker applications.

The vacuum circuit breaker offers the advantages of frequent switching operations, low life-cycle costs, excellent arc quenching capabilities, and nearly maintenance-free operation. These characteristics make vacuum circuit breaker suitable for frequent compensation of reactive power in the power system. However, the switching of reactive power supplied by capacitor banks proves to be a challenging task for vacuum circuit breakers. For example, restrikes are potentially dangerous as they could cause overvoltage during capacitive current switching duties [4–7]. A reliable capacitive current switching performance of vacuum circuit breakers is favorable to the power systems, especially for transmission voltage levels.

Capacitive current switching performance of vacuum circuit breakers is influenced by many factors, including prestriking and inrush current, contact design and magnetic field, contact materials, and contact conditioning et al. Dullni et al. [8] found that the restrike phenomenon was correlated with the prestrike process during the corresponding making operation. Körner et al. [9] showed that a pair of spiral type transverse magnetic field contacts had a better dielectric behavior than that of butt type contacts. Yokokura et al. [10] found that CuW contact material was the best material for the capacitive current switching duty compared with CuTeSe, CuBi, and CuCr. This chapter addresses four influencing factors of capacitive current switching in vacuum. The first one is prestriking phenomena and inrush current in making operations. The second one is the magnetic field of contacts. The third one is the contact

Z. Liu et al., *Switching Arc Phenomena in Transmission Voltage Level Vacuum Circuit Breakers*, https://doi.org/10.1007/978-981-16-1398-2_5

materials. The fourth one is a high-frequency high-voltage impulse conditioning technology. For general purpose vacuum circuit breakers, the capacitive current switching is quite difficult to achieve restrike free, especially when vacuum circuit breakers are approaching transmission voltage levels. By considering the four influencing factors, a special purpose vacuum circuit breaker for capacitive current switching is developed, which is a nonsynchronous double break type vacuum circuit breaker. Type test demonstrated that it has a good back-to-back capacitor bank switching performance.

5.1 Prestriking Phenomena and Inrush Current

Capacitive current switching operations consist of CO operations. During a making operation, the contacts of vacuum interrupter approach each other continually. Once the electric field strength exceeds the dielectric strength between the contacts, a prestrike arc occurs in the vacuum gap. Then an inrush current flows through the vacuum interrupter. At energization of a capacitor bank with one or more capacitor banks already in service ("back-to-back switching"), the already energized bank (or banks) provides a very large inrush current during the prestrike arc before contacts touch. The inrush current could reach 20-kA peak and 4250-Hz frequency according to IEC standard for back-to-back capacitor banks switching. The contact surfaces are possible to melt locally by the prestrike arc. At last the two contacts could be welded together as the contacts mate. During a subsequent breaking operation, the welding area is ruptured by an operating mechanism. Then protrusions, craters and particles could be generated on the contact surfaces. These damages may directly influence the dielectric strength of the vacuum gap. So, it is difficult for vacuum interrupters to withstand the following recovery voltage after current zero with the damaged contact surfaces. The peak values of the recovery voltage can reach up to 2 times or 2.5 times of the rated system voltage, which may result in occasional restrikes during the breaking operation. Thus, the inrush current in making operation can significantly affect the breaking performance in capacitive current switching.

5.1.1 Capacitive Current Switching of Vacuum Interrupters and Inrush Currents

Inrush currents have a significant influence on the capacitive current switching in vacuum. Kamikawaji et al. [11] revealed an influence of the inrush current on the restrike probability in a vacuum chamber. Doen et al. [12] found that the breakdown phenomena and field emission currents were influenced by the inrush current during capacitor switching. Juhsasz et al. [13] found that the restrike occurring frequency was dependent on the prestrike arc energy. Koochack Zadeh et al. [14] observed that the amplitude of the inrush current had a significant influence on the field emission

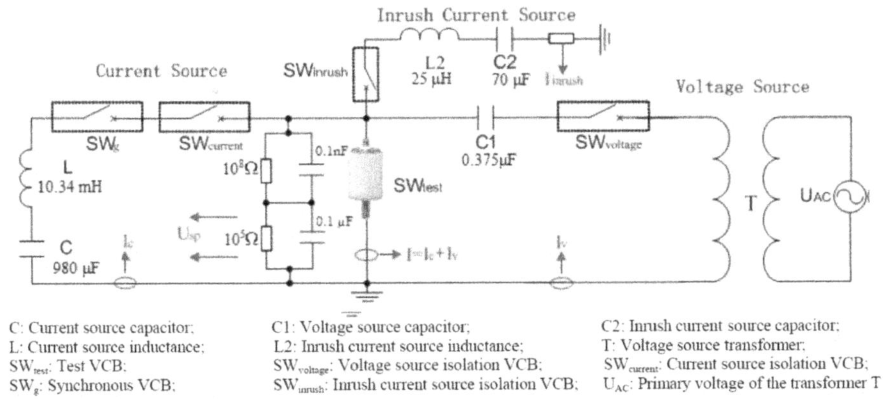

Fig. 5.1 A synthetic test circuit for capacitive current switching

current. Delachaux et al. [15] and Smeets et al. [16] found the inrush current interruption phenomena in vacuum interrupters during the prestrike process. However, the details of the prestrike and restrike characteristics in vacuum scattered in various literatures. This section experimentally determines an impact of the inrush current amplitude on the prestrike and restrike characteristics of vacuum interrupters during capacitive current switching under comparable conditions. The amplitudes of the inrush currents are set to four levels (0, 2, 5 and 10 kA). First, it investigates the influence of the inrush current amplitude on the prestrike gaps during making operations. Then, it analyzes the inrush current interruption phenomenon. Finally, it shows the restrike characteristics, including the restrike probability, the restrike type and the restrike occurring time.

Figure 5.1 shows a synthetic test circuit that is used to evaluate the capacitive current switching performance of vacuum interrupters. The test circuit is composed of three sub-circuits: an inrush current source circuit, a current source circuit and a voltage source circuit. The inrush current source circuit supplies an inrush current I_{inrush} up to a peak value of 20 kA at a frequency of 3800 Hz by discharging a capacitor C2 through an inductance L2 after closing a switch SW_{inrush}. The current source circuit provides a power frequency current I_c up to 2 kA (50 Hz) by discharging a capacitor C through an inductance L after closing a switch SW_{current} and a switch SW_g. The voltage source circuit produces a recovery voltage of $(1-\cos\omega t)$ wave shape through a superposition of a cosine waveform generated by a high voltage transformer T to a DC voltage generated by a capacitor C1.

The test operation is composed of a making operation and a breaking operation. Before starting the test operation, the inrush current source capacitor C2 and the power frequency current source capacitor C were pre-charged to a required voltage. When the phase angle of U_{AC} was zero, a signal triggered a sequence of operations by controlling the switches. The operation sequence was as following: First, the switches SW_{inrush} and SW_{test} were closed in sequence. I_{inrush} flowed through a test vacuum interrupter. Next, the switches $SW_{current}$ and $SW_{voltage}$ were closed. The voltage source current I_v flowed through the test vacuum interrupter. Then, the switch SW_{inrush} was opened to isolate the inrush current source circuit from the other two circuits. Thereafter, the switch SW_g closed to synchronize the power frequency currents I_c and I_v in the same phase. Finally, the switches SW_{test} and $SW_{current}$ were opened. When the switching current I $(= I_c + I_v)$ was interrupted, the recovery voltage was applied to the test vacuum interrupter.

In the test, eleven 7.2 kV vacuum interrupters were used. The test vacuum interrupters have not been conditioned by any electrical methods before, such as spark conditioning and current conditioning. Butt type contact was used. The contact diameter was 30 mm. The contact material was CuCr25 (25% weight of Cr). The rated contact gap d was 4 mm. Moreover, a permanent magnetic operating mechanism was adopted. The average opening velocity was 0.8 m/s from the closing position to 75% of d. The average closing velocity was 0.8 m/s from 2/3 d to the closing position.

Figure 5.2 shows typical oscillograms of making and breaking operations. Figure 5.2a shows a making operation. The upper trace was the contact travel. The lower trace was the high frequency inrush current. Figure 5.2b shows a breaking operation. The upper trace was a recovery voltage. The lower trace was the current. It successfully interrupted the current without any restrike. This was indicated by the recovery voltage applied to the test vacuum interrupter.

To evaluate the capacitive current switching performance of the test vacuum interrupters, a series of 80 CO operations were executed for each test vacuum interrupter. The details of test conditions are shown in Table 5.1. Four levels of the inrush current amplitudes of 0 kA (no inrush current), 2, 5 and 10 kA corresponded to four test series a, b, c and d, respectively. Their corresponding pre-charged voltages of C2 were 0, 1.2, 3 and 6 kV, respectively. In test series a–c, three test vacuum interrupters were used in each test series. In test series d, only two test vacuum interrupters were used because of contact welding. One test vacuum interrupter cannot be opened by the operating mechanism after five test operations and the second test vacuum interrupter cannot be opened after three test operations. The switching current was less than 1 A, which was considered as a rigorous test condition. In contrast, the switching current with hundreds of amperes would have a conditioning effect. Moreover, the arcing time was chosen close to 0 ms to simulate the breaking operations with short arcing time. The peak value of the recovery voltage was about 15 kV (2 U_{peak}). The full period of recovery voltage application was 500 ms.

The prestrike gap d_{pre} is an instantaneous contact gap which is measured at the occurring of the first prestrike (see Fig. 5.2a). In test series a, the inrush current was 0 kA. So the making operation was made as a no-load operation. The prestrike phenomenon does not exist. In test series d, the inrush current was 10 kA. The contact welding became so serious that the distribution of d_{pre} cannot be accounted. Figures 5.3 and 5.4 show the distribution of the prestrike gap d_{pre} in test series b and test series c, respectively. Figure 5.5 shows the cumulative probability distribution of the prestrike gap d_{pre} given in Figs. 5.3 and 5.4. It shows that the cumulative probability distribution followed a 3-parameter Weibull distribution, as shown in Eq. (5.1).

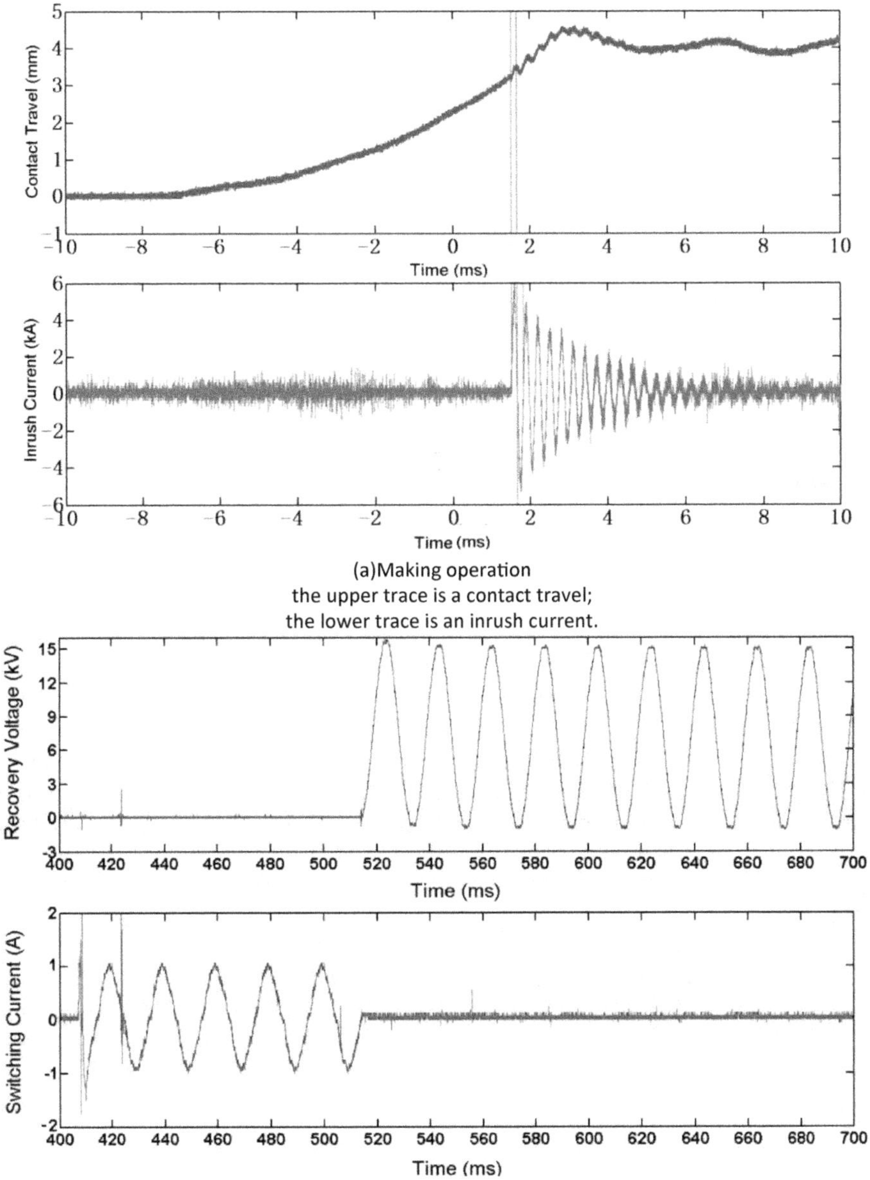

(a)Making operation
the upper trace is a contact travel;
the lower trace is an inrush current.

(b)Breaking operation
the upper trace is a recovery voltage;
the lower trace is a switching current.

Fig. 5.2 Typical oscillograms of making and breaking operations

Table 5.1 Conditions for capacitive current switching test

Inrush current	Series a	Series b	Series c	Series d
	0 kA	2 kA	5 kA	10 kA
	Frequency: 3800 Hz			
Switching current	<1 A			
Recovery voltage	$U_{trv} = U_{peak} \times (1 - \cos\omega t)$			
	$U_{peak} = \frac{7.2kV \times \sqrt{2}}{\sqrt{3}} \times 1.25 = 7.35kV$			
	Duration: 500 ms			
Test duty	C–O: 80 times			
Test VI	VI_No01	VI_No21	VI_No51	VI_No101
	VI_No02	VI_No22	VI_No52	VI_No102
	VI_No03	VI_No23	VI_No53	

$$F(d_{pre}) = 1 - \exp\left[-\left(\frac{d_{pre} - d_o}{\Theta - d_o}\right)^m\right] \quad (5.1)$$

where F—The cumulative probability; d_0—The location parameter; m—The shape parameter; Θ—The scale parameter.

Table 5.2 shows the values of the Weibull parameters. The R-squared is used to indicate how well the fitting degree between F and d_{pre} for the 3-Parameter Weibull distribution. R-squared is in the range of 0 and 1. When it is approaching 1 it is well fitting. As shown in Table 5.2, all values of R-squared are above 0.98. Therefore, it is reasonable to assume that the cumulative probability of the d_{pre} followed the 3-parameter Weibull distribution. Furthermore, the scattering of the prestrike gap can be evaluated by the scale ($\Theta - d_0$). With increase of the inrush current from 2 kA in test series b to 5 kA in test series c, the average scale ($\Theta - d_0$) increased from 0.2 to 0.6 mm.

Table 5.3 shows the 10% prestrike gap d_{10}, 50% prestrike gap d_{50} and 90% prestrike gap d_{90}, which can also be obtained from Eq. (5.1). The average values of d_{10}, d_{50} and d_{90} in test c ($I_{inrush} = 5$ kA) were 0.7, 1.0 and 1.5 mm, respectively. They were 40.0, 42.9 and 50.0% higher than that in test b ($I_{inrush} = 2$ kA), respectively.

The inrush current was interrupted frequently during the prestrike processes both in test series b and test series c. This phenomenon caused a short period of no inrush current passing during the prestrike process. Figure 5.6 shows two typical oscillograms of the inrush current interruption. Figure 5.6a shows that the inrush current is interrupted once. The corresponding occurring contact gap was 1.2 mm. The duration of the inrush current interruption was 255 μs. Then, the inrush current reignited when the two contacts were approaching each other further. When the prestrike gap was large, the inrush current may be interrupted twice or more, as shown in Fig. 5.6b. It shows that, when the inrush current was interrupted at first current zero, the corresponding occurring contact gap was 1.9 mm. After a duration

Fig. 5.3 Prestrike gap d_{pre} versus number of making operations in test series b for VI_No21 to VI_No23

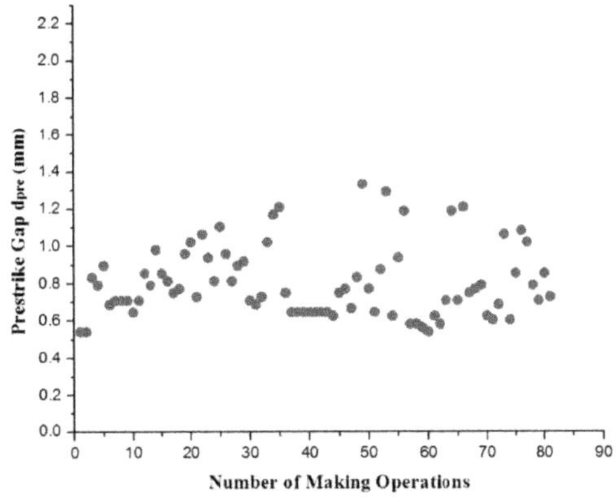

(a) VI_No21

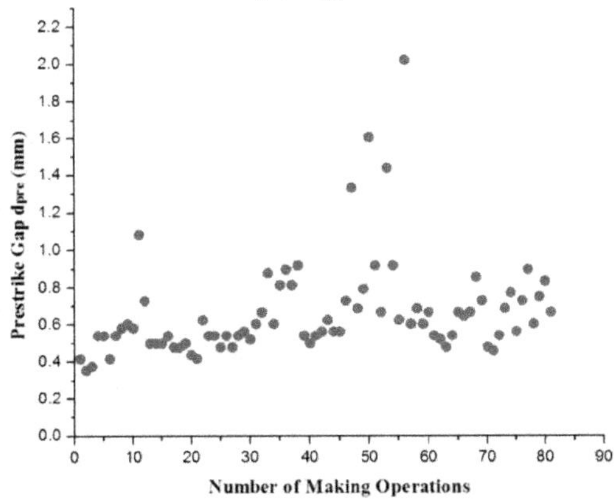

(b) VI_No22

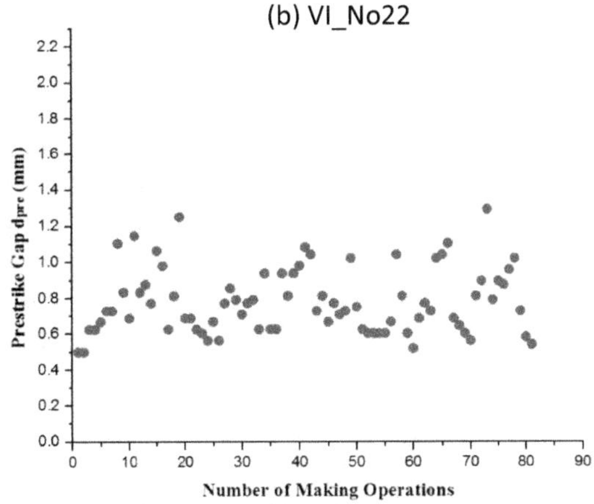

(c) VI_No23

Fig. 5.4 Prestrike gap d_{pre} versus number of making operations in test series c for VI_No51 to VI_No53

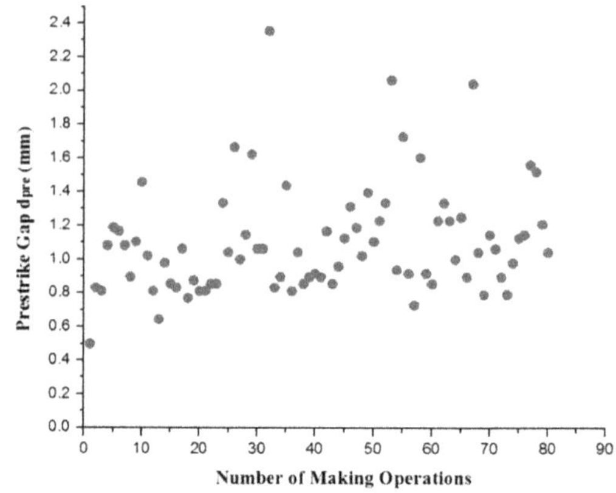

(a) VI_No51

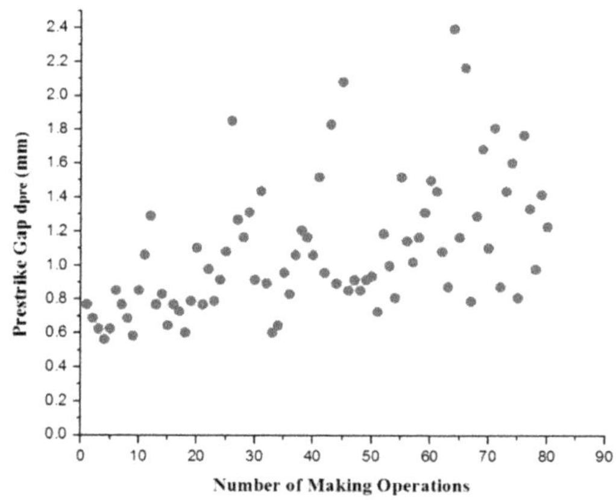

(b) VI_No52

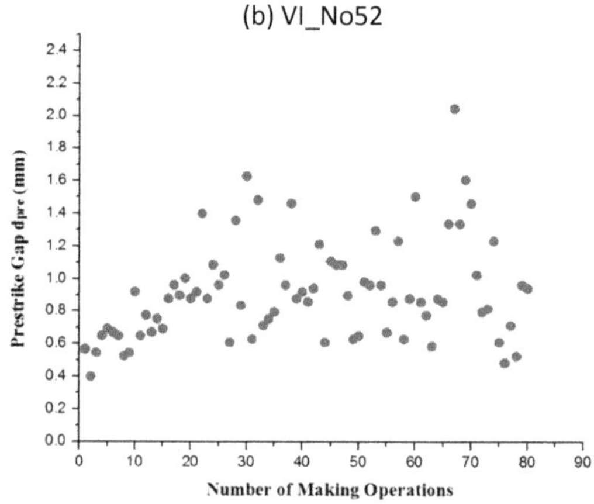

(c) VI_No53

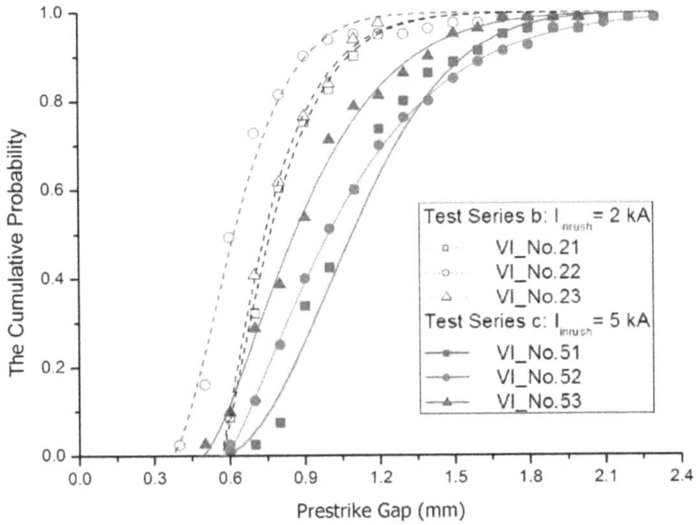

Fig. 5.5 Cumulative probability of prestrike gaps for test series b (VI_No.21–VI_No.23) and test series c (VI_No.51–VI_No.53)

Table 5.2 Values of the Weibull parameters for the 3-parameter Weibull distribution

Test VI	m	d_0 (mm)	Θ (mm)	$\Theta - d_0$ (mm)	R-squared
No.21	1.3	0.6	0.8	0.2	0.996
No.22	1.6	0.4	0.7	0.3	0.988
No.23	1.2	0.6	0.8	0.2	0.998
No.51	2.0	0.6	1.2	0.7	0.981
No.52	1.3	0.6	1.1	0.5	0.999
No.53	1.4	0.5	1.0	0.5	0.996

Table 5.3 10% prestrike gap d_{10}, 50% prestrike gap d_{50} and 90% prestrike gap d_{90} in Test Series b and Test Series c

Test series	Test VI	d_{10} (mm)	d_{50} (mm)	d_{90} (mm)
b $I_{inrush} = 2$ kA	No.21	0.6	0.8	1.1
	No.22	0.4	0.6	0.9
	No.23	0.6	0.7	1.1
Average value		0.5	0.7	1.0
c $I_{inrush} = 5$ kA	No.51	0.8	1.1	1.5
	No.52	0.7	1.0	1.6
	No.53	0.6	0.9	1.3
Average value		0.7	1.0	1.5

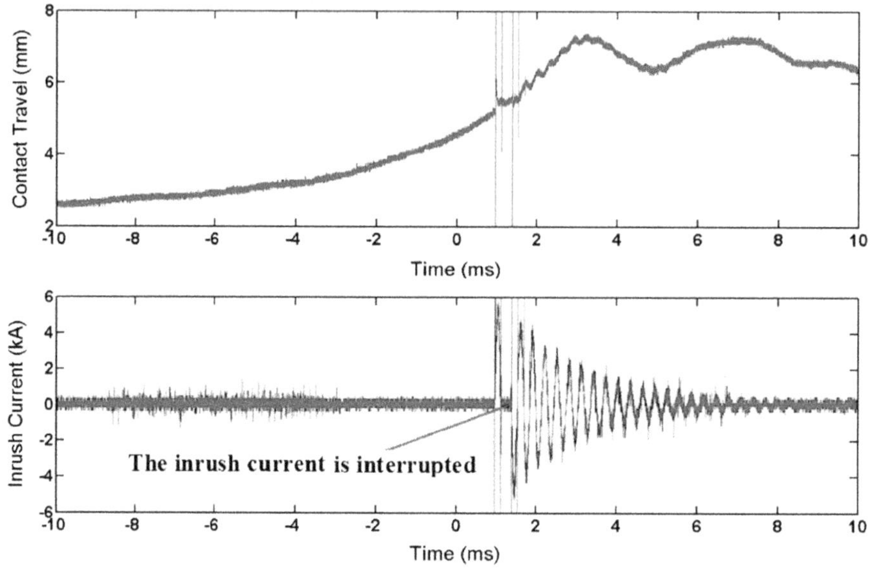

(a) The inrush current is interrupted once

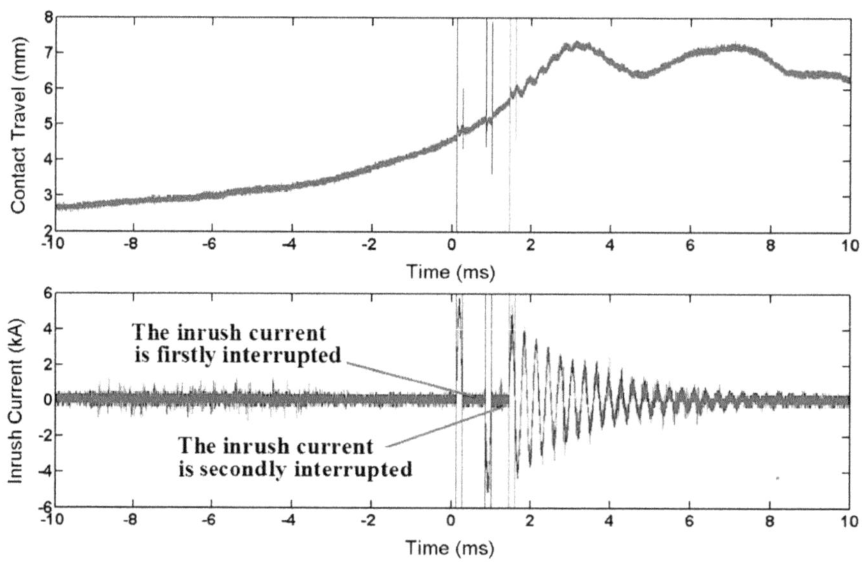

(b) The inrush current is interrupted twice

Fig. 5.6 Typical oscillograms of the inrush current interruption phenomena during making operations of 570 μs, the inrush current continued to flow through the contact gap. The second current interruption occurred at the contact gap of 1.5 mm. The duration of the second inrush current interruption lasted 430 μs.

Figure 5.7 shows a relationship between the inrush current interruption and the prestrike gap d_{pre} in test series b and test series c. It shows that the inrush current tended to be interrupted at a higher prestrike gap. Moreover, the occurrence frequency of the inrush current interruption in the case of $I_{inrush} = 5$ kA was 68.3% (test series c), which was 37.9% higher than that in the case of $I_{inrush} = 2$ kA (test series b).

Figure 5.8 shows a relationship between the duration of the inrush current interruption and the prestrike gap d_{pre} in test series b and test series c, respectively. It found that the duration of the inrush current interruption became longer at a higher prestrike gap both in test series b and test series c. Figure 5.8a shows that the duration of the inrush current interruption has a 99% possibility to be lower than 400 μs in test series b. It is also indicated a possibility of 20.7% of the inrush currents interruption for more than once. Figure 5.8b shows that the duration of the inrush current interruption has a 99% possibility to be lower than 800 μs in test series c. For test series c, it indicated a possibility of 26.2% of the inrush currents interruption for more than once.

Restrikes and non-sustained disruptive discharges (NSDDs) are two significant phenomena when vacuum interrupters are withstanding the recovery voltage during capacitive current switching. In the power system, restrikes are defined as a resumption of the current, whereas the NSDDs do not result in the current in the main load. The restrike is referred to the breakdown which occurs after (1/4)·T of the recovery voltage (Here $T = 20$ ms, which is a cycle of the recovery voltage). While, the breakdown occurs before (1/4)·T of the recovery voltage is referred as a reignition. Here, the capacitive current switching tests were executed by the synthetic test circuit. Both of the restrikes and NSDDs were regarded as the restrikes, which followed that of Delachaux et al. [15].

It found that the inrush currents have an influence on the features of the restrikes. Figure 5.9 shows three types of the restrikes in vacuum interrupters. Figure 5.9a shows restrike type I. It shows that one restrike occurred in the first cycle and another restrike occurred in the seventh cycle of the recovery voltage. This type of the restrike seemed to be a “disturbance” to the recovery voltage. There was no influence on the next cycle of the voltage waveform. When this type of the restrike occurred, there was a high frequency current appearing at the instant. Figure 5.9b shows restrike type II. The restrike occurred in the first cycle of the recovery voltage, which influenced the voltage waveform of the next cycle. The restrike current consisted of a high frequency current and a power frequency current. The power frequency current established after the high frequency current. Figure 5.9c shows restrike type III. The restrike occurred at the crest of the third circle of the recovery voltage with a following voltage reversal. There were also several restrikes occurring later. It seemed to be a “full breakdown” under the recovery voltage. The amplitude of the restrike current was higher than that in the type II.

Table 5.4 shows the occurrence frequency of the three restrike types. It found that the inrush current has a significant influence on the distribution probability of

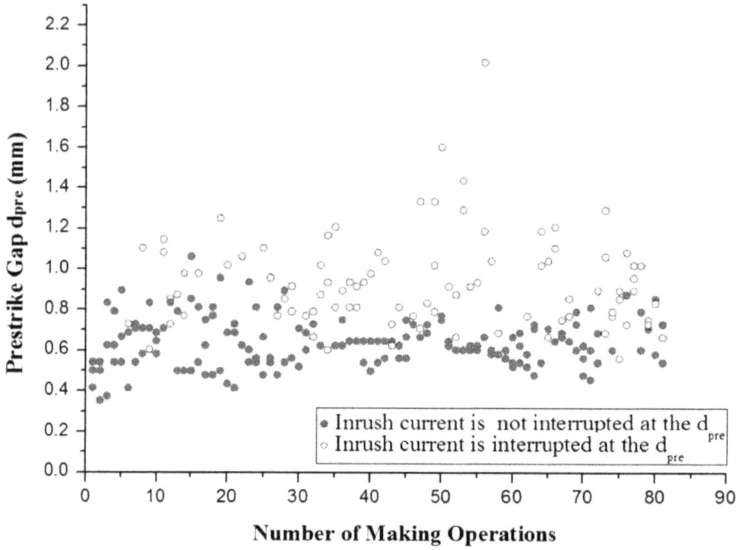

(a) Test series b

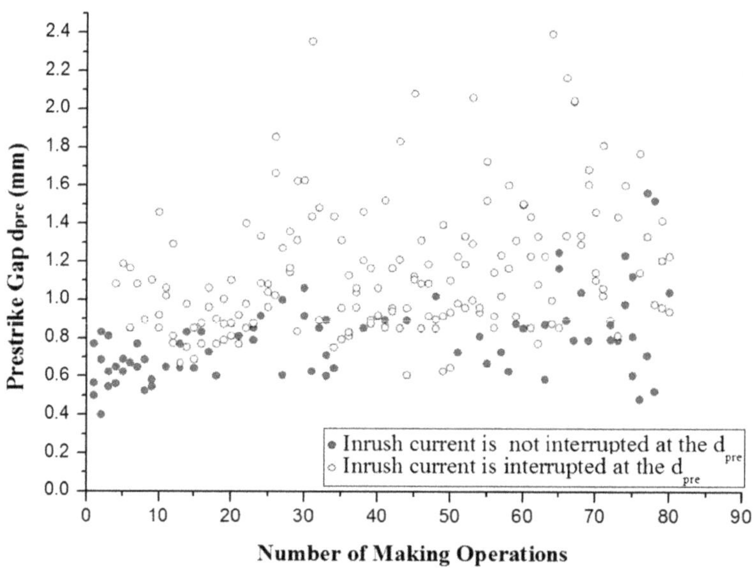

(b) Test series c

Fig. 5.7 Relationship between the inrush current interruption and the prestrike gap in test series b and test series c

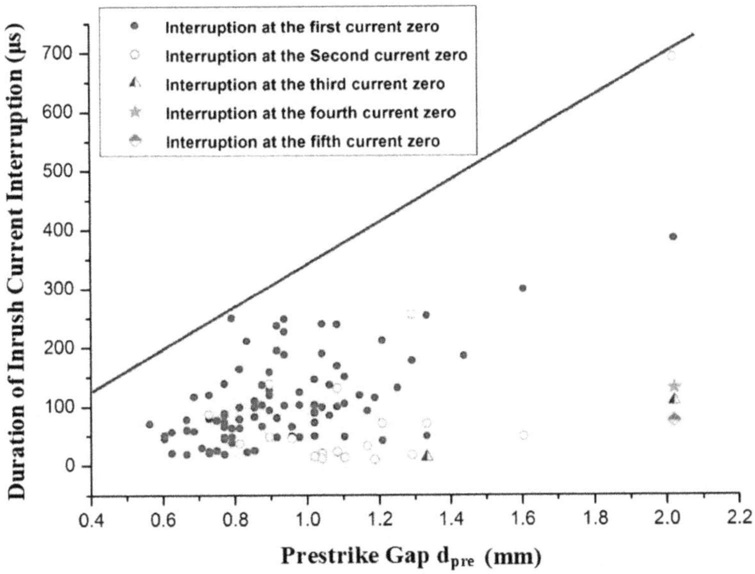

(a) Test series b

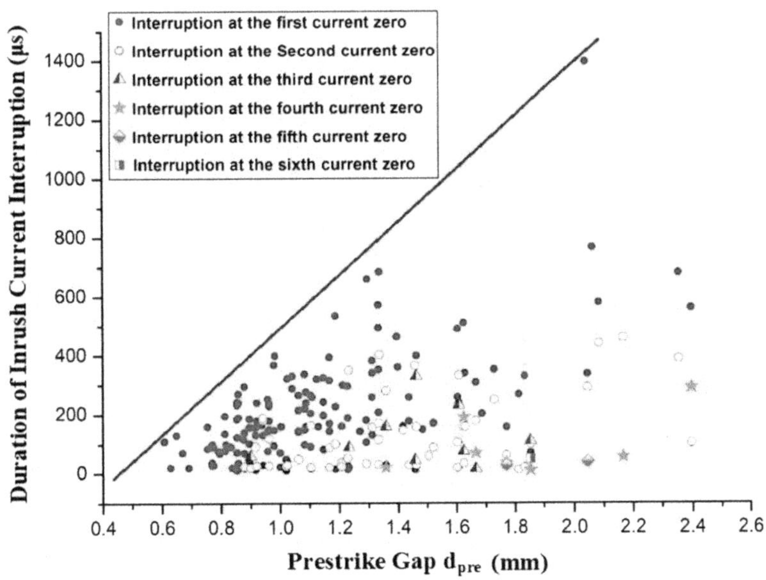

(b) Test series c

Fig. 5.8 Duration of inrush current interruption versus prestrike gaps

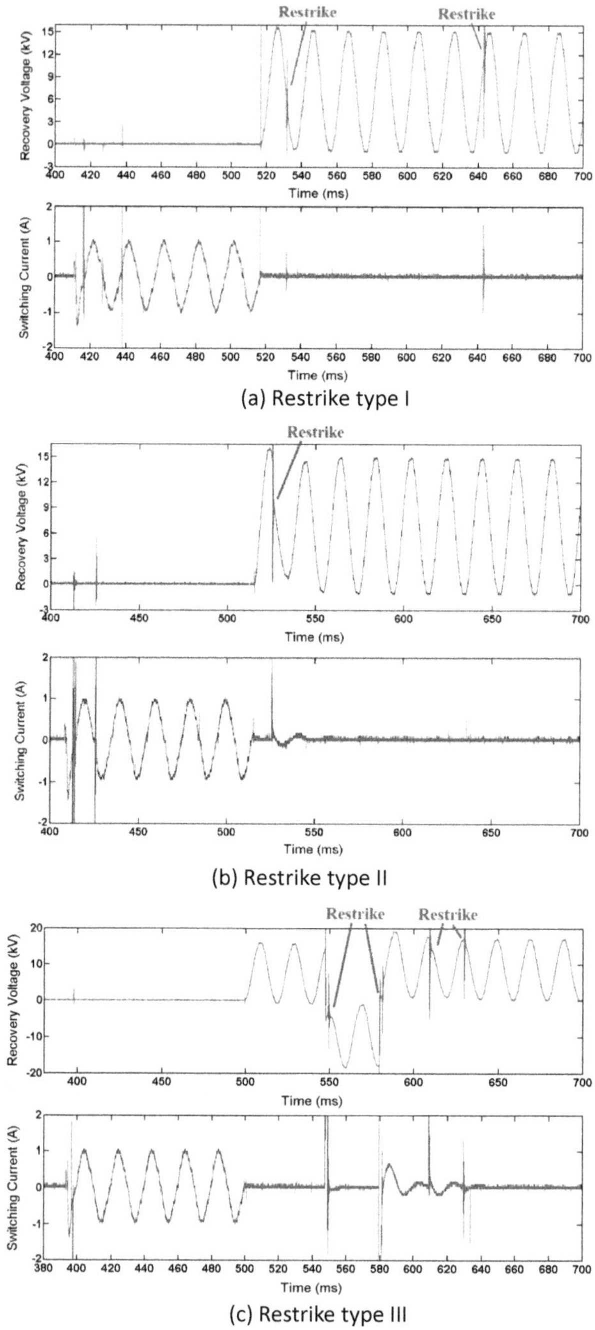

Fig. 5.9 Three types of restrikes in vacuum interrupters

Table 5.4 Occurrence frequency of the three restrike types

Test series	Restrike type I (%)	Restrike type II (%)	Restrike type III (%)
a $I_{inrush} = 0$ kA	75.0	16.7	8.3
b $I_{inrush} = 2$ kA	62.8	25.6	11.6
c $I_{inrush} = 5$ kA	20.5	20.5	59.0

Table 5.5 Restrike probability for test series a, b, c and d

Test series	Test VI	Restrike probability	Average value
a $I_{inrush} = 0$ kA	No.01 No.02 No.03	3.8% 2.5% 7.5%	4.6%
b $I_{inrush} = 2$ kA	No.21 No.22 No.23	13.8% 20.0% 20.0%	17.9%
c $I_{inrush} = 5$ kA	No.51 No.52 No.53	41.3% 28.8% 21.3%	30.4%
d $I_{inrush} = 10$ kA	No.101 No.102	Welding	

the restrike types. When the inrush current increased from 0 to 2 kA and 5 kA, the occurrence frequency of restrike type I decreased from 75.0 to 62.8% and further to 20.5% respectively. In contrast, the occurrence frequency of restrike type III increased from 8.3 to 11.6% and further to 59.0%, respectively.

Table 5.5 shows the results of the restrike probabilities of the test vacuum interrupters in test series a, b, c and d, respectively. It is obviously found that the restrike probability was closely related to the amplitude of the inrush current. The higher inrush current corresponded to more restrike phenomena. In test series a, where $I_{inrush} = 0$ kA, the average restrike probability was 4.6%. In test series b, where $I_{inrush} = 2$ kA, the average restrike probability was 17.9%. When I_{inrush} increased to 5 kA in test series c, the average restrike probability reached 30.4%. It was 1.7 times of that in test series b. In test series d, the restrike probability cannot be calculated as only a few operations have been executed because of contact welding of the test vacuum interrupters.

Figure 5.10 shows the occurring time distribution of the reignition and the restrike in test series a, b and c. T (= 20 ms) is a cycle of the recovery voltage. The zero point is the instant when the recovery voltage applied to the test vacuum interrupter. It is found that the occurring time was strongly influenced by the inrush current amplitude. Figure 5.10a shows that a majority of the restrikes occurred after 5·T in the case of no inrush current applied. Figure 5.10b shows that the restrike occurring time mainly

Fig. 5.10 Occurring time distribution of the ignition and restrike under different amplitudes of the inrush currents

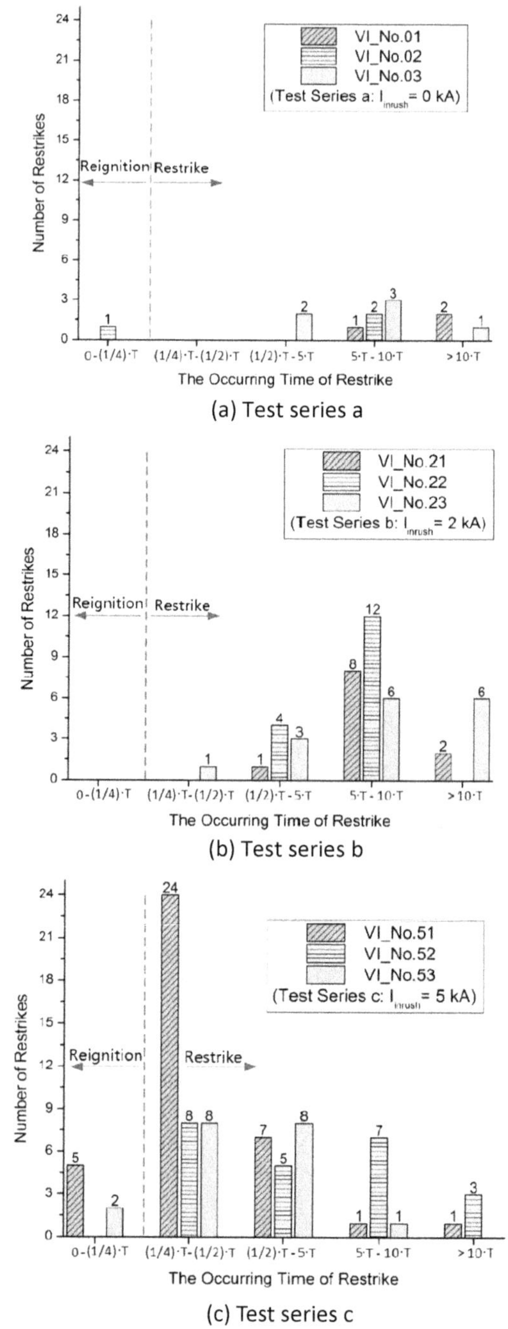

distributed between from 5 to 10 T. However, there were a few restrikes occurring before 5 T when I_{inrush} was 2 kA. Figure 5.10c shows that a majority of restrikes occurred before (1/2)·T when I_{inrush} increased to 5 kA. The restrike occurred earlier with the increase of the inrush current. Moreover, the reignitions became significant as I_{inrush} reached 5 kA.

The contact surfaces were seriously damaged by the prestrike arcs of inrush current. When the inrush current was 0 kA in test series a, there was no damage found on the contact surfaces. Figures 5.11 and 5.12 show the contact surfaces after 80 CO operations in test series b (VI_No21, 22 and 23) and in test series c (VI_No51, 52 and 53), respectively. Each test vacuum interrupter showed damaged contact surfaces with a macroscopic protrusion or crater. It suggested that virtually all the prestrike

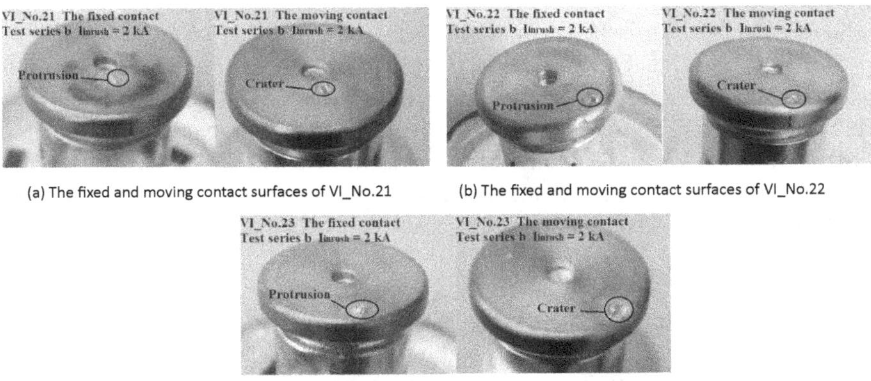

(a) The fixed and moving contact surfaces of VI_No.21 (b) The fixed and moving contact surfaces of VI_No.22

(c) The fixed and moving contact surfaces of VI_No.23

Fig. 5.11 Photos of the contact surfaces after 80 CO operations in test series b with $I_{inrush} = 2$ kA

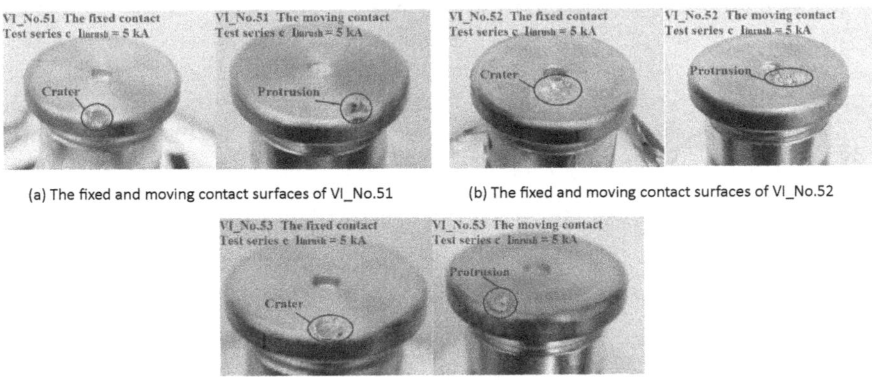

(a) The fixed and moving contact surfaces of VI_No.51 (b) The fixed and moving contact surfaces of VI_No.52

(c) The fixed and moving contact surfaces of VI_No.53

Fig. 5.12 Photos of the contact surfaces after 80 operations in test series c of $I_{inrush} = 5$ kA phenomena occurred in the same area of the contact surface during the making operations. Meanwhile, a crater corresponding to the macroscopic protrusion formed on the opposite contact surface. The macroscopic protrusions in test series c were not only significantly higher but also larger in size than that in test series b. In test series d, the contact surface condition was not analyzed as only a few CO operations have been applied.

Figure 5.13 shows the protrusion height on one contact surface and the crater area on the opposite contact surface after the 80 CO operations. Figure 5.13a shows the protrusion height. It shows that in test series b, the heights of the protrusions were 0.3, 0.7 and 1.1 mm for VI_No21, VI_No22 and VI_No23, respectively. In test series c, the heights of the protrusions were 2.4, 1.3 and 2.2 mm for VI_No51, VI_No52 and VI_No53, respectively. So the average height of protrusions was 2.0 mm in test series c. It was 2.8 times of that in test series b. Figure 5.13b shows the crater area. It shows that in test series b, the areas of the craters were 7.5, 10.6 and 11.3 mm^2 for VI_No21, VI_No22 and VI_No23, respectively. So the average area was 9.8 mm^2 in test series b. In test series c, the areas of the craters were 17.0, 31.0 and 22.9 mm^2 for VI_No51, VI_No52 and VI_No53, respectively. So the average area was 23.6 mm^2 in test series c, which was 2.4 times of that in test series b. The results suggested the damages on the contact surfaces had a strong dependence on the amplitude of the inrush current. The damages on the contact surfaces became larger with the increase of inrush currents.

The prestrike process plays an important role in capacitive current switching, as it can influence the restrikes of vacuum interrupters. When an inrush current flows through a vacuum interrupter during a making operation, a prestrike arc melts a local area of the contact surface. Then the two contacts could be welded after they finally mate each other. During a following breaking operation, the welding area would be ruptured. The rupture of the welding could cause serious damages on the contact surfaces (see Figs. 5.11 and 5.12). The next prestrike process may be influenced by the deterioration of the contact surface condition. It means that the prestrike could be determined by the previous history of the contacts [9]. Thus, the prestrike gaps d_{pre} fluctuated with the number of the inrush current application in each test series because of the deteriorating conditions of the contact surfaces. When I_{inrush} increased from 2 to 5 kA, the prestrike gap increased accordingly. The d_{10} increased from 0.5 to 0.7 mm. The d_{50} increased from 0.7 to 1.0 mm. The d_{90} increased from 1.0 to 1.5 mm. It is obvious that the higher pre-charge voltage of C2 resulted in the higher prestrike gap. Besides the pre-charge voltage, the contact surface condition could also influence the prestrike gap. It found that more significant damages (higher protrusions and larger size craters) generated on the contact surfaces by higher amplitude of the inrush current. It suggests that the high macroscopic protrusion causes the early prestrikes. Furthermore, the scattering of the prestrike gap in test series c was larger than that in test series b. It is strongly related to the deteriorating conditions of the contact surfaces.

During the prestrike process, it observed that the inrush current can be interrupted frequently at current zeros. When the contacts approach each other further, the inrush current can be established again. So there is a short no current period during the

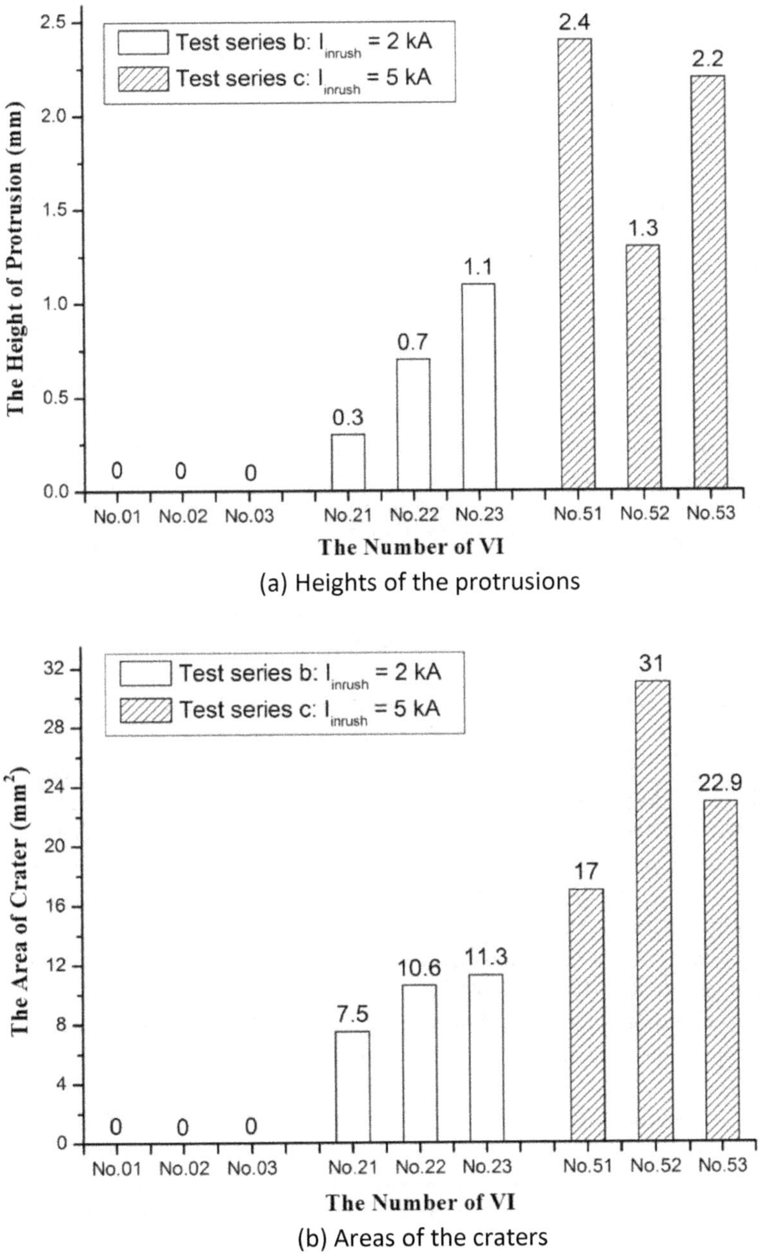

Fig. 5.13 The protrusion height on one contact surface and the crater area on the opposite contact surface after the 80 CO operations

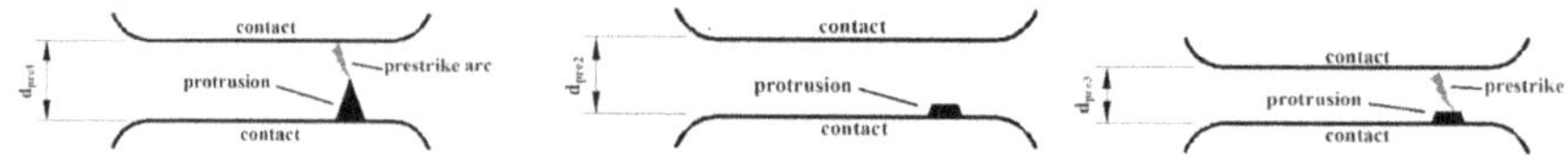

Fig. 5.14 A proposed physical process of the prestrike process: **a** a prestrike occurs on a protrusion at d_{pre1}; **b** the protrusion is eroded and the inrush current is interrupted at current zero as the dielectric strength of the vacuum interrupter increases ($d_{\text{pre2}} \approx d_{\text{pre1}}$); **c** the inrush current reignites when the moving contact approaches to the fixed one more closely ($d_{\text{pre3}} < d_{\text{pre2}}$)

prestrike process. This phenomenon is also found by Smeets et al. [16] and Delachaux et al. [15]. One can find that the amplitude of the inrush current has an influence on the inrush current interruption phenomenon. The occurrence frequency of this phenomenon in the case of $I_{\text{inrush}} = 5$ kA was 68.3%, which was higher than that of 37.9% in the case of $I_{\text{inrush}} = 2$ kA. The duration of the inrush current interruption increased with the increase of I_{inrush}. The duration of the inrush current interruption has a 99% possibility to be lower than 400 μs at $I_{\text{inrush}} = 2$ kA and to it was lower than 800 μs at $I_{\text{inrush}} = 5$ kA. Moreover, the inrush current can be interrupted more than once at higher prestrike gaps.

An obvious reason, resulting in this inrush current interruption phenomenon, is an inherent property of "vacuum" that is capable of interrupting current of very high di/dt [16]. However, the phenomenon of the inrush current interruption may have a close relationship with contact surface conditions. As is shown in Fig. 5.8, the inrush current cannot be interrupted in the first few operations both in test series b and test series c. The inrush current interruption phenomena usually occurred at the higher prestrike gaps. Furthermore, the inrush current can be interrupted twice or more at high prestrike gaps, which is probably correlated with high macroscopic protrusions. So, the conditions of the contact surfaces before the prestrike occurs would play an important role on the inrush current interruption phenomenon.

We propose that the macroscopic protrusion generated by the previous applications of inrush current may be a main reason that results in the inrush current interruption phenomenon. Figure 5.14 shows a proposed physical process of a prestrike during a making operation. As shown in Fig. 5.14a, when a prestrike occurs at d_{pre1}, a prestrike arc establishes on a macroscopic protrusion generated by the previous making operations. As the frequency of the inrush current is quite high (3800 Hz), the moving contact only moves several hundreds of micrometers at the first inrush current zero when the contact gap reaches to d_{pre2}, as shown in Fig. 5.14b. So, it means that d_{pre2} is approximately equal to d_{pre1}. But the height of the macroscopic protrusion is in the order of several millimeters. The protrusion would be eroded by the prestrike arc significantly. Then the dielectric strength of the vacuum interrupter increases to withstand the applied voltage. So the inrush current can be interrupted at an inrush current zero. Figure 5.14c shows that the inrush current reignites a moment later when the moving contact approaches to the fixed one more closely (d_{pre3} is much smaller than d_{pre2}).

In the same way, the inrush current can be interrupted twice or more during the prestrike process. According to the above proposed mechanism, the inrush current interruption phenomenon would occur more frequently with a slower closing speed. So, it can explain the experimental results of Delachaux et al. [15] well, where a slower closing speed led to inrush current interruption phenomenon occurring more frequently. The mechanism can also predict that the inrush current interruption phenomenon occurs more frequently with a higher frequency inrush current. Considering that the higher inrush current causes higher protrusion height on the contact surface (see Fig. 5.13), the inrush current interruption phenomenon occurs more frequently at $I_{\text{inrush}} = 5$ kA than $I_{\text{inrush}} = 2$ kA. More research work indicated that the vacuum interrupters tended to interrupt the inrush current at each inrush current zero. If the contact gap can not withstand the "recovery voltage" at the current zero when the inrush current is interrupted, a breakdown occurs and the inrush current continues to flow. Otherwise, the inrush current is interrupted. More detail about inrush current interruption can be found in Sect. 5.1.5.

The damages on the contact surfaces exhibit strong dependence on the inrush current amplitude. There are no arc erosions on the contact surfaces when I_{inrush} is 0 kA. The protrusion and crater are generated from the rupture of the contact welding caused by the inrush current. As shown in Fig. 5.13, when I_{inrush} increases from 2 to 5 kA, the average height of the protrusion increases from 0.7 to 2.0 mm, and the average area of the crater increases from 9.8 to 23.6 mm^2. When I_{inrush} increases to 10 kA, the vacuum interrupters cannot be opened because of the serious contact welding.

With the increase of the inrush current amplitude, the prestrike arc energies injected into the contact surfaces become higher during the making operations. So, the prestrike arc with the higher energy makes more severe contact welding. The pre-charged energy in C2 is 193 J for the case of $I_{\text{inrush}} = 5$ kA, while the pre-charged energy is only 31 J for the case of $I_{\text{inrush}} = 2$ kA. Therefore, the prestrike gaps of $I_{\text{inrush}} = 5$ kA are higher than that of $I_{\text{inrush}} = 2$ kA, as shown in Figs. 5.3 and 5.4. When the prestrike occurs at a high contact gap, the prestrike arc can last for a long period of time. The inrush current can provide more energy to contact surface. Thus, more serious damages generate on the contact surfaces in the case of $I_{\text{inrush}} = 5$ kA than that in the case of $I_{\text{inrush}} = 2$ kA.

It is obvious that the dielectric strengths of vacuum interrupters are strongly depended on the conditions of the contact surfaces in capacitive current switching. The contact surfaces can be damaged by the prestriking arcs of the inrush currents. The average heights of the protrusions reach 0.7 mm ($I_{\text{inrush}} = 2$ kA) and 2.0 mm ($I_{\text{inrush}} = 5$ kA), respectively, which are comparable to the rated contact gap of 4 mm. So, the macroscopic protrusion can significantly influence the electric field distribution in contact gap. The restrike occurs if this contact gap cannot withstand the recovery voltage. Moreover, the crater and the particles generated by the rupture of the contact welding can also lead to restrikes. As the damages generated on the contact surfaces become more serious in the case of $I_{\text{inrush}} = 5$ kA, its average restrike probability reaches to 30.4%, which is 1.7 times of that in the case of $I_{\text{inrush}} = 2$ kA and 6.6 times of that in the case of $I_{\text{inrush}} = 0$ kA.

It is well known that the electric field between two contacts is not determined by U/d, where U is the voltage applied across a contact gap d, but it is given by $\beta U/d$. The β is an enhancement factor which is composed of a microscopic enhancement factor β_m and a geometric enhancement factor β_g. It is observed that the amplitude of the inrush current has a significant influence on the field emission current [12, 14, 16]. In the case of a higher inrush current, a higher field emission current was observed after the capacitive current [14]. It is obviously that the enhancement factor β would increase significantly due to the growing damage generated on the contact surfaces. Therefore, the electric field on the protrusion could increase significantly. The high electric field may cause a strong field emission under the recovery voltage. The field emission current may have an influence on the occurring time of the restrike. It shows that the restrikes almost occur later than 5 T (T = 20 ms) in the case of $I_{inrush} = 0$ kA. With the increase of the inrush current amplitude, the electric field may increase significantly because of the damages on the contact surfaces. So, the field emission current becomes higher to generate restrikes that occur earlier during the recovery voltage period. In the case of $I_{inrush} = 2$ kA, the occurring time of restrikes mainly resides from 5·T to 10·T. When I_{inrush} reaches up to 5 kA, the restrikes mainly occur before (1/2)·T and the reignitions become more obvious.

Additionally, it is found that the inrush current amplitude has a significant influence on the restrike type. When I_{inrush} is 0 kA, 75.0% of the restrikes belong to restrike type I. Restrike type I seems to be a "disturbance" to the recovery voltage. It has no influence on the next cycle of the voltage waveform. In this case, the field emission current should be relatively low as no arc damage is generated on the contact surfaces. So, the main reason leading to restrike type I may be due to the original particles inside the vacuum interrupter and the mechanical erosion. Restrike type II can influence the voltage waveform of the next cycle. A power frequency current is established after a high frequency current. Restrike type III seems to be a "full breakdown" of the recovery voltage. The amplitude of the restrike current is higher than that of restrike type II. With the increase of the inrush current amplitude, restrike type II and restrike type III become more frequently. In the case of $I_{inrush} = 2$ kA, 25.6% of the restrikes belong to restrike type II and 11.6% of that belong to restrike type III. In the case of $I_{inrush} = 5$ kA, 20.5% of the restrikes belong to restrike type II and 59.0% of that belong to restrike type III. Restrike type II and restrike Type III could be related to the higher field emission current. The higher inrush current leads to the higher protrusion and the broader crater on the contact surfaces (see Fig. 5.13), which can increase the enhancement factor β. As a result of the deterioration of the contact surface condition, the dielectric strength of the vacuum interrupter decreases and the corresponding field emission current would increase. Koochack Zadeh et al. [14] showed the higher inrush current led to the higher field emission current. Doen et al. [12] showed that the higher field emission current resulted in more restrike phenomena. So it is assumed that the field emission current takes an important role leading to restrike type II and restrike type III.

5.1.2 Asymmetrical AC Field Emission Current Characteristics of Vacuum Interrupters Subjected to Inrush Current

In Sect. 5.1.1, it revealed that the inrush current has a significant impact on the prestrike and restrike characteristics of vacuum interrupters during capacitive current switching. It also assumed that the field emission current takes an important role leading to restrike type II and III. Field emission—also called Fowler–Nordheim tunneling—is a process whereby electrons tunnel through a barrier in the presence of a high electric field [17]. The numerous electrons emitted from the cathode by an electrostatic field compose the field emission current. The field emission current characteristics of vacuum interrupters in capacitive current switching have been widely studied. Smeets et al. [16] revealed that a longer arc duration for the normal capacitive load current reduced the field emission current level after load current interruption. Koochack Zadeh et al. [14] showed that the higher inrush current led to the higher field emission current. Sandolache et al. [18] found that there could be no one-to-one relationship between field emission current levels measured during the recovery period and late breakdown occurrence in capacitive current switching. The field emission current measurements mentioned above were all performed during a recovery voltage phase, which is unidirectional with DC characteristics as a (1-cosωt) wave shape. However, Figs. 5.11 and 5.12 showed that the CuCr contact exhibited an unsymmetrical welding rupture with protrusions/craters characteristics after suffering inrush currents. A measurement of field emission current under bipolar voltage could reflect the conditions of both contact surfaces after the vacuum interrupters suffer the inrush currents. Therefore, more details of the unsymmetrical protrusion/crater emission properties could be disclosed by measuring the field emission current under power frequency voltage. This section is to determine an impact of the inrush currents during back-to-back capacitor bank making operations on the field emission current characteristics of vacuum interrupters under AC power frequency voltage. It considered the number of making operations and the inrush current amplitude. The results would help to discover the restrike mechanism for vacuum interrupters in capacitive current switching.

Figure 5.15a shows an inrush current circuit for back-to-back capacitor bank making operations. The inrush current circuit can supply a current by discharging capacitors C2 through inductances L2. The test vacuum interrupter SW_{test} was initially kept in an open position. C2 was pre-charged to a voltage U_r. After closing switch SW_{inrush} and then closing SW_{test}, C2 discharged through L2. During the making operation of SW_{test}, a prestriking occurs. The inrush current started to flow. Because of the circuit resistance, the inrush current damped with oscillation to zero. The frequency and amplitude of the inrush current can be adjusted by changing capacitance of C2, inductance of L2, or charging voltage U_r. Figure 5.15b shows a typical oscillogram of the inrush current in the making operation with an amplitude of 20 kA and a frequency of 3800 Hz.

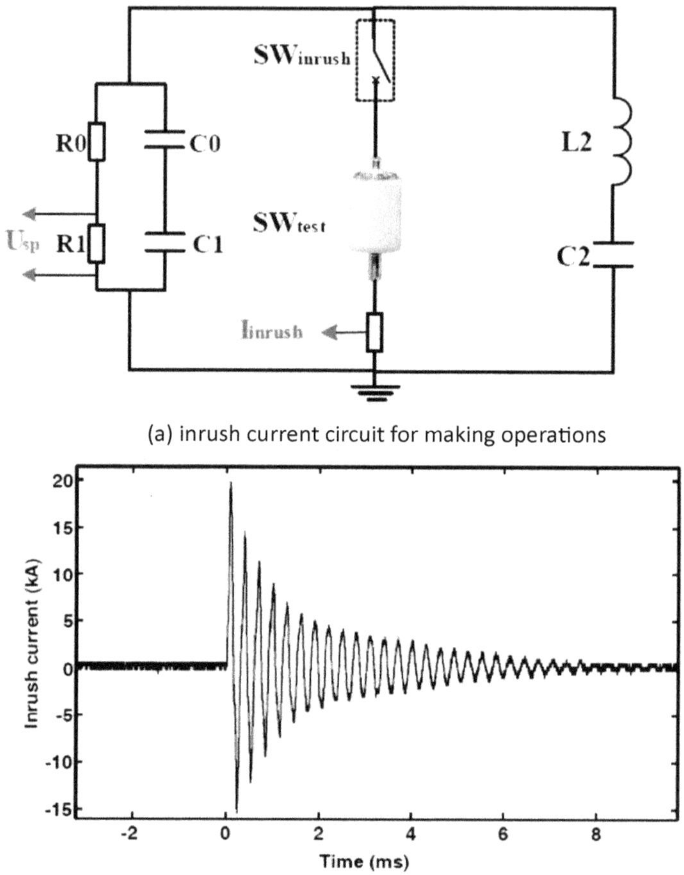

(a) inrush current circuit for making operations

(b) a typical oscillogram of the inrush current

Fig. 5.15 A generation of inrush current

Four 7.2 kV vacuum interrupters were adopted to evaluate the impact of the inrush current on the emission current characteristics. The contact was butt type. The contact diameter was 30 mm. The contact material was CuCr25 (25% weight of Cr). The full contact gap was 4 mm. Table 5.6 shows the experimental conditions. The four tested vacuum interrupters were divided into two groups of A and B according to the inrush current, which was 10 and 20 kA, respectively. Each group included 2

Table 5.6 Experimental conditions

Samples		A1, A2	B1, B2
Inrush current	Amplitude	10 kA	20 kA
	Frequency	3800 Hz	
Closing-measurement operation		Twice	

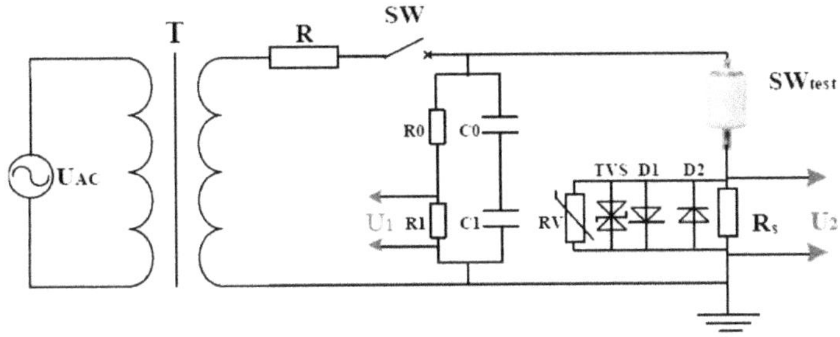

Fig. 5.16 Field emission current measurement circuit. U_{AC}: AC voltage source; T: Voltage regulator and transformer; R: Current-limiting resistance; SW: Switch; resistances and capacitances of the voltage divider: R0 = 100 MΩ, R1 = 0.1 MΩ, C0 = 104 pF, C1 = 0.1 μF; SW_{test}: Test vacuum interrupter; Noninductive resistance $R_s = 100\ \Omega$; D1, D2: Rectifiers; TVS: Transient voltage suppressor; RV: Voltage-dependent resistor

vacuum interrupters, which were labeled 1 or 2. The four vacuum interrupters were conditioned before the application of inrush current. In the conditioning process, the four vacuum interrupters were subjected to a power frequency voltage rising from 0 to 15 kV gradually. In this way, the initial dielectric strength of the test vacuum interrupters reached 15 kV at a contact gap of 1 mm. Thereafter, the test included two steps. First, a making operation was applied with an inrush current of 10 kA (group A) or 20 kA (group B). Second, the field emission current of the test vacuum interrupter was measured under a power frequency voltage (closing-measurement operation). The closing-measurement operation was performed twice for each group.

Figure 5.16 shows a circuit for field emission current measurement. At the beginning of the measurement, SW_{test} was in the open position with a contact gap of 1 mm. Before the measurement of field emission current, both the two vacuum interrupters in group A underwent an inrush current of 10 kA peak. Both vacuum interrupters in group B underwent an inrush current of 20 kA peak. After closing the switch SW, a power frequency voltage was applied across SW_{test} through a transformer T. The voltage rose from 2 to 30 kV in a step of 2 kV. The power frequency voltage was applied with a duration of 5 min and it lasted 500 ms for each step. When the power frequency voltage across SW_{test} reached a certain level, the field emission current flowing between the contacts can be recorded. A current-limiting resistance R was used to limit the current to a few tenths of a milli-ampere once a breakdown occurred to avoid a severe polishing effect due to a restrike on the contact surfaces. The applied voltage was measured by a voltage divider. The field emission current was measured using a 100-Ω non-inductive resistance R_s, which was protected by a combination of a transient voltage suppressor, a voltage dependent resistor, and a pair of anti-parallel diodes from overvoltages in case of a restrike. The protection was proposed by Koochack Zadeh et al. [19]. The turn-on voltage of the anti-parallel diodes was 0.5 V. The 0.5 V divided by 100 Ω presented a current limit of 5 mA. That meant that a restrike current greater than 5 mA would flow through the anti-parallel diodes D1 and D2. However, the field emission current within μA to mA level (less than the limited 5 mA) must flow through the measurement element of the non-inductive resistance R_s rather than through the diodes [20]. Additionally, Koochack Zadeh et al. [19] reported a quite large range of emission current amplitudes (10 A to 5 mA), which were observed in more than 1,000 measurements. Here the field emission current limit was set as 5 mA, thus it will not lead to breakdowns in the vacuum interrupters.

Figure 5.17 shows the measured field emission current of vacuum interrupter A1 at a contact gap of 1 mm. Figure 5.17a shows the field emission current waveform with a local enlargement. The vacuum interrupter A1 suffered an inrush current of 10-kA peak twice before the measurement of the field emission current. The applied power frequency voltage was 14 kV. As shown in Fig. 5.17a, the sinusoidal curve is the applied power frequency voltage, whose duration is 500 ms. The unsymmetrical curve is the measured field emission current, varying with the applied power frequency voltage. The enlargement waveform shows that the field emission current was not symmetrical at each polarity, even under a symmetrical power frequency voltage. In contrast, the field emission currents before application of inrush currents were almost symmetrical under the same symmetrical power frequency voltage. Figure 5.17a shows the field emission current of positive polarity was significantly larger than that of negative polarity. This polarity effect of the field emission current was the most significant characteristic. The field emission current did not increase immediately after the applied power frequency voltage increased from its zero-crossing points, but it rose after the voltage exceeded a certain value. Figure 5.17b shows a dependence of the positive polarity field emission current on the applied power frequency voltage. The spots are the measured field emission current. Figure 5.17b revealed a nonlinear relationship between the field emission current and the applied voltage. It shows that the positive polarity field emission current increased sharply when the applied voltage exceeded 12 kV. When the field emission current reached 5 mA (the value limited by the diodes), the applied voltage was close to 20 kV. In addition, as breakdowns might significantly influence the contact surface conditions, breakdowns should be avoided as much as possible. A field emission current limit of 5 mA proved to be a reasonable value to avoid breakdowns.

Figure 5.18 shows the measured positive polarity field emission current for test samples A1, A2, B1, and B2 as the applied voltage increases. As can be seen, not only the amplitudes of inrush current but also the numbers of making operations had a significant impact on the field emission current characteristics. Higher inrush current amplitudes and the second making inrush current generated higher field emission currents under the power frequency voltage.

Figure 5.19 shows the contact surfaces photos of vacuum interrupters after suffering the inrush currents. Figure 5.19a shows the photos for vacuum interrupter A1. It suffered inrush current of 10 kA peak twice. The light spots on the contact surfaces probably originated from the conditioning procedure before the application of inrush currents. It shows that the damage caused by the 10 kA peak inrush current during the making operation formed a macroscopic protrusion on the moving contact surface and a macroscopic crater on the fixed contact surface. Figure 5.19b shows the photos for vacuum interrupter B1. It suffered inrush current of 20 kA peak

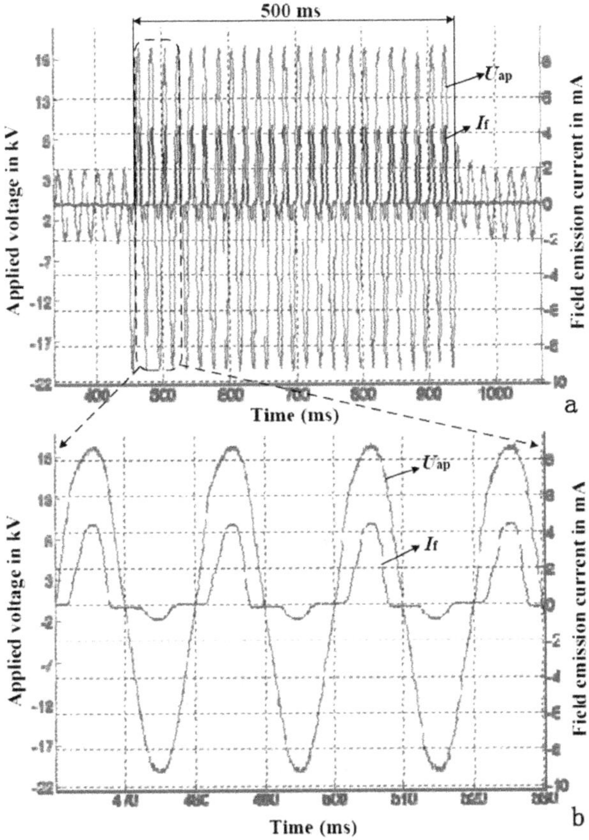

(a) field emission current waveform with a local enlargement

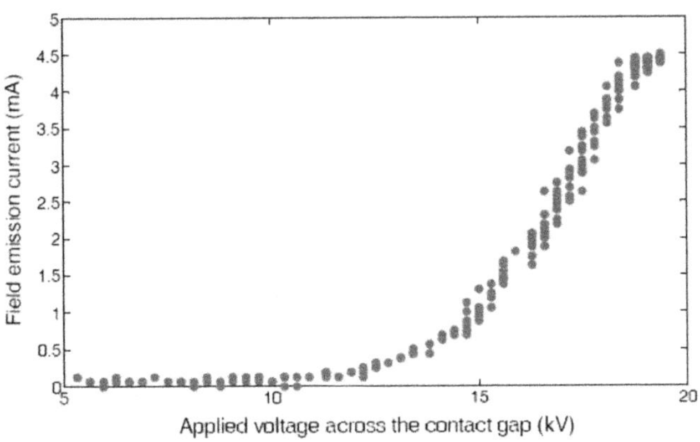

(b) positive polarity field emission current and applied voltage

Fig. 5.17 Field emission current measurement of vacuum interrupter A1 at a contact gap of 1 mm

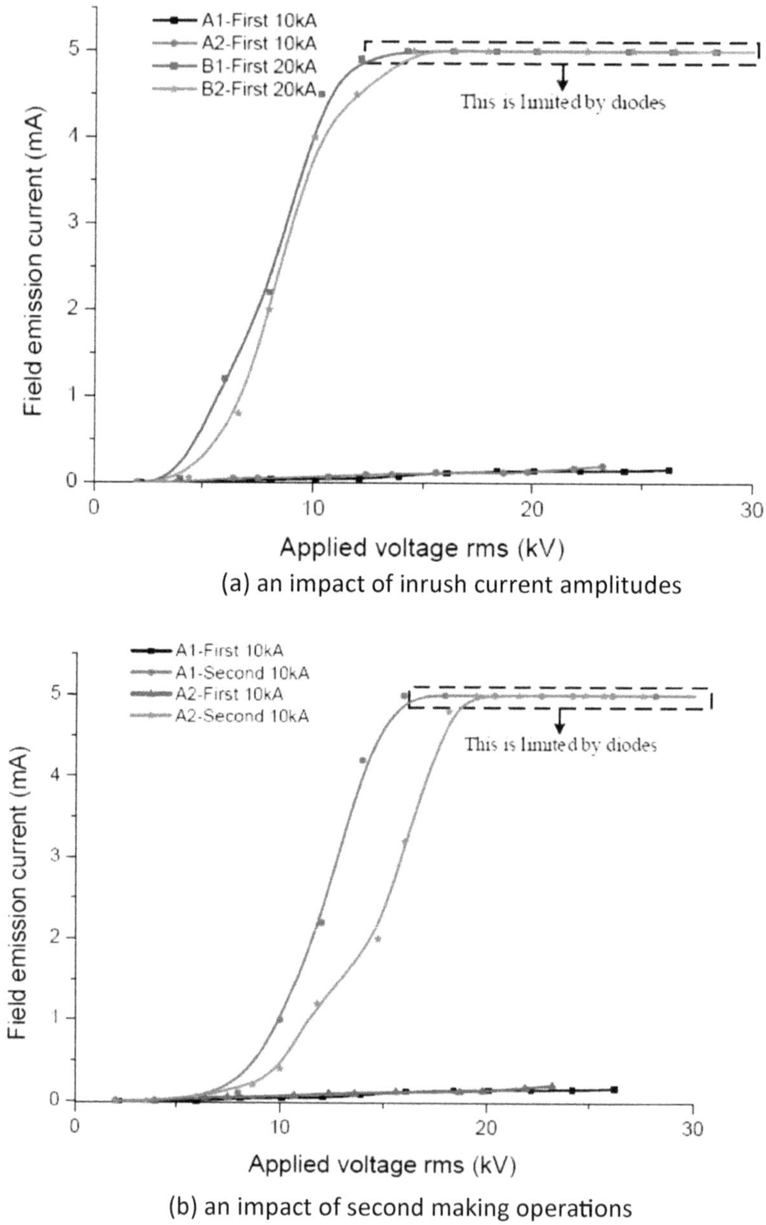

Fig. 5.18 The measured positive polarity field emission current for test samples A1, A2, B1, and B2 as the applied voltage varies

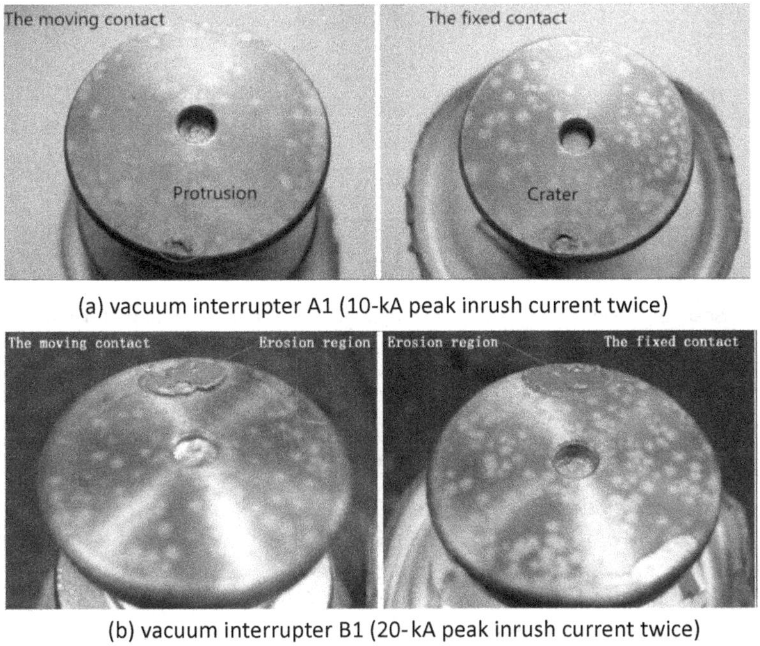

Fig. 5.19 Contact surface photos of the vacuum interrupters after suffering inrush currents

twice. It shows that a local CuCr contact material layer was torn out, exposing the copper substrate on both the moving contact and the fixed contact. Compared with Fig. 5.19a, the damage on the contact surfaces caused by the 20 kA peak inrush current was more serious than that of the 10 kA peak inrush current. Figure 5.19 indicated that the CuCr contacts were seriously damaged by the high energy inrush current during the prestrike process. This was in accordance with Figs. 5.11 and 5.12. After the separation of the welded contacts, a crater corresponding to a macroscopic protrusion formed on the contact surfaces.

Figure 5.20 shows scanning electron microscope photos of the protrusion and crater shown in Fig. 5.19a. Figure 5.20a shows there was a micro-projection tip on the macroscopic protrusion surface. Its scale bar is 3.33 μm. The micro-projection tip is closely related to field emission. Figure 5.20b shows the microscopic image of the macroscopic crater. The scale bar is 33.3 μm. The microscopic image of the macroscopic crater did not indicate any field emission structures.

The electric field between contacts is not determined by U/d, where U is the voltage applied across the contact gap d, but is given by $\beta U/d$ [21], where β is an enhancement factor. The electric field enhancement is induced by the surface conditions of the electrodes. The Fowler–Nordheim-equation [22] explains the field emission current quantitatively. One form of the equation is shown in Eq. (5.2):

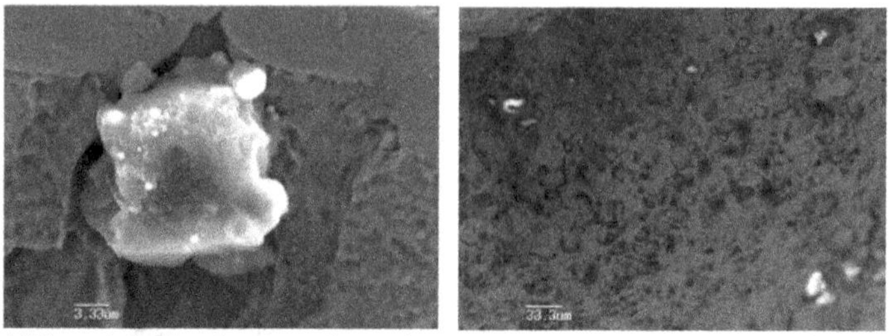

(a) Micro-projection tip on the macroscopic protrusion (b) Microscopic image of the macroscopic crater

Fig. 5.20 Scanning electron microscope photographs of the macroscopic protrusion and macroscopic crater shown in Fig. 5.19

$$j_e = 1.54 \times 10^{-6} \frac{E_m^2}{\varphi t^2(y)} \exp\left(-6.83 \times 10^7 \frac{\varphi^{1.5} v(y)}{E_m}\right) \tag{5.2}$$

where j_e is the field emission current density in A/cm^2; $E_m = \beta U/d$ is the actual field strength in V/cm; U is the applied voltage across the electrodes; d is the contacts gap distance; φ is the work function of the contact material; $t(y)$ and $v(y)$ are slowly varying functions related to the field strength E_m and the work function φ. As copper and chrome are not chemically bonded with each other in copper-chromium alloy, it is only a physically bonded material, where each material retains its own physical property. The work functions of copper and chrome are 4.65 and 4.5 eV, respectively. The work function of the CuCr25 electrodes was estimated to be 4.6 eV, considering their composition ratio.

If one puts $I_e = j_e A_e$ in Eq. (5.2), where I_e is the emission current; A_e is the emission area and $E_m = \beta U/d$ for substitution of E_m and j_e, it appears Eq. (5.3).

$$\log_{10}(I_e U^{-2}) = \log_{10}\left(\frac{1.5 \times 10^{-6} A_e \beta^2}{\varphi d^2}\right) - \frac{2.966 \times 10^9 \varphi^{1.5} d}{\beta} U^{-1}. \tag{5.3}$$

According to Eq. (5.3), if the current I_e is measured as a function of U, a straight line can be plotted between $\log_{10}(I_e/U^2)$ and U^{-1}. The slope and the intercept of the straight line will allow β and A_e to be calculated.

Figure 5.21 shows the field emission current characteristics of vacuum interrupter A1 at a contact gap of 1 mm. Figure 5.21a choose one of the cycles of the waveforms of the applied power frequency voltage and the field emission current shown in Fig. 5.17a. Figure 5.21a shows that the field emission current curve can be divided into five phases: (I) positive rising phase; (II) positive falling phase; (III) negative rising phase; (IV) negative falling phase; and (V) cooling phase, where the field emission current is approximately zero. Based on Eq. (5.3), Fig. 5.21b shows the Fowler–Nordheim plot of $\log_{10}(I_e/U^2)$ versus U^{-1} with the five phases. The markers 'o' and 'x' represent the data during the positive voltage phase and negative voltage

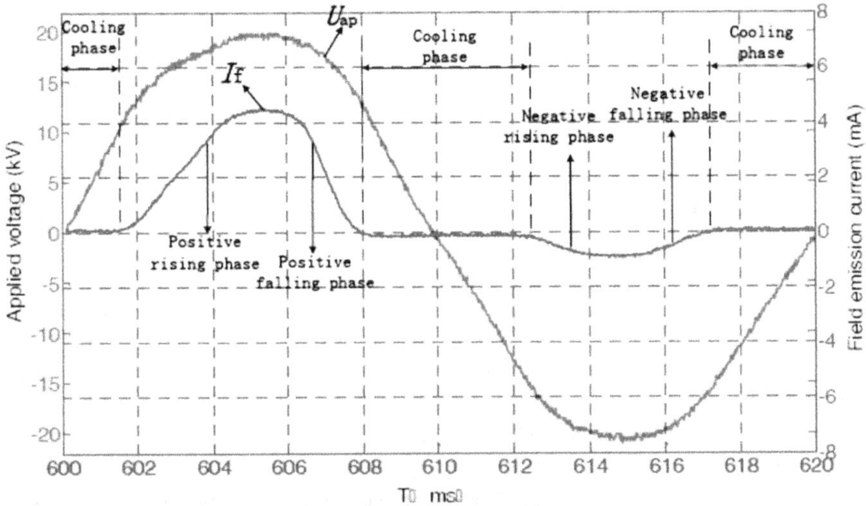

(a) one cycle of power frequency voltage and field emission current

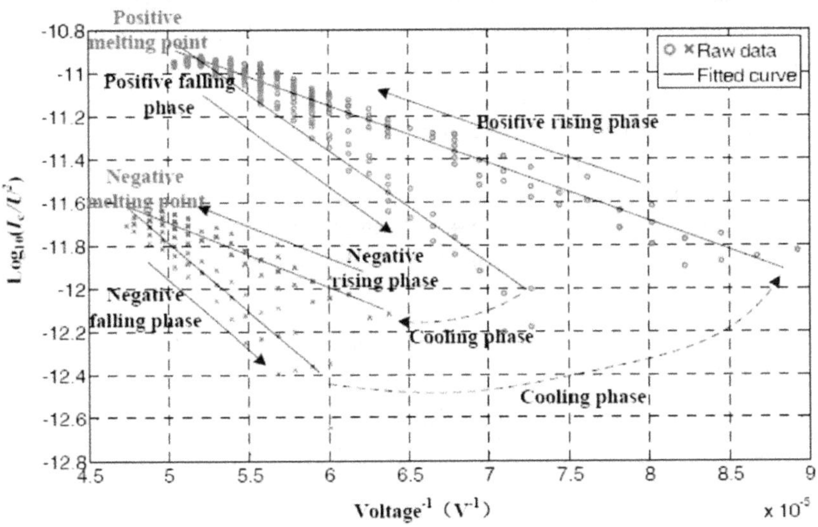

(b) Fowler-Nordheim plot for the data

Fig. 5.21 Field emission current characteristics of vacuum interrupter A1 at a contact gap of 1 mm

phase, respectively. As can be seen, the markers 'o' consists of two parts: the data during the positive rising phase and the data during the positive falling phase. The markers 'x' also consists of two parts: the data during the negative rising phase and the data during the negative falling phase. The straight lines are the linear fitting curves. According to Eq. (5.3), the field enhancement factor β and the effective emission area A_e can be calculated using the slope of the straight lines and the intersection with the y

axis. A lower slope of the fitted line corresponds to a higher field enhancement factor β. During the positive rising phase, the straight line slowly rose with the increasing applied voltage, which meant that β was high. The high β value corresponded to the sharp micro-projection tip shown in Fig. 5.20a. During the positive falling phase, the straight line decreased sharply, which meant a smaller β. The high field emission current during the positive rising phase heats the micro-projection tips, causing them to melt and leading to a lower β. The negative voltage phase seems similar to the positive voltage phase. It shows a slow rising slope in the negative rising phase and a steep decreasing slope in the negative falling phase. A significant relevant parallel relationship can be found when making a comparison with the four fitted lines shown in Fig. 5.21b. In the positive rising phase and the negative rising phase, the two lines were almost parallel. Similarly, in the positive falling phase and the negative falling phase, the two lines were nearly parallel.

Table 5.7 shows the calculated β and A_e values for phases I–IV of vacuum interrupter A1 at 14 kV after suffering 10-kA peak inrush current twice. According to Table 5.7, the slopes of the straight lines for the positive rising phase and the positive falling phase in Fig. 5.21 were -2.68×10^4 and -5.23×10^4, which corresponded to field enhancement factors of 1093 and 559, respectively. Therefore, the field enhancement factor during the positive rising phase doubled that of the positive falling phase. The same happened in the negative voltage phase. The slopes of the negative rising phase and negative falling phase were -3.02×10^4 and -6.41×10^4, respectively, which corresponded to field enhancement factors of 970 and 457. The field enhancement factor β had similar values in the voltage rising stages or in the falling stages for both the positive and negative phases, i.e., 1093 to 970 in the positive and negative rising phases, respectively, and 559 to 457 during the positive and negative falling phases, respectively. Table 5.8 shows the field emission parameters of vacuum inter-

Table 5.7 Calculated field emission parameters of vacuum interrupter A1 at 14 kV after suffering 10-kA peak inrush current twice

Parameter	Positive rising	Positive falling	Negative rising	Negative falling
Slope	-2.68×10^4	-5.23×10^4	-3.02×10^4	-6.41×10^4
y intercept	−9.54	−8.22	−10.18	−8.58
β	1093	559	970	457
A_e/m^2	7.18×10^{-16}	5.77×10^{-14}	2.08×10^{-16}	3.76×10^{-14}

Table 5.8 Calculated field emission parameters of vacuum interrupter A2 at 14 kV after suffering 10-kA peak inrush current twice

Parameter	Positive rising	Positive falling	Negative rising	Negative falling
Slope	-3.15×10^4	-6.34×10^4	-3.54×10^4	-6.72×10^4
y intercept	−9.27	−7.76	−9.70	−8.33
β	928	462	827	435
A_e/m^2	1.87×10^{-15}	2.43×10^{-13}	8.76×10^{-16}	7.41×10^{-14}

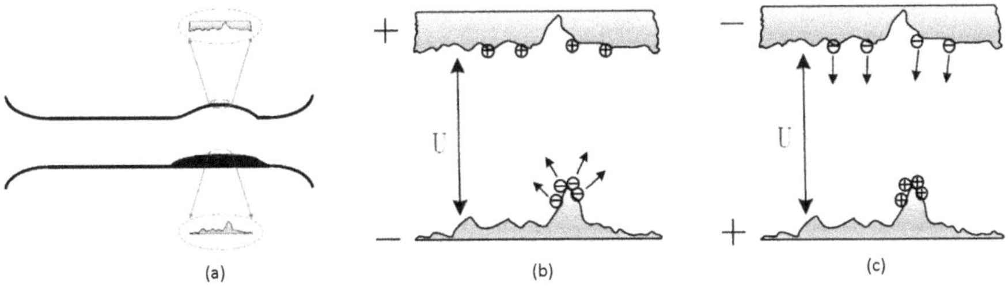

Fig. 5.22 Physical explanation for the unsymmetrical field emission current phenomenon

rupter A2 at 14 kV after suffering 10-kA peak inrush current twice. According to Table 5.8, the same trend of the field emission parameters can be found during the four phases in a cycle of the applied power frequency voltage.

As can be seen from Fig. 5.18, the field emission currents of vacuum interrupters A1 and A2 were approximately zero after suffering the first inrush current with a 10 kA peak. With so low field emission current, the calculation of the field emission parameters was not feasible. In addition, according to Fig. 5.19b, because a local CuCr contact material layer of vacuum interrupters B1 and B2 was totally worn-out, the analysis of field emission parameters for vacuum interrupters B1 and B2 did not make sense.

Figure 5.22 shows a proposed physical model for the unsymmetrical field emission current phenomenon. According to the microscopic observation results in Fig. 5.20, the micro-projection tip located on the protrusions, shown in Fig. 5.22a. In Fig. 5.22b, the microscopic crater is the anode and the protrusion is the cathode. In this case, the electric field is primarily concentrated on the micro tips. The electric field E_{m} near the micro tips surface is far higher than the E simply calculated by $E = U/d$. The high electric field would trigger a strong electron emission, causing a field emission current. However, after the change of voltage polarity, the case shown in Fig. 5.22c appears. In this case, the protrusion is the anode and the crater is the cathode. Because the crater is primarily composed of numerous micro-craters, the electric field around the micro-crater cathode would be much lower than the case shown in Fig. 5.22b. Then, with a lower local electric field at the cathode, when applying the same voltage, the field emission effect becomes lower, which also means that the field emission current will be lower than that shown in Fig. 5.22b.

Figure 5.23 shows a proposed electrical heating–cooling cycle of the emission sites by joule heating effect of the field emission current, which may explain the results shown in Fig. 5.21b. During the positive rising phase, with an increase of the applied voltage, the field emission current would significantly increase. The growing field emission current heats the emission sites. The emission sites are likely to melt at the positive melting point labeled in Fig. 5.21b. As Fig. 5.23 shows, the previous cold and sharp micro-tips melted, causing a smoother tip with a larger field emission area at the field emission sites. The changes in the field emission microstructure and

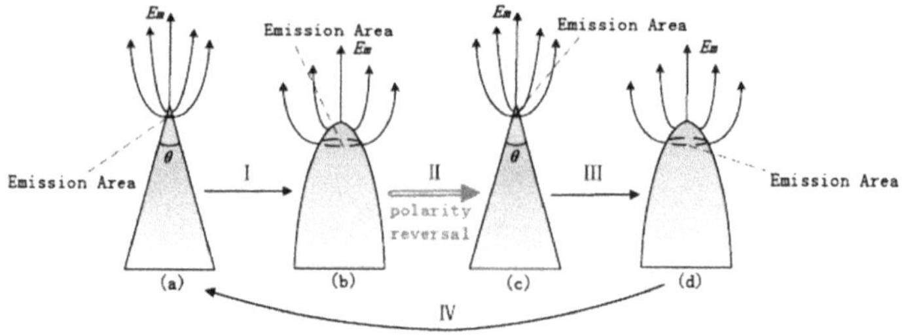

Fig. 5.23 A proposed heating–cooling cycle of the field emission sites. **a** The condensed cold state of the emission sites; **b** The melted state of the emission sites; **c** The condensed cold state of the field emission sites on the opposite contact; **d** The melted state of the emission sites on the opposite contact. E_m: The electric field around the field emission sites; I: Positive rising phase; II: Positive falling phase and cooling phase; III: Negative rising phase; IV: Negative falling phase and cooling phase

the field enhancement factor β at the application of high alternating voltage were also found by Frohlich et al. [23]. The actual electric field E_m is given by $\beta U/d$. A sharper tip will lead to a higher electric field around the micro-projections, and it also means a high field enhancement factor β, while the smoother tips have the opposite effect. Then, after the positive melting point, the decrease of the field enhancement factor β and the increase of the field emission area A_e become reasonable. As shown in Fig. 5.17a, after the applied voltage fell below a critical value, the field emission current was approximately zero, which was labeled as the cooling phase in Fig. 5.21a, b. During the cooling phase, there would be no joule heating occurring on the field emission sites. Because of the thermal diffusion and radiation process, the melted tips cool down and re-crystallize to form field emission sites again. The structure of the new emission sites was determined by the crystal structure of the material [24]. They tend to form a similar shape as the previous ones during the positive rising phase. During the negative rising phase, the field enhancement factor β of the opposite contact would be similar to that of the positive rising phase. The same situation would apply for the negative falling phase as that of the positive falling phase. This explains the parallel F–N plot straight lines shown in Fig. 5.21b. The waveform shown in Fig. 5.21a can be explained by the heating–cooling cycle, which includes a heating process in the positive voltage phase, a crystallizing process during the positive falling phase and cooling phase, a re-heating process in the negative voltage phase, and a re-crystallizing process during the next cooling phase. The periodic waveforms of the applied voltage and field emission current shown in Fig. 5.17a are a repetition of this pattern.

5.1.3 Prestrike Electric Field Characteristics When Making Inrush Current in 40.5 kV Vacuum Interrupters

In Sect. 5.1.1, it discussed the impact of the inrush current amplitude on the prestrike and restrike characteristics of 7.2 kV vacuum interrupters during capacitive current switching. Then Sect. 5.1.2 discussed the impact of the inrush currents during back-to-back capacitor bank making operations on the field emission current characteristics of 7.2 kV vacuum interrupters under AC power frequency voltage. Even though it evidenced a significant impact of the inrush current on the prestriking phenomena of vacuum interrupters during capacitive current making operations, only 7.2 kV vacuum interrupter was tested. More evidence is desired, especially on higher voltage, such as 40.5 kV vacuum interrupters. This section is to determine an impact of inrush current on prestrike electric field characteristics for 40.5 kV vacuum interrupters in capacitive current making operations.

The 40.5 kV vacuum interrupters were installed in a commercially available 40.5 kV vacuum circuit breaker, which was equipped with a permanent magnetic actuator. The contact was butt type design with a diameter of 50 mm. Two kinds of contact materials were used. One was infiltration CuCr50 (50% of Cr) and the other was arc-melted CuCr40 (40% of Cr). The full contact gap was 20 mm. The average closing velocity in the full contact gap was 0.77 m/s. All the vacuum interrupters experienced current and voltage conditioning.

The experiment was carried out with a *L-C* oscillating circuit, which provided an inrush current of 207 Hz with 6.48 kA peak. The chosen of inrush current frequency and amplitude considered a 35 kV reactive power compensation in 500 kV extra-high-voltage substations in China. The 35 kV capacitor units are Y connected with isolated neutral. The capacitor banks are with reactive power Qc = 60 Mvar at 35 kV, which are connected to the substation busbar through a series reactor whose damping factor is 6%. Figure 5.24 shows the *L-C* oscillating circuit. In the circuit

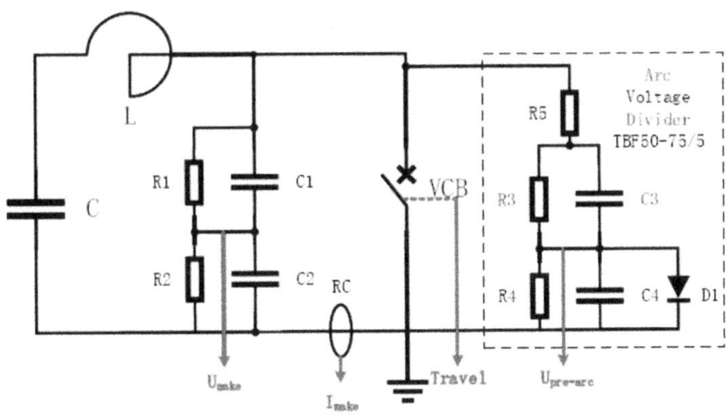

Fig. 5.24 *L-C* oscillating circuit for capacitive making operation

the capacitor C was pre-charged and then it discharged through the reactor L and test vacuum circuit breaker. Then an inrush current was initiated. The pre-charged voltage of C was 1.4 per unit (p.u.), which is the capacitive voltage factor for breaking of capacitor banks with isolated neutral. In order to understand the prestrike electric field characteristics, a series of 80 making operations was executed for six 40.5 kV vacuum interrupters. Table 5.9 shows the conditions for the 80 making operations. Three vacuum interrupters were with the infiltration CuCr50 contact material and the other three were with arc-melted CuCr40 contact material. No load current switching was carried out, because it simulated a more serious condition, by which no current conditioning takes effect.

Figure 5.25 shows a typical oscillogram in the making operations. In Fig. 5.25, traces "Umake", "Imake", and "Travel" represent the making voltage across the vacuum interrupter, the inrush current passing through the vacuum interrupter, and the displacement of the moving contact, respectively. "Upre-arc" indicates the instant when the contacts physically mate. The bouncing of the vacuum circuit breaker was

Table 5.9 Capacitive making test conditions

Test VIs	Arc-melted CuCr40			Infiltration CuCr50		
	No.B1	No.B2	No.B3	No.C1	No.C2	No.C3
Inrush current	6.48 kA peak and 207 Hz					
Switching current	0 A (no load)					
Applied voltage	$40.5 \times \sqrt{2}/\sqrt{3} \times 1.4$ kV					
Full stroke	20 mm					
Switching test duty	C–O: 80 times					

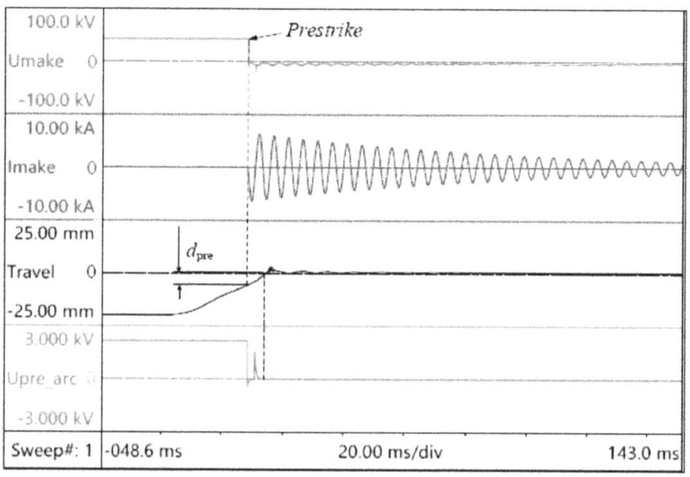

Fig. 5.25 A typical oscillogram in making operations

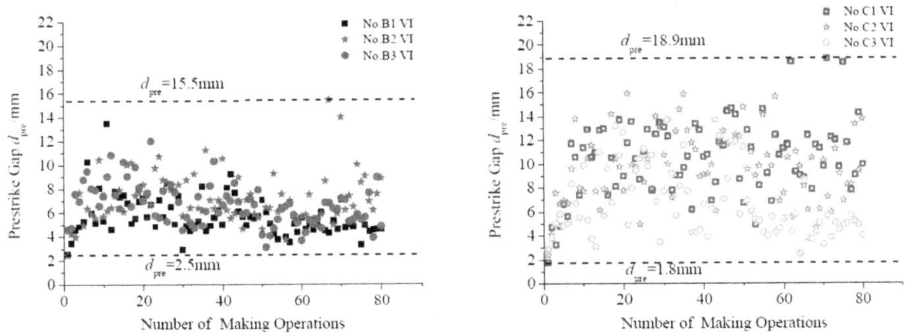

(a) Prestrike gaps of test series B (b) Prestrike gaps of test series C

Fig. 5.26 Distributions of prestrike gaps versus the number of making operations for six test vacuum interrupters

eliminated by adjusting the permanent magnetic actuator. The prestrike gap d_{pre} is defined as the distance between the contacts when prestrike occurs.

Figure 5.26 shows the distributions of prestrike gaps versus the number of making operations for the six test vacuum interrupters. As mentioned in Sect. 5.1.1, the cumulative probability distribution of the prestrike gap d_{pre} can be evaluated by a 3-parameter Weibull distribution function. Table 5.10 summarizes the parameters of the Weibull parameters. R-squared is used to indicate the goodness of the fit. As all the R-squared are above 0.95, it is reasonable to assume that the complementary cumulative probability of the d_{pre} follows the 3-parameter Weibull distribution. In Table 5.10, d_{100} is the location parameter that represents the prestrike gap below which the prestrike probability is 1. m is the Weibull shape or shape factor. Θ is the scale parameter.

The prestrike electric field E_{pre} is the instantaneous electric field that impressed across the contact gap when prestrike occurs. U_{make} and d_{pre} is measured in the experiment. Thus, E_{pre} is defined as Eq. (5.4).

$$E_{\text{pre}} = U_{make}/d_{\text{pre}} \tag{5.4}$$

Table 5.10 Parameters of the complementary Weibull distribution in test series B and C

Test VIs	m/mm	d_{100}/mm	Θ/mm	Θ−d_{100}/mm	R-squared
No.B1	2.1	3.0	6.5	3.5	0.951
No.B2	2.8	2.5	7.5	5.0	0.988
No.B3	2.5	3.1	7.1	4.0	0.996
No.C1	3.1	3.0	11.9	8.9	0.996
No.C2	2.7	2.4	10.5	8.1	0.992
No.C3	1.6	2.6	8.8	6.2	0.992

In addition, Slade showed a compendium relationship of the DC vacuum breakdown voltage U_B as a function of the contact gap d [21]. As the applied voltage was also DC voltage in this experiment, Eq. (5.5) would be applicable, which was suggested by Slade for 0.4 mm < d < 40 mm.

$$U_B = 58d^{0.58} \tag{5.5}$$

There is a difference between the experiments mentioned by Slade and the prestrike experiments here. In the experiments mentioned by Slade, the contacts were fixed, while the contacts were approaching to each other in this prestrike experiment. Thus a theoretical static state breakdown field strength E_B at an instantaneous prestrike gap d_{pre} was defined as Eq. (5.6).

$$E_B = \frac{U_B}{d_{pre}} = 58d_{pre}^{-0.42} \tag{5.6}$$

Figure 5.27 shows the measured prestrike field strength E_{pre} and theoretical static state breakdown field strength E_B at each instantaneous prestrike gap d_{pre} versus the making operations for the six test vacuum interrupters. E_{pre} and E_B were calculated according to Eqs. (5.4) and (5.6), respectively. As shown in Fig. 5.27, E_{pre} was lower than E_B. The lower E_{pre} may be caused by inrush current arc erosions, as discussed in Sects. 5.1.1 and 5.1.2.

If one sorts the electric field strengths shown in Fig. 5.27 as an ascending order of prestrike gaps, it shows curves like Fig. 5.28. The upper curve is the theoretical static state breakdown field strength E_B and the lower curve is the measured prestrike electric field strength E_{pre}. The measured prestrike electric field strength curve followed a function of Eq. (5.7), where A and B are constants. Table 5.11 shows the fitting parameters A and B for the six test vacuum interrupters. R-squared is used to indicate the goodness of the fit. As all the R-squared are above 0.96, it is reasonable to assume that prestrike field strengths follow Eq. (5.7). Therefore, the term $-Ad_{pre}^{-B}$ quantified the influence of inrush current arc erosions on the vacuum insulation level from the original static state level.

$$E_{pre} = 58d_{pre}^{-0.42} - Ad_{pre}^{-B} \tag{5.7}$$

5.1.4 Impact of Inrush Current on Field Emission Current and Voltage Distribution for Double Vacuum Interrupters in Series

Previous Sects. 5.1.1–5.1.3 discuss the impact of inrush current on the prestrike characteristics of single vacuum interrupter, which include 7.2 kV and 40.5 kV vacuum

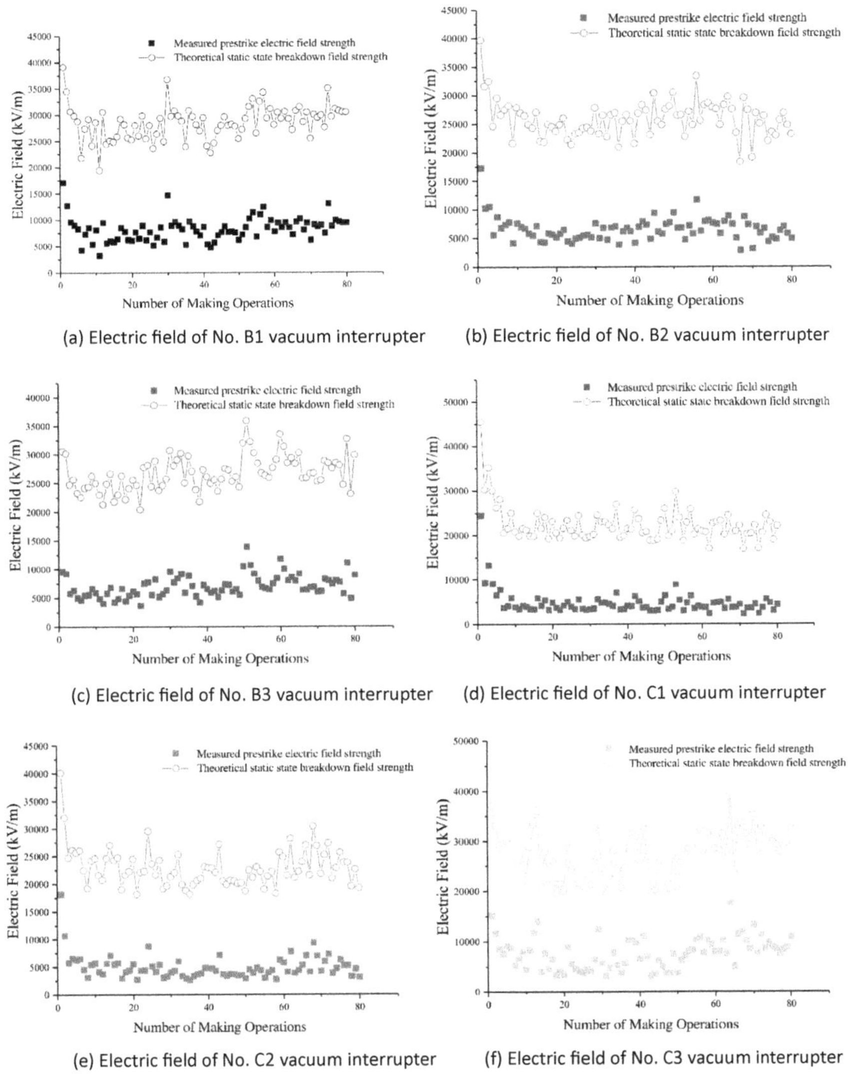

Fig. 5.27 Measured prestrike electric field strength and theoretical static state breakdown field strength of the six test vacuum interrupters in 80 making operations

interrupters. However, the impact of inrush current on the double vacuum interrupters in series is still unclear. This section experimentally determines the impact of inrush current in making operation on the prestrike characteristics of double vacuum interrupters in series, including field emission current and voltage distribution.

Two 12 kV vacuum interrupters in series were tested. The contact was horse-shoe type axial magnetic field contact (see Sect. 5.2.2). The contact diameter was 50 mm,

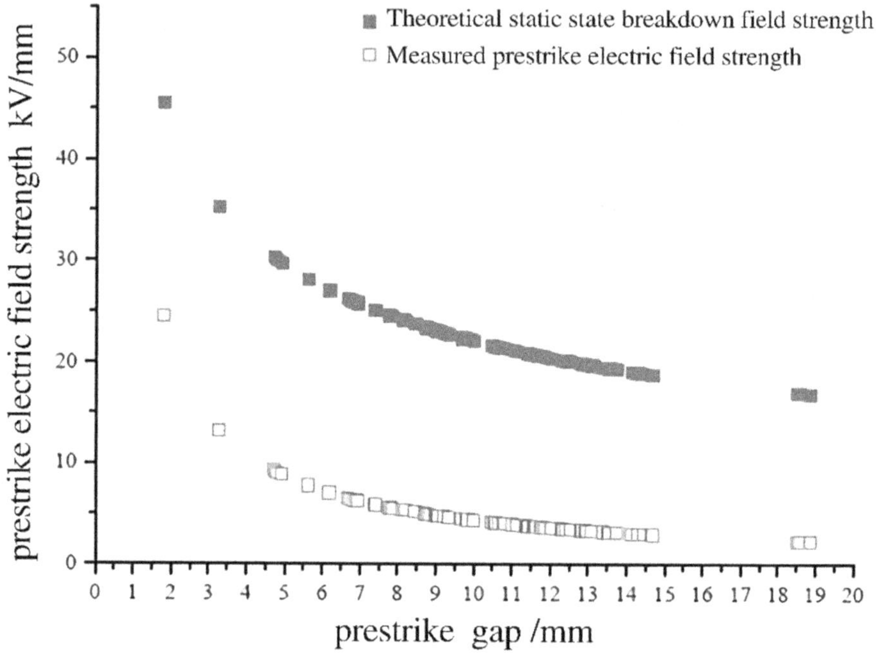

Fig. 5.28 Relationship between prestrike electric field strengths and prestrike gaps in vacuum interrupter C1

Table 5.11 Fitting parameters of Eq. (5.7)

Test vacuum interrupters		A	B	R-squared
Arc-melted CuCr40 contact material	B1	27.82	0.18	0.99
	B2	29.19	0.21	0.98
	B3	29.07	0.21	0.99
Infiltration CuCr50 contact material	C1	27.50	0.19	0.96
	C2	29.56	0.22	0.97
	C3	28.62	0.21	0.99

and the contact material was CuCr40 (40% of Cr). The test program included two parts, which were part I and part II, respectively.

Table 5.12 shows the test arrangement of part I. Three vacuum interrupters A1, A2 and N1 were tested. A1 and A2 were used to make an inrush current of 20 kA peak with a frequency of 3800 Hz. The inrush current was generated by discharging a capacitor through an inductor (see Fig. 5.15). The full contact gap was 9 mm. The contacts were opened with no load by a permanent magnetic actuator after a making operation, which avoided a conditioning effect from the switching current. Only one CO operation cycle was executed during the test. N1 was a new vacuum interrupter.

Table 5.12 Test arrangement in test part I

Test sample		A1	A2	N1
Inrush current	Amplitude	20 kA	20 kA	0
	Frequency	3800 Hz		
Breaking current		0 A, no load opening		
Operation cycle		1 CO		

Table 5.13 shows the test arrangement of part II. Part II measures field emission current I_e under a DC voltage. Figure 5.29a shows a schematic diagram of the measurement circuit. First, a capacitor C2 was charged to a certain value U_{C2}. Then U_{C2} applied to a test vacuum interrupter by closing an isolated switch SW_2 after an interval T_a. Two anti-parallel diodes D_{21} and D_{22} were used to protect the oscilloscope from the high voltage caused by possible breakdowns of vacuum interrupter. Five diode modules (SanRex DD200KB) were connected in series. A resistance R_e of 100–500 Ω was used to measure the field emission current. In order to keep the leakage current of D_{21} and D_{22} less than 0.5 mA, the maximum value of I_e was limited to less than 25 mA. Figure 5.29b shows a typical oscillogram of field emission current. The upper curve is the applied DC voltage and the lower curve is the field emission current. The same three vacuum interrupters A1, A2 and N1 in part I were measured in five groups. In group 1, the contact gap was 2 mm. For each vacuum interrupter, I_e was measured when U_{C2} was applied to the moving terminal and the fixed terminal, respectively. Furthermore, each terminal was tested twice at $T_a = 100$ ms and $T_a = 10$ ms, respectively. In group 2, vacuum interrupters A1 and A2 were in series by connecting their fixed terminals. Then I_e was measured when U_{C2} was applied to each moving terminal. In group 3, vacuum interrupters A1 and N1 were in series. In group 4, vacuum interrupters A2 and N1 were in series. The test procedure in group 3 and group 4 were as same as that of group 2. In group 5, the contact gaps of vacuum interrupters A1 and A2 were set to 4 mm. The test procedure was as same as that of group 1. For group 2 to group 5, T_a was set to 10 ms.

In group 1 the field emission current measurement was carried out for vacuum interrupter A1, A2 and N1, respectively. Figure 5.30 shows a dependence of the field emission current on the applied voltage for vacuum interrupter A1, A2 and N1. It was found that I_e of A1 and A2 were much higher than that of N1 at same applied voltage. In addition, I_e was extremely low when the applied voltage was less than a certain value U_0. Once the applied voltage was higher than U_0, I_e increased significantly. Equation (5.8) describes a dependence of I_e on the applied voltage U_{C2}.

$$I_e = a \times e^{b \times U_{C2}} + c \tag{5.8}$$

where a, b and c are constants, which were fitted as shown in Table 5.14. Well-fitting was achieved as the R-squared values were above 0.98. A common phenomenon observed was that I_e decreased in test 2 for all vacuum interrupters. This means that

Table 5.13 Test arrangement in test part II

Group1	A1		A2		N1	
d	2 mm	2 mm	2 mm	2 mm	2 mm	2 mm
U_{c2}	Moving	Fixed	Moving	Fixed	Moving	Fixed
Grounded	Fixed	Moving	Fixed	Moving	Fixed	Moving
Test 1	$T_a = 100$ ms					
Test 2	$T_a = 10$ ms					
Group2	A1	A2	A1	A2		
d	2 × 2 mm		2 × 2 mm			
U_{c2}	Moving	–	–	Moving		
Grounded	–	Moving	Moving	–		
in Series	Fixed terminal connected		Fixed terminal connected			
Test 1	$T_a = 10$ ms					
Group3	A1	N1	A1	N1		
d	2×2 mm		2×2 mm			
U_{c2}	Moving	–	–	Moving		
Grounded	–	Moving	Moving	–		
in Series	Fixed terminal connected					
Test 1	$T_a = 10$ ms					
Group 4	A2	N1	A2	N1		
d	2 × 2 mm		2 × 2 mm			
U_{c2}	Moving	–	–	Moving		
Grounded	–	Moving	Moving	–		
in Series	Fixed terminal connected					
Test 1	$T_a = 10$ ms					
Group5	A1		A2			
d	4 mm	4 mm	4 mm	4 mm		
U_{c2}	Moving	Fixed	Moving	Fixed		
Grounded	Fixed	Moving	Fixed	Moving		
Test 1	$T_a = 10$ ms					

I_e had a conditioning effect of the contact surface. Moreover, it could cause different degrees of conditioning to the prestrike arc deteriorated vacuum interrupter and new vacuum interrupter. For vacuum interrupter A1 and A2, the voltage difference between test 1 and test 2 was less than 3 kV. For vacuum interrupter N1, the voltage difference was 6 to 8 kV. In order to limit the conditioning impact, T_a decreased from 100 to 10 ms from test group 2 to test group 5.

Table 5.15 shows four calculated voltages U_0, U_5, U_{10} and U_{20} with $I_e = 0.01$, 5, 10 and 20 mA for vacuum interrupter A1 and A2. Table 5.16 shows four calculated voltage U_0, U_{01}, U_{02} and U_{03} with $I_e = 0.01$, 0.1, 0.2 and 0.3 mA for vacuum

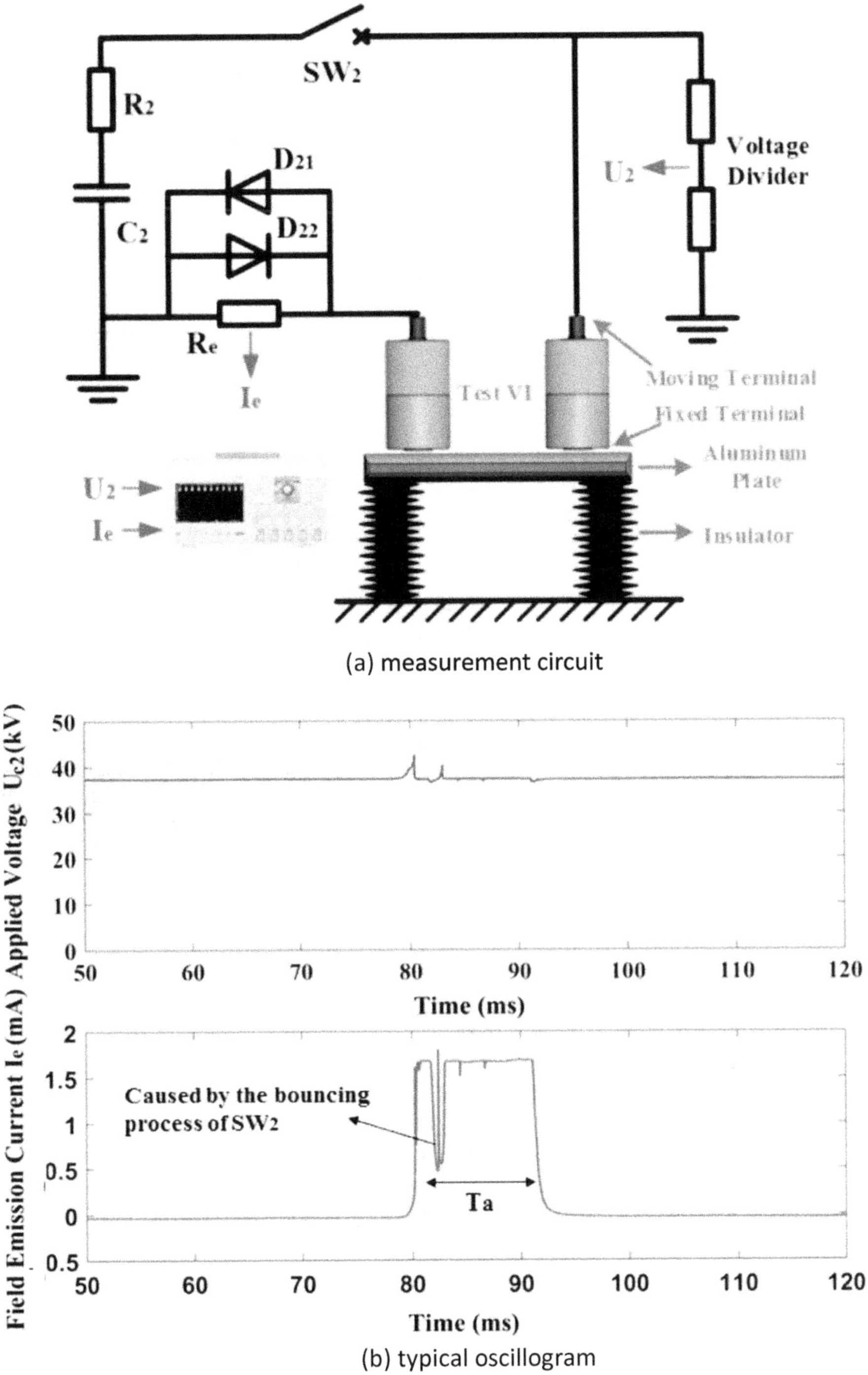

Fig. 5.29 Measurement of field emission current

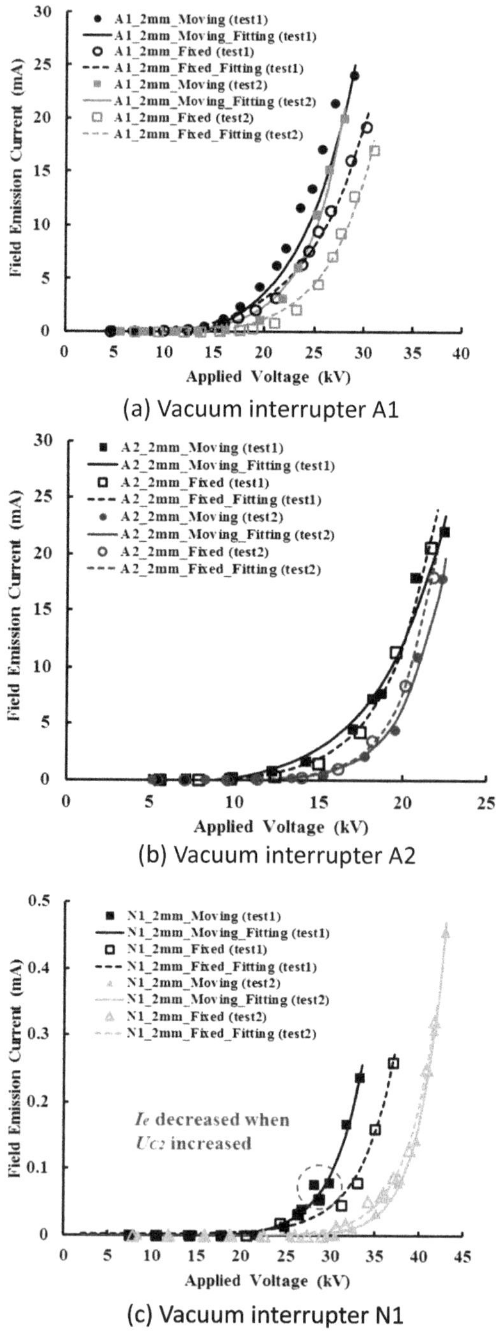

Fig. 5.30 A dependence of field emission current on the applied voltage for the three vacuum interrupters in group 1. Contact gap was 2 mm and $T_a = 100$ ms (test 1) or 10 ms (test 2)

Table 5.14 Fitting parameters of Eq. (5.8) for test group 1

VI	T_a(ms)	U_{c2}	a	b	c	R-square
A1	100	Moving Fixed	3.60E-01 1.68E-01	1.50E-01 1.60E-01	−1.82E-00 −1.03E-00	0.98 0.99
	10	Moving Fixed	1.47E-02 2.34E-02	2.60E-01 2.15E-01	−2.91E-01 −5.44E-01	0.99 0.99
A2	100	Moving Fixed	9.28E-02 1.34E-02	2.48E-01 3.41E-01	−9.55E-01 −2.73E-01	0.99 0.99
	10	Moving Fixed	1.10E-03 5.90E-04	4.36E-01 4.75E-01	−1.95E-01 −8.37E-02	0.99 0.99
N1	100	Moving Fixed	1.71E-05 1.74E-05	2.87E-01 2.59E-01	−2.61E-02 2.33E-03	0.99 0.98
	10	Moving Fixed	1.46E-07 1.75E-06	3.49E-01 2.90E-01	−9.54E-04 −5.37E-04	0.99 0.99

Table 5.15 Calculation of applied voltage with $I_e = 0.01, 5, 10$ and 20 mA for vacuum interrupter A1 and A2 (contact gap = 2 mm)

I_e (mA)				0.01	5	10	20
VI	Test	T_a(ms)	U_{c2}	U_0(kV)	U_5(kV)	U_{10}(kV)	U_{20}(kV)
A1	Test 1	100	Moving Fixed	10.8 11.3	19.6 22.4	23.2 26.2	27.3 30.2
	Test 2	10	Moving Fixed	11.6 14.7	22.7 25.4	25.2 28.4	27.8 31.5
A2	Test 1	100	Moving Fixed	9.45 8.94	16.8 17.5	19.3 19.5	21.9 21.5
	Test 2	10	Moving Fixed	12.0 10.7	19.4 19.1	21.0 20.5	22.5 22.0

Table 5.16 Calculation of applied voltage with $I_e = 0.01$, 0.1, 0.2 and 0.3 mA for vacuum interrupter N1 (contact gap = 2 mm)

I_e (mA)				0.01	0.1	0.2	0.3
VI	Test	T_α(ms)	U_{c2}	U_0(kV)	U_{01}(kV)	U_{02}(kV)	U_{03}(kV)
N1	Test 1	100	Moving Fixed	23.0 23.5	30.3 33.3	32.7 36.0	34.1 37.6
	Test 2	10	Moving Fixed	32.2 30.0	38.6 37.8	40.6 40.1	41.7 41.5

interrupter N1. The voltage calculation was based on Eq. (5.8). It was found that the emission currents of vacuum interrupter N1 were significantly lower than A1 and A2, even the applied voltage for N1 was higher than that for A1 and A2.

After testing group 1, two vacuum interrupters in series were tested in the following groups, which was from test group 2 to group 4. The test samples were still the same three vacuum interrupters A1, A2 and N1 in group 1. Each contact gap was set as 2 mm. The time interval T_a was set to 10 ms. During these tests, the fixed terminals of two vacuum interrupters were electrically connected. Then I_e was measured by applying U_{C2} on the moving terminal of each vacuum interrupter. At last, in test group 5 single vacuum interrupter was tested. The contact gaps of vacuum interrupter A1 and A2 were set to 4 mm in group 5. For each vacuum interrupter, I_e was measured when U_{C2} was applied to the moving terminal and the fixed terminal, respectively.

Figure 5.31 shows a dependence of field emission current on U_{C2} in test group 2 and test group 5. It was found that I_e of vacuum interrupter A1 and A2 in series was higher than that of single vacuum interrupter A1, but it was lower than that of single vacuum interrupter A2. The field emission current also followed Eq. (5.8). Table 5.17 shows calculated applied voltage U_0, U_5, U_{10} and U_{20} with $I_e = 0.01$, 5, 10 and 20 mA for test group 2 and test group 5 by using Eq. (5.8). It found that the four voltages of vacuum interrupter A1 and A2 in series were all higher than that of single vacuum interrupter A2, but they were lower than that of vacuum interrupter A1. So a 2 mm plus 2 mm vacuum gap was not necessary dielectrically stronger than a single vacuum gap of 4 mm when both vacuum interrupters were deteriorated by the inrush current. Table 5.18 shows the fitting parameters *a*, *b* and *c* in Eq. (5.8) for test group 2 to test group 5. Well-fitting was achieved as the *R*-squared values were above 0.9.

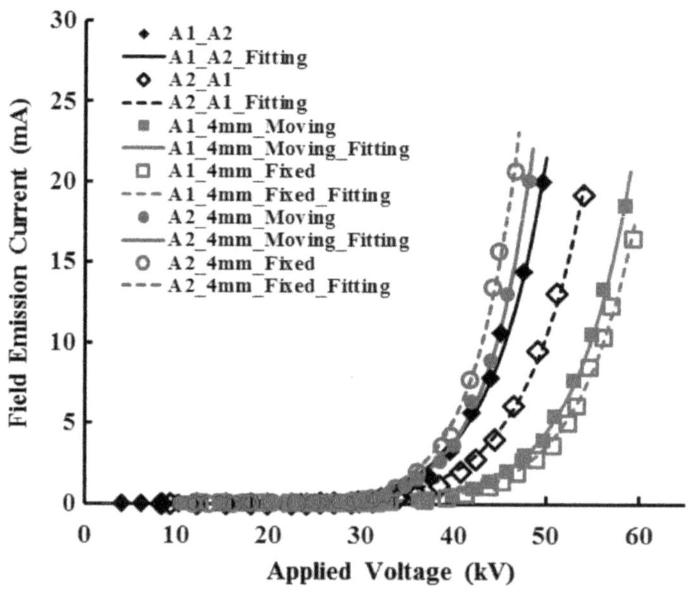

Fig. 5.31 A dependence of field emission current on U_{C2} in test group 2 and test group 5. "A1_A2" and "A2_A1" refer to Table 5.17 "A1 A2 in series". "A1_4mm_Moving" and "A1_4mm_Fixed" refer to Table 5.17 "A1". "A2_4mm_Moving" and "A2_4mm_Fixed" refer to Table 5.17 "A2"

Table 5.17 Calculated applied voltage with $I_e = 0.01$, 5, 10 and 20 mA in test group 2 and test group 5

I_e (mA)				0.01	5	10	20
VI	d (mm)	U_{c2}	Grounded	U_0(kV)	U_5(kV)	U_{10}(kV)	U_{20}(kV)
A1A2 in series	2 × 2	Al Moving	A2 Moving	23.9	41.6	45.5	49.6
		A2 Moving	Al Moving	26.4	45.4	49.7	54.0
A1	4	Moving	Fixed	29.2	50.8	54.8	58.9
		Fixed	Moving	29.8	52.1	56.2	60.4
A2	4	Moving	Fixed	22.2	41.3	44.6	48.0
		Fixed	Moving	23.1	39.8	43.1	46.3

Table 5.18 Fitting parameters of Eq. (5.8) for test group 2 to 5

VI	D (mm)	U_{c2}	Grounded	a	b	c	R-square
A1A2 in series	2 × 2	Al Moving	A2 Moving	4.57 E-03	1.69 E-01	−2.54 E-01	0.99
		A2 Moving	Al Moving	4.19 E-03	1.57 E-01	−2.56 E-01	0.99
AlN1 in series	2 × 2	Al Moving	N1 Moving	2.97 E-05	1.44 E-01	1.68 E-03	0.94
		N1 Moving	A1 Moving	1.34 E-04	1.10 E-01	−6.94 E-04	0.91
A2N1 in series	2 × 2	A2 Moving	N1 Moving	7.57 E-05	1.35 E-01	−4.93 E-03	0.97
		N1 Moving	A2 Moving	6.65 E-06	1.80 E-01	−1.70 E-03	0.99
Al	4	Moving	Fixed	9.56 E-04	1.69 E-01	−1.24 E-01	0.99
		Fixed	Moving	8.96 E-04	1.66 E-01	−1.16 E-01	0.99
A2	4	Moving	Fixed	1.07 E-03	2.05 E-01	−9.11 E-02	0.99
		Fixed	Moving	1.15 E-03	2.11 E-01	−1.40 E-01	0.99

Figure 5.32 shows the field emission current dependence on the applied voltage U_{C2} in test group 3 and test group 4. Compared to Fig. 5.31, it was found that I_e was relatively low when a prestrike arc deteriorated vacuum interrupter and a new vacuum interrupter were in series (A1 and N1 in series, and A2 and N1 in series). When U_{C2} reached about 60 kV, I_e was still lower than 0.4 mA. Table 5.19 shows four calculated voltages U_0, U_{01}, U_{02} and U_{03} when $I_e = 0.01, 0.1, 0.2$ and 0.3 mA. It was found that the U_{01}, U_{02} and U_{03} of vacuum interrupters A1 and N1 in series were higher than that of vacuum interrupters A2 and N1 in series, even though their

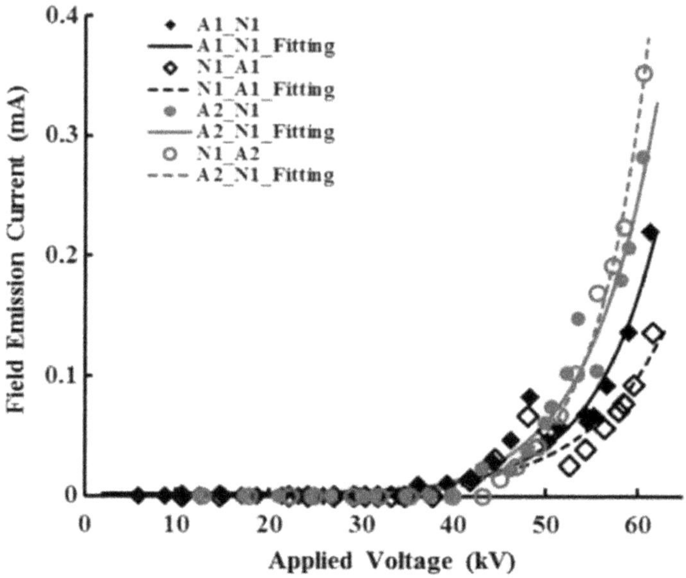

Fig. 5.32 A dependence of field emission current on U_{C2} in test group 3 and test group 4. Traces of A1_N1 and N1_A1 refer to Table 5.19 "A1 N1 in series". Traces of A2_N1 and N1_A2 refer to Table 5.19 "A2 N1 in series"

Table 5.19 Calculated applied voltage with $I_e = 0.01, 0.1, 0.2$ and 0.3 mA in test group 3 and test group 4

I_e (mA)				0.01	0.1	0.2	0.3
VI	d (mm)	U_{c2}	Grounded	U_0(kV)	U_{01}(kV)	U_{02}(kV)	U_{03}(kV)
AlNl in series	2 × 2	Al Moving	N1 Moving	38.9	56.3	61.1	64.0
		Nl Moving	Al Moving	40.1	60.2	66.5	70.1
A2Nl in series	2 × 2	A2 Moving	N1 Moving	39.1	53.5	58.4	61.3
		N1 Moving	A2 Moving	41.6	53.6	57.4	59.7

values of U_0 were almost the same. The reason could be due to the dielectric strength of single vacuum interrupter A1 was better than that of A2, as indicated in Table 5.15.

When a high voltage is applied to two vacuum interrupters in series, it is difficult to measure their voltage distributions by voltage divider, because the impedance of voltage divider is not high enough, which is typically less than hundreds MΩ. This section proposes a method to estimate the voltage distribution between two vacuum interrupters in series. The proposed method estimates the voltage distribution through I_e. From Eq. (5.8) one can get Eq. (5.9).

Table 5.20 Equation (5.9) constants for single vacuum interrupter

VI	Test	T_a (ms)	U_{c2}	a′	b′	c′
Al	Test 2	10	Moving Fixed	3.85E+00 4.64E+00	6.80E+01 4.28E+01	2.91E-01 5.44E-01
A2	Test 2	10	Moving Fixed	2.29E+00 2.10E+00	9.14E+02 1.69E+03	1.95E-01 8.37E-02
N1	Test 2	10	Moving Fixed	2.87E+00 3.45E+00	6.83E+06 5.72E+05	9.54E-04 5.37E-04

$$U_{C2} = a' \times ln\left[b' \times (I_e + c')\right] \tag{5.9}$$

where a′, b′and c′are constants, which can be calculated from *a*, *b* and *c*. By using Eq. (5.9), one can obtain a dependence of the voltage across single vacuum interrupter or double vacuum interrupters in series on the field emission current. Table 5.20 shows the constants in Eq. (5.9) for single vacuum interrupter A1, A2 and N1 when the contact gap was 2 mm. Therefore, the voltage across a single vacuum interrupter can be estimated if I_e is known. Table 5.21 shows the constants in Eq. (5.9) for two vacuum interrupters in series with each contact gap of 2 mm. It includes vacuum interrupters A1 and A2 in series, A1 and N1 in series, and A2 and N1 in series.

One can determine the voltage distribution of double vacuum interrupters in series with the following method. The method is shown in Fig. 5.33. In Fig. 5.33, vacuum interrupters A1 and A2 were in series. Therefore, their emission currents were identical. The voltage across each vacuum interrupter can be determined by a calculation with Eq. (5.9) and Table 5.20 if I_e is known. The solid line in Fig. 5.33 is the experimental curve for double vacuum interrupters A1 and A2 in series, which is based on Eq. (5.9) and Table 5.21. The dashed line in Fig. 5.33 is the curve that is the sum of voltage of A1 and voltage of A2. It showed that the dashed curve well fitted the experiment data. Thus, with a measured I_e, the voltage distribution for double

Table 5.21 Equation (5.9) constants for vacuum interrupters in series

VI	d (mm)	U_{c2}	Grounded	a′	b′	c′
A1A2 in series	2 × 2	A1 Moving A2 Moving	A2 Moving Al Moving	5.90 E+00 6.37 E+00	2.19 E+02 2.39 E+02	2.54 E-01 2.56 E-01
AlN1 in series	2 × 2	A1 Moving N1 Moving	N1 Moving A1 Moving	6.94 E+00 9.09 E+00	3.36 E+04 7.46 E+03	−2.00 E-03 1.00 E-03
A2N1 in series	2 × 2	A2 Moving N1 Moving	N1 Moving A2 Moving	7.39 E+00 5.56 E+00	1.32 E+04 1.50 E+05	4.93 E-03 1.70 E-03

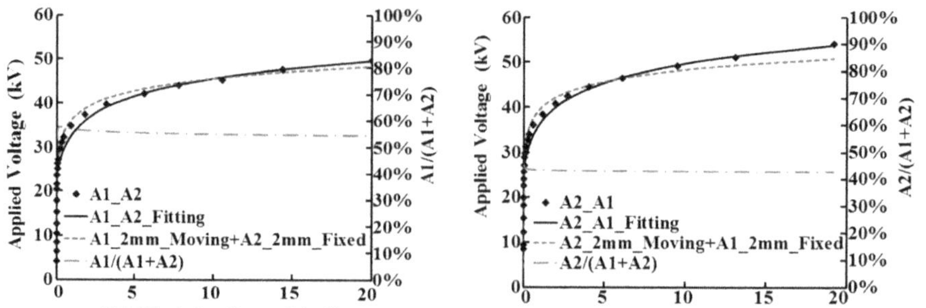

(a) the high voltage applied to the moving terminal of A1 (b) the high voltage applied to the moving terminal of A2

Fig. 5.33 A method to determine voltage distribution for double vacuum interrupters in series. Traces '—' represents vacuum interrupter 'A1 and A2 in series'. It refers to Table 5.21. Traces '-----' represents the sum of voltage of A1 and voltage of A2. The voltage of each vacuum interrupter refers to Table 5.20. Traces '–·–·' represents the ratio of vacuum interrupter A1 or A2 to the total voltage

vacuum interrupters in series can be determined by the voltage of each vacuum interrupter at this I_e. In Fig. 5.33a, the high voltage was applied to the moving terminal of vacuum interrupter A1. The ratio of voltage across vacuum interrupter A1 was about 55% of the total voltage. In Fig. 5.33b, the high voltage was applied to the moving terminal of vacuum interrupter A2. The ratio of voltage across vacuum interrupter A2 was about 44% of the total voltage.

With the same method, it determined the voltage distribution of double vacuum interrupters for vacuum interrupter A1 and N1 in series, and vacuum interrupters A2 and N1 in series. Figure 5.34 shows the situation of vacuum interrupters A1 and N1 in series. Figure 5.35 shows the situation of vacuum interrupters A2 and N1 in series. It showed that the sum of voltages on each vacuum interrupter (trace '---') was close to that of the experimental curve for double vacuum interrupters in series (trace '—'). The voltage distribution (trace '–·–·') of the two vacuum interrupters is also

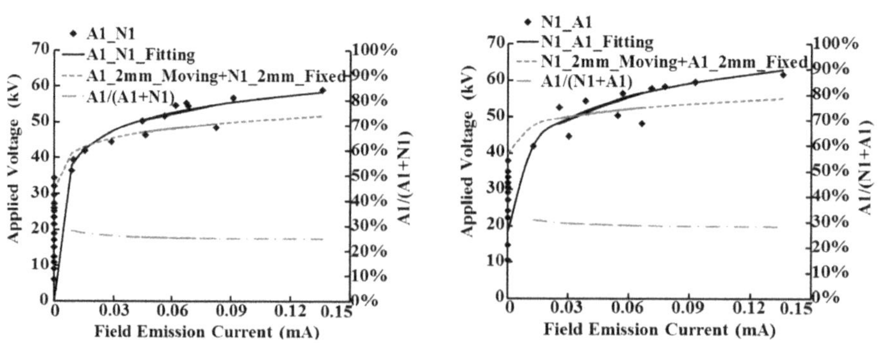

(a) the high voltage was applied to the moving terminal of A1 (b) the high voltage was applied to the moving terminal of N1

Fig. 5.34 Determination of voltage distribution with vacuum interrupters A1 and N1 in series

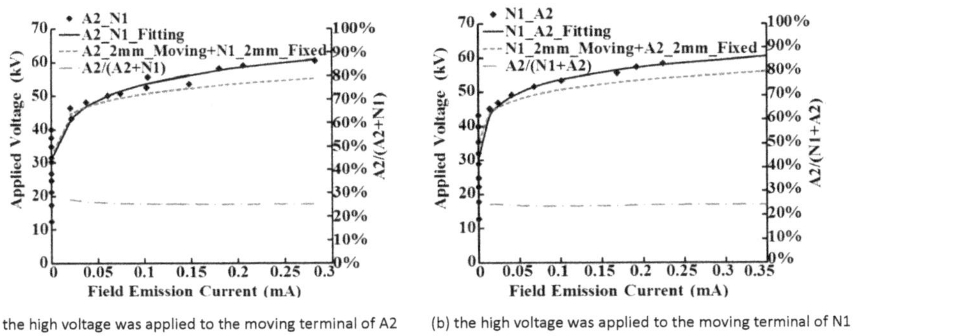

Fig. 5.35 Determination of voltage distribution with vacuum interrupters A2 and N1 in series

shown. For vacuum interrupter A1 and N1 in series, the voltage shared by A1 was about 25% when the high voltage was applied to the moving terminal of A1. When the high voltage was applied to the moving terminal of N1, the voltage shared by A1 was about 29%. For vacuum interrupter A2 and N1 in series, the voltage shared by vacuum interrupter A2 was about 25% when the high voltage was applied to the moving terminal of A2. When the high voltage was applied to the moving terminal of N1, the voltage shared by A2 was about 24%. Thus, the new vacuum interrupter N1 shared more than 70% of the total voltage when it was in series with a prestrike arc deteriorated vacuum interrupter.

As mentioned in previous Sects. 5.1.1–5.1.3, the main reason for restrikes during capacitive current switching could be related to the contact damages caused by inrush current, especially by that of high inrush current (see Figs. 5.12 and 5.19). The damaged contact surfaces are difficult to withstand the following recovery voltage. It found that the field emission currents of vacuum interrupters A1 or A2 (prestrike arc deteriorated by making inrush current) were much higher than that of new vacuum interrupter N1. The field emission current I_e is strongly dependent on the contact surface condition and the applied voltage [21]. The electrical field strength between the two contacts $E = \beta_m \times \beta_g \times U/d$ plays an important role on field emission current, where β_m is micro-enhancement factor and β_g is geometric enhancement factor. For large protrusions and craters, β_m and even β_g can increase significantly. So strong electrical filed would appear on the damaged contact surface area and further cause high emission current.

When a high voltage is applied to two vacuum interrupters in series, it is difficult to precisely measure the voltage distribution by a voltage divider. As the impedance of voltage divider (hundreds MΩ) is much lower than that of vacuum interrupter, the voltage divider in parallel changes the voltage distribution of the two vacuum interrupters in series. This section proposes a voltage distribution estimation method based on I_e. It found that a new vacuum interrupter shared more than 70% of the total voltage when it is in series with a prestrike arc deteriorated vacuum interrupter. However, if two prestrike arc deteriorated vacuum interrupters in series, each one shared about a half of the total voltage.

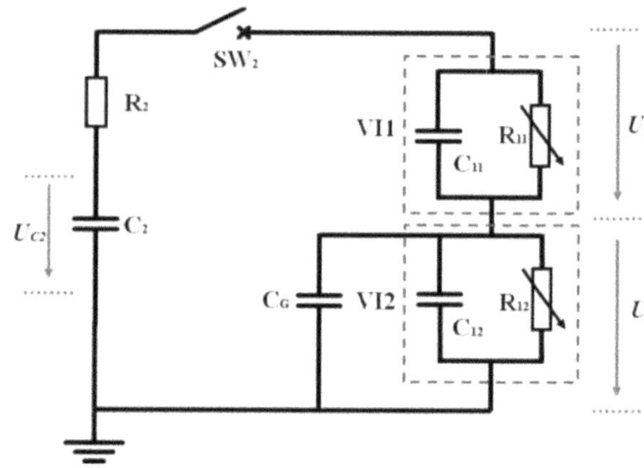

Fig. 5.36 An equivalent circuit of two vacuum interrupters in series. C_2: Capacitor 0.5 μF; R_2: Current-limiting resistance 10 kΩ; SW2: Isolated switch; C_G: Ground stray capacitance; C_{11}, C_{12}: Equivalent capacitance; R_{11} R_{12}: Equivalent resistance

Figure 5.36 shows an equivalent circuit for two vacuum interrupters in series. It explains the voltage distribution between the two vacuum interrupters. C_{11} and C_{12} are equivalent capacitance for the two vacuum interrupters. C_G is the ground stray capacitance. C_{11}, C_{12} and C_G are about several to tens of pF. So the voltage distribution was firstly determined by the three capacitance. Once the isolated switch SW_2 was closed, the equivalent resistance R_{11} and R_{12} varied with the shared voltage (U_1 or U_2). The middle terminal is floating until a voltage ratio was achieved that provided the same field emission current through R_{11} and R_{12}. The condition of the contact surfaces determined the field emission current, which further impacted the voltage distribution of two vacuum interrupters in series.

Moreover, the field emission current of a prestrike arc deteriorated vacuum interrupter and a new vacuum interrupter in series was significantly lower than that of two prestrike arc deteriorated vacuum interrupters in series. It suggested that if a good dielectric strength is desired for two vacuum interrupters in series, it was better to keep at least one of the vacuum interrupter's contact surface fine. For traditional double break vacuum circuit breakers, the two vacuum interrupters usually operate simultaneously. The two vacuum interrupters would make inrush current simultaneously. So that the contact surfaces would be both damaged by the prestrike arcs. Therefore, the dielectric performance decreased. For back-to-back capacitor banks switching, a specific operation sequence is proposed for two vacuum interrupters in series. Figure 5.37 shows the operation sequence. There was a time intervalΔT_c between the two vacuum interrupters VI_1 and VI_2 during a making operation. VI_1 closes at instant T1 and VI_2 closes at instant T2. The prestrike arc tends to occur in VI_2 as VI_1 already closed. So the prestrike arc damages would mainly generate on the contact surfaces of VI_2. During the opening operation, there is a time intervalΔT_o between the two vacuum interrupters. VI_1 opens first at instant T3, which has a relatively high insulation strength to withstand the recovery voltage, because its contact surfaces are minimized for prestrike arcs. It suggests that ΔT_o is less than arcing time because hundreds of amperes of the switching current would condition the damaged contact surface of VI_2. Besides, to achieve a better current

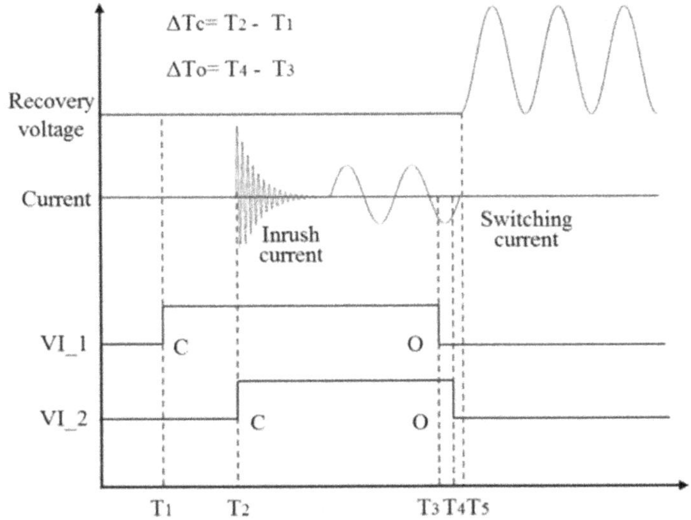

Fig. 5.37 A proposed operation sequence for two vacuum interrupters in series

switching performance, VI_1 can choose a strong breaking capability contact material to interrupt short circuit current, such as CuCr. VI_2 can choose a strong anti-welding contact material, such as CuW. Section 5.5 shows the proposed technologies in details.

5.1.5 Inrush Current Interruption and Contact Gap Breakdown

Section 5.1.1 discussed inrush current interruption phenomena in the capacitive current making operations of vacuum interrupters. This section is to understand the inrush current interruption phenomena in details. As the phenomena became significant at a large contact gap (see Sect. 5.1.1), the capacitive current making experiments were carried out with 40.5 kV vacuum interrupters, which had a larger contact gap than the 7.2 kV vacuum interrupters. Firstly, 80 times capacitive current making operations were carried out. Then, the inrush current interruption phenomena and contact gap breakdown phenomena during the capacitive current making were observed and analyzed. The results would help to get insight to prestrike phenomena in vacuum.

Table 5.22 shows the experimental conditions for the capacitive current making operations. In the experiment, the inrush current was 6.48 kA peak and the frequency was 207 Hz. Such choice of inrush current parameters was based on a service condition of 500 kV substations in China, in which the capacitor banks were used for reactive power compensation. The capacitor bank was typically with a reactive power

Table 5.22 The experimental conditions for the capacitive current making operations

Test VIs	Arc-melted CuCr40			Infiltration CuCr50		
	No.B1	No.B2	No.B3	No.C1	No.C2	No.C3
Inrush Current	6.48 kA peak and 207 Hz					
Switching Current	0 A (No Load)					
Applied Voltage	$40.5 \times \frac{\sqrt{2}}{\sqrt{3}} \times 1.4$kV					
Full Stroke	20 mm					
Switching Test Duty	C–O: 80 times					

of $Q_c = 60$ Mvar at $U_r = 35$ kV, which was connected to the substation bus-bar through a series reactor whose damping factor was 6%. The experiment was carried out with a commercial 40.5 kV vacuum circuit breaker equipped with permanent magnetic actuators. There were six vacuum interrupters prepared for the experiment. The diameter of the contact was 50 mm with butt type design. Three vacuum interrupters were with the contact material of infiltration CuCr50 (50 wt% of Cr) and the other three vacuum interrupters were with arc-melted CuCr40 (40 wt% of Cr). The full contact gap was 20 mm. The average closing velocity was 0.77 m/s which was measured from 2/3 gap to the closing position. Before the capacitive making test, all the six vacuum interrupters experienced current and voltage conditioning as commercial vacuum interrupters would experience.

Figure 5.38a shows a capacitive current making test circuit. The test vacuum circuit breaker was integrated in a *L-C* oscillating circuit, which provided an inrush current with a first peak of 6.48 kA at 207 Hz. The charging voltage of the capacitor was 46.3 kV, which was 1.4 per unit (p.u.). The capacitive voltage factor of 1.4 was considered for capacitor banks with isolated neutral [25, 26]. In the capacitive current making test, the capacitor *C* was pre-charged and then it discharged through a reactor *L*. Thus an inrush current passed through the test vacuum circuit breaker. The bounce of the test vacuum circuit breaker was eliminated by adjusting the permanent magnetic actuator. A series of 80 making operations was executed for each of the six 40.5 kV vacuum interrupters. The breaking operations were performed under no-load so as to minimize the conditioning effect of the switching current. Figure 5.38b shows a typical making operation oscillogram. In the Figure, traces U_{make}, I_{make}, Travel and $U_{\text{pre-arc}}$ represent the making voltage across the vacuum interrupter, the inrush current passing through the vacuum interrupter, the displacement curve of the moving contact and the prestriking arc voltage, respectively. The prestriking arc voltage was measured by a special arc voltage measurement circuit. The measurement circuit uses a TBF50-75/5 type voltage divider whose ratio is 100:1. Its power frequency withstand voltage was 50 kV while BIL was 85 kV. The output of the voltage divider was protected by a transient voltage suppressor (TVS) whose action voltage was

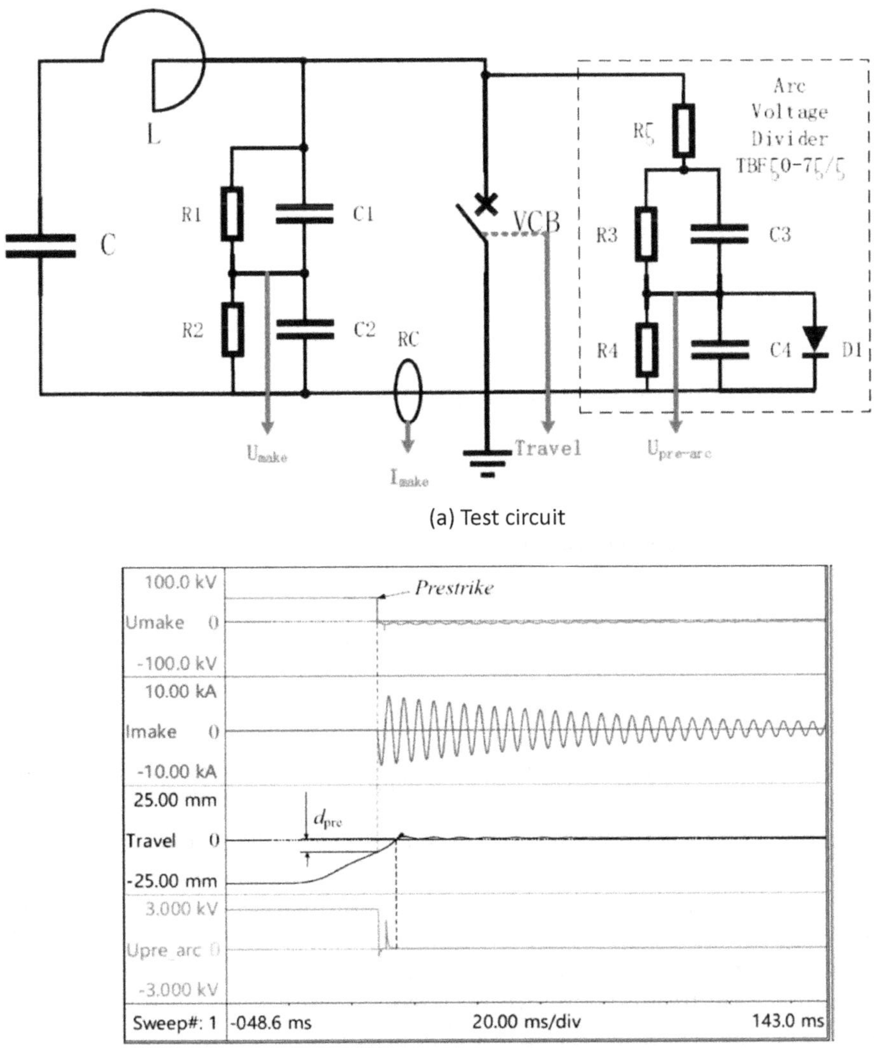

Fig. 5.38 Capacitive current making test circuit and a typical oscillogram

2500 V, which means that the TVS would cut off the voltage higher than 2500 V. $U_{\text{pre-arc}}$ was also used to indicate the instant when the contacts physically touch. The prestrike gap d_{pre} was defined as the distance between the contacts when prestrike first occurs.

Figure 5.39 shows a complementary cumulative probability distribution (CCDF) of the prestrike gaps. As the probability of prestrike increased with a decrease of the prestrike gap, it was found the CCDF of the prestrike gaps would be evaluated by a

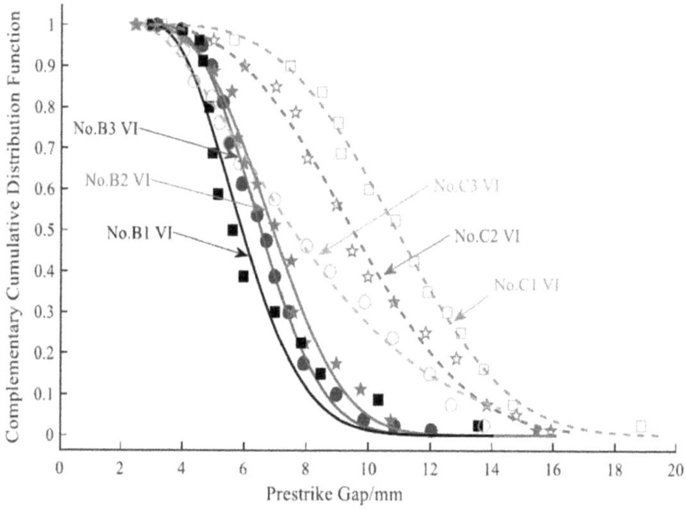

Fig. 5.39 Complementary cumulative probability distribution of prestrike gap for test series B and C

3-parameter Weibull distribution function as Eq. (5.10), in which d_{100} is the location parameter that means the prestrike gap below which the prestrike probability is 1. m is the Weibull shape or shape factor. Θ is the scale parameter.

$$F(d_{\text{pre}}) = \exp\left\{-\left[(d_{\text{pre}} - d_{100})/(\Theta - d_{100})\right]^{m}\right\} \tag{5.10}$$

Table 5.23 shows the Weibull distribution parameters. R-squared is in a range of 0–1 and it indicates the goodness of the fit. It shows all the R-squared are above 0.95, so that it is reasonable to assume that the complementary cumulative probability of the d_{pre} follows the 3-parameter Weibull distribution. It found that although the contact materials and preparing technologies are different, the prestrike gap d_{100} below which prestrike would always occur is close to each other, which is from 2.4 to 3.1 mm.

Figure 5.40 shows a typical prestrike oscillogram with inrush current interrup-

Table 5.23 Parameters of the 3-parameter Weibull distribution in test series B and C

Test VIs	m/mm	d_{100}/mm	Θ/mm	R-squared
No.B1	2.1	3.0	6.5	0.951
No.B2	2.8	2.5	7.5	0.988
No.B3	2.5	3.1	7.1	0.996
No.C1	3.1	3.0	11.9	0.996
No.C2	2.7	2.4	10.5	0.992
No.C3	1.6	2.6	8.8	0.992

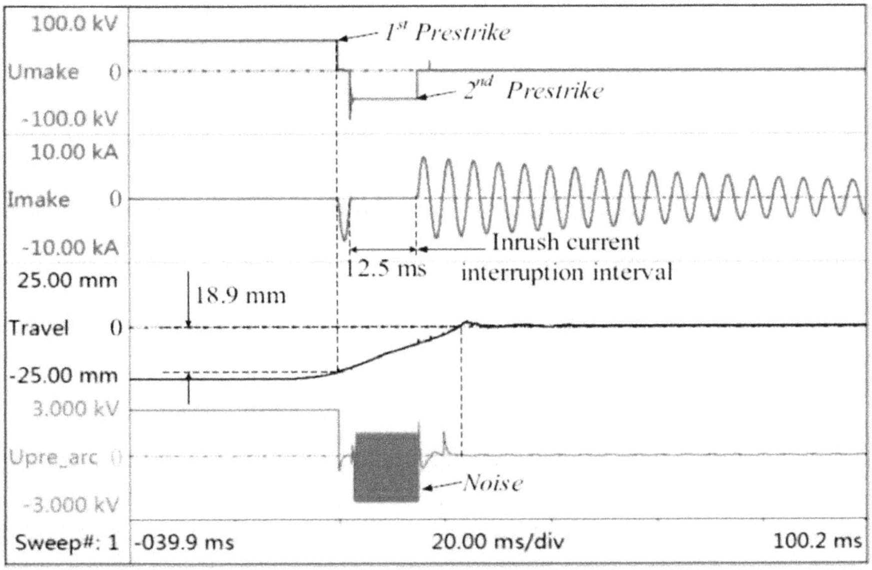

(a) An inrush current interruption with 12.5 ms interval

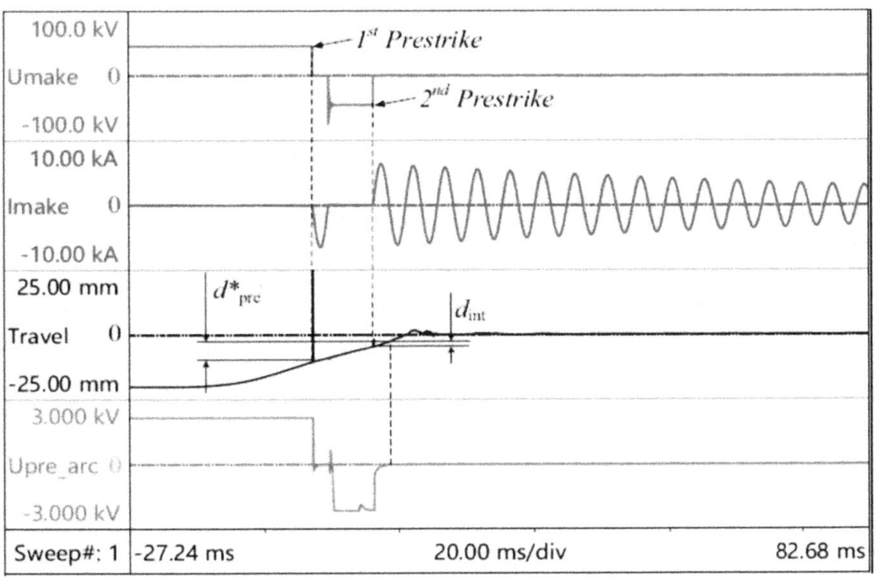

(b) The definition of d_{int} and d^*_{pre} when inrush current interruption occurs

Fig. 5.40 A typical prestrike oscillogram with inrush current interruption *Note* The $U_{\text{pre-arc}}$ noise shown in Fig. 5.40a is due to high frequency noise generated by the voltage suppressor in the arc voltage measuring circuit tion. Figure 5.40a shows that the prestrike gap could reach 18.9 mm that meant that prestrike occurred nearly at the moment the moving contact started to move, because the full contact stroke was 20 mm. In this case there was an inrush current interruption and its interval was 12.5 ms. This interval was between an interruption of an initial inrush current and a later inrush current initiation in the making operation. Figure 5.40b shows a definition of inrush current interruption re-ignition gap d_{int}. It is the contact gap at which the second prestrike occurs after an inrush current interruption. Figure 5.41 shows the distributions of inrush current interruption re-ignition gaps d_{int} versus the number of making operations for the six test vacuum interrupters. Here d_{int} equal to zero means the inrush current interruption did not occur. It shows that, during the capacitive current making operations, inrush current interruption was a common phenomenon.

It was found that the inrush current interruption is the other side of the contact gap breakdown phenomenon. Figure 5.42 shows a typical prestrike oscillogram of contact gap breakdowns. This phenomenon can be observed in all the six test vacuum interrupters. Figure 5.42 shows five events of contact gap breakdown. It indicated that the inrush current was interrupted at each inrush current zero, then a recovery voltage rose. The contact gap cannot withstand the recovery voltage, which led to a contact gap breakdown and inrush current continued flowing. The five breakdown voltages were −45.4, 44.0, −38.0, 50.7 and −29.6 kV for the No.1–No.5 spikes, respectively.

It found that, there are three types of contact gap breakdown. Figure 5.43 shows the three types of contact gap breakdowns. Figure 5.43a shows the contact gap breakdown type I. In type I breakdown occurred only once. Figure 5.43b shows the contact gap breakdown type II. In type II there are two consecutive positive breakdowns. Figure 5.43c shows the contact gap breakdown type III. In type III there are two consecutive negative breakdowns. The breakdown voltage is not polarity dependent. It is associated with the polarity of the inrush current at that current zero. The contact gap breakdown type II has a decreasing breakdown voltage while the contact gap breakdown type III has an increasing breakdown voltage. In Fig. 5.43, the small current at contact gap breakdown might be originated from a discharge of parasitic capacitances in the vicinity of the vacuum interrupter.

The possible mechanism of the inrush current interruption phenomenon is explained as following. During the 80 times making-breaking operations, the prestriking arc melt the contact surface locally and then the two contacts would weld when they mated each other. Figure 5.44 shows the photos of the contact surfaces after 80 making operations for the arc-melted CuCr40 contact material and infiltration CuCr50 contact material. Then in the following no-load breaking process, the welded area would rupture and macroscopic protrusion would be generated on the contact surface, while micro-particles might be released in the contact gap. Figure 5.45 shows a physical explanation of inrush current interruption. In making operation, the 1st prestrike might be initiated between the moving contact and the nearby particles, as shown in Fig. 5.45a. At the next inrush current zero the prestriking arc would easily to be extinguished because of the strong interrupting capability of the vacuum interrupters. As the contacts approach each other more closely, the 2nd prestrike is

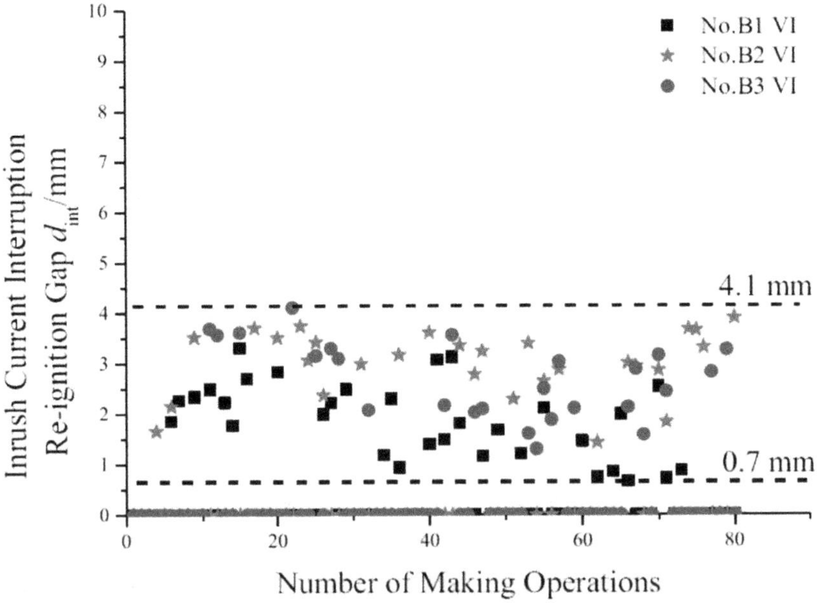

(a) test series B with arc-melted CuCr40 contact material

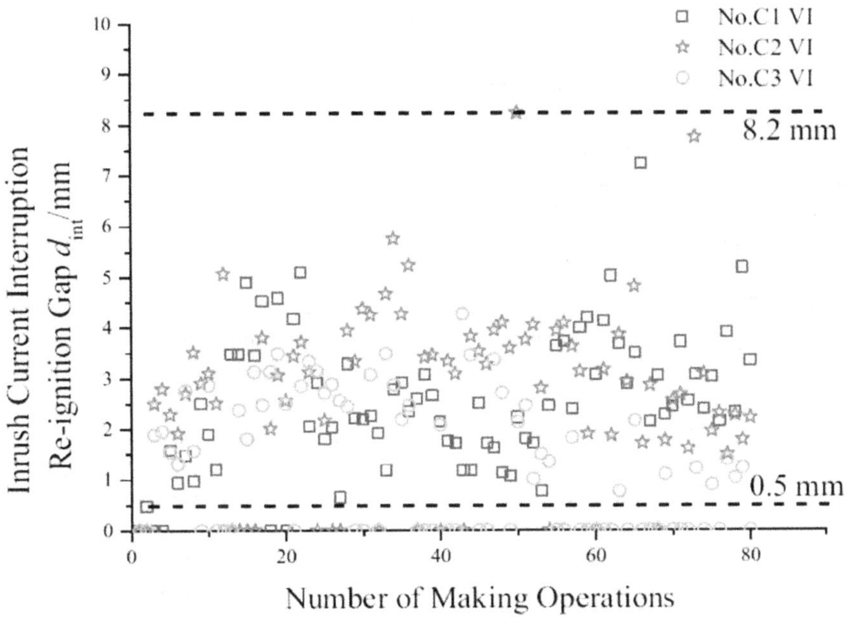

(b) test series C with Infiltration CuCr50 contact material

Fig. 5.41 Distributions of inrush current interruption re-ignition gaps versus the number of making operations for the six test vacuum interrupters

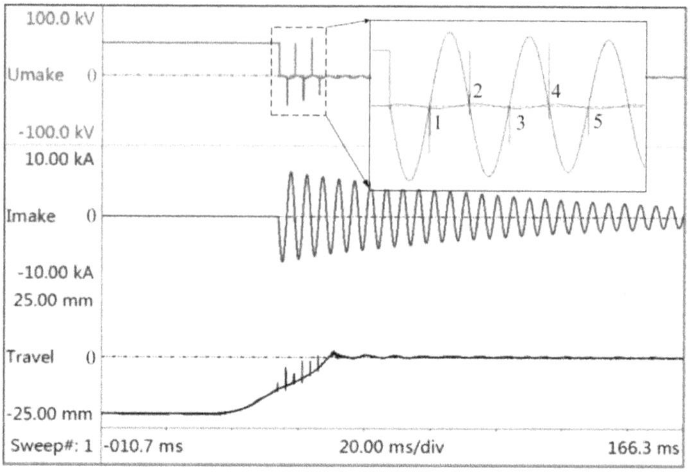

Fig. 5.42 A typical prestrike oscillogram of contact gap breakdowns

possible to be initiated between the contacts that would be modified with the microscopic protrusion, as shown in Fig. 5.45b. The distance between the moving contact and the micro particle(s) or the stationary contact may determine the prestrike gaps.

It is found that, after prestrike initiating inrush current during the making operations, it would always tend to interrupt inrush current at each current zero before contacts mate each other. This phenomenon of course results from the inherent characteristics that vacuum interrupters would interrupt the current with very high di/dt. If the inrush current was successfully interrupted, the inrush current interruption phenomenon would occur. But in most cases, contact gap breakdown occurs and inrush current is not interrupted, as shown in Fig. 5.42. Figure 5.43 presents three types of contact gap breakdowns during capacitive making operations. The contact gap breakdowns are the other side of inrush current interruption.

Figure 5.46 proposes possible physical processes of contact gap breakdowns for type I, type II, and type III, respectively. The vacuum insulation community presently considers two mechanisms for the origin of vacuum breakdown: field emission dominated breakdown and micro-particles induced breakdown [17]. The contact gap breakdowns are possibly related to not only generated protrusions but also micro-particles released by the welding rupture in the no-load breaking process. Figure 5.46a explains the type I contact gap breakdowns. At the inrush current zeros, prestriking arc would extinguish, then the dielectric strength of the contact gap would increase to withstand the increasing voltage. A discharge might occur between a contact and the opposite protrusion if the applied voltage is higher than the dielectric strength of the contact gap, as shown in a process of type I (a). This discharge may initiate a prestriking arc again immediately, which could continue the inrush current flowing, as shown in a process of type I (b). Figure 5.46b explains the type II contact gap breakdowns. A protrusion with a relatively high height might appear. This also explains why the contact gap breakdown voltage in type I is higher than

Fig. 5.43 Three types of contact gap breakdowns during making operations

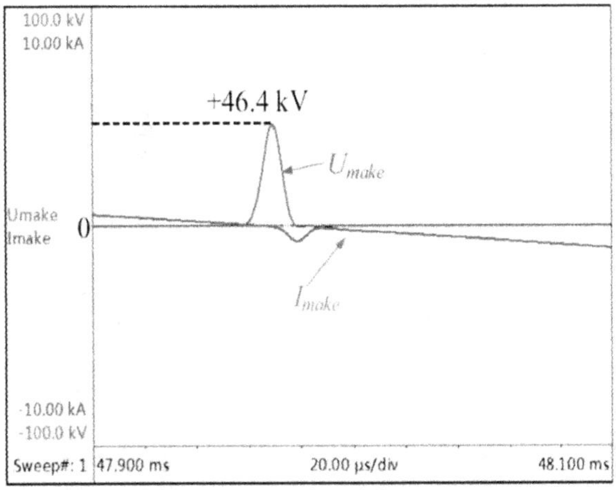

(a) Contact gap breakdown type I

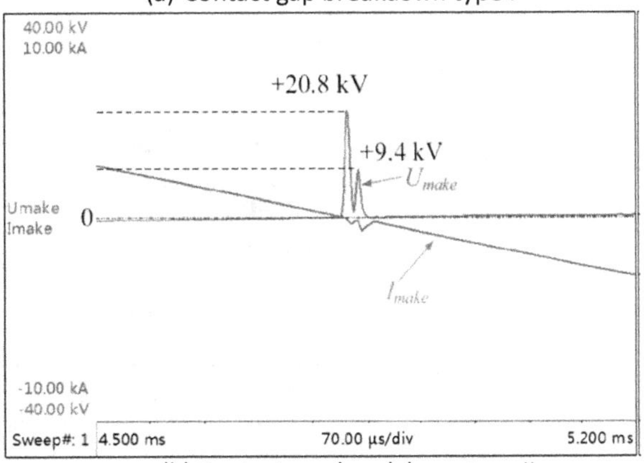

(b) Contact gap breakdown type II

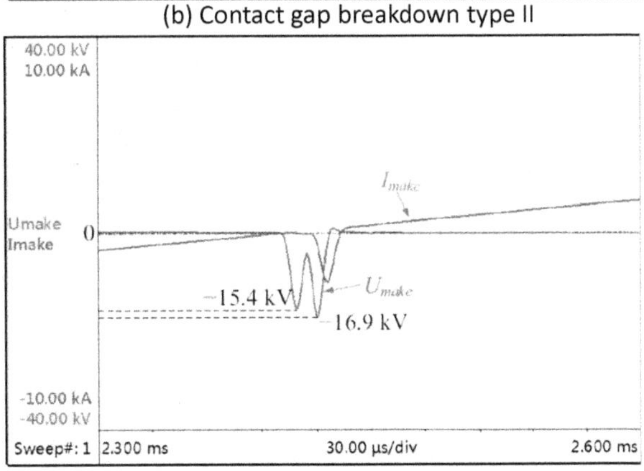

(c) Contact gap breakdown type III

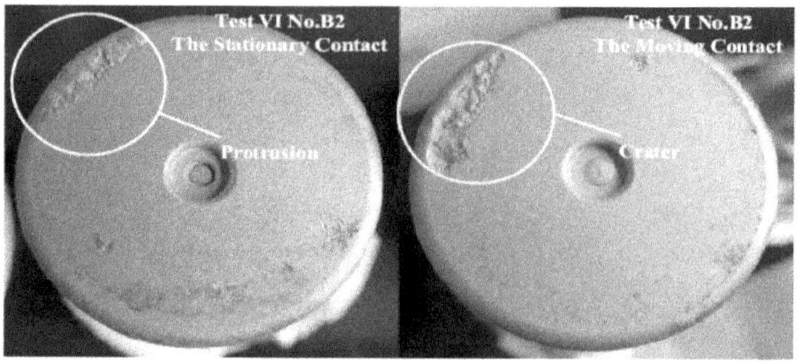

(a) Arc-melted CuCr40 contact material (No.B2)

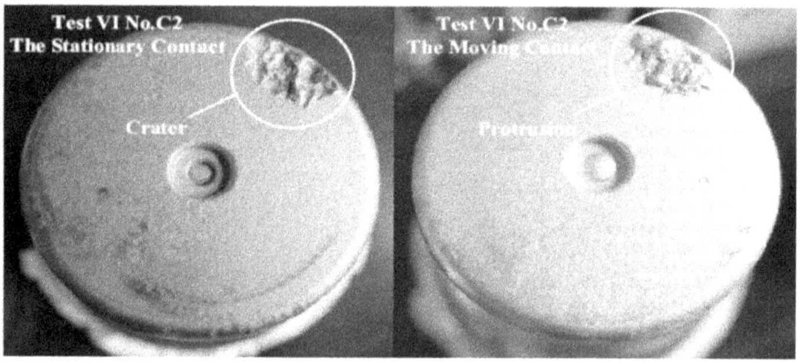

(b) Infiltration CuCr50 contact material (No.C2)

Fig. 5.44 Photograph of the contact surfaces after 80 making operations

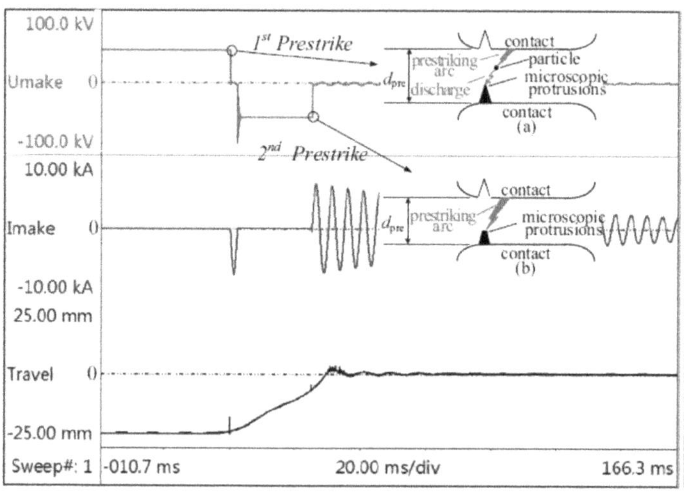

Fig. 5.45 Physical explanation of inrush current interruption. **a** The 1st prestrike initiated between the contact and the micro particle; **b** The 2nd prestrike initiated between the contact and the microscopic protrusion

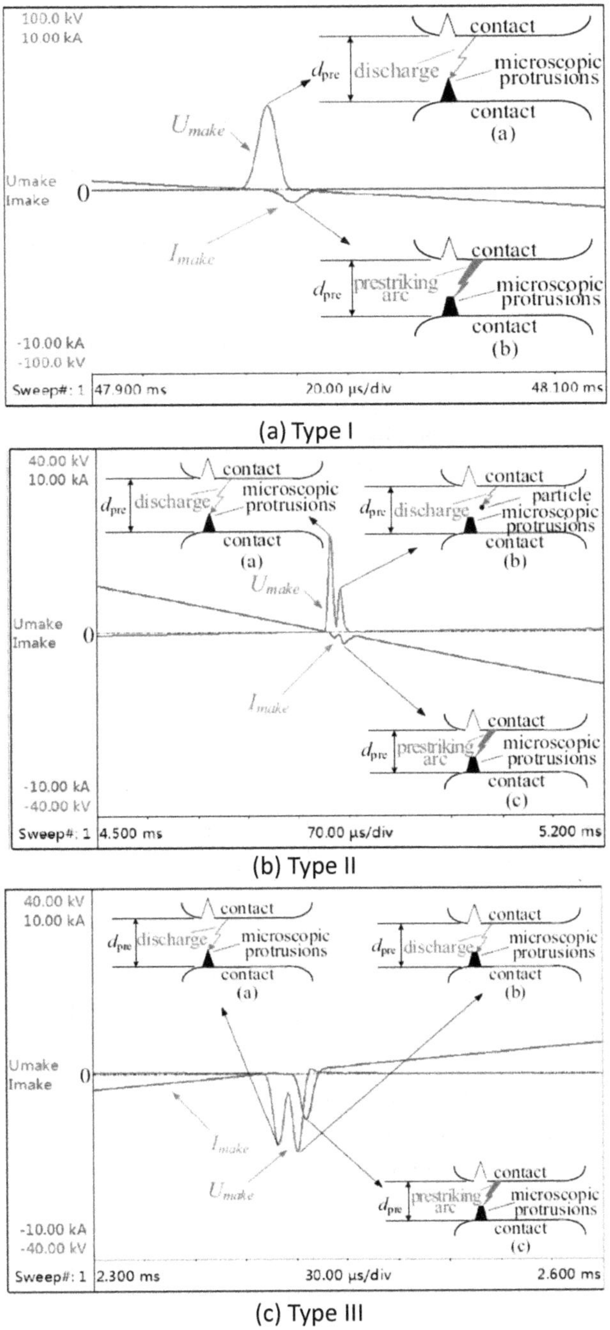

Fig. 5.46 Physical explanation of contact gap breakdowns

that in type II. At the inrush current zeros, the prestriking arc would extinguish, then the dielectric strength of the contact gap increased. As the applied voltage increases, the first discharge might initiate between a contact and the opposite protrusion if the applied voltage is higher than the dielectric strength of the contact gap, as shown in a process of type II (a). In such case the tip of the protrusion might be eroded and micro-particles might be released. Then the second discharge might initiate between the contact and the released particles, as shown in a process of type II (b). The second discharge may transform into a prestriking arc and inrush current would flow again subsequently, as shown in a process of type II (c). Figure 5.46c explains the type III contact gap breakdowns. Again, at the inrush current zeros, prestriking arc would extinguish and the dielectric strength of the contact gap would increase. As shown in a process of type III (a), if the applied voltage is higher than the dielectric strength of the contact gap, the first discharge may occur. The tip of the protrusion might be eroded significantly. The second discharge might occur between the contact and the opposite eroded protrusion, as shown in a process of type III (b). That is a possible reason why the amplitude of the first breakdown voltage is lower than that of the second one. This secondary discharge may transform into a prestriking arc immediately and inrush current would continue flowing subsequently, as shown in a process of type III (c).

5.2 Capacitive Current Switching and Magnetic Field Between Contacts

5.2.1 Prestrike Inrush Current Arc Behaviors in Vacuum Interrupters Subjected to a Transverse Magnetic Field and an Axial Magnetic Field

This section is to determine the influences of magnetic field on the prestrike inrush current arc behaviors in vacuum interrupters during capacitive current making processes. In the experiment, the influences of transverse magnetic field and axial magnetic field were investigated. The amplitudes of the inrush current were set as three levels, which were 5, 10, and 20 kA, respectively. The distributions of the prestrike gaps and prestrike arc characteristics were observed. The prestrike gap d, corresponding to pre-arcing durations, reflected the actual dielectric strength between the contacts. In addition, the prestrike arc characteristics (i.e., diffuse arc or intense arc) influenced the conditions of the contact surfaces. The experimental results could help to choose appropriate magnetic field contact to improve the capacitive current switching performance of vacuum interrupters.

The inrush current was generated via a L-C discharging circuit. Figure 5.47 shows the L-C discharging circuit. The discharging circuit can generate inrush current up to 20-kA peak value at a frequency of 3800 Hz. In the beginning of each test, according to a required inrush current amplitude, the capacitor $C1$ was pre-charged to a certain

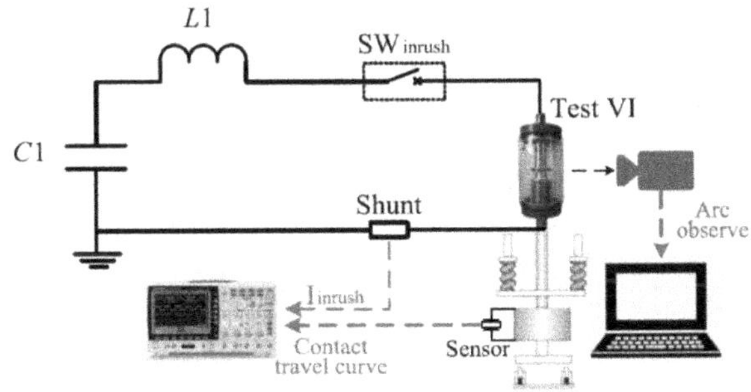

Fig. 5.47 *L–C* discharging circuit for inrush current generation. $C1 = 74.51\ \mu F$: inrush current source capacitor; $L1 = 21.34\ \mu H$: inrush current source inductance; Shunt: 75 $\mu\Omega$; SW_{inrush}: inrush current making switch; Test VI: test vacuum interrupter

voltage U_r. U_r was 10.7 kV to generate an inrush current of 20-kA peak. After closing a switch SW_{inrush} and then closing test vacuum interrupter, an inrush current flowed in the circuit, during which process a prestrike arc generated in the test vacuum interrupter before the contacts mechanically mated. The waveform of the inrush current was measured through a shunt. The contact closing travel curve was recorded by a displacement sensor. The test vacuum interrupter was driven by a permanent magnetic actuator, which provided an average closing velocity of 0.8 m/s for the moving contact. The prestrike arc behaviors during the making process were recorded by a charge-coupled device (CCD) video camera. The CCD video camera had a recording speed of 30 000 frames/s and the exposure time was set to 2 μs.

Figure 5.48 shows a typical oscillogram in a making operation. Inrush current was 5-kA peak and its frequency was 3800-Hz. The upper trace corresponds to contact travel, and the lower trace corresponds to inrush current. The prestrike moment and the prestrike gap *d* were determined by comparing the inrush current waveform with the contact travel curve.

In the test, six 12-kV vacuum interrupters were adopted to understand the influence of the magnetic field on the prestrike arc characteristics during capacitive current

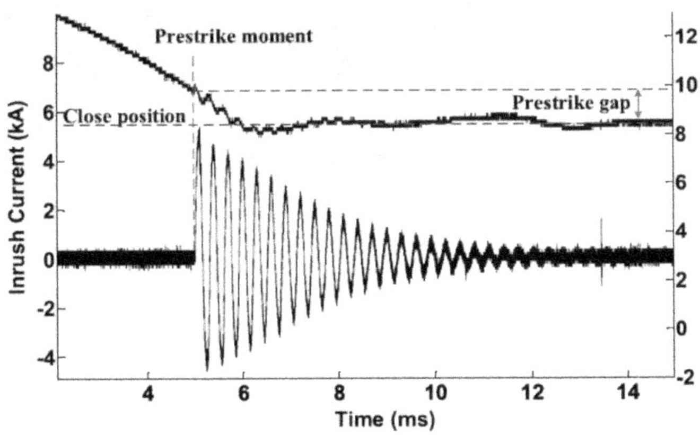

Fig. 5.48 A typical oscillogram in a making operation

making operations. Each vacuum interrupter composed of a glass envelope and no metal vapor shields were equipped, which enabled an optical observation of the prestrike arcs. The contact diameter was 48 mm and the contact material was CuCr50 (50% of Cr). The contact materials were produced via a sintering method. Three vacuum interrupters were with cup-type transverse magnetic field contact, while the other three vacuum interrupters were with cup-type axial magnetic field contact. The contact plates were all manufactured in the same way to make sure a similar state of surface roughness. Figure 5.49 shows the photos of the transverse magnetic field and axial magnetic field contacts. Besides, all the test vacuum interrupters were conditioned by a power frequency voltage to ensure the same rated dielectric strength.

Table 5.24 shows the experiment conditions. The six test vacuum interrupters were divided into two groups, which was group A and B. Group A were with cup type transverse magnetic field contacts, and group B were with cup-type axial magnetic field contacts. The vacuum interrupters in each group were labeled 1, 2, and 3. They corresponded to the making inrush currents of 5-, 10-, and 20-kA peaks, respectively. A series of 30 making operations was executed for the vacuum interrupters A1 and B1 by using the inrush current of 5-kA peak and of 3800-Hz frequency. While the prestrike gap d was measured and the prestrike arc behaviors were observed. The vacuum interrupters A2 and B2 experienced an inrush current of 10-kA peak and of 3800-Hz frequency for 30 operations, while the prestrike arc behaviors and

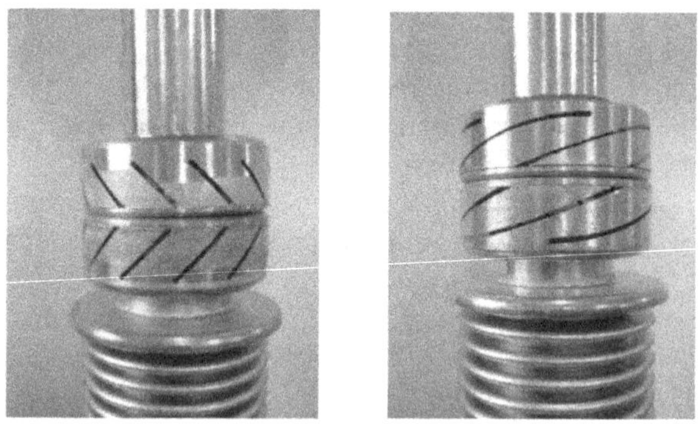

(a) cup-type transverse magnetic field contact

(b) cup-type axial magnetic field contact

Fig. 5.49 Transverse magnetic field and axial magnetic field contacts in test vacuum interrupters

Table 5.24 The experiment conditions

Test VIs		A1	A2	A3	B1	B2	B3
Inrush	Amplitude (kA)	5	10	20	5	10	20
Current	Frequency (Hz)	3800					
Contact material		CuCr50					
Contact structure		Cup-type TMF			Cup-type AMF		
Making operations		30	30	15	30	30	15

prestrike gap were measured. The vacuum interrupters A3 and B3 experienced an inrush current of a 20-kA peak and of 3800-Hz frequency for 15 operations. After each making operation of the inrush current, the contact welding forces of vacuum interrupters A3 and B3 were measured. The details of measurement of welding forces refers to Sect. 5.3.1. The breaking operation was performed with no load and without any applied voltage.

As mentioned above, 30 making operations was conducted on vacuum interrupters A1, A2, B1, and B2, and 15 making operations was conducted for vacuum interrupters A3 and B3. Moreover, after each making operation of the inrush current for A3 and B3, the welding force between the contacts were measured. Therefore, the effect of cup-type transverse magnetic field contact and cup-type axial magnetic field contact on the prestrike arc behaviors could be observed and statistically evaluated. During the prestrike process, a special attention was given to the prestrike arc behaviors, including the prestrike gaps and the prestrike arc behaviors observations. In addition, the contact welding force statistics, and hence, the contact surface conditions after the inrush current, were also shown.

Figure 5.50 shows the distributions of the prestrike gap *d* for the test vacuum interrupters versus the number of making operations. The prestrike gaps are plotted in the order of the making operations. Figure 5.50a shows the prestrike gap *d* of each making operation for vacuum interrupters of A1 and B1. Figure 5.50b shows the prestrike gap *d* of each making operation for vacuum interrupters of A2 and B2. Figure 5.50c shows the prestrike gap *d* of each making operation for vacuum interrupters of A3 and B3. In Fig. 5.50, the triangle marker and the square marker represent the prestrike gaps of transverse magnetic field contact and axial magnetic field contact, respectively. In order to study the data, a probability theory was applied.

The probability of prestrike gaps increased with a decrease in the contact gap. Thus, a complementary cumulative distribution function (CCDF) was introduced to study the prestrike gap *d*. The complementary cumulative distribution function was defined as Eq. (5.11).

$$\overline{F}(d) = 1 - F(d) \tag{5.11}$$

where $F(d)$ is the cumulative distribution function that presents the probability that the prestrike takes at a gap less than or equal to *d*. Therefore, $\overline{F}(d)$ presents the probability that the prestrike takes at a gap above *d*. Figure 5.51 shows the complementary cumulative distribution function of the prestrike gap *d*. It showed that the complementary cumulative distribution function followed a complementary three-parameters Weibull distribution function well. The Weibull function is shown in Eq. (5.12)

$$\overline{F}(d) = 1 - F(d) = \exp\left[-\left(\frac{d - d_{100}}{\Theta - d_{100}}\right)^m\right] \tag{5.12}$$

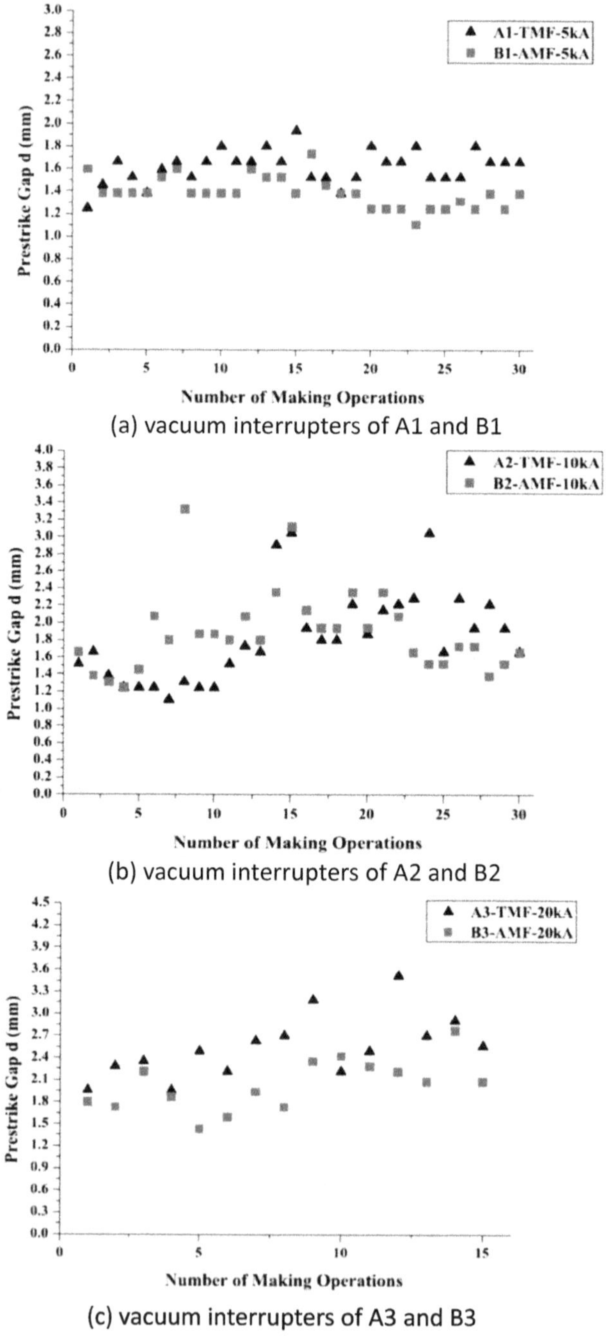

Fig. 5.50 Distributions of the prestrike gap *d* for the test vacuum interrupters versus the number of making operations

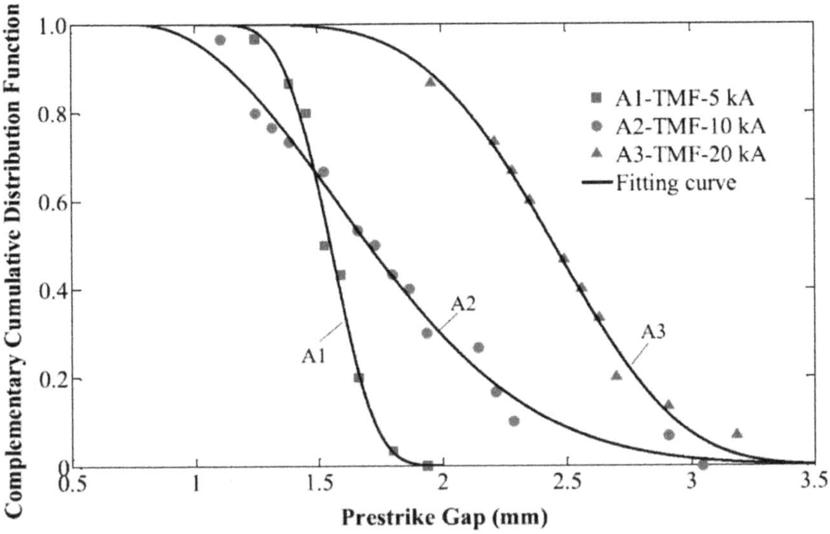

(a) vacuum interrupters A1–A3.

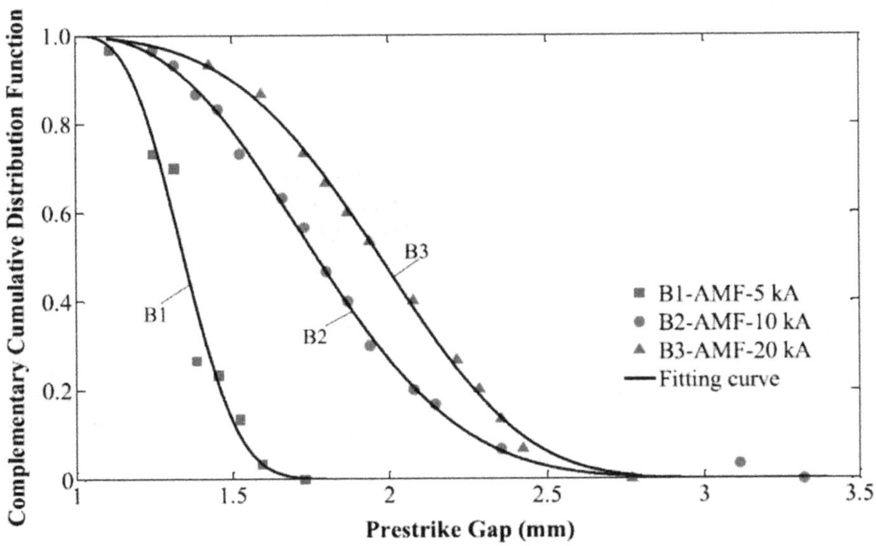

(b) vacuum interrupters B1–B3.

Fig. 5.51 Complementary cumulative probability functions of the prestrike gaps for test series A and B

Table 5.25 The fitting parameters of the complementary three-parameters Weibull distribution function for test series A and B

Test VIs	m	$d_{\overline{100}}$/mm	Θ/mm	$\Theta - d_{\overline{100}}$/mm	R-squared
Al	3.93	1.06	1.60	0.54	0.984
A2	2.07	0.76	1.89	1.13	0.987
A3	3.76	1.15	2.59	1.44	0.992
B1	2.89	1.00	1.39	0.39	0.967
B2	2.48	0.99	1.89	0.90	0.998
B3	4.02	0.70	2.10	1.40	0.998

where $d_{\overline{100}}$ is the location parameter that represents the prestrike gap below which the prestrike probability is 1. m is the Weibull shape or the shape factor. Θ is the scale parameter.

Table 5.25 shows the fitting parameters of the complementary Weibull distribution function for test series A and B. The R-squared parameter is used to indicate the fitting similarity between the F–d plot. It showed that the R-squared values were all above 0.96. Therefore, it is reasonable to assume that the complementary cumulative distribution function of the prestrike gap d followed the Weibull function. Furthermore, the scatters of the prestrike gap d could be evaluated by the scale of $(\Theta - d_{\overline{100}})$. Table 5.25 shows that, with the increase of the inrush current from 5 up to 10 and finally to 20-kA peak, the scales of $(\Theta - d_{\overline{100}})$ increased significantly. Therefore, the scattering of d increased with the increase of the inrush current from 5 up to 10 kA and finally to 20-kA peak in the test vacuum interrupters of both series A and B. The increased prestrike gap scattering could be related to the deteriorating conditions of the contact surface. In addition, it showed that vacuum interrupters A1, A2, and A3 have higher scatters of prestrike gaps than vacuum interrupter B1, B2, and B3, respectively.

Figure 5.52 shows typical prestrike arc characteristics in the making operations. Figure 5.52a shows the vacuum interrupters of A2 (cup-type transverse magnetic field contact) and B2 (cup-type axial magnetic field contact) with CuCr50 contact material during the making operation of 10-kA peak inrush current. The photos (a)–(d) were during the first half-wave (approximately 132 μs) of the inrush current after the prestrike occurred. The labeled time under each photograph presents a duration from the prestrike moment to the instant at which the photo was taken. As shown in Fig. 5.52a, for vacuum interrupter A2 of cup-type transverse magnetic field contact, the initial prestrike arcs originated at one point between the contacts. The prestrike arcs rapidly evolved into intense arcs at a short prestrike gap with a fast increase of the inrush current. Subsequently, the prestrike arcs quickly diffused in a short duration when the inrush current approached zero. For vacuum interrupter B2 of cup-type axial magnetic field contacts, the ignition of arcs also started at one point between the contacts. However, with an increase of the inrush current up to the peak of 10 kA, the prestrike arcs tend to be diffuse arcs with an expanding distribution and the arcs kept diffuse until current zero. Figure 5.52b shows typical

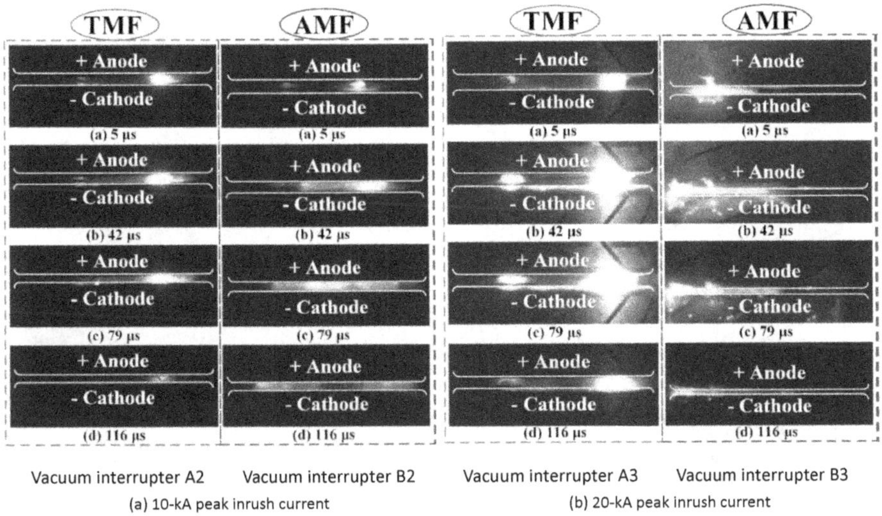

Fig. 5.52 Prestrike arc characteristics in making operation of capacitive current

prestrike arc characteristics of vacuum interrupters A3 and B3 during the making operation of 20-kA peak inrush current. The photos (a)–(d) were also taken during the first half-wave of the inrush current. It shows the initial prestrike arcs in vacuum interrupter A3 originated from two points between the contacts. They rapidly evolved to be intense arcs and kept in intense arcs until current zero. The prestrike arcs in vacuum interrupter B3 originated from one point between the contacts, and then the arc expanded rapidly, during which a fierce emission of a strong light with metal droplets or particles could be observed. The larger arcing areas and reduced arcing brightness in vacuum interrupter B3 indicated the arc was more diffuse than that in A3. In addition, the intense arcs during the prestriking process were more significant in cup-type transverse magnetic field contacts; whereas the prestrike arcs in cup-type axial magnetic field contacts tended to be diffuse.

The welding forces were measured after each making operation for vacuum interrupters of A3 and B3. The welding force was measured for 15 times. Figure 5.53 shows the welding force measurement results. The measured forces are marked by circles, of which the filled circles refer to "welded" cases, whereas the empty circles refer to "un-welded" cases. In the case of vacuum interrupter A3, the welding force was within a range of 980–3880 N. The average welding force for the 15 operations was 2039 N. The welding force for vacuum interrupter B3 was within a range of 620–2880 N for the 12 welded cases. The average welding force of the 12 operations was 1518 N.

The contact surfaces were observed after tests. Figure 5.54 shows the photos of the contact surfaces after the inrush current making operations for test series A and B. It found that each contact surface had a macroscopic arcing erosion area, whereas the rest of the surface area is definitely "clean". This finding suggested that virtually all

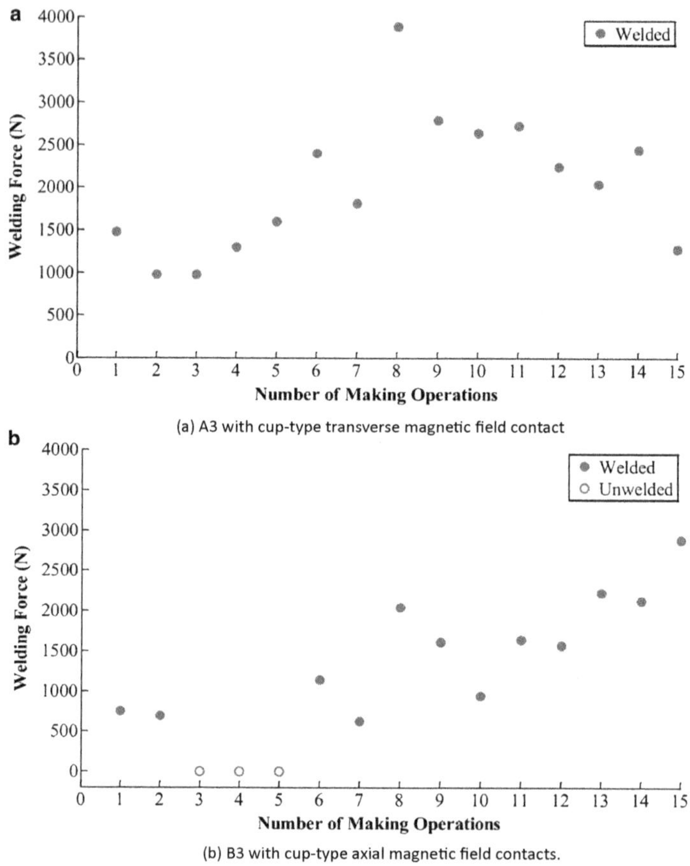

Fig. 5.53 Welding force measurement results for vacuum interrupter A3 and B3

the prestrike phenomena occurred in a specific area of the contact surface during the making operations. In addition, with the increase of making inrush current amplitude, the macroscopic arcing erosion areas become larger in size in both test series A and B, indicating more serious damage on the contact surfaces.

Figure 5.55 shows a measurement of the arcing erosion areas on the contact surfaces of test series A and B. It shows that, the areas of arcing erosion on the contact surfaces become larger in size with the increase of inrush currents in both series A and B. In test series A, the average arcing erosion areas of moving and fixed contacts were 8.2, 30.7, and 120.1 mm^2 for vacuum interrupters of A1, A2, and A3, respectively. In test series B, the average arcing erosion areas of moving and fixed contacts were 14.1, 33.0, and 163.1 mm^2 for vacuum interrupters of B1, B2, and B3, respectively. Therefore, with the increase of inrush currents from 5 up to 10 and finally to 20 kA, the average arcing erosion areas of moving and fixed contacts

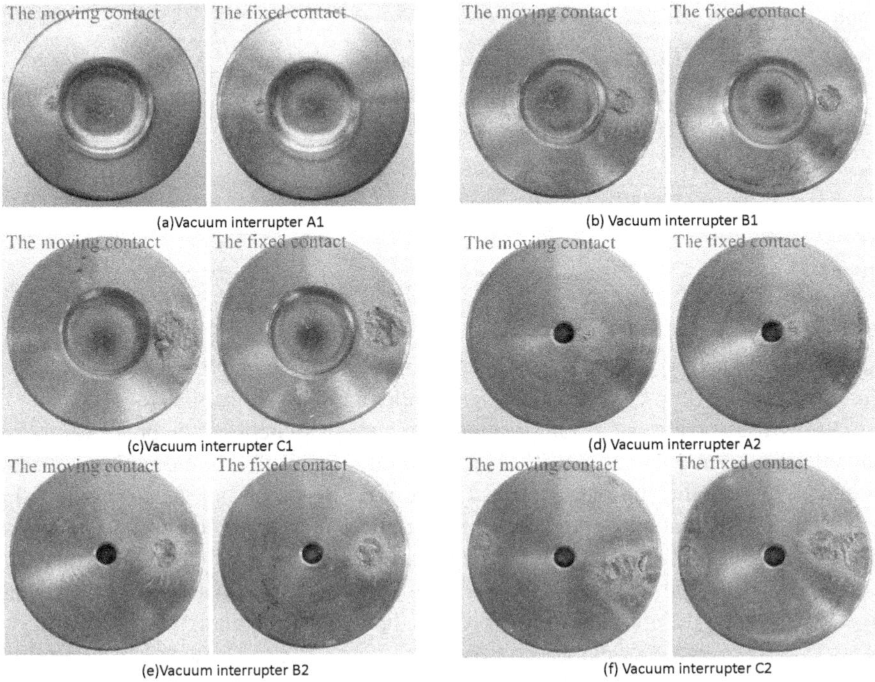

Fig. 5.54 Contact surface photos of test series A and B

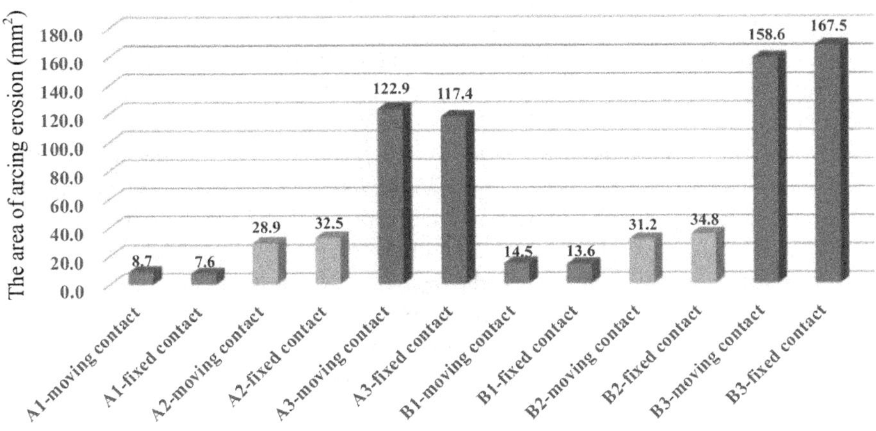

Fig. 5.55 Arcing erosion areas on the contact surfaces of test series A and B

Table 5.26 $d_{\overline{90}}$, $d_{\overline{50}}$, and $d_{\overline{10}}$ for test series A and B

Test VIs	Contact structure	Inrush current I_{inrush}	$d_{\overline{90}}$/mm	$d_{\overline{50}}$/mm	$d_{\overline{10}}$/mm
A1	Cup-type TMF	5-kA peak	1.4	1.6	1.7
A2		10-kA peak	1.1	1.7	2.5
A3		20-kA peak	1.9	2.5	2.9
B1	Cup-type AMF	5-kA peak	1.2	1.3	1.5
B2		10-kA peak	1.4	1.8	2.3
B3		20-kA peak	1.5	2.0	2.4

in series A (cup-type transverse magnetic field contact) are significantly lower than those in series B (cup-type axial magnetic field contact), respectively.

During the making operation of capacitive current switching, the prestrike arcs is capable to melt at a local contact surface, which may result in welding between the contacts. During the following contact separation, these partial welds break and a subsequent deformation of the contact material results in protrusions and craters at the local contact surface. This result affects the dielectric strength of the vacuum interrupter. Therefore, the performance of vacuum circuit breakers in capacitive switching depends more on the prestrike arcing process than on the current interrupting process [11]. The prestrike gap d can be considered to appraise the surface condition of the contact. This consideration allows an evaluation of the dielectric strength between the contacts affected by the last breaking operation and previous inrush current stress. Thus, the influence of prestrike gaps on restrike characteristics of vacuum interrupters is significant. Therefore, the prestrike gaps analysis takes an important role.

As shown in Fig. 5.50, the prestrike gap d scattered, which may due to the deteriorated conditions of the contact surfaces. As the complementary cumulative distribution function of the prestrike gaps followed the Weibull function, one can calculate $d_{\overline{90}}$ (CCDF 90% prestrike gap), $d_{\overline{50}}$ (CCDF 50% prestrike gap), and $d_{\overline{10}}$ (CCDF 10% prestrike gap) by using Eq. (5.12). Table 5.26 shows the values of $d_{\overline{90}}$, $d_{\overline{50}}$, and $d_{\overline{10}}$ for test series A and B. It shows that, vacuum interrupters B1, B2, and B3 have generally lower prestrike gaps than those of vacuum interrupter A1, A2, and A3, respectively. Compared to vacuum interrupter A1, the $d_{\overline{90}}$, $d_{\overline{50}}$, and $d_{\overline{10}}$ of vacuum interrupter B1 decreased by 14, 19, and 12%, respectively. For vacuum interrupter B3, the $d_{\overline{90}}$, $d_{\overline{50}}$, and $d_{\overline{10}}$ decreased by 27, 20, and 17% than vacuum interrupter A3, respectively. Therefore, the cup-type axial magnetic field contact had a lower prestrike gap than the cup-type transverse magnetic field contact.

Therefore, the vacuum interrupters with cup-type axial magnetic field contact showed a better dielectric behavior than the cup-type transverse magnetic field contact during the prestrike process, if one considers the cup-type axial magnetic field contact exhibited lower prestrike gaps, larger arcing erosion areas, and lower scatterings of prestrike gaps. For cup-type transverse magnetic field contacts, the intense arc that forms under a high current should be forced on a circular motion by a self-induced magnetic field generated by the conducting current. The constricted vacuum arc is driven by the Lorentz force under a self-induced transverse magnetic field and it should lead to a uniform contact erosion. However, the arc movement by the driving force is a physical motion, which has a time delay. Kimura et al. [27] revealed an immobility or slow arc period for the arc motion under transverse magnetic field contacts, and the slow arc period reached a few milliseconds with a gap distance of several millimeters. The prestrike inrush current arcs locally originate under a short gap distance (1–2 mm) and last for a short duration (1–2 ms). With the time delay of the transverse magnetic field, the prestrike arc motion proved to be not significant under the short arc duration. Figure 5.52 shows that no obvious arc motions are observed for vacuum interrupters in test series A with the cup-type transverse magnetic field contacts during the making operation. In contrast, the cup-type axial magnetic field contacts are capable of enhancing the threshold current level of the intense arcs formation [28], causing the prestrike arc to be more diffuse. This effect is presented irrespective of the short gap distance and short duration. Then, the axial magnetic field diffusing effect is considered to be faster than the transverse magnetic field driving effect during the prestrike process with a high frequency of inrush current and short gap distance. The diffusing effect leads to a wider distribution of the eroded area and reduces erosion of the contacts, thereby helping to decrease the damage caused by the inrush current on the local contact surfaces.

5.2.2 *Capacitive Current Switching Performance of Vacuum Interrupters with Three Kinds of Contacts*

This section compares the capacitive current switching performance of vacuum circuit breakers by using three commercially available contacts: cup-type axial magnetic field contact, horse-shoe axial magnetic field type contact, and spiral-type transverse magnetic field contact. A synthetic test circuit was used to evaluate the capacitive current switching performance of vacuum circuit breaker.

Figure 5.56a shows the synthetic test circuit. It contains two parts: an inrush current source and a power frequency source. The inrush current source provides an inrush current by discharging a capacitor C_2 through an inductance L_2. The inrush current can reach 20 kA peak with a frequency of 4250 Hz. SW_{test} is the test vacuum circuit breaker. The power frequency source provides a power frequency current I_{sp} for SW_{test}. Its frequency and the amplitude were decided by a capacitor C_4 and an inductance L_4. A DC recovery voltage can be generated by the power frequency source. The recovery voltage was determined by capacitor C_5 and resistor R_5, together with C_4 and L_4.

Figure 5.56b shows a C-O test operation sequence. Before starting the test operation, the capacitor C_2 and C_4 were pre-charged to an appropriate voltage. Vacuum circuit breakers SW_{test}, SW_{inrush} and SW_{mc} were all opened, expected for SW_{mo}, which was closed. Firstly, SW_{inrush} and SW_{test} were closed in sequence, then an inrush current I_{inrush} was generated by discharging of C_2 through L_2. Thereafter, vacuum circuit breaker SW_{mg} and SW_{mc} were closed, by which a power frequency

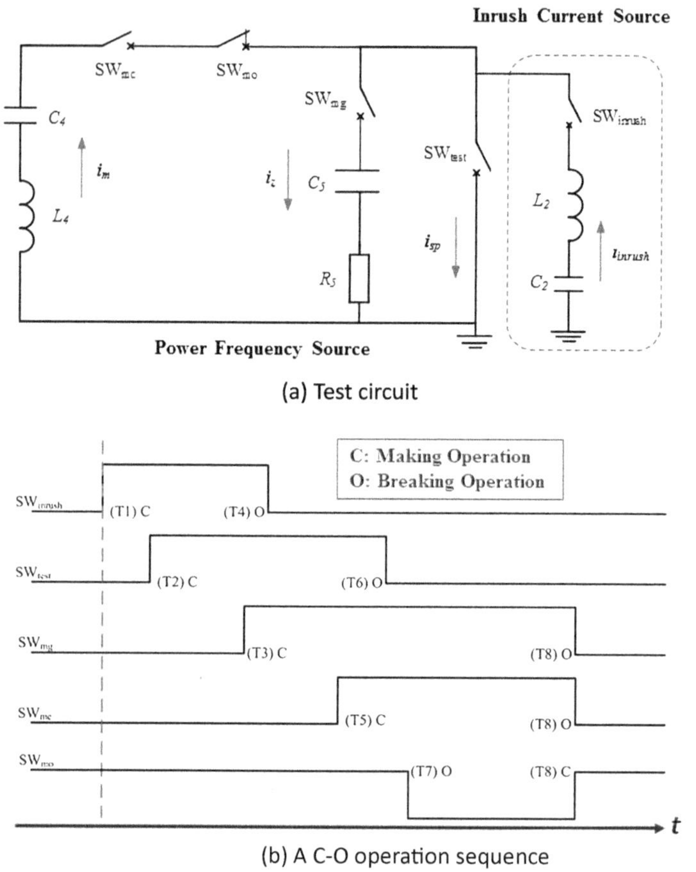

Fig. 5.56 A synthetic test circuit for the back-to-back capacitor bank switching and C-O operation sequence for circuit breakers

current I_{sp} with an amplitude of several hundred amperes flowed through the test vacuum circuit breaker SW_{test}. Then SW_{test} opened. After I_{sp} passed current zero, C_4 continued to charge C_5 through L_4 and R_5. So the charging voltage of C_5 was the recovery voltage of SW_{test}. Then SW_{mo} opened. The recovery voltage was a DC voltage and it maintained at a certain value after the branch current I_c passed current zero. At last, all vacuum circuit breaker restored to their initial status after the recovery voltage applying duration.

Figure 5.57 shows a C-O operation for a back-to-back capacitor bank switching. Figure 5.57a shows a typical oscillogram of a making operation. The upper trace was the contact travel. The lower trace was the high frequency inrush current. Figure 5.57b shows a typical oscillogram of a failed breaking operation. The upper trace was the DC recovery voltage. The rising time from 10 to 90% of the maximum voltage

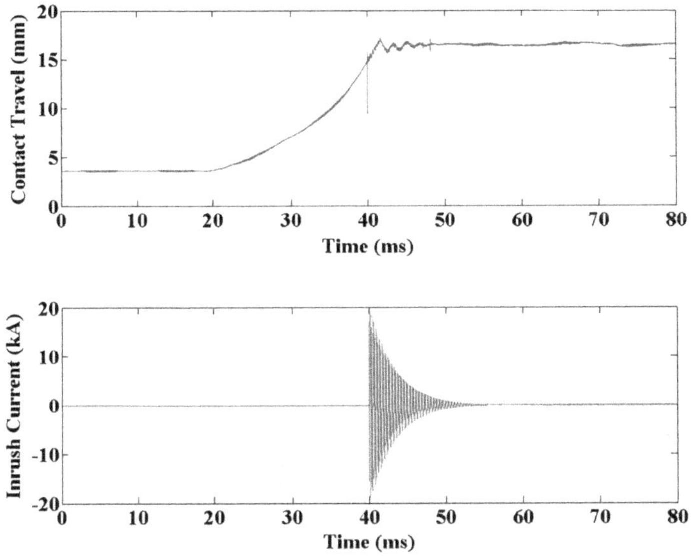

(a) a making operation

Fig. 5.57 A C-O operation for a back-to-back capacitor bank switching

was about 3 ms. The lower trace was the switching current I_{sp}. It showed a restrike occurred at about 48 ms after current zero.

Six 24 kV vacuum circuit breakers with 3 kinds of contacts were tested. VCB_11 and VCB_12 were with cup-type axial magnetic field contact. VCB_21 and VCB_22 were with spiral-type transverse magnetic field contact. VCB_31 and VCB_32 were with horse-shoe type axial magnetic field. All contact materials were CuW. The rated contact gaps were 12 ± 1 mm. To evaluate back-to-back capacitor bank current switching performance, the inrush current was set to 20 kA. The switching current was 400 A and the peak value of recovery voltage was 55 kV. A series of 80 CO operations were executed for each test vacuum circuit breaker.

Table 5.27 shows the restrike probabilities of each test vacuum circuit breaker. For cup-type axial magnetic field contact, the restrike probabilities of VCB_11 and VCB_12 were 5.0 and 15.0%, respectively. For spiral-type transverse magnetic field contact, the restrike probabilities of VCB_21 and VCB_22 were 17.5 and 3.8%, respectively. For horse-shoe type axial magnetic field contact, the restrike probabilities of VCB_31 and VCB_32 were 1.3 and 10%, respectively. So, the average value with horse-shoe type axial magnetic field contact was the lowest among the three kinds of contacts.

Figure 5.58 shows the distribution of the restrike occurring time for each test vacuum circuit breaker. For VCB_11 and VCB_12, 56% of the restrikes occurred within 10 ms after current zero during the recovery voltage period; 25% of the restrike occurred between 10 and 100 ms; the remained 19% restrikes occurred later than 100 ms. For VCB_21 and VCB_22, 24% restrikes occurred within 10 ms after current

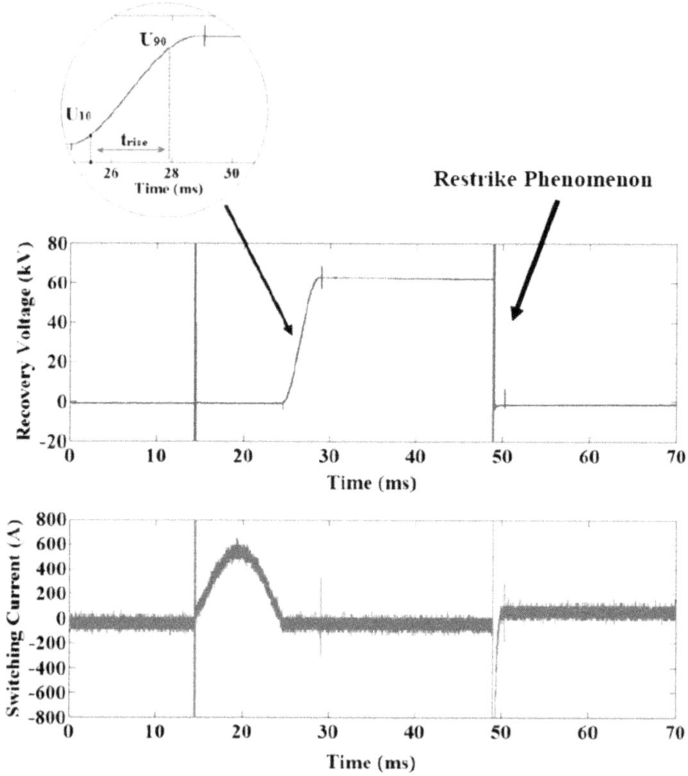

(b) a failed breaking operation

Fig. 5.57 (continued)

Table 5.27 The restrike probabilities of each test vacuum circuit breaker

Contact structure	Test sample	Restrike probability P (%)	Average value (%)
Cup type axial magnetic field contact	VCB_11	5.0	10.0
	VCB_12	15.0	
Spiral type transverse magnetic field contact	VCB_21	17.5	10.6
	VCB_22	3.8	
'Horse-shoe' type axial magnetic field contact	VCB_31	1.3	5.6
	VCB_32	10.0	

zero during the recovery voltage period; 29% of the restrikes occurred between 10 and 100 ms; 47% of the restrikes occurred later than 100 ms. For VCB_31 and VCB_32, almost all restrike occurred around 100 ms, excepted one occurred close to 10 ms.

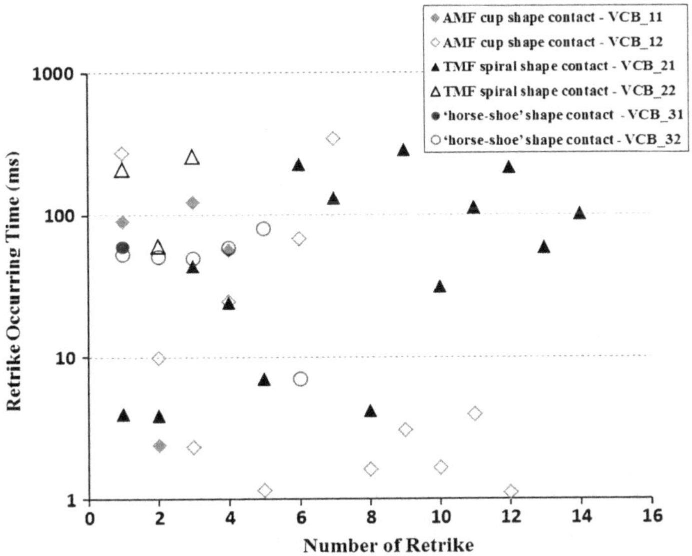

Fig. 5.58 The distribution of the restrike occurring time for each test vacuum circuit breaker

The above restrike characteristics suggested that the contact structure and the magnetic field played an important role in capacitor bank switching. The reason would be the significant impact of the inrush current on contact surface erosion. The horse-shoe type axial magnetic field contact showed better performance than the other two kinds of contacts for lower restrike probability and nearly no restrike occurred during the initial stage of the recovery voltage. Therefore, vacuum interrupters with the horse-shoe type axial magnetic field contact were chosen for back-to-back capacitor bank switching technology, which is discussed in Sect. 5.5.

5.3 Capacitive Current Switching and Contact Materials

5.3.1 Inrush Current Prestrike Arc Behaviors of Contact Materials CuCr50/50 and CuW10/90

Section 5.1 suggested that the inrush current has a significant impact on the prestrike arc erosions of the contact surfaces during capacitive making operations. Obviously, contact materials would take an important role in the contact erosion. Many studies investigated the role of contact materials in the contact erosion. Kamikawaji et al. [11] showed that CuCr was one of the most popular contact materials for vacuum circuit breakers because of its high current interrupting and high voltage withstanding ability. Slade et al. [29]. revealed the different effects of CuCr and CuW contact materials during making fault current in vacuum. Yokokura et al. [10] found that the CuW contact material was the best material for the capacitive current switching duty compared with CuTeSe, CuBi, and CuCr. However, more work should be done to understand how the contact materials impact the inrush current prestrike arc behaviors and the contact erosions in vacuum during capacitive making operations.

This section investigates an influence of contact materials on the prestrike arc behaviors and the contact erosions in vacuum interrupters during capacitive making operations. The amplitudes of the inrush current are set to three levels, which are 5-, 10-, and 20-kA, respectively. The contact materials include CuCr50/50 and CuW10/90. The concerned behaviors include the distributions of the prestrike gaps, prestrike arc behaviors, contact welding forces, and the contact surface erosion conditions. The results could help in improving the capacitive switching performance of vacuum interrupters.

Figure 5.59a shows a test circuit for capacitive inrush current making operation. The inrush current circuit can supply a current up to 20-kA peak value at a frequency of 3800 Hz. The switches SW_{inrush} and SW_{test} were set open initially. The capacitor $C1$ was pre-charged to a voltage U_r. After closing SW_{inrush} and then SW_{test}, the capacitor $C1$ was discharged through the inductor $L1$. During the making

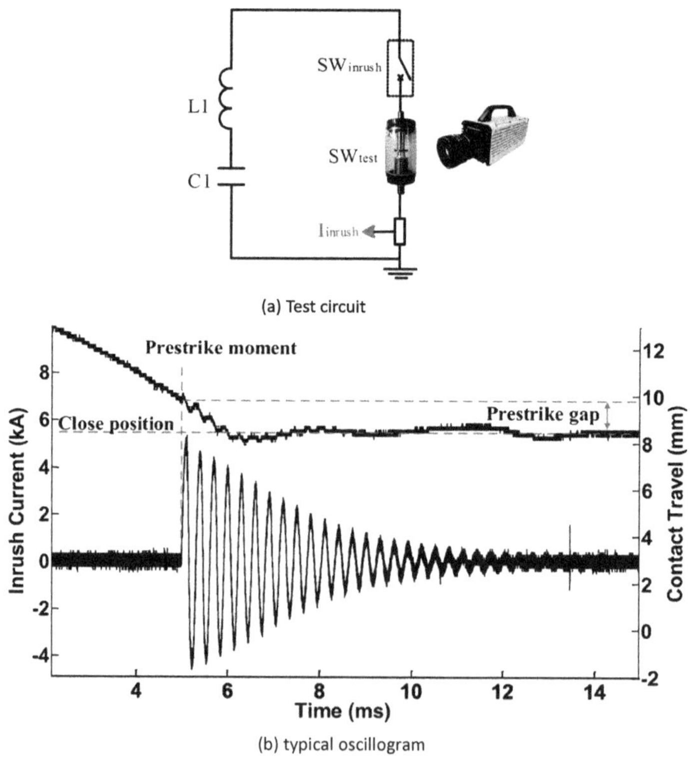

Fig. 5.59 Test circuit for capacitive inrush current making operation and a typical oscillogram

Table 5.28 The experimental conditions for capacitive current making operations

Test VIs		Al	A2	A3	B1	B2	B3
Inrush	Amplitude (kA)	5	10	20	5	10	20
Current	Frequency (Hz)	3800					
Contact material		CuCr50/50			CuW10/90		
Making operations		30	30	15	30	30	15

of SW_{test}, the electric field strength between its contacts increased continuously, while the contacts approached each other. After the electric field strength exceeded the dielectric strength between the contacts, a prestrike occurred. Then, an inrush current started to flow at a certain gap distance *d* through vacuum arcs. Because of the circuit resistance, the inrush current damped with oscillations. The amplitude of the inrush current can be adjusted by changing the pre-charged voltage value U_r of capacitor *C*1. The frequency of the inrush current can be adjusted by the capacitance of *C*1 and inductance of *L*1. Figure 5.59b shows a typical oscillogram of a making operation. The inrush current was 5-kA peak with a frequency of 3800 Hz. The upper trace is contact closing displacement curve and the lower trace is the high-frequency inrush current. By comparing the inrush current waveform with the contact closing displacement curve, the prestrike moment and the prestrike gap *d* could be determined. Simultaneously, the evolution of the prestrike arcs during the prestrike process was recorded by a charge-coupled device (CCD) video camera. The CCD video camera had a recording speed of 30 000 frames/s and the exposure time was set to 2 μs.

Table 5.28 shows the experimental conditions. Six specially designed 12-kV vacuum interrupters with butt-type contacts were adopted. They were used to evaluate the impact of the inrush currents and contact materials on the prestrike arc characteristics during capacitive current making operations. The vacuum interrupters were composed of a glass envelope and they were equipped without metal vapor shields to enable an optical observation of prestrike arc behaviors. The contact diameter was 48 mm. The full contact gap was 9 mm. In the six vacuum interrupters, three were made of CuCr50/50 contact material, while the other three were made of CuW10/90 contact material. All the contact materials were made in sintered method. A permanent magnetic operating mechanism was adopted. The average closing velocity was 0.8 m/s from 2/3 full contact gap to the closing position. The six test vacuum interrupters were divided into groups A and B. The vacuum interrupters in group A were made of CuCr50/50 contact material, and the vacuum interrupters in group B were made of CuW10/90 contact material. Every vacuum interrupter in each group was labeled as 1, 2, or 3, which corresponds to the making inrush currents with 5- 10-, and 20-kA peaks, respectively. A series of 30 making operations were executed for the A1 and B1 vacuum interrupters by using the inrush current with a 5-kA peak and a 3800 Hz frequency, while the prestrike gap was measured and the prestrike arc features were observed. The A2 and B2 vacuum interrupters suffered an inrush current with a 10-kA peak and a 3800 Hz frequency for 30 times, while the prestrike arc behaviors and prestrike gap were recorded. Furthermore, the A3 and B3 vacuum interrupters suffered an inrush current with a 20-kA peak for 15 times. After each making operation of the inrush current, the contact welding forces of A3 and B3 were measured.

Figure 5.60a shows an experimental setup for the measurement of contact welding forces. A force sensor, which is a strain gauge, was connected to the stationary rod of a test vacuum interrupter. It measured contact separation forces during opening operations, which is subsequent to a contact welding. While the upper movable rod of the vacuum interrupter was manually driven up, a steady increased force can be acquired until the instant of contact separation [30]. A 3 V dc trigger circuit was used to trace the fracture moment of the welding points between the contact surfaces. Figure 5.60b shows a typical oscilogram of a welding force measurement. The upper curve is the trigger signal and the lower curve is the measured force. It

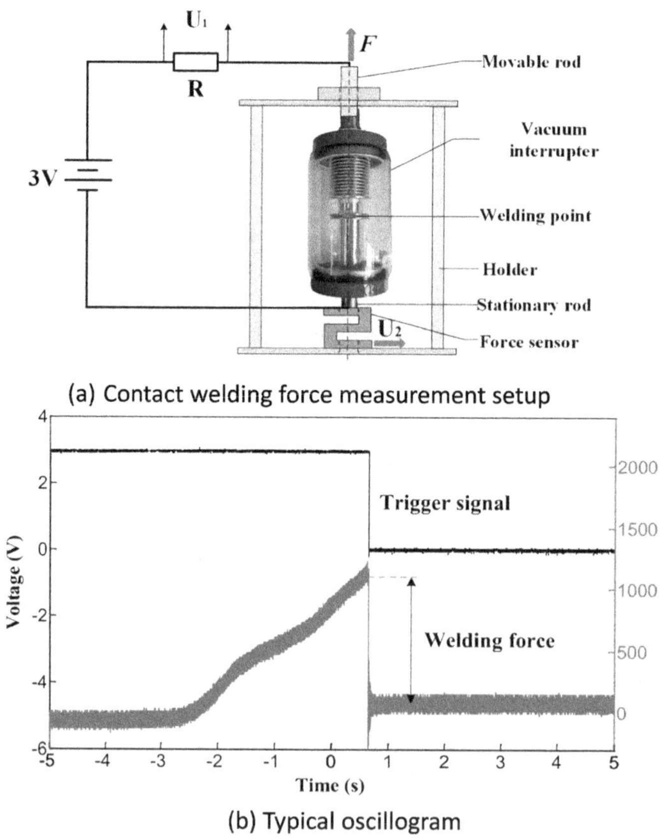

Fig. 5.60 Contact welding force measurement and typical oscillogram

shows that the loop resistance of the vacuum interrupter before the separation of the welded contacts was quite low. The voltage across the resistor R was approximated to 3 V (dc battery voltage). After opening the contacts of the vacuum interrupter, the voltage across R jumped to 0 V, which generated a falling trigger signal of welding fracture to the oscilloscope. At the same time, the welding forces that were initiated by the making operation of the 20-kA peak inrush current were measured by the force sensor. As a slowly increasing drawing force F was applied to the movable contact rod, it resulted in a steady increase of the force until it breaks the welding spots between the contact surfaces. Then the contacts were separated. Moreover, the trigger signal dropped from 3 to 0 V indicated the fracture of the welding points. The welding force kept 0 N after the fracture of the welding points. Therefore, the force at the instant of contact separation was the welding force. The force measurement accuracy was 10 N.

Figure 5.61 shows the cumulative probability distribution of the prestrike gap d. The cumulative probability distribution followed the three-parameter Weibull distribution function, as shown in Eq. (5.13)

$$F(d) = 1 - \exp\left[-\left(\frac{d - d_o}{\Theta - d_o}\right)^m\right] \tag{5.13}$$

where $F(d)$ is the cumulative probability at a prestrike gap d, which represents the probability of the prestrike gaps that take on a value less than and equal to d; d_0 is the location parameter; m is the Weibull shape or "shape" factor; and Θ is the scale parameter. Table 5.29 shows the values of the Weibull parameters for series A and B. The R-squared value is used to indicate the fitting similarity. All values of R-squared were above 0.96. Therefore, it was reasonable to assume that the cumulative probability of the d value followed the three-parameter Weibull distribution. The scattering of the prestrike gaps can be evaluated by the scale of $(\Theta - d_0)$.

According to Eq. (5.13), one can calculate the 10% prestrike gap d_{10}, the 50% prestrike gap d_{50}, and the 90% prestrike gap d_{90}. Table 5.30 shows the calculated d_{10}, d_{50}, and d_{90}. Generally speaking, the d_{10}, d_{50}, and d_{90} of test series A (CuCr 50/50 contact material) was close to that of test series B (CuW 10/90 contact material), and the prestrike gap increased as inrush current increasing. However, an exception appeared for CuW 10/90 contact material at inrush current of 10 kA peak, at which the prestrike gap was lower than that of inrush current 5 kA peak. Maybe this came from randomness of the test vacuum interrupter.

Figure 5.62 shows the prestrike arc characteristics of vacuum interrupter A3 with the CuCr50/50 contact material during the making operation of 20-kA inrush current. The photos of Fig. 5.62a–h were during the first two periods (a duration of approximately 600 μs) of the inrush current after the prestrike initiated. The labeled time under each photo represents a duration from the prestrike occurring to the instant at which the photo was taken. Figure 5.62i shows the waveform of the inrush current in the making operation. The photos show that the initial prestrike arcs originated at one point between the contacts before they mated. Then the prestrike arcs appeared

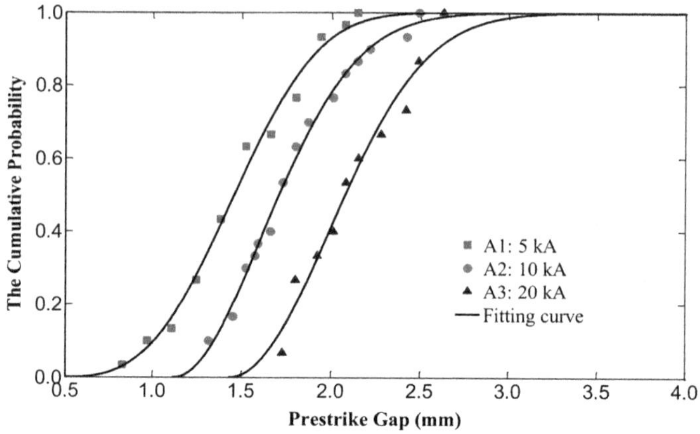

(a) Vacuum interrupter A1, A2, and A3

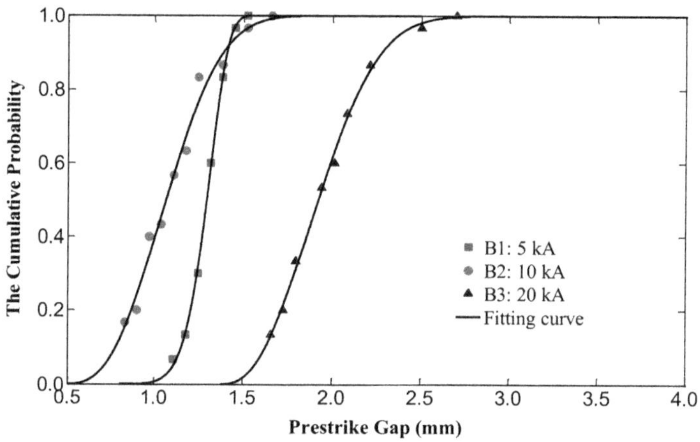

(b) Vacuum interrupter B1, B2, and B3

Fig. 5.61 Cumulative probability distribution of the prestrike gaps for test series A and B

Table 5.29 The fitting parameters of the three-parameters Weibull distribution function for test series A and B

Test VIs	m	d_0/mm	Θ/mm	$\Theta - d_0$/mm	R-squared
A1	3.1	0.5	1.6	1.1	0.991
A2	2.0	1.1	1.8	0.7	0.993
A3	2.0	1.4	2.2	0.8	0.965
B1	7.3	0.7	1.4	0.7	0.998
B2	2.7	0.5	1.1	0.6	0.985
B3	2.4	1.4	2.0	0.6	0.997

Table 5.30 The 10% prestrike gap d_{10}, the 50% prestrike gap d_{50}, and the 90% prestrike gap d_{90} calculated for test series A and B

Test VIs	Contact material	Inrush current I_{inrush}	d_{10}	d_{50}	d_{90}
A1	CuCr50/50	5-kA peak	1.0	1.4	1.9
A2		10-kA peak	1.3	1.7	2.2
A3		20-kA peak	1.7	2.1	2.6
B1	CuW10/90	5-kA peak	1.1	1.3	1.4
B2		10-kA peak	0.8	1.1	1.4
B3		20-kA peak	1.6	1.9	2.3

as an intense arc. The arc brightness varied with the instantaneous value of inrush current.

Figure 5.63 shows the prestrike arc characteristics of vacuum interrupter B1 with the CuW10/90 contact material during the making operation of 5-kA inrush current. The arc photos during the making operation are shown in Fig. 5.63a–h, which were during the first period (a duration of approximately 300 μs) of the inrush current after the prestrike occurred. It shows that the prestrike arcs initially originated at two points between the contacts. They rapidly evolved into intense arcs. Then a significant ejection of metal vapor or particles can be observed at the moment of the contacts mating. The phenomena of multiple initial arcing points during the prestrike process were mostly observed in test series B with the CuW10/90 contact material. While the prestrike arcs in test series A with CuCr50/50 contact material observed mainly one initial arcing point. Moreover, the arc occupying areas in series B with the CuW10/90 contact material were observed larger than that of test series A with CuCr50/50 contact material.

The welding forces were measured 15 times after each making operation for vacuum interrupter A3 and B3. Figure 5.64 shows the measured welding forces. The solid circles are "welded" cases, while empty circles are "un-welded" cases. Figure 5.64a shows the measured results of vacuum interrupter A3. It shows that 9 out of 15 making operations welded. The welding force was in a range of 40–2070 N. The welding forces showed no correlation with the previously performed making operations. Figure 5.64b shows the measured results of vacuum interrupter B3. Only one case of welding occurred. The welding force was 210-N.

Figure 5.65 shows the contact surface photos after inrush current making operations. The left two columns show test series A (A1, A2, and A3) with the CuCr50/50 contact material and the right two columns show test series B (B1, B2, and B3) with the CuW10/90 contact material. It shows that the welding fracture surfaces in test series A were relatively rough. Moreover, the fracture surfaces suggested that the prestrike phenomena occurred at a local area of the contact surfaces in vacuum interrupters A1 and A2, because there was only a single macroscopic crater on one contact, corresponding to a macroscopic protrusion on the opposite contact. For test vacuum interrupter A3, there were separate macroscale welding areas distributed on the contact surfaces instead of a single protrusion or crater. Moreover, the welding

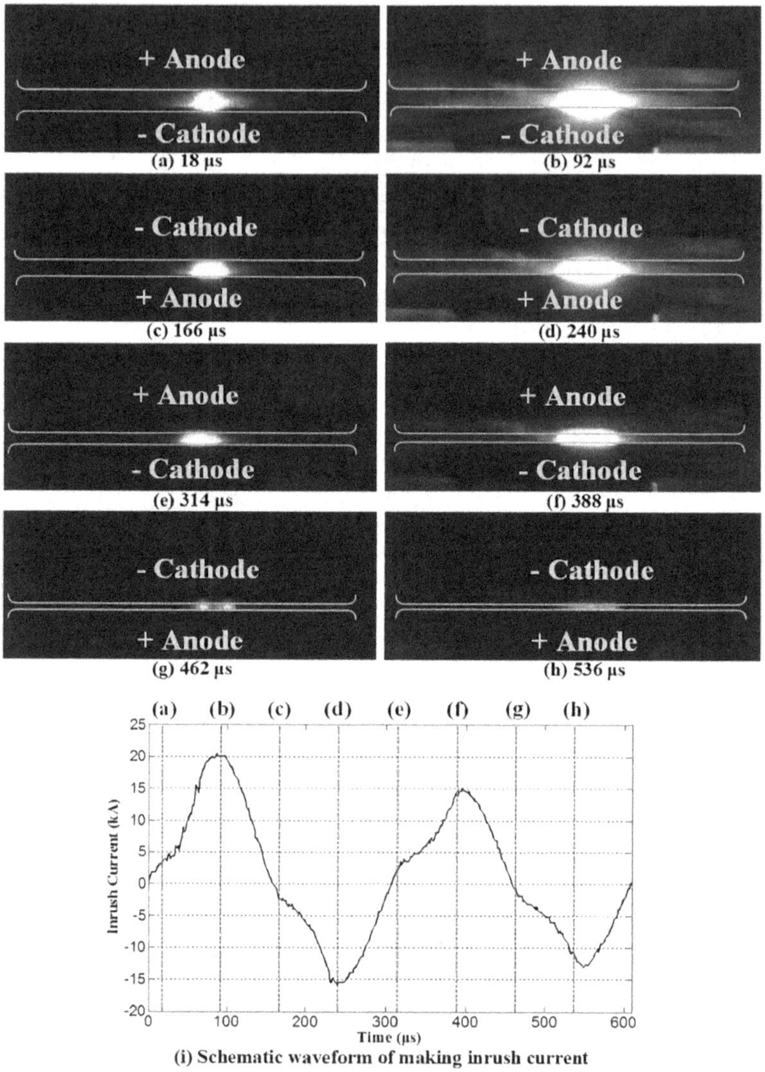

Fig. 5.62 Prestrike arc characteristics of vacuum interrupter A3 during the making operation of 20-kA inrush current

areas on contact surfaces of test vacuum interrupter A3 were significantly larger in size than those of vacuum interrupters A1 and A2. In test series B with the CuW10/90 contact material, there were no obvious macroscopic craters and protrusions. There were only some bright erosion spots appearing on the contact surfaces.

The experimental results suggested that higher inrush currents would cause severer damage, as shown in Fig. 5.65. It shows that the contact surfaces of vacuum interrupter A3 were most seriously damaged in test series A, which showed more welding spots

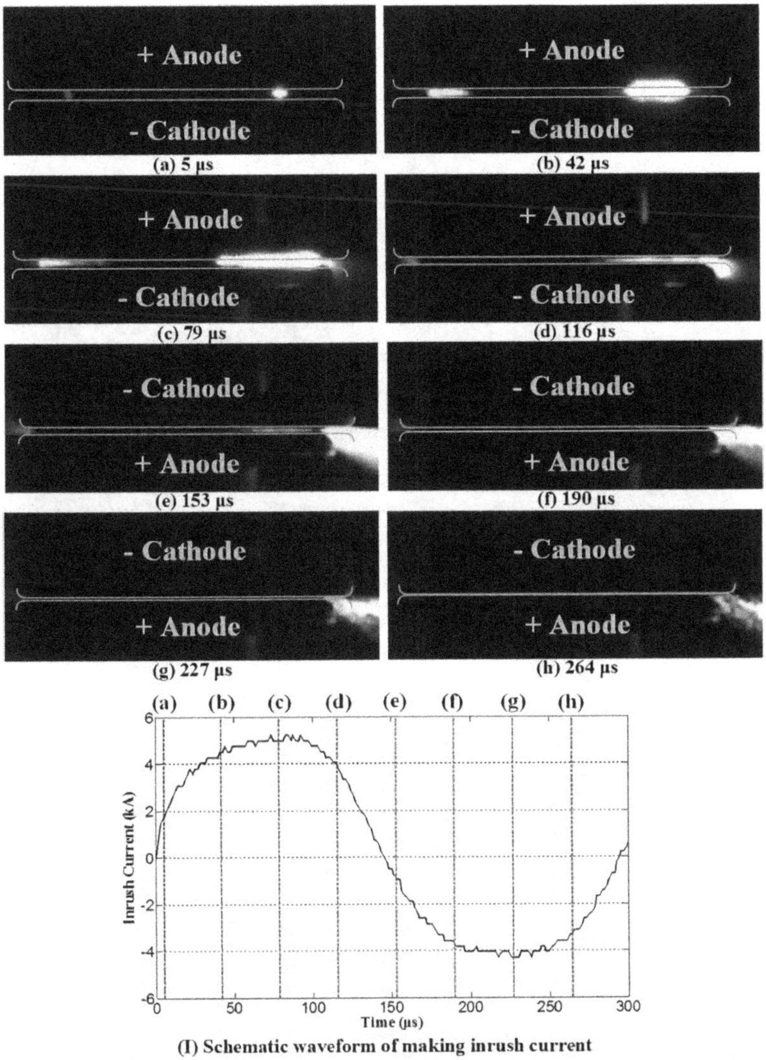

Fig. 5.63 Prestrike arc characteristics of vacuum interrupter B1 during the making operation of 5-kA inrush current

and larger erosion areas after 15 making operations of the 20-kA peak inrush current. The severer damage may be related to higher arc energy. A higher prestrike gap corresponds to a longer duration of the prestrike arcs. Therefore, a higher arc energy would be injected into the contact surfaces to melt local contact surfaces and cause more serious damage on the contact surfaces.

Figures 5.62, 5.63, 5.64, and 5.65 suggested that the CuW10/90 contact material had a better anti-welding property compared to the CuCr50/50 contact material. The

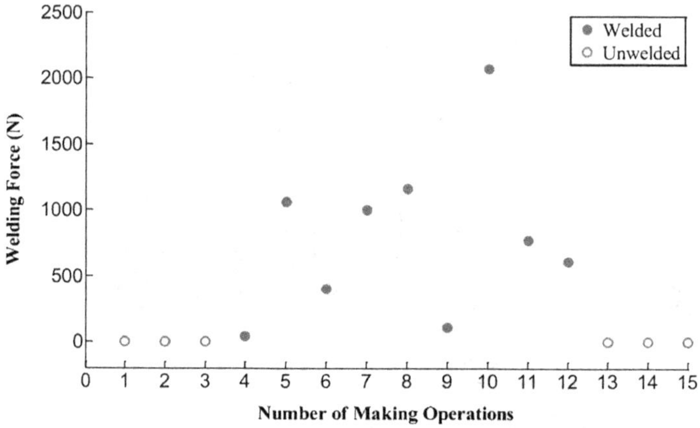

(a) Vacuum interrupter A3

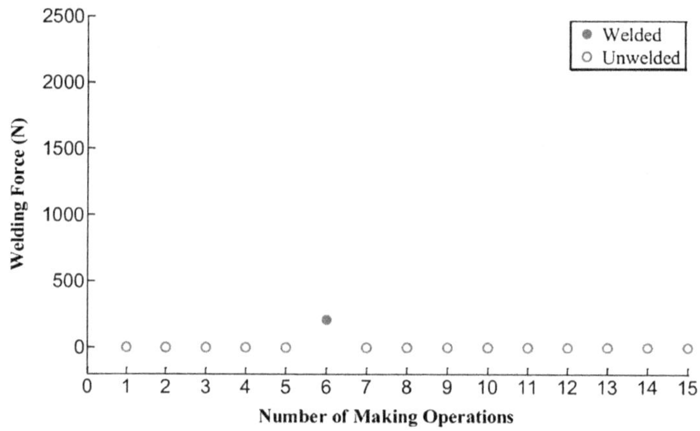

(b) Vacuum interrupter B3

Fig. 5.64 Welding force measurement

CuCr contacts was easier to weld together, and protrusions, craters, and particles formed more easily. The prestrike arcs with the CuCr contact material tended to concentrate in one location, while the prestrike arcs were prone to distribute widely on the contact surfaces of CuW contact material. However, despite a very good resistance to welding, the CuW contact material has a lower electrical conductivity and a low current interrupting capability. The low interrupting ability of CuW contact material could be that the W is an excellent thermionic emitter and as the current increases, the contact surface is likely to emit electrons to initiate a restrike.

Figure 5.66 suggests a physical explanation of the contact erosions for the CuCr50/50 and CuW10/90 contact material during the prestrike process. Figure 5.66a show that of CuCr50/50 contact material. When an inrush current flows through the vacuum arcs during the prestrike process, the prestrike arc melts a local area of the

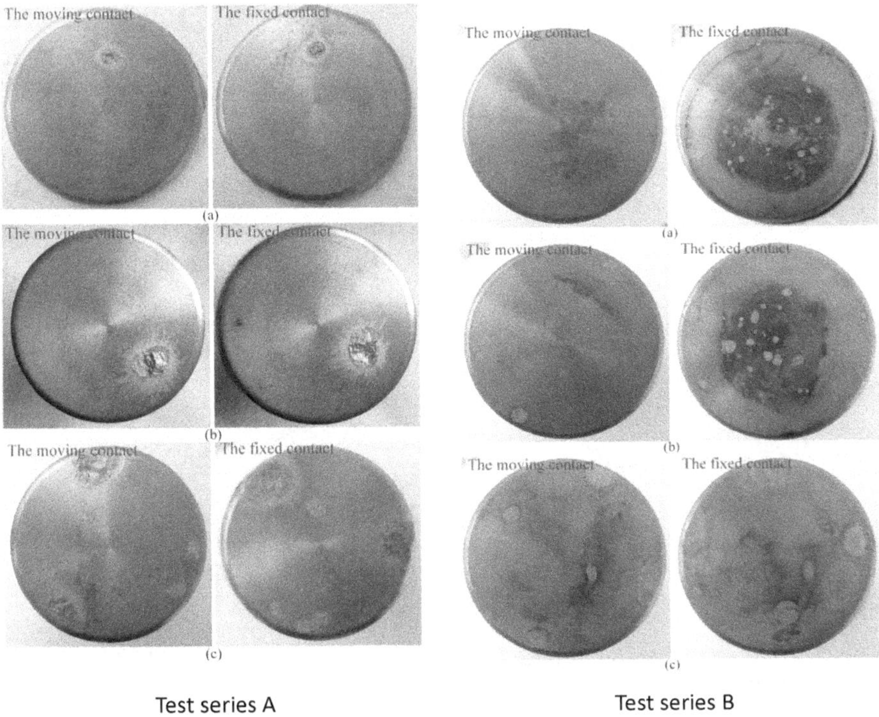

Fig. 5.65 Contact surface photos. **a** vacuum interrupter A1 and B1, 30 making operations of $I_{\text{inrush}} = 5$ kA. **b** vacuum interrupter A2 and B2, 30 making operations of $I_{\text{inrush}} = 10$ kA. **c** vacuum interrupter A3 and B3, 15 making operations of $I_{\text{inrush}} = 20$ kA

contact surface. This accompanied by a concentrated heating from the arc to the local contacts. The prestrike arc erosion starts upon liquid formation in a confined region. As inrush current increases and the arc becomes more concentrated, the region affected by the arc erosion grows larger. The temperature at the local erosion region is well above the CuCr melting point, so liquid contact material can be observed on the CuCr50/50 local contact surfaces. With continuous heating effects of the prestrike vacuum arcs, an amount of the copper and chromium metal vapor mixture evaporates from the liquid areas on the contacts, since the metal copper and chromium have similar boiling points (2868 K for Cu and 2955 K for Cr). In addition, the splashing of the liquid metal also occurs by the mechanical impacting effect of the contacts. Then, the two contacts may be welded together after they finally mate together. This is consistent with the welding force results of test vacuum A3, as shown in Fig. 5.64. Figure 5.66b shows the physical explanation of the contact erosion for the CuW10/90 contact material during the prestrike process. With a prestrike occurs, the prestrike arcs heat the contact surfaces to a high temperature and local contact surface areas melt. The high temperature of the vacuum arcs would be equalized by the liquid flow to redistribute heat energy and metal evaporation. Moreover, the projection of

Fig. 5.66 A suggested physical explanation of the contact erosions during the prestrike process

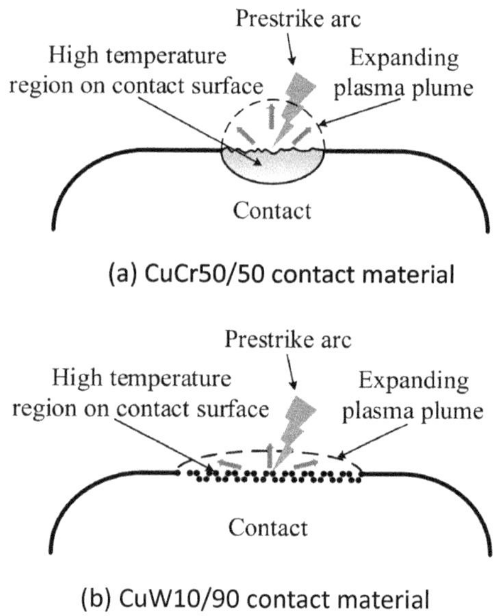

droplets from the metal liquid may also contribute to the cooling of the contacts, as observed in the prestrike arc photos shown in Fig. 5.63. Tungsten has a high melting point of 3695 K, while the melting point of copper is 1356 K. Therefore, it is likely that a certain temperature between the melting points of copper and tungsten could be established on the contacts, at which copper can be easily evaporated, while tungsten remains in the solid state. The solid tungsten may act as a thermal sheath to prevent the contacts from experiencing further serious damage. This can account for the rare welding phenomena observed on vacuum interrupter B3. After the condensation of the metal liquid, a rich copper zone may be formed on the contact surfaces, which would be the bright spots shown in Fig. 5.65.

5.3.2 *Capacitive Current Switching Performance of Vacuum Interrupters with Three Contact Materials*

Section 5.1 suggested that the capacitive current switching performance of a vacuum interrupter is mainly determined by a condition of the contact surfaces modified during a prestrike process. Section 5.3.1 showed more evidence on the prestrike arc behaviors and the arc erosions on the contacts with two kinds of contact materials. So that the contact material and its fabrication process play an important role on the capacitive current switching behaviors in vacuum. This section is to determine the capacitive current switching performance of vacuum interrupters with three different contact materials and fabrication process, which are infiltration CuCr 50 (50% weight of Cr), vacuum induction melting CuCr40Te0.005 (40% weight of Cr and 0.005%

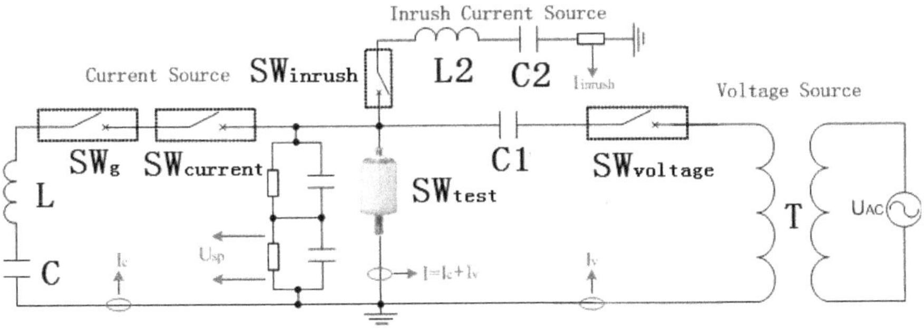

Fig. 5.67 A synthetic circuit for back-to-back capacitor bank switching

weight of Te), and vacuum arc melting CuCr 50 (50% weight of Cr). The results would be helpful for the chosen of contact materials for the back-to-back capacitor bank switching of vacuum interrupters.

A synthetic circuit is used to evaluate the capacitive current switching performance of vacuum interrupters. Figure 5.67 shows the synthetic circuit. The circuit is composed of three sub-circuits: an inrush current source circuit, a current source circuit and a voltage source circuit. SW_{test} is the test vacuum interrupter. The inrush current source circuit supplies an inrush current I_{inrush} at a frequency of 3800 Hz. The inrush current is generated by discharging a capacitor $C2$ through an inductance $L2$ with closing a switch SW_{inrush}. The current source circuit can provide a power frequency current I_c (50 Hz) by discharging a capacitor C through an inductance L through closing a switch $SW_{current}$ and a switch SW_g. The voltage source circuit can produce a recovery voltage of (1-cosωt) wave shape through a superposition of a cosine waveform generated by a high voltage transformer T and a DC voltage generated by a capacitor $C1$.

The test vacuum interrupters endured CO operations. The CO operation was composed of a making operation and a subsequent breaking operation. Figure 5.68 shows typical oscillograms of a CO operation. Figure 5.68a shows a making operation. The upper trace is a contact travel and the lower trace is an inrush current. The inrush current was 8 kA peak. The contact gap corresponded to the inrush current occurring was the prestrike gap. Figure 5.68b shows a successful breaking operation oscillogram. The lower trace is a switching current. After a current zero, a recovery voltage (the upper trace) was applied to the test vacuum interrupter and it withstood the voltage. Figure 5.68c shows a failed breaking oscillogram. After a current zero, a recovery voltage was applied. Then the first restrike occurred at the peak of the first circle of the recovery voltage. The second restrike occurred at the peak of the fifth circle of the recovery voltage. A high frequency restriking current appeared at the instant when the restrike occurred.

In the experiment, nine 12 kV vacuum interrupters were tested. The test vacuum interrupters were not conditioned by any electrical methods before. The contact materials were infiltration CuCr50 (50% weight of Cr), vacuum induction melting CuCr40Te0.005 (40% weight of Cr and 0.005% weight of Te) and vacuum arc melting

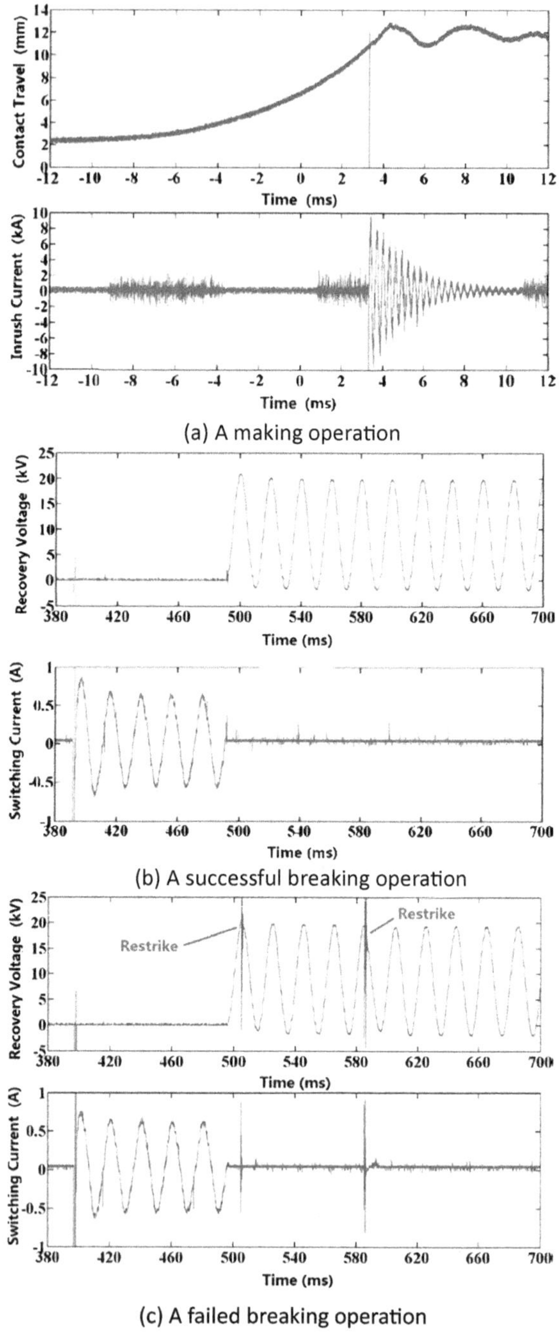

Fig. 5.68 Typical oscillograms of a CO operation of a test vacuum interrupter

Table 5.31 Three test series and the experimental conditions

Test VIs	Test series I	Test series II	Test series III
	A1	B1	C1
	A2	B2	C2
	A3	B3	C3
Fabrication process	Infiltration CuCr 50	Vacuum induction melting CuCr40Te0.005	Vacuum arc melting CuCr 50
Inrush	Peak value: 8 kA		
Current	Frequency: 3800 Hz		
Switching current	<1 A		
Recovery voltage	$U_{trv} = U_{peak} \times (1 - \cos\omega t)$		
	$U_{peak} = \frac{12kV \times \sqrt{2}}{\sqrt{3}} = 9.8kV$		
	Duration: 500 ms		
Test duty	C–0: 80 times		

CuCr50 (50% weight of Cr). The full contact gap was 9 mm. To evaluate the capacitive current switching performance, a series of 80 CO operations were executed for each test vacuum interrupter. Table 5.31 shows the experimental conditions. The amplitude of the inrush current was set as 8 kA. Three test series were carried out. Test series I was with the contact material of infiltration CuCr50. Test series II was with the contact material of vacuum induction melting CuCr40Te0.005. Test series III was with the contact material of vacuum arc melting CuCr50. In each test series, three test vacuum interrupters were used. The switching current was less than 1 A, which was a severe test condition. In contrast, the switching current with hundreds of amperes would have a conditioning effect. The peak value of the recovery voltage was about 20 kV ($2U_{peak}$). The full period of recovery voltage application was 500 ms.

The restrike is referred to the breakdown which occurs after (1/4)·T of the recovery voltage, where T is a cycle of the recovery voltage. Here T is 20 ms. The reignition is referred to a breakdown which occurs before (1/4)·T of the recovery voltage. Table 5.32 shows the restrike probabilities of the test vacuum interrupters. In test series I, the contact material was infiltration CuCr50/50. The average value of the restrike probabilities of the vacuum interrupters (A1, A2 and A3) was 6.7%. In test series II, the contact material was vacuum induction melting CuCr40Te0.005. The average value of the restrike probabilities of the vacuum interrupters (B1, B2 and B3) was 5.8%. In test series III, the contact material was vacuum arc melting CuCr50/50. The average value of the restrike probabilities of the vacuum interrupters (C1, C2 and C3) was 8.3%. Thus, in capacitive current switching, the vacuum interrupters with vacuum induction melting CuCr40Te0.005 contact material showed the best performance among the three kinds of contact materials.

Figure 5.69 shows the distribution of the restrike occurring instant. Figure 5.69a–c show the results with contact material infiltration CuCr50/50, vacuum induction

Table 5.32 Restrike probabilities of the test vacuum interrupters

Test series (contact material)	Test VI	Restrike probability (%)	Average value (%)
Test series I (infiltration CuCr 50)	Al	5.0	6.7
	A2	10.0	
	A3	5.0	
Test series II (vacuum induction melting CuCr40Te0.005)	B1	3.6	5.8
	B2	6.3	
	B3	8.0	
Test series III (vacuum arc melting CuCr 50)	C1	8.8	8.3
	C2	7.5	
	C3	8.8	

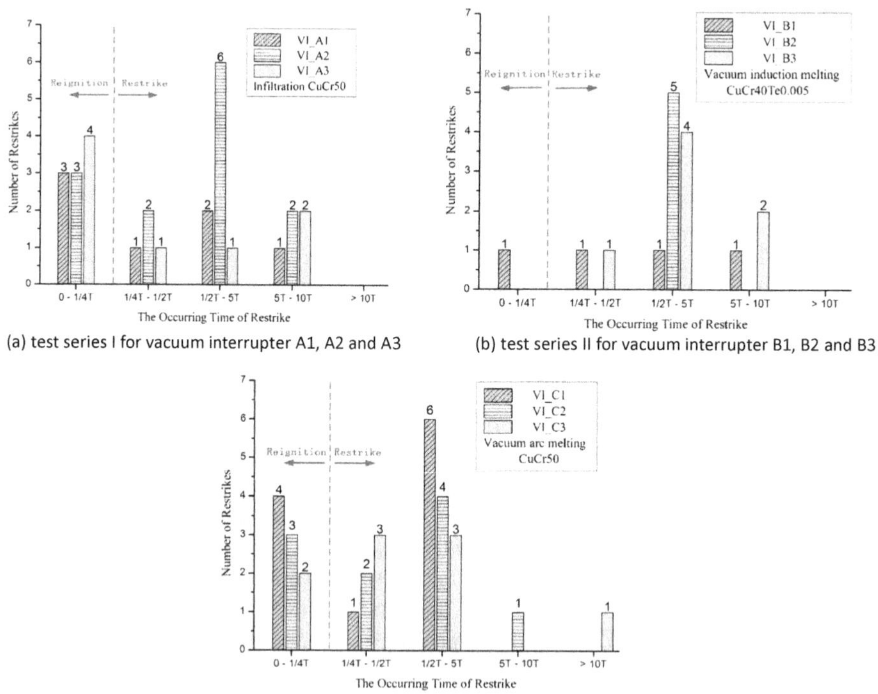

(a) test series I for vacuum interrupter A1, A2 and A3

(b) test series II for vacuum interrupter B1, B2 and B3

(c) test series III for vacuum interrupter C1, C2 and C3

Fig. 5.69 The occurring instant of the restrikes

melting CuCr40Te0.005 and vacuum arc melting CuCr50/50, respectively. The zero point is the instant when a recovery voltage is applied. Figure 5.69a shows there were 28 breakdowns in test series I, in which 18 restrikes occurred after (1/4)·T and 10 reignitions occurred before (1/4)·T. Figure 5.69b shows there are 15 restrikes and only one reignition in test series II. Figure 5.69c shows there are 21 restrikes and

9 reignitions in test series III. Figure 5.69a–c shows almost all the restrikes reside between (1/4) T and 10·T for each vacuum interrupter. The number of the restrikes of vacuum induction melting CuCr40Te0.005 contact materials was the least among the three kinds of contact materials. At the same time, the vacuum induction melting CuCr40Te0.005 also had the least reignitions.

The vacuum induction melting CuCr40Te0.005 proved better capacitive current switching performance than that of infiltration CuCr50/50 and vacuum arc melting CuCr50/50 contact material. The average values of the restrike probabilities followed the order of vacuum induction melting CuCr40Te0.005 (5.8%) < infiltration CuCr50/50 (6.7%) < vacuum arc melting CuCr50/50 (8.3%). The number of the reignitions followed the order of vacuum induction melting CuCr40Te0.005 (1 time) < vacuum arc melting CuCr50/50 (9 times) < infiltration CuCr50/50 (10 times).

5.4 High-Frequency High-Voltage Impulse Conditioning and Capacitive Current Switching of Vacuum Interrupters

Section 5.1 shows that a high inrush current in making operation tended to cause contact welding. The contact welding rupture damages the contact surfaces by resulting protrusions, craters and particles. They reduce the dielectric strength of the vacuum gap. Finally, the it may lead the restrikes during the breaking operations. The insulation weak points on the deteriorated contact surface are potentially responsible to the restrikes. It is well known that conditioning is an effective way to enhance the high-voltage withstanding capability of vacuum interrupters by eliminating the insulation weak points on the contact surfaces [31]. Traditional conditioning technologies include spark conditioning using high-voltage ac power supply, spark conditioning using high-voltage pulse, and current conditioning et al. [31]. We proposed a high-frequency high-voltage impulse conditioning technology [32]. In each conditioning operation, a vacuum interrupter is conditioned by several hundreds of sub-microsecond high-voltage impulses. If a breakdown occurs at any voltage impulse, a high frequency conditioning current of several kA (peak value) flows through the vacuum interrupter. The high frequency current is likely to be interrupted by the vacuum interrupter. Then the sub-microsecond high-voltage impulses resume. Therefore, the high-frequency high-voltage impulse conditioning technology is a mixture of spark conditioning using high-voltage pulse and current conditioning. This section determines an influence of the high-frequency high-voltage impulse conditioning on the capacitive current switching performance of vacuum interrupters.

Figure 5.70 shows the proposed high-frequency high-voltage impulse conditioning technology. Figure 5.70a shows the conditioning circuit. $C1$ is a main capacitor for storing energy. It supplies for a high voltage transformer T. The voltage of $C1$ determines the peak value of each impulse, which is 100 kV peak in the test. $C2$ is charged by $C1$ through a reactor $L1$ and a resistance R_s. $C2$ discharges after

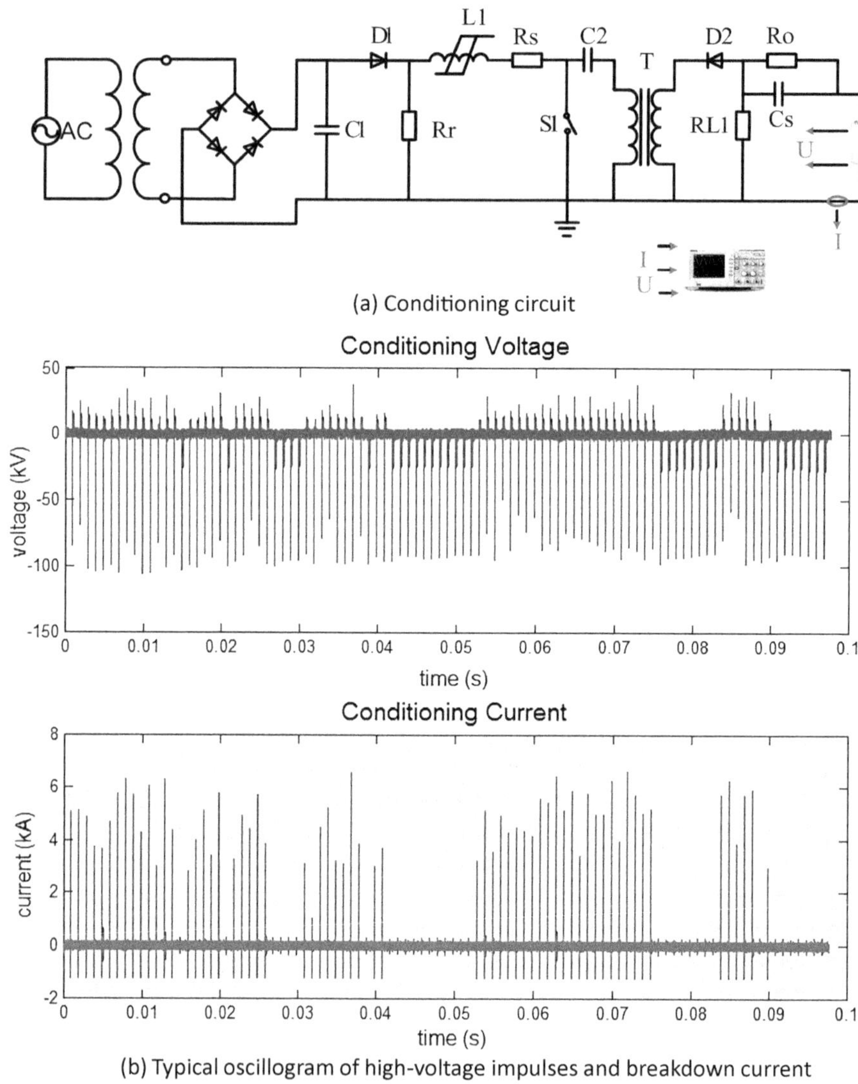

Fig. 5.70 A high-frequency high-voltage impulse conditioning technology

a hydrogen thyratron *S*1 is triggered. Thereafter a high voltage impulse is applied to the test vacuum interrupter through the pulse transformer *T*. If a breakdown in the vacuum interrupter occurs, a high frequency conditioning current of several kA peak flows. The conditioning energy injected into the vacuum interrupter can be controlled, which is from 0 to 3.5 J, by the charged voltage and the capacitance of *C*2. The frequency and duration of a series of high voltage impulses depend on an operating frequency of *S*1, which can be controlled as 1–2000 Hz and 0.1–10.0 s.

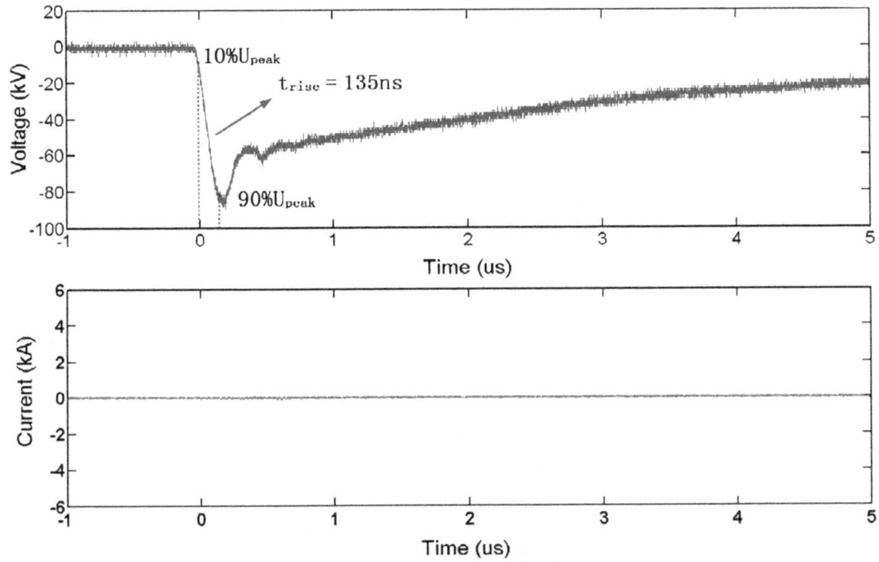

(c) A high voltage impulse without breakdown

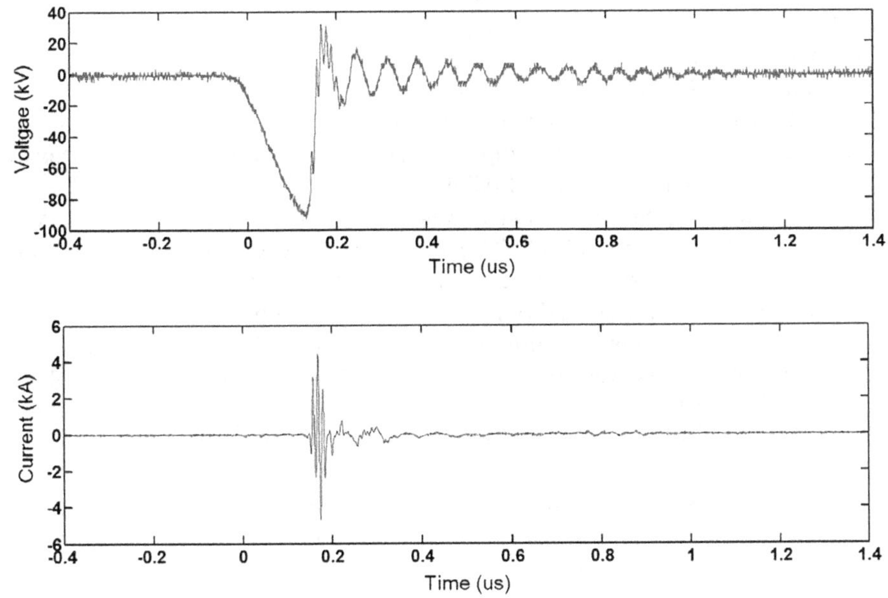

(d) A high voltage impulse with a breakdown

Fig. 5.70 (continued)

The size of the conditioning apparatus is 100 cm × 80 cm × 106 cm. Figure 5.70b shows a typical oscillogram of a series of high voltage impulses. Here nearly half of the voltage impulses cause breakdowns. Figure 5.70c shows a typical oscillogram of a high voltage impulse without breakdown, which has a rising time t_{rise} of 135 ns from 10 to 90% U_{peak}. Figure 5.70d shows a typical oscillogram of a high voltage impulse with breakdown. A high frequency conditioning current (several kA) flows through the vacuum interrupter when it cannot withstand the impulse voltage. This conditioning technology is similar to the high frequency conditioning proposed by ABB and RWTH Aachen University [33]. However, the injected energy at each shot of impulse voltage application would be lower than the previous one, because of the shorter rising time. The low injected energy is beneficial to the conditioning process; otherwise high injected energy would cause de-conditioning effects.

In the experiment, nine identical 7.2-kV vacuum interrupters were used. The pressure of the vacuum interrupters kept on the order of 10^{-5} Pa. The contact was butt type with a diameter of 30 mm. The contact material was CuCr25 (25% of Cr). The full contact gap was 4 mm. A permanent magnetic operating mechanism was adopted in the experiments. The average closing velocity was 0.8 m/s, which was from 2/3 rated contact gap to the closing position.

Before the switching test, each test vacuum interrupter was conditioned by a series of 400 high-frequency high-voltage impulses, in which 200 high-voltage impulses were applied with the stationary contact earthing and the other 200 voltage impulses were applied with the moving contact earthing. The frequency of the impulses was set to 1000 Hz. The duration of each voltage impulse series was set to 0.1 s. Therefore, there were about 100 high-voltage impulses in each series. The vacuum interrupters were conditioned at a contact gap of 0.8 mm.

The capacitive current switching test was carried out by the synthetic test circuit shown in Fig. 5.1. A series of 80 CO operations were executed for each vacuum interrupter. Table 5.33 shows the test conditions. The amplitudes of the inrush currents I_{inrush} were set to 0-, 2-, and 5-kA peak, which corresponded to test series A, B and C, respectively. The inrush current frequency was 3800 Hz. In each test series, the vacuum interrupters were preconditioned by the high-frequency high-voltage impulses. The switching current I_{c} was set to be less than 1 A. The recovery voltage u_{trv} satisfied $u_{\mathrm{trv}} = U_{\mathrm{peak}} \cdot (1 - \cos\omega t)$. The peak value of the recovery voltage was set at 15 kV. The duration of the recovery voltage was about 500 ms.

The prestrike gap d_{pre} was determined as an instantaneous contact gap at the occurrence of the first prestrike during a making operation. In test series A, the inrush current was 0 kA. Therefore, the making operation was a no-load operation. In this case, there was no prestrike phenomenon, and d_{pre} of test series A did not exist. Figures 5.71 shows the distribution of d_{pre} versus the number of making operations in test series B (I_{inrush} 2-kA peak) and test series C (I_{inrush} 5-kA peak). In test series B, d_{pre} ranged between 0.44 and 0.77 mm. In test series C, d_{pre} ranged between 0.45 and 1.95 mm. Thus, the scatter of d_{pre} in test series B was lower than that in test series C.

Since the probability of prestrike increased with a decrease of the contact gap, a complementary cumulative distribution function (CCDF) was introduced for d_{pre}, as

Table 5.33 Capacitive current switching test conditions

Test VIs	Test series A No. 01c VI No. 02c VI No. 03c VI	Test series B No. 21c VI No. 22c VI No. 23c VI	Test series C No. 51c VI No. 52c VI No. 53c VI
Inrush current I_{inrush}	0 kA	2 kA	5 kA
	Frequency: 3800 Hz		
Switching current I_c	<1 A		
Contact gap d	4 mm		
Recovery voltage u_{trv}	$u_{trv} = U_{peak}(1 - cons\omega t)$ $U_{peak} = \frac{7.2kV \times \sqrt{2}}{\sqrt{3}} \times 1.25$		
	Duration: 500 ms		
Test duty	C-O: 80 times		

Note No.**c VI means the vacuum interrupter was conditioned by the high-frequency high-voltage impulse conditioning technology

shown in Eq. (5.11). Figure 5.72 shows the CCDF of d_{pre} in test series B and C. The CCDF was evaluated by a complementary Weibull distribution function as shown in Eq. (5.12). Table 5.34 shows the complementary Weibull parameters values for all of the nine vacuum interrupters. It shows that the R-squared are all above 0.98. Therefore, the CCDF of d_{pre} followed the complementary Weibull distribution function. The shape factor m decreased significantly with the increase of the inrush current. The location parameter $d_{\overline{100}}$ represents the prestrike gap below which the prestrike probability is 1 (CCDF 100%). $d_{\overline{100}}$ increased with the increase of inrush current. The scatters of the prestrike gap can be evaluated by the scale $(\Theta - d_{\overline{100}})$. With the increase in the inrush current from 2- to 5-kA peak, the scale $(\Theta - d_{\overline{100}})$ increased. Therefore, the scattering of d_{pre} increased with the increase of inrush current. Table 5.35 compares the average values of the prestrike gaps of $d_{\overline{10}}$, $d_{\overline{50}}$, $d_{\overline{90}}$, and $(\Theta - d_{\overline{100}})$ with and without the high-frequency high-voltage impulse voltage conditioning. The data without conditioning came from the average values shown in Table 5.3. The high-frequency high-voltage impulse voltage conditioning decreased $d_{\overline{10}}$, $d_{\overline{50}}$, $d_{\overline{90}}$, and $(\Theta - d_{\overline{100}})$ by 39, 19, 0 and 52% respectively for test series B (I_{inrush} = 2-kA peak). For test series C (I_{inrush} = 5-kA peak), the conditioning decreased $d_{\overline{10}}$, $d_{\overline{50}}$, $d_{\overline{90}}$, and $(\Theta - d_{\overline{100}})$ by 19, 12, 4.0 and 28%, respectively. The high-frequency high-voltage impulse voltage conditioning significantly decreased not only the prestrike gaps, but also the prestrike gap scattering.

Figure 5.73 shows a typical oscillogram of the restrike. In the experiment, only one type of restrike phenomenon was observed. The restrike seemed as a disturbance to the recovery voltage. It had no influence on the next cycle of the voltage waveform. Moreover, there was a high-frequency current at the instant when the restrike occurred. This is the restrike type I shown in Sect. 5.1.1. Table 5.36 shows the restrikes characteristics with and without the high-frequency high-voltage impulse

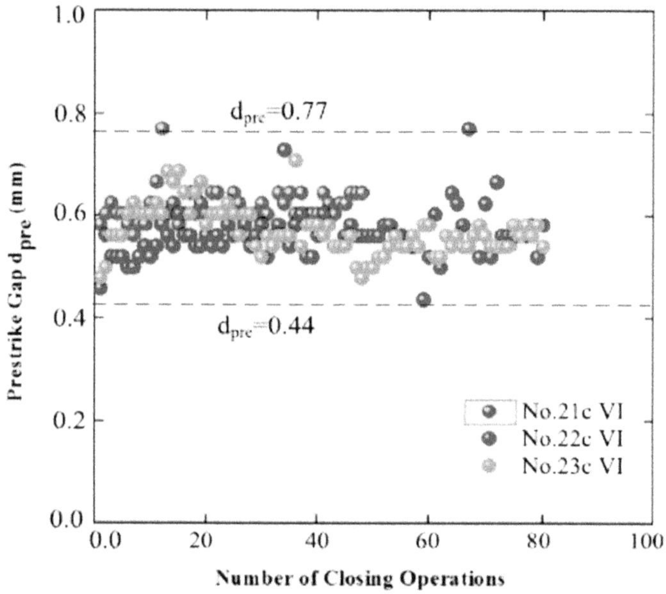

(a) test series B (I_{inrush} = 2-kA peak)

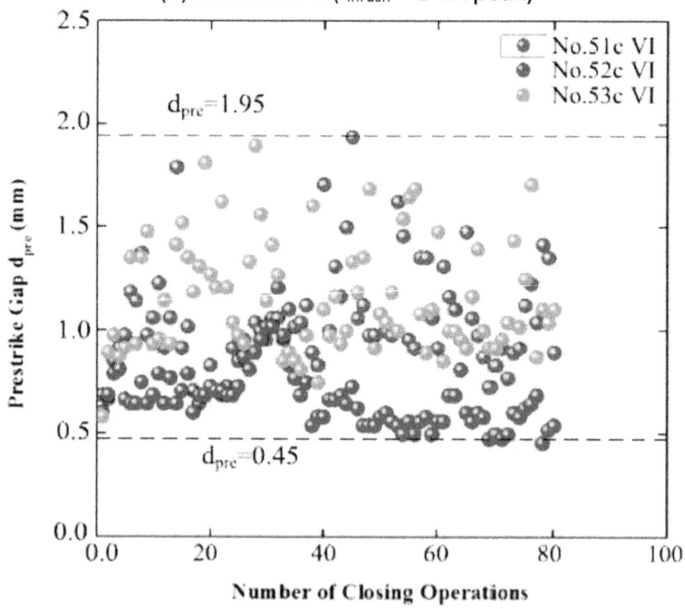

(b) test series C (I_{inrush} 5-kA peak)

Fig. 5.71 Prestrike gap d_{pre} versus number of making operations

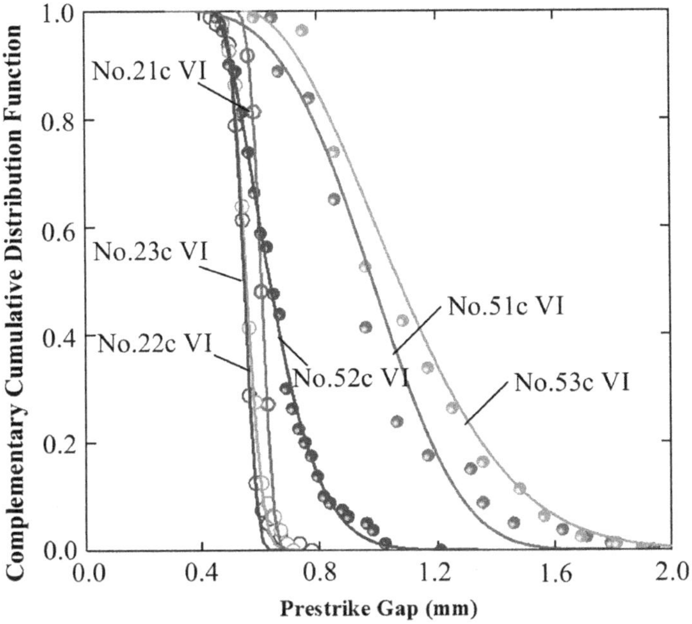

Fig. 5.72 Complementary cumulative probability functions of prestrike gap d_{pre} for test series B ($I_{inrush} = 2$ kA peak) and test series C ($I_{inrush} = 5$-kA peak)

Table 5.34 Complementary Weibull distribution parameters values in test series B and C

VIs	m	$d_{\overline{100}}$/mm	Θ/mm	$\Theta - d_{\overline{100}}$/mm	R-squared
No.21c	3.84	0.51	0.61	0.10	0.99
No.22c	3.95	0.44	0.56	0.12	0.99
No.23c	2.59	0.47	0.57	0.10	0.99
No.51c	1.54	0.66	1.03	0.37	0.98
No.52c	1.72	0.44	0.68	0.23	0.99
No.53c	1.98	0.54	1.18	0.63	0.99

conditioning. The data without conditioning came from Table 5.4. For vacuum interrupters without the conditioning, with increasing of inrush current, the restrikes tended to evolve into restrike type II and restrike type III. By contrast, for vacuum interrupters with the conditioning, all of the restrikes followed restrike type I, no matter what the amplitude of inrush current was.

Table 5.37 shows the restrike probabilities P_{res} in the 80 CO operations for the test vacuum interrupters in test series A, B and C. In test series A ($I_{inrush} = 0$ kA), the restrike probability was 5.0% for each of the three vacuum interrupters (No. 01c, No. 02c, and No. 03c). The average restrike probability was 5.0%. In test series B ($I_{inrush} =$ 2 kA peak), the restrike probabilities were 6.3, 11.3, and 11.3% for the three vacuum

Table 5.35 An influence of the high-frequency high-voltage impulse voltage conditioning on the prestrike gap characteristics

Test VIs	Conditioning	$d_{\overline{10}}$/mm	$d_{\overline{50}}$/mm	$d_{\overline{90}}$/mm	$\Theta - d_{\overline{100}}$/mm
Series B (2-kA peak)	Without conditioning	1.00	0.70	0.53	0.23
	High frequency high voltage conditioning	0.61	0.57	0.53	0.11
	Relative variational rate (%)	−39	−19	0	−52
Series C (5-kA peak)	Without conditioning	1.50	1.00	0.70	0.57
	High frequency high voltage conditioning	1.21	0.88	0.67	0.41
	Relative variational rate (%)	−19	−12	−4	−28

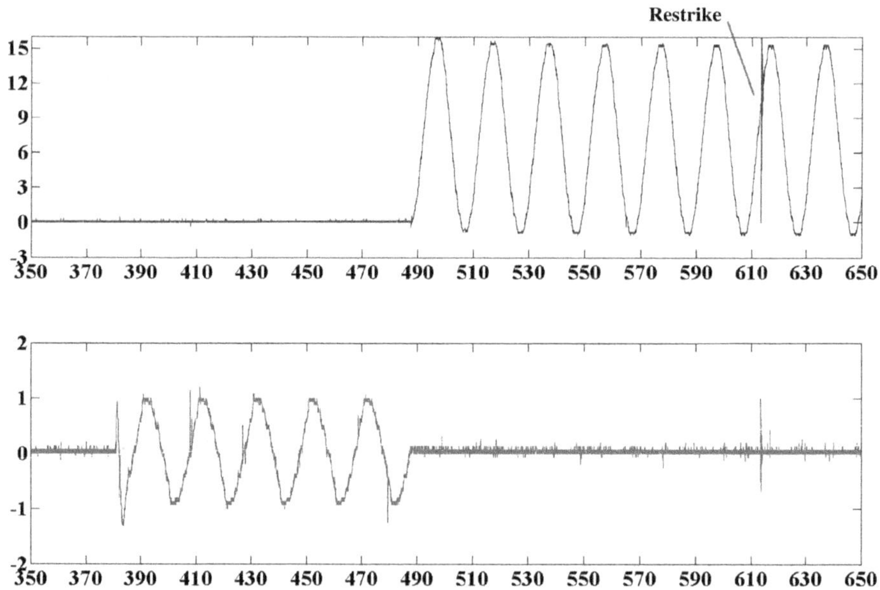

Fig. 5.73 A typical oscillogram of the restrike

interrupters (No. 21c, No. 22c, and No. 23c). The average restrike probability was 9.6%. In test series C ($I_{\text{inrush}} = 5$ kA peak), the restrike probabilities were 7.5, 13.8 and 2.5% for the three vacuum interrupters. The average restrike probability was 7.9%. Table 5.38 compares the restrike probabilities with and without the conditioning. The data without conditioning came from Table 5.5. It shows that the restrike probability almost unchanged in the case of $I_{\text{inrush}} = 0$ kA with and without the conditioning. However, when the inrush current increased to 2-kA peak and further to 5-kA peak, the restrike probability decreased by 8.3 and 22.5%, respectively by undergoing the

Table 5.36 An influence of the high-frequency high-voltage impulse voltage conditioning on the restrikes characteristics

Test series	Conditioning	Restrike times	Restrike Type I (%)	Restrike Type II (%)	Restrike Type III (%)
A (0-kA peak)	Without conditioning	11	75.0	16.7	8.3
	High frequency high voltage conditioning	12	100	0	0
B (2-kA peak)	Without conditioning	43	62.8	25.6	11.6
	High frequency high voltage conditioning	23	100	0	0
C (5-kA peak)	Without conditioning	73	20.5	20.5	59.0
	High frequency high voltage conditioning	19	100	0	0

Table 5.37 Restrike probability for test series A, B, and C

Test series	VIs	P_{res} (%)	Average value $\overline{P}_{res}$ (%)
A (I_{inrush} 0-kA peak)	No01.c	5.0	5.0
	No02.c	5.0	
	No03.c	5.0	
B (I_{inrush} 2-kA peak)	No.21c	6.3	9.6
	No.22c	11.3	
	No.23c	11.3	
C (I_{inrush} 5-kA peak)	No.51c	7.5	7.9
	No.52c	13.8	
	No.53c	2.5	

high-frequency high voltage impulse conditioning. Thus, the high-frequency high voltage impulse conditioning technology had a significant influence on reducing the restrike probabilities, especially at higher inrush currents.

Figure 5.74 shows the distribution of the restrike occurring time t_{res} for test series A, B, and C. In the figure, T is a cycle of the recovery voltage (20 ms). The zero point is the instant when the recovery voltage applied to the test vacuum interrupter. For test series A ($I_{inrush} = 0$ kA), the majority of the restrikes (75%) distributed between 0.5·T and 5·T (75%). For test series B (I_{inrush} = 2-kA peak), the restrike occurring

Table 5.38 An influence of the high-frequency high-voltage impulse voltage conditioning on the restrikes probabilities

Test series	Conditioning	$\overline{P}_{res}$ (%)	$\Delta\overline{P}_{res}$ (%)
A (I_{inrush} 0-kA peak)	High frequency high voltage conditioning	5.0	0.4
	Without conditioning	4.6	
B (I_{inrush} 2-kA peak)	High frequency high voltage conditioning	9.6	−8.3
	Without conditioning	17.9	
C (I_{inrush} 5-kA peak)	High frequency high voltage conditioning	7.9	−22.5
	Without conditioning	30.4	

time became more scattering. The majority of the restrikes (82%) distributed between 0.5·T and 10·T. For test series C (I_{inrush} = 5-kA peak), the restrike occurring time became even more scattering, in which 16% ranging between 0.25 and 0.5 T, 47% ranging between 0.5 and 5 T, 11% ranging between 5 and 10 T, 26% ranging above 10 T. Comparing to the no conditioning case shown in Fig. 5.10, in which the restrike occurred earlier with the increase in the inrush current, the higher inrush current tended to cause the restrikes occurring more scattered in the conditioned vacuum interrupters.

Figures 5.75, 5.76, and 5.77 show the contact surfaces photos after the 80 CO operations in test series A, B, and C, respectively. In test series A, the prestrike phenomenon did not exist. Figure 5.75 shows there was no damages generated by the prestrike arc on the contact surfaces. On the other side, the conditioning prints by the high-frequency high-voltage impulses can be found all over the contact surface. Furthermore, the conditioning prints uniformly distributed over the contact surfaces. Figures 5.76 and 5.77 show that the conditioning prints also uniformly distributed over the contact surfaces. At the same time, with the high-frequency high-voltage impulse conditioning, the welding spots scattered on the contact surfaces for each vacuum interrupter in test series B and C. In comparison, the no conditioned case, as shown in Figs. 5.11 and 5.12, the welding concentrated at one location between the contacts. There was a single macroscopic protrusion on one contact surface and a corresponding macroscopic crater on the opposite contact surface. The scattered welding spots on the conditioned contact surfaces may be related to the molten metal layer caused by the high-frequency high-voltage impulse conditioning. With the layer, the prestrike arc energies may not concentrate on a small area, but on more spots of the contact surface. The distributed prestrike arc energy on contact surfaces would benefit the capacitive current switching performance by reducing the restrike probability, because the dielectric strength of the vacuum gap would increase under the recovery voltage.

Fig. 5.74 Restrike occurring time. T is a cycle of the recovery voltage (20 ms)

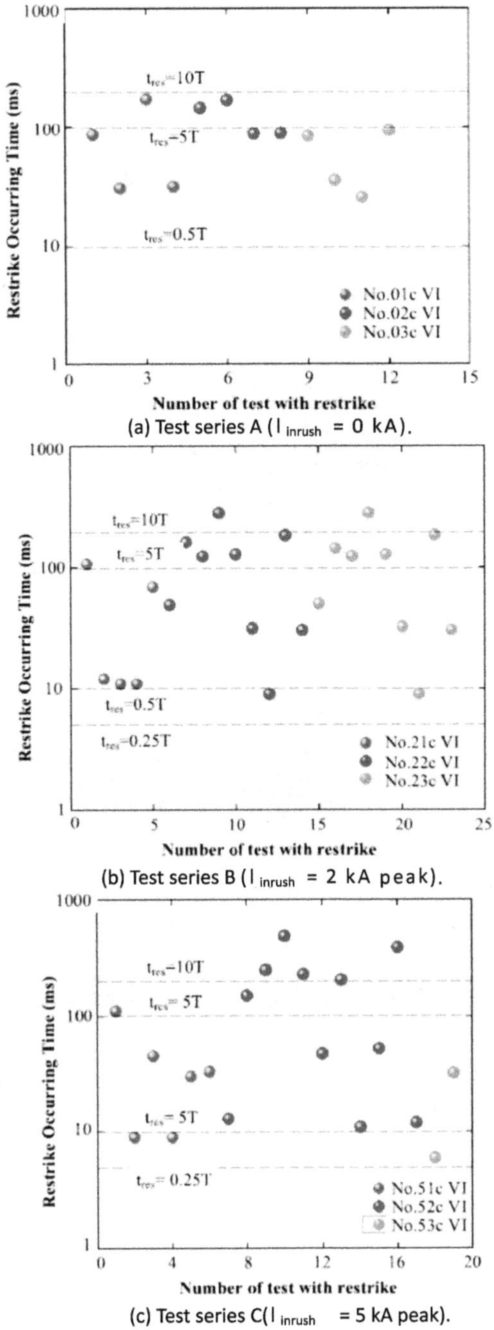

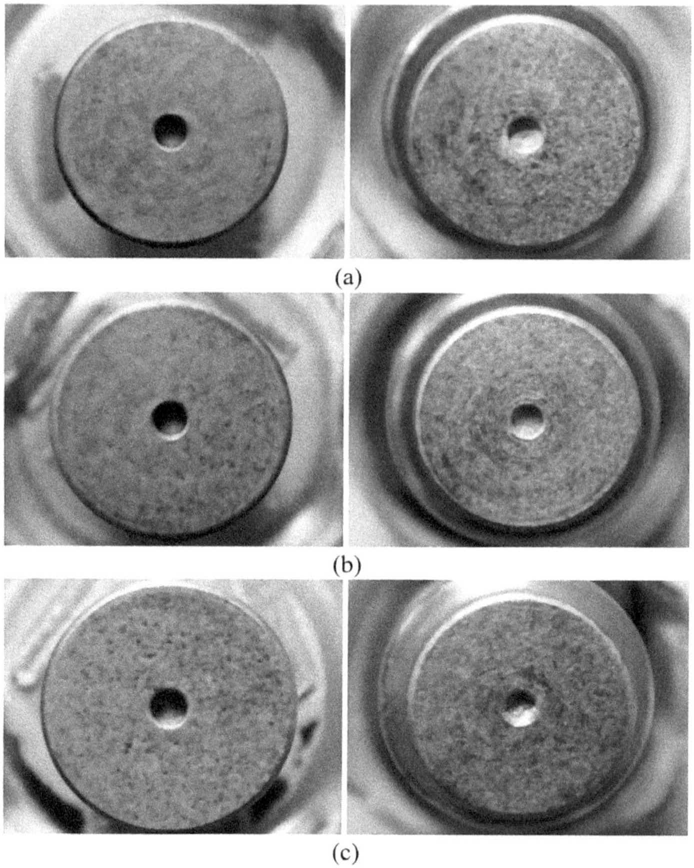

Fig. 5.75 Contact surface photos of test series A ($I_{inrush} = 0$-kA). Left: fixed contact. Right: moving contact. **a** No. 21c VI. **b** No. 22c VI. **c** No. 23c VI

5.5 Back-To-Back Capacitor Bank Switching Technology

In most of extra high voltage 500 kV substations in China, 72.5 kV SF_6 circuit breakers are used to energize or de-energize the capacitor banks for reactive power compensation, although the voltage level for reactive power compensation is only 35 kV [34]. Why this happened? It is to achieve low restrike probability. Vacuum circuit breakers have the advantages including long mechanical and electrical life, low cost, and environmentally friendly. However, restrikes occasionally occur in vacuum interrupters when switching capacitor banks because of modification for contact surface generated by prestrike arcs, especially for back-to-back capacitor banks switching (see Sect. 5.1). Therefore, it is meaningful to improve the conditions of contact surfaces and to reduce the probability of restrikes. Section 5.1.4 suggested a specific operation sequence for two vacuum interrupters in series to keep at least

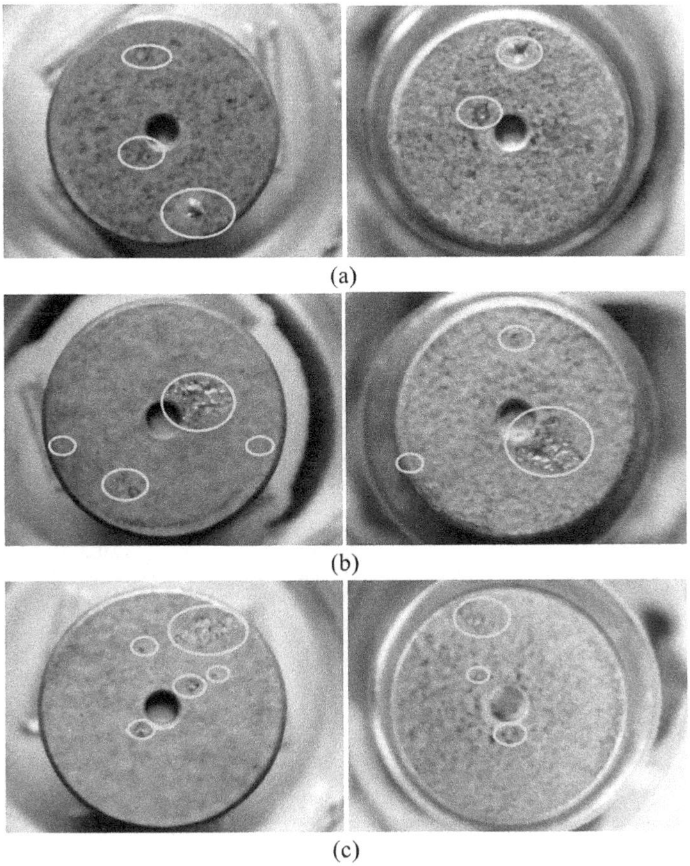

Fig. 5.76 Contact surface photos of test series B (I_{inrush} = 2-kA peak). Left: fixed contact. Right: moving contact. **a** No. 01c VI. **b** No. 02c VI. **c** No. 03c VI

one of the vacuum interrupters' contact surface fine, as shown in Fig. 5.37. In this way, a good dielectric strength would be achieved. Section 5.2.2 choose horse-shoe type axial magnetic field contact for back-to-back capacitor bank switching for lower restrike probability than that of cup-type axial magnetic field contact and spiral-type transverse magnetic field contact. Such ideas inspired a new vacuum circuit breaker technology for back-to-back capacitor bank switching by using series double breaks, which is introduced in the two following sections in details. The proposed technology adopts two vacuum interrupters in series. The two vacuum interrupters operate as sequence shown in Fig. 5.37. The two vacuum interrupters use horse-shoe type axial magnetic field contact. One vacuum interrupter is responsible for inrush current making and it use CuW contact material. The other vacuum interrupter is responsible for capacitive current and/or short circuit current breaking, which use CuCr contact material. Based on the proposed technology, two 40.5 kV vacuum circuit breaker

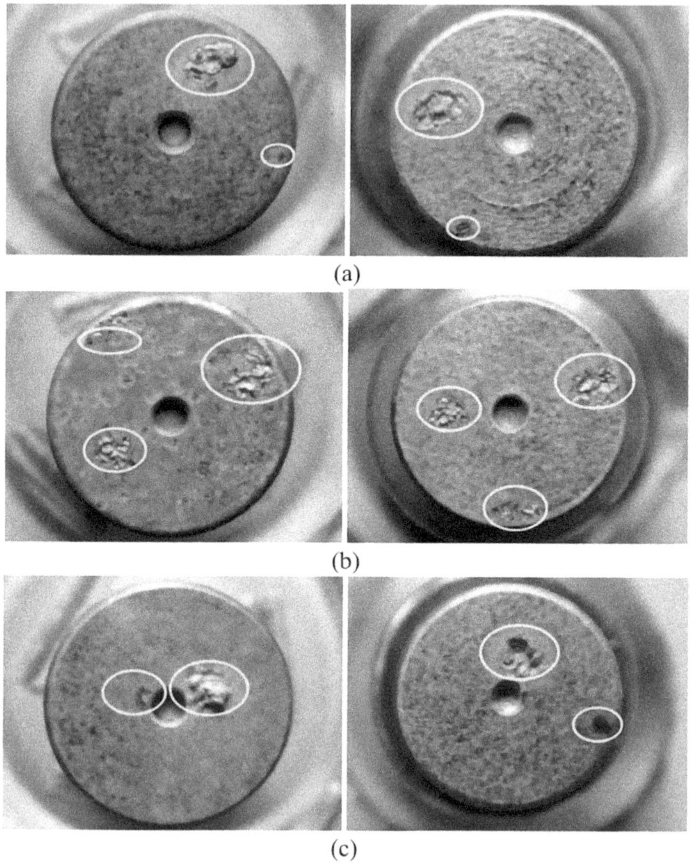

Fig. 5.77 Contact surface photos of test series C (I_{inrush} = 5-kA peak). Left: fixed contact. Right: moving contact. **a** No. 51c VI. **b** No. 52c VI. **c** No. 53c VI

products were developed for back-to-back capacitor bank switching. Both of them were certificated as class C2 by type test. One was an outdoor 40.5 kV/2500 A/31.5 kA vacuum circuit breaker with back-to-back capacitor bank switching current of 1500 A. The other was an indoor 40.5 kV/2500 A/31.5 kA vacuum circuit breaker with back-to-back capacitor bank switching current of 1200 A. The mechanical life of both products was 30,000 cycles.

5.5.1 An Outdoor Double-Break 40.5 kV Vacuum Circuit Breaker for Back-To-Back Capacitor Bank Switching

Figure 5.78a shows the outdoor 40.5 kV double-break vacuum circuit breaker devel-

Fig. 5.78 A class C2 40.5 kV/2500 A/31.5 kA outdoor vacuum circuit breaker with back-to-back capacitor bank switching current of 1500 A (Courtesy of Eaton)

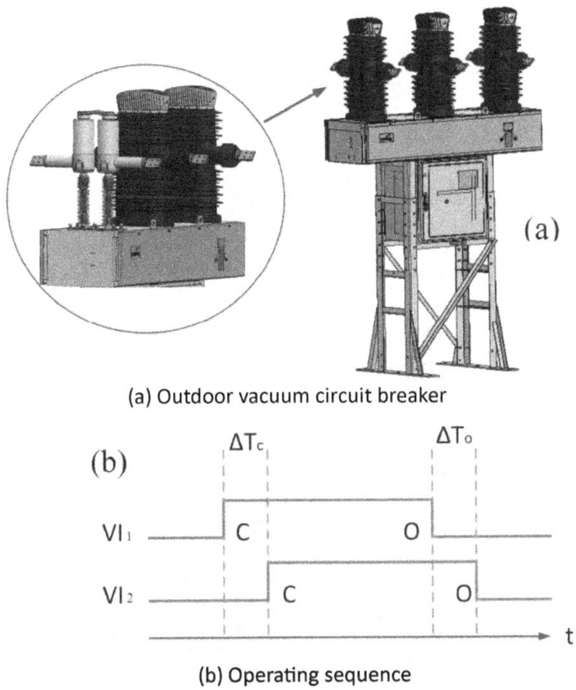

oped for back-to-back capacitor bank switching. Its nominal rated current is 2500 A and its short circuit breaking current is 31.5 kA. For each phase, there are two vacuum interrupters (VI_1 and VI_2) embedding in an epoxy pole. Both of the vacuum interrupters chose horse-shoe type axial magnetic field contact. The contact material of VI_1 is CuCr and the contact material of VI_2 is CuW. Figure 5.78b showed its operating sequence. VI_1 closes and opens before VI_2. The time ΔT_c and ΔT_o represent two operation intervals between the two vacuum interrupters. Type test results proved it as a class C2 circuit breaker with back-to-back capacitor bank switching current of 1500 A.

A direct three-phase test circuit was used for type testing of the back-to-back capacitor bank switching capability of the vacuum circuit breaker. Figure 5.79 shows a schematic diagram of the three phase back-to-back capacitor bank switching direct test circuit. The test power is provided by short-circuit generators CF through a short-circuit transformer DB. C_2 is the load capacitor bank, which determines the value of breaking current. C_1 is the back-to-back capacitor bank, which determines the value of inrush current during making operations. Voltage transformers YH_1/YH_2 measure the power frequency voltages. Current transformers LH_1/LH_2 measure circuit currents. Voltage dividers FY measure the recovery voltage of the test vacuum circuit breaker SW_{test}.

Table 5.39 shows the type test conditions of back-to-back capacitor banks switching. The type test followed IEC 62271-100 standard [25] and GB 1984-2014

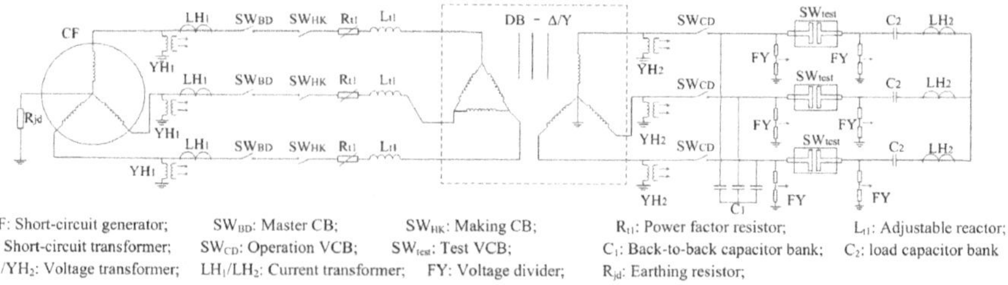

Fig. 5.79 A schematic diagram of direct test circuit for three-phase back-to-back capacitor bank switching

Table 5.39 Type test conditions of back-to-back capacitor banks switching for the 40.5 kV outdoor vacuum circuit breaker

Test duty	Operation	Circuit voltage	Breaking current
T60	O	–	18.9 kV %dc ≤ 20%
BC2	80 CO	40.5 kV	1500 A
BC1	24 O	40.5 kV	150–600 A

standard [35]. A test duty T60 was carried out as a pre-conditioning test before the back-to-back capacitor banks switching test. Then a test duty BC2 that contained a series of 80 "CO" operations was carried out. Figure 5.80 shows a typical current and voltage oscillogram of the back-to-back capacitor banks switching test. An inrush current of 20kA peak and frequency of 4250 Hz flowed through the test vacuum

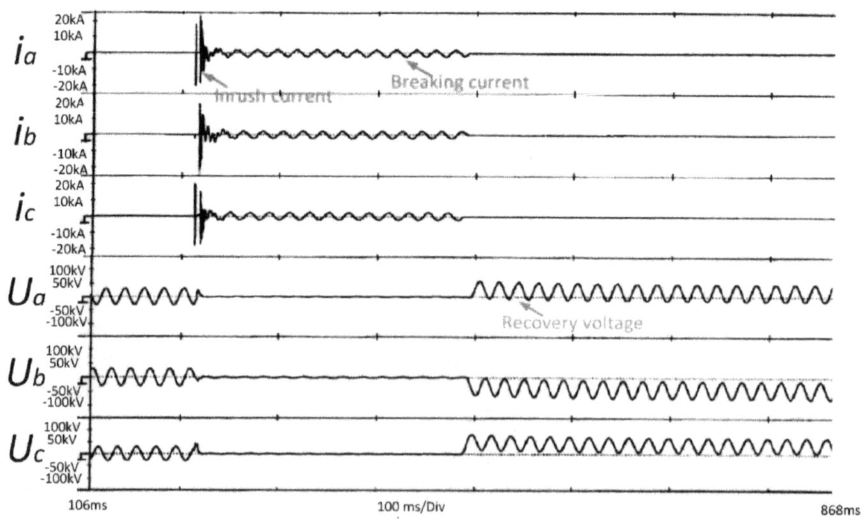

Fig. 5.80 A typical current and voltage oscillogram of the back-to-back capacitor banks switching test

circuit breaker during the making operation. Then a recovery voltage with a peak value of 85 kV was applied to the test vacuum circuit breaker after interrupting a capacitive current of 1500 A rms during the breaking operation. After that, a test duty BC1 was carried out for the test vacuum circuit breaker. It contained 24 "O" operations which meant the test vacuum circuit breaker do not suffer inrush current. The capacitive current in the BC1 duty is 150–600 A.

Figure 5.81 showed the peak values of inrush current during the making operations in test duty BC2. The peak values of inrush current varied according to the making phase angles relative to the source side voltage. The peak values of inrush current varied dramatically. Some reached 50 kA and some only were about 5 kA. The average values were 24, 24 and 22 kA for Phase A, Phase B and Phase C, respectively. So the inrush current in this test were even more severe than that required in IEC62271-100 standard [25].

Table 5.40 shows the restrike probability of the 40.5 kV outdoor vacuum circuit breaker during the back-to-back capacitor banks switching test. For both BC2 and BC1 test duties, there was no restrike and no NSDD phenomenon occurring during interrupting the capacitive current. It meant that the vacuum circuit breaker reached C2 class as clarified in the IEC standard [25].

Figure 5.82 shows the contact surfaces photos of the 40.5 kV outdoor vacuum

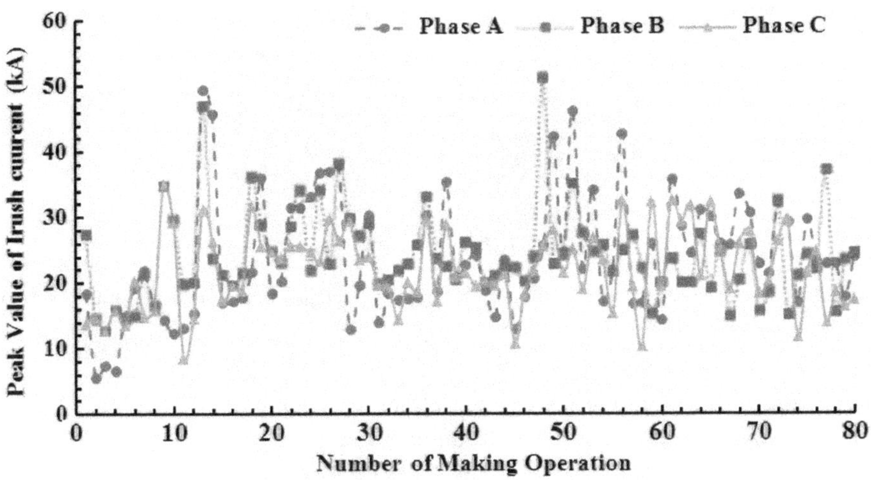

Fig. 5.81 Peak values of inrush current in test duty BC2

Table 5.40 Restrike probability of the 40.5 kV outdoor vacuum circuit breaker during the back-to-back capacitor banks switching test

Test duty	Operation	Restrike probability (%)	NSDD
BC2	80 CO	0	0
BC1	24 O	0	0

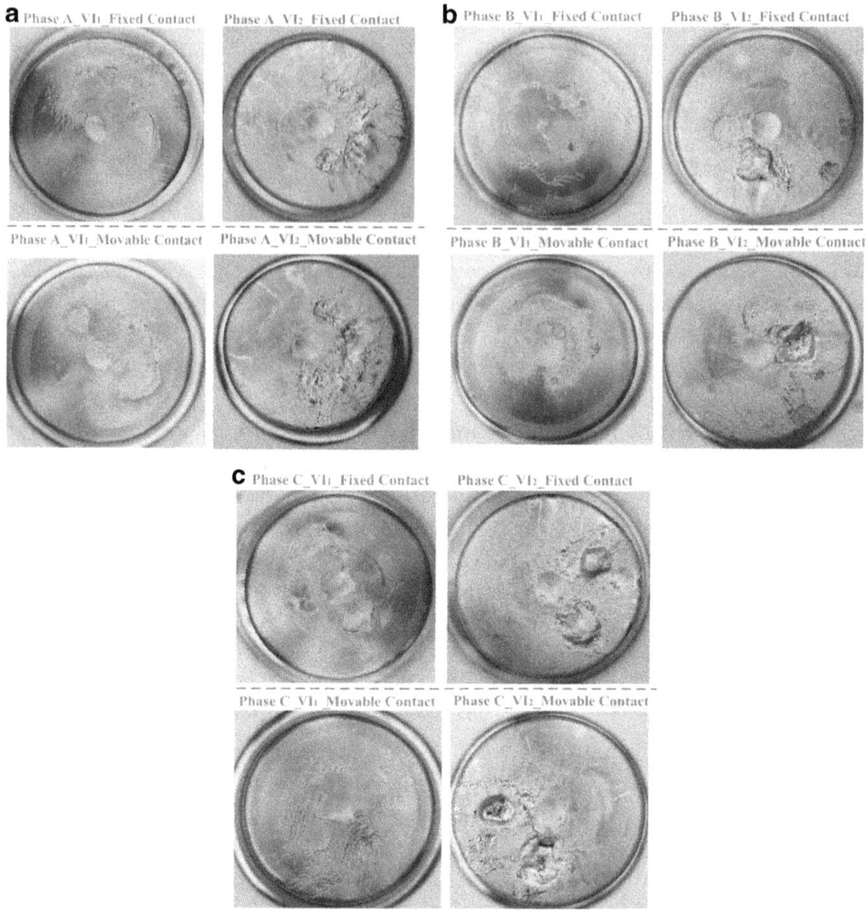

Fig. 5.82 Contact surface photos of the 40.5 kV outdoor vacuum circuit breaker after the back-to-back capacitor banks switching test

circuit breaker after the back-to-back capacitor banks switching test. It found that the contact surface conditions of VI_2 were quite different from that of VI_1. Destructive effects appeared on the CuW contact surfaces of VI_2. Significant welding erosion areas can be found on the three phases of contact surfaces of VI_2. Moreover, there were protrusions appearing on the stationary contact surface and carters appearing on the movable contact surface. A 3D contour measurement showed that the height of the protrusions on the stationary contacts of phase A, B, and C were 1.5, 3.0, and 2.8 mm respectively. The depth of the carters on the movable contacts of phase A, B, and C were 1.7, 3.5, and 3.7 mm, respectively. In contrast, the damage on the CuCr contact surfaces of VI_1 were insignificant. There were only light erosions with a depth ranging from 0.24 to 0.58 mm appearing.

The 1500 A back-to-back capacitor bank switching type test passing proves that the 40.5 kV outdoor vacuum circuit breaker follows class C2. It also showed that the proposed technology is an effective approach for minimizing restrikes during capacitive current switching. The proposed vacuum circuit breaker electrically connected two vacuum interrupters (VI_1 and VI_2) in series for each phase. During the making operation, VI_2 closes after VI_1. The time interval ΔT_c ensures that the prestriking primarily occurs inside VI_2. So that the prestrike arc of inrush current mainly erodes the contacts of VI_2. In order to increase the anti-welding ability of VI_2, CuW contact material was chosen for VI_2. CuW contact material proved excellent anti-welding performance, even though serious damages can be found on the contact surfaces of VI_2, such as significant welding erosion areas, high protrusions and craters with large size. VI_1 was largely prevented from serious welding even the inrush current peak value reached 50 kA. During the following breaking operation, a capacitive current of below 2 kA is usually not difficult for vacuum circuit breakers to interrupt. Instead, contact surfaces damages, such as protrusions, craters and particles, generated by the rupture of welding area are critical factors to influence the dielectric strength of the vacuum gap. It meant that vacuum interrupter with damaged contact surfaces is easy to interrupt the capacitive current, but it is difficult to withstand the $1-\cos(\omega t)$ shape recovery voltage. In the proposed technology, another time interval ΔT_o determines VI_2 opens later than VI_1 during the breaking operation. VI_1 is to interrupt the capacitive current (1500 A for BC2 test duty or 600 A for BC1 test duty) and then to withstand the initial stage of recovery voltage ($0 - \Delta T_o$). For normal vacuum circuit breakers, breakdowns tended to occur in the initial recovery stage while the ruptures of the welding area deteriorated the dielectric strength (see Fig. 5.10). However, the situation was different for the proposed technology, as the contact surfaces of VI_1 were virtually not damaged by the inrush current arcs. Thus, the VI_1 was easily to withstand the recovery voltage. After ΔT_o, the VI_2 opened. Therefore, the two vacuum interrupters in series collaborated to withstand the following recovery voltage. It was an effective way to prevent later breakdowns from ΔT_o up to hundreds of ms.

5.5.2 *An Indoor 40.5 kV Vacuum Circuit Breaker for Back-To-Back Capacitor Bank Switching*

Figure 5.83a shows the indoor 40.5 kV vacuum circuit breaker, which was developed to meet the requirement of class C2 with a back-to-back capacitor bank switching current of 1200 A. In order to facilitate convenient retrofitting, the main dimensions were kept as same as the currently installed indoor SF_6 circuit breakers. For each phase, two vacuum interrupters with horse-shoe axial magnetic field contact (VI_1 and VI_2) were embedded in an epoxy pole (see Sect. 5.2.1). The contact material of VI_1 was CuCr and the contact material of VI_2 was CuW. For long mechanical life, the two vacuum interrupters were operated by two separated actuators following the

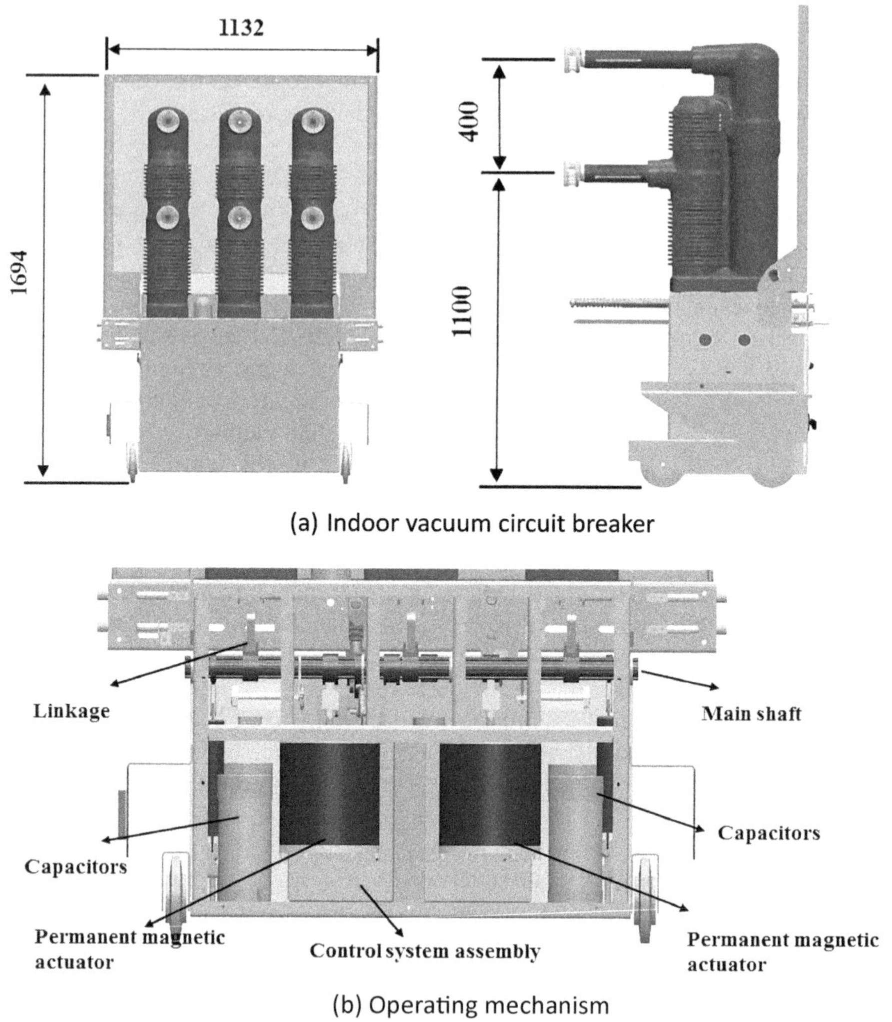

Fig. 5.83 A class C2 40.5 kV/2500 A/31.5 kA indoor vacuum circuit breaker with back-to-back capacitor bank switching current of 1200 A (Courtesy of Eaton)

sequence as shown in Fig. 5.78b. In this operating sequence, VI_1 closes and opens before VI_2. The time intervals ΔT_c and ΔT_o were the operation intervals between two vacuum interrupters. Figure 5.83b shows the operating mechanism of this indoor vacuum circuit breaker. Two permanent magnetic actuators were mechanically linked to two independent linkages and finally they drove the movement of the two vacuum interrupters.

Table 5.41 showed the back-to-back capacitor banks switching type test conditions according to IEC62271-100 standard [25] and GB1984-2014 standard [35]. The

Table 5.41 Type test conditions of back-to-back capacitor banks switching for the 40.5 kV indoor vacuum circuit breaker

Test duty	Operation	Circuit voltage	Breaking current
T60	O	–	18.9 kV %dc ≤ 20%
BC2	80 CO	40.5 kV	1200 A
BC1	24 O	40.5 kV	350 A

Table 5.42 Restrike probability of the 40.5 kV indoor vacuum circuit breaker during the back-to-back capacitor banks switching test

Test duty	Operation	Restrike probability (%)	NSDD
BC2	80 CO	0	0
BC1	24 O	0	0

back-to-back capacitor banks switching type test circuit was as same as that shown in Fig. 5.79. A test duty T60 was carried out as a pre-conditioning test before the back-to-back capacitor banks switching test. Then a test duty BC2 that contained a series of 80 "CO" operations was carried out. The typical current and voltage oscillogram was similar to that shown in Fig. 5.80. An inrush current of 20 kA peak with a frequency of 4250 Hz flowed through the test vacuum circuit breaker during making operation. Then the recovery voltage with a peak value of 85 kV was applied to the test vacuum circuit breaker after interrupting a capacitive current of 1200 A rms during breaking operation. After that, a test duty BC1 was carried out for the test vacuum circuit breaker. It contained 24 "O" operations which meant the test vacuum circuit breaker did not suffer inrush current. The capacitive current in the BC1 duty was 350 A.

Table 5.42 shows the restrike probability of the test vacuum circuit breaker during the back-to-back capacitor banks switching test. For both BC2 and BC1 test duties, there was no restrike and even no NSDD phenomenon occurred during interrupting the capacitive current. It meant that this test vacuum circuit breaker reached the requirement of class C2, as clarified in IEC and GB standards [25, 35].

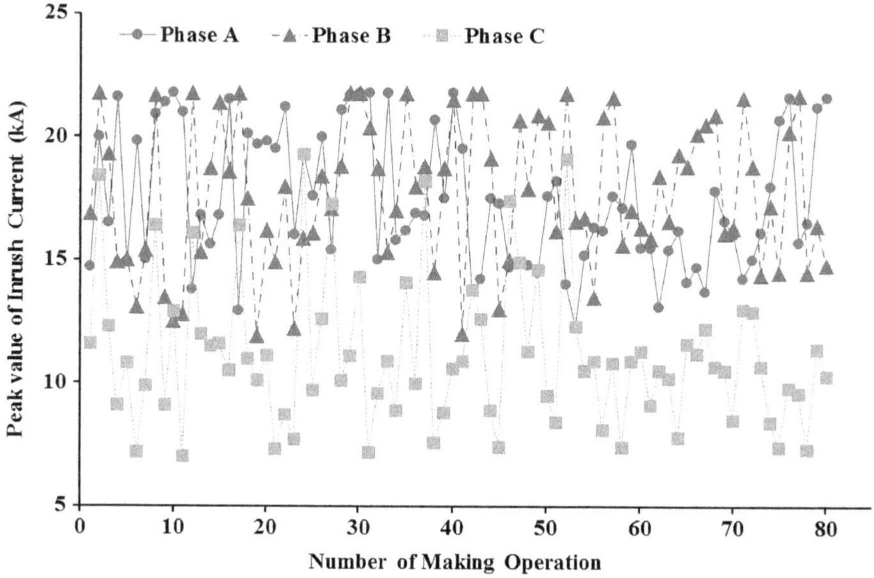

Fig. 5.84 Peak values of inrush current in test duty BC2

As the BC2 test was executed by the direct three-phase test circuit, the peak values of inrush current varied according to the making phase angles relative to the source side voltage. Figure 5.84 shows the peak values of the inrush current in making operations during BC2 test. Some reached more than 20 kA and some were lower than 10 kA. The average values were 24, 17 and 18 kA for Phase A, Phase B and Phase C, respectively.

Figure 5.85 shows the contact surfaces photos. For the CuCr contacts in VI_1, there was only a slight arc erosion. There was no welding between the CuCr contacts caused by inrush current because of the time interval ΔT_c. For the CuW contact in VI_2, there were significant damages caused by inrush current. However, benefitting from the strong anti-welding ability of CuW contact material, no serious welding was found on the contact surfaces.

During the breaking operations, normal commercial vacuum circuit breakers would be difficult to withstand the $1-\cos(\omega t)$ shape recovery voltage with restrikes free. The contact surface damages caused by inrush current, such as protrusions, craters and particles, had negative impacts on the dielectric strength of the vacuum gap. The proposed technology set the time interval ΔT_o, by such setting VI_1 interrupts the capacitive current and withstand the initial stage of recovery voltage ($0 - \Delta T_o$ ms). Breakdowns tended to occur frequently in the initial stage of the recovery voltage because the ruptures of the welding area deteriorated the dielectric strength. However, the situation was improved in this vacuum circuit breaker, as the contact surfaces of VI_1 were virtually not damaged by the inrush current. Thus, the VI_1 was easily to withstand the recovery voltage. The VI_2 opened after ΔT_o ms. Therefore,

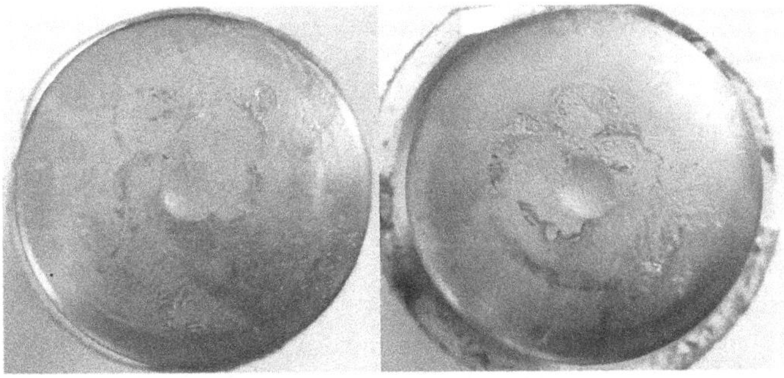

(a) CuCr contact in VI_1 (left: movable contact, right: fixed contact)

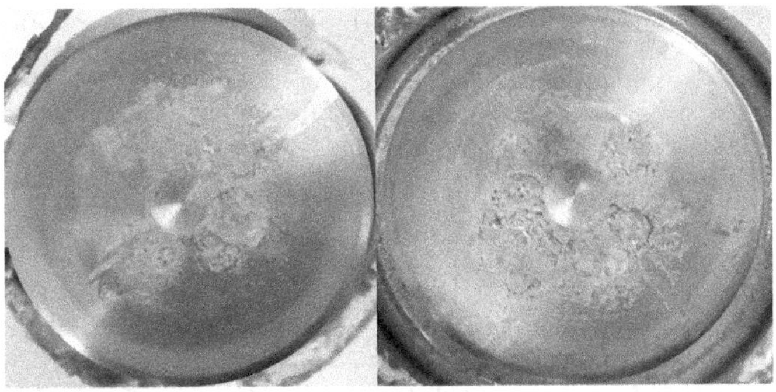

(b) CuW contact in VI_2 (left: movable contact, right: fixed contact)

Fig. 5.85 Contact surface photos of the 40.5 kV indoor vacuum circuit breaker after the back-to-back capacitor banks switching test

the two vacuum gaps in series to withstand the remaining recovery voltage. It was an effective way to prevent later breakdowns.

References

1. Bonfanti I, Guilloux C, Damstra G and Smeets R, Shunt Capacitor Bank Switching Stresses and Test Methods, Electra, No. 183, pp. 12–44, 1999.
2. Slade P G, The Vacuum Interrupter: Theory. Design and Application, CRC Press Taylor & Francis Group, 2008, pp. 446–450.
3. Griot S and Moreau A, Vacuum Circuit Breaker's Electrical Life for Shunt Capacitor Switching, 24th International Symposium on Discharges and Electrical Insulation in Vacuum, Braunschweig, Germany, pp. 194–197, 2010.
4. Smeets R and Lathouwers A, Capacitive Currents Switching Duties of High-Voltage Circuit Breakers: Background and Practice of New IEC Requirements," IEEE Power Engineering Soc. Winter Meeting, Singapore, Vol. 3, pp. 2123–2128, 2000.
5. Glinkowski M, Greenwood A, Hill J, Mauro R and Varneckes V, Capacitance Switching with Vacuum Circuit Breakers - A Comparative Evaluation," IEEE Trans. Power Delivery, Vol. 6, No. 3, pp. 1088–1095, 1991.

6. Schellekens H, Capacitor Bank Switching with Vacuum Circuit Breaker, 23rd Int'l. Sympos. Discharges and Electr. Insul. Vacuum, Braunschweig, pp. 157–160, 2008.
7. Voshall R E and Holmes F A, An Indirect Test Circuit for Testing Vacuum Interrupters for Capacitor Switching Duty, IEEE Trans. Power App. Syst., Vol. 99, No. 3, pp. 1276–1279, 1980.
8. Dullni E, Shang W, Gentsch D, Kleberg I and Niayesh K, Switcing of Capacitive Currents and the correlation of Restrike and Pre-ignition Behavior, IEEE Trans. Dielectr. Electr. Insul., Vol. 13, No. 1, pp. 65–71, 2006.
9. Körner F, Lindermayer M, Kurrat M and Gentsch D, Switching Behavior of Different Contact Materials under Capacitive Switching Conditions, 23rd International Symposium on Discharges and Electr. Insul. Vacuum, Bucharest, Romania, pp. 202–205, 2008.
10. Yokokura K, Matsuda M, Atsumi K, Miyazawa T, Sohma S, Kaneko E and Ohshima I, Capacitor Switching Capability of Vacuum Interrupter with CuW Contact material, IEEE Trans. Power Delivery, Vol. 10, No. 2, pp. 804–809, 1995.
11. Kamikawaji T, Shioiri T, Funahashi T, Satoh Y, Kaneko E and Ohshima I, An Investigation into Major Factors in Shunt Capacitor Switching Performances by Vacuum Circuit Breakers with CuCr contacts, IEEE Trans. Power Delivery, Vol. 8, No. 4, pp. 1789–1795, 1993.
12. Doen T, Tsukima M, Sato S and Yoshida T, Investigation of correlation between vacuum breakdown phenomena and Field Emission Current during Shunt Capacitor Switching, 25th International Symposium on Discharges and Electrical Insulation in Vacuum, Tomsk, pp. 161–164, 2012.
13. Juhsasz A and Rieder W, Capacitive Switching with Vacuum Interrupters, 10th International Conference on Gas Discharges and Their Applications, pp. 62–65, 1992.
14. Koochack Zadeh M, Hinrichsen V, Smeets R and Lawall A, The Impact of Capacitive Bank Inrush Current on Field Emission Current in Vacuum, 24th International Symposium on Discharges and Electrical Insulation in Vacuum, Braunschweig, Germany, pp. 210–213, 2010.
15. Delachaux T, Rager F, Gentsch D, Study of Vacuum Circuit Breaker Performance and Weld Formation for Different Drive Closing Speeds for Switching capacitive current, 24th International Symposium on Discharges and Electrical Insulation in Vacuum, Braunschweig, Germany, pp. 241–244, 2010.
16. Smeets R, Kuivenhoven S, Chakraborty S and Sandolache G, Field Electron Emission Current in Vacuum Interrupters after Large Inrush Current, 25th International Symposium on Discharges and Electrical Insulation in Vacuum, Tomsk, Russia, pp. 157–160, 2012.
17. Latham R V, High voltage vacuum insulation: basic concepts and technological practice, Academic Press, 1995. pp.1–59
18. Sandolache G, Chakraborty S, Gaches L, Smeets R, Kuivenhoven S, Novak P, Beer P, An investigation into late breakdown phenomena during capacitor switching performances in relation with vacuum interrupter design and field emission current, 25th International Symposium on Discharges and Electrical Insulation in Vacuum, Tomsk, Russia, pp. 441–444, 2012.
19. Koochack Zadeh M, Hinrichsen V, Smeets R, and Lawall A, Field Emission Currents in Vacuum Breakers after Capacitive Switching, IEEE Trans. Dielectr. Electr. Insul, vol. 18, pp. 910–917, Jun 2011.
20. Koochack Zadeh M, Hinrichsen V, and Kuivenhoven S, Measurement of field emission current during switching of capacitive current in vacuum, 24th International Symposium on Discharges and Electrical Insulation in Vacuum, Braunschweig, Germany, pp. 249–252, 2010.
21. Slade P G, The vacuum interrupter: theory, design, and application, CRC press, 2008, pp. 29–49.
22. Fowler R H and Nordheim L, Electron emission in intense electric fields, Proc. R. Soc. London, Ser. A, Containing Papers of a Mathematical and Physical Character, pp. 173–181, 1928.
23. Frohlich K and Widl W, Breakdown and field-emission behaviour of differently prestressed vacuum interrupter copper contacts, IEE Proceedings C - Generation, Transmission and Distribution, vol. 128, issue 4, pp. 243–249, 1981.
24. Xu S, Hidaka K, Kumada A, Kaneko E, Ikeda H, and Liu P, Research the late discharge in VCB using point-to-plane electrodes, 26th International Symposium on Discharges and Electrical Insulation in Vacuum, Mumbai, India, pp. 9–12, 2014.

25. IEC International standard, High voltage Alternating Current Circuit Breakers, IEC 62271–100, 2008.
26. IEC International standard, Synthetic Testing of High Voltage Alternating Current Circuit Breakers, IEC 62271–101, 2012.
27. Kimura T, Sawada A, Koyama K, Koga H, and Yano T, Influence of vacuum arc behavior on current interrupting limit of spiral contact, in Proceedings of 19th International Symposium on Discharges and Electrical Insulation in Vacuum, vol. 2, Xi'an, China, Sep. 2000, pp. 443–446.
28. Ma H, Zhang Z, Liu Z, Wang Z, Geng Y, and Wang J, Anode current density distribution measurements for different vacuum arc modes subjected to axial magnetic field, IEEE Trans. Plasma Sci., vol. 43, no. 8, pp. 2335–2344, Aug. 2015.
29. Slade P G, Smith R K, and Taylor E D, The effect of contact closure in vacuum with fault current on prestrike arcing time, contact welding and the field enhancement factor, in Proc. 53rd IEEE Holm Conf. Elect. Contacts, Sep. 2007, pp. 32–36.
30. Korner F, Lindmayer M, Kurrat M, and Gentsch D, Contact behavior in vacuum under capacitive switching duty, IEEE Trans. Dielectr. Electr. Insul., vol. 14, no. 3, pp. 643–648, Jun. 2007.
31. Slade P G, The vacuum interrupter: theory, design, and application, CRC press, 2008, pp. 82–88.
32. Yang H, Geng Y, Liu Z, Zhang Y, Zhai X, Li Y, Back-to-back capacitor switching performance of vacuum interrupters after high frequency sub-microsecond voltage impulse conditioning, 2nd International Conference on Electric Power Equipment – Switching technology (ICEPE-ST), Matsue, Japan, pp. 1–4, 2013.
33. Fink H, Gentsch D, Heil B, Humpert C, and Schnettler A, Conditioning of series vacuum interrupters (VIs) for medium voltage by applying high-frequency (HF) current to increase the dielectric strength of VIs, IEEE Trans. Plasma Sci., vol. 35, no. 4, pp. 873-878, 2007.
34. Wang H, Lin J, Li X, Liu N, Geng Y, Liu Z, Prestrike characteristics when switching on inrush current in 40.5 kV VIs with axial magnetic field, 4th International Conference on Electric Power Equipment – Switching technology (ICEPE-ST), Xi'an, China, pp.133–136, 2017.
35. High-Voltage Alternating-Current Circuit-Breakers, GB standard 1984–2014, 2014.

CPSIA information can be obtained
at www.ICGtesting.com
Printed in the USA
LVHW080348290521
688315LV00017BC/146

9 789811 613975